Communications and Control Engineering

Springer
*London
Berlin
Heidelberg
New York
Barcelona
Hong Kong
Milan
Paris
Singapore
Tokyo*

Other titles published in this series:

Stability and Stabilization of Infinite Dimensional Systems with Applications
Zheng-Hua Luo, Bao-Zhu Guo and Omer Morgul

Nonsmooth Mechanics (2nd edition)
Bernard Brogliato

Nonlinear Control Systems II
Alberto Isidori

L2-Gain and Passivity Techniques in nonlinear Control
Arjan van der Schaft

Control of Linear Systems with Regulation and Input Constraints
Ali Saberi, Anton A. Stoorvogel and Peddapullaiah Sannuti

Robust and H∞ Control
Ben M. Chen

Computer Controlled Systems
Efim N. Rosenwasser and Bernhard P. Lampe

Dissipative Systems Analysis and Control
Rogelio Lozano, Bernard Brogliato, Olav Egeland and Bernhard Maschke

Control of Complex and Uncertain Systems
Stanislav V. Emelyanov and Sergey K. Korovin

Robust Control Design Using H∞ Methods
Ian R.Petersen, Valery A. Ugrinovski and Andrey V. Savkin

Model Reduction for Control System Design
Goro Obinata and Brian D.O. Anderson

Control Theory for Linear Systems
Harry L. Trentelman, Anton Stoorvogel and Malo Hautus

Functional Adaptive Control
Simon G. Fabri and Visakan Kadirkamanathan

Positive 1D and 2D Systems
Tadeusz Kaczorek

Identification and Control Using Volterra Models
F.J. Doyle III, R.K. Pearson and B.A. Ogunnaike

Non-linear Control for Underactuated Mechanical Systems
Isabelle Fantoni and Rogelio Lozano

Robust Control (Second edition)
Jürgen Ackermann

M. Vidyasagar

Learning and Generalisation

With Applications to Neural Networks

With 38 Figures

Springer

M. Vidyasagar, PhD
Executive Vice President, Tata Consultancy Services, 1-2-10 Sardar Patel Road, Secunderabad 500 003, India

Series Editors
E.D. Sontag •M. Thoma • A. Isidori • J. van Schuppen

British Library Cataloguing in Publication Data
Vidyasagar, M. (Mathukumalli), 1947-
   Learning and generalisation : with applications to neural
   networks. - 2nd ed. - (Communications and control
   engineering)
   1.Machine learning 2.Neural networks (Computer science)
   I.Title
   006.3'1
   ISBN 1852333731

Library of Congress Cataloging-in-Publication Data
Vidyasagar, M. (Mathukumalli), 1947-
   Learning and generalisation : with applications to neural networks / M. Vidyasagar.--
   2nd ed.
       p. cm. -- (Communications and control engineering, ISSN 0178-5354)
     ISBN 1-85233-373-1 (alk. paper)
      1. Machine learning. 2. Control theory. 3. Neural networks (Computer science)
   I. Title. II. Series.
   Q325.5 .V53 2002
   006.3'1--dc21   2002070674

Apart from any fair dealing for the purposes of research or private study, or criticism or review, as permitted under the Copyright, Designs and Patents Act 1988, this publication may only be reproduced, stored or transmitted, in any form or by any means, with the prior permission in writing of the publishers, or in the case of reprographic reproduction in accordance with the terms of licences issued by the Copyright Licensing Agency. Enquiries concerning reproduction outside those terms should be sent to the publishers.

Communications and Control Engineering Series ISSN 0178-5354

ISBN 1-85233-373-1 Springer-Verlag London Berlin Heidelberg
a member of BertelsmannSpringer Science+Business Media GmbH
http://www.springer.co.uk

© Springer-Verlag London Limited 2003
Printed in Great Britain

The use of registered names, trademarks etc. in this publication does not imply, even in the absence of a specific statement, that such names are exempt from the relevant laws and regulations and therefore free for general use.

The publisher makes no representation, express or implied, with regard to the accuracy of the information contained in this book and cannot accept any legal responsibility or liability for any errors or omissions that may be made.

Typesetting: Electronic text files prepared by author
Printed and bound at The Cromwell Press Ltd, Trowbridge, Wiltshire
69/3830-543210 Printed on acid-free paper SPIN 10775958

Dedicated with affection and gratitude to
My father, Professor Mathukumalli Venkata Subbarao
On the occasion of his eightieth birthday, and
To my mother, Mrs. Suseela Subbarao

Table of Contents

Preface to the Second Edition xiii

Preface to the First Edition xvii

1. **Introduction** 1
2. **Preliminaries** 13
   2.1 Pseudometric Spaces, Packing and Covering Numbers 13
       2.1.1 Pseudometric Spaces 13
       2.1.2 Packing and Covering Numbers 14
       2.1.3 Compact and Totally Bounded Sets 16
   2.2 Probability Measures 17
       2.2.1 Definition of a Probability Space 17
       2.2.2 A Pseudometric Induced by a Probability Measure ... 18
       2.2.3 A Metric on the Set of Probability Measures 19
       2.2.4 Random Variables 21
       2.2.5 Conditional Expectations 23
   2.3 Large Deviation Type Inequalities 24
       2.3.1 Chernoff Bounds 24
       2.3.2 Chernoff-Okamoto Bound 26
       2.3.3 Hoeffding's Inequality 26
   2.4 Stochastic Processes, Almost Sure Convergence 29
       2.4.1 Probability Measures on Infinite Cartesian Products .. 29
       2.4.2 Stochastic Processes 29
       2.4.3 The Borel-Cantelli Lemma and Almost Sure Convergence 30
   2.5 Mixing Properties of Stochastic Processes 33
       2.5.1 Definitions of Various Kinds of Mixing Coefficients ... 34
       2.5.2 Inequalities for Mixing Processes 36
3. **Problem Formulations** 43
   3.1 Uniform Convergence of Empirical Means 43
       3.1.1 The UCEM Property 43
       3.1.2 The UCEMUP Property 52

viii Table of Contents

		3.1.3	Extension to Dependent Input Sequences	54
	3.2	Learning Concepts and Functions		55
		3.2.1	Concept Learning	55
		3.2.2	Function Learning	64
		3.2.3	Extension to Dependent Input Sequences	65
		3.2.4	Assumptions Underlying the Model of Learning	66
		3.2.5	Alternate Notions of Learnability	70
	3.3	Model-Free Learning		76
		3.3.1	Problem Formulation	76
		3.3.2	Relationship to the Uniform Convergence of Empirical Means	81
	3.4	Preservation of UCEMUP and PAC Properties		83
		3.4.1	Preservation of UCEMUP Property with Beta-Mixing Inputs	84
		3.4.2	Law of Large Numbers Under Alpha-Mixing Inputs	89
		3.4.3	Preservation of PAC Learning Property with Beta-Mixing Inputs	94
		3.4.4	Preservation of PAC Learning Property with Beta-Mixing Inputs: Continued	95
		3.4.5	Replacing $\mathcal{P}$ by its Closure	97
	3.5	Markov Chains and Beta-Mixing		100
		3.5.1	Geometric Ergodicity and Beta-Mixing	100
		3.5.2	Beta-Mixing Properties of Markov Sequences	105
		3.5.3	Mixing Properties of Hidden Markov Models	110
4.	**Vapnik-Chervonenkis, Pseudo- and Fat-Shattering Dimensions**			**115**
	4.1	Definitions		115
		4.1.1	The Vapnik-Chervonenkis Dimension	115
		4.1.2	The Pseudo-Dimension	120
		4.1.3	The Fat-Shattering Dimension	122
	4.2	Bounds on Growth Functions		123
		4.2.1	Growth Functions of Collections of Sets	123
		4.2.2	Bounds on Covering Numbers Based on the Pseudo-Dimension	128
		4.2.3	Metric Entropy Bounds for Families of Functions	132
		4.2.4	Bounds on Covering Numbers Based on the Fat-Shattering Dimension	139
	4.3	Growth Functions of Iterated Families		141
5.	**Uniform Convergence of Empirical Means**			**149**
	5.1	Restatement of the Problems Under Study		149
	5.2	Equivalence of the UCEM and ASCEM Properties		153
	5.3	Main Theorems		155
	5.4	Preliminary Lemmas		161

	5.5	Theorem 5.1: Proof of Necessity 173
	5.6	Theorem 5.1: Proof of Sufficiency 178
	5.7	Proofs of the Remaining Theorems 190
	5.8	Uniform Convergence Properties of Iterated Families 194
		5.8.1 Boolean Operations on Collections of Sets 195
		5.8.2 Uniformly Continuous Mappings on Families of Functions... 196
		5.8.3 Families of Loss Functions 200
6.	**Learning Under a Fixed Probability Measure** 207	
	6.1	Introduction 207
	6.2	UCEM Property Implies ASEC Learnability 209
	6.3	Finite Metric Entropy Implies Learnability 216
	6.4	Consistent Learnability................................... 224
		6.4.1 Consistent PAC Learnability 224
		6.4.2 Consistent PUAC Learnability 226
	6.5	Examples... 230
	6.6	Learnable Concept Classes Have Finite Metric Entropy...... 236
	6.7	Model-Free Learning 242
		6.7.1 A Sufficient Condition for Learnability 244
		6.7.2 A Necessary Condition 248
	6.8	Dependent Inputs 250
		6.8.1 Finite Metric Entropy and Alpha-Mixing Input Sequences 250
		6.8.2 Consistent Learnability and Beta-Mixing Input Sequences 251
7.	**Distribution-Free Learning**................................ 255	
	7.1	Uniform Convergence of Empirical Means 255
		7.1.1 Function Classes 256
		7.1.2 Concept Classes 258
		7.1.3 Loss Functions 261
	7.2	Function Learning 263
		7.2.1 Finite P-Dimension Implies PAC and PUAC Learnability 264
		7.2.2 Finite P-Dimension is not Necessary for PAC Learnability 267
	7.3	Concept Learning 269
		7.3.1 Improved Upper Bound for the Sample Complexity ... 269
		7.3.2 A Universal Lower Bound for the Sample Complexity . 273
		7.3.3 Learnability Implies Finite VC-Dimension 278
	7.4	Learnability of Functions with a Finite Range 280

8. Learning Under an Intermediate Family of Probabilities .. 285
8.1 General Families of Probabilities 287
8.1.1 Uniform Convergence of Empirical Means 287
8.1.2 Function Learning 288
8.1.3 Concept Learning 292
8.2 Totally Bounded Families of Probabilities................... 297
8.3 Families of Probabilities with a Nonempty Interior 308

9. Alternate Models of Learning............................. 311
9.1 Efficient Learning 312
9.1.1 Definition of Efficient Learnability 313
9.1.2 The Complexity of Finding a Consistent Hypothesis .. 317
9.2 Active Learning 326
9.2.1 Fixed-Distribution Learning 329
9.2.2 Distribution-Free Learning.......................... 332
9.3 Learning with Prior Information: Necessary and Sufficient Conditions.. 335
9.3.1 Definition of Learnability with Prior Information 335
9.3.2 Some Simple Sufficient Conditions 337
9.3.3 Dispersability of Function Classes 341
9.3.4 Connections Between Dispersability and Learnability WPI.. 344
9.3.5 Distribution-Free Learning with Prior Information 348
9.4 Learning with Prior Information: Bounds on Learning Rates . 352

10. Applications to Neural Networks 365
10.1 What is a Neural Network? 366
10.2 Learning in Neural Networks............................... 369
10.2.1 Problem Formulation 369
10.2.2 Reprise of Sample Complexity Estimates 372
10.2.3 Complexity-Theoretic Limits to Learnability 377
10.3 Estimates of VC-Dimensions of Families of Networks........ 381
10.3.1 Multi-Layer Perceptron Networks................... 382
10.3.2 A Network with Infinite VC-Dimension 388
10.3.3 Neural Networks as Verifiers of Formulas 390
10.3.4 Neural Networks with Piecewise-Polynomial Activation Functions 396
10.3.5 A General Approach 402
10.3.6 An Improved Bound 406
10.3.7 Networks with Pfaffian Activation Functions 410
10.3.8 Results Based on Order-Minimality 413
10.4 Structural Risk Minimization 415

11. Applications to Control Systems ... 421
11.1 Randomized Algorithms for Robustness Analysis ... 421
11.1.1 Introduction to Robust Control ... 421
11.1.2 Some NP-Hard Problems in Robust Control ... 424
11.1.3 Randomized Algorithms for Robustness Analysis ... 426
11.2 Randomized Algorithms for Robust Controller Synthesis: General Approach ... 429
11.2.1 Paradigm of Robust Controller Synthesis Problem ... 429
11.2.2 Various Types of "Near" Minima ... 432
11.2.3 A General Approach to Randomized Algorithms ... 435
11.2.4 Two Algorithms for Finding Probably Approximate Near Minima ... 436
11.3 VC-Dimension Estimates for Problems in Robust Controller Synthesis ... 438
11.3.1 A General Result ... 438
11.3.2 Robust Stabilization ... 438
11.3.3 Weighted H_∞-Norm Minimization ... 441
11.3.4 Weighted H_2-Norm Minimization ... 444
11.3.5 Sample Complexity Considerations ... 445
11.3.6 Robust Controller Design Using Randomized Algorithms: An Example ... 449
11.4 A Learning Theory Approach to System Identification ... 453
11.4.1 Problem Formulation ... 453
11.4.2 A General Result ... 455
11.4.3 Sufficient Conditions for the UCEM Property ... 458
11.4.4 Bounds on the P-Dimension ... 461

12. Some Open Problems ... 465

Preface to the Second Edition

In the roughly five years since the first edition of this book was published, several significant advances have taken place in statistical learning theory. New approaches have been successfully evolved, and some of the open problems stated in the first edition have been solved. In view of these developments, it has been decided to bring out a second edition of the book.

Compared to the first edition, here are some of the specific changes that have been made in the book. First, the substantial changes:

- At the time the first edition was published, practically all of statistical learning theory was based on the assumption that the samples to the learning algorithm were independent and identically distributed. Clearly the assumption that the learning samples are statistically independent is a very serious restriction, that deserved to be removed at the earliest opportunity. In the present edition, the notion of independence is replaced by the weaker notion of *mixing*, and it is shown that most of the main results of statistical learning theory continue to hold under this weaker hypothesis. Thus it becomes of interest to study whether stochastic processes of "practical" interest have this mixing property. It is shown that state sequences of Markov chains and output sequences of hidden Markov models both possess the mixing property, under appropriate conditions. Most of the relevant material is introduced in Chapters 2 and 3; however, in almost all chapters, the consequences of replacing i.i.d. input sequences by mixing input sequences are explored.
- The application of statistical learning theory to control systems was in a nascent state when the first edition was written. Since that time, there have been some major advances in this area. Two such advances are highlighted in the present edition. The first pertains to the use of "randomized" algorithms to provide probabilistic solutions to controller synthesis problems that are NP-hard in their deterministic form. If one insists on finding an algorithm that works all the time (i.e., a deterministic algorithm that is guaranteed to find a solution for every problem instance), then many simple-looking problems in robust controller synthesis are by now known to be NP-hard, and thus intractable unless P = NP (which most people don't believe). On the other hand, if one is willing to settle for an algorithm that "works reasonably well most of the time" (i.e., a randomized algorithm),

then all of these problems become tractable in the sense that there exist efficient (polynomial-time) randomized algorithms. The second theme is studying the problem of system identification as a problem in statistical learning theory. System identification is a mature and well-established discipline, and one might wonder why a new approach is needed at all. The reason is that, by tradition, system identification theory is addressed to the derivation of *asymptotic* results, that tell us what happens in the limit, i.e., as the number of samples approaches infinity. However, if one is interested in combining system identification with robust control, then it is essential to have *finite-time* estimates, of the kind provided by statistical learning theory. Thus by recasting the problem of identifying an unknown system as a learning problem, it becomes possible to derive finite-time estimates for the rate at which the identified model converges to the unknown system that is being identified. These estimates can then be used to design robust controllers.

Now for more minor changes:

- The concept of "fat-shattering dimension" is introduced, and its application to learning real-valued functions is explained.
- In Chapter 9, the section on learning with prior information has been expanded to include recent results, which provide both necessary and sufficient conditions for learning with prior information.
- The chapter on neural networks has been modified to reflect recent advances.
- Since several of the open problems stated in Chapter 12 have since been solved, this chapter is thoroughly revamped.

As a result of all these changes, the pedagogical approach of the book has been substantially altered from the first edition. In that work, I attempted to build up the level gradually, whereby the first two chapters (after the Introduction) could be skipped by experts, who could proceed directly to the later chapters. However, in the present edition advanced material can be found in *every* chapter. I had to choose between introducing each new result in a place that appeared most natural to me, and retaining the monotonicity of the "difficulty function." I opted for the former approach.

In view of the length of the book, I made a conscious decision to leave out a discussion of support vector machines, which represent an important advance in machine learning. The interested reader is referred to [46, 172, 49].

Since the publication of the first edition, I have been fortunate to have initiated a continuing collaboration with Rajeeva Karandikar of the Indian Statistical Institute, Delhi. I would like to thank him for furthering my education in probability theory, and for replacing my "seat of the pants" approach to probability theory with something more rigorous. I would also like to thank him for reading Chapters 2 and 3, and also for permitting the inclusion of several previously unpublished results in these chapters. Any remaining errors are of course my own responsibility.

As always I would like to thank my wife Shakunthala for her consistent support throughout my career. I would also like to thank my employers, Tata Consultancy Services, especially my CEO, Mr. S. Ramadorai, for having the vision to encourage an activity of this kind in a software company.

I take this opportunity to dedicate this book to my parents, who have made me what I am. My father, Professor M. V. Subbarao, continues to be an active researcher in number theory even at the age of eighty. From his example I learnt to aspire to a career in research when I was still a boy. My mother, Mrs. Suseela Subbarao, not only played a major role in my precocity, but also passed on to me a wonderful religous lineage, whose value cannot be measured by any worldly yardstick.

Hyderabad, India
March 2002

Preface to the First Edition

The objective of this book is to present a comprehensive treatment of some recent developments in statistical learning theory, and their applications to analyzing the ability of neural networks to generalize; in addition, some potential applications to control systems are also sketched, and some problems for future research are indicated. The book is aimed at engineers, computer scientists, and applied mathematicians who have an interest in the broad area of machine learning. The background required to read and understand this book consists primarily of a basic understanding of the elements of probability theory. It should be emphasized that, while learning theory as discussed here uses the *formalism* of probability theory, most of the deep concepts in probability are *not* used. Chapter 2 gives a summary of the background required of the reader. The book can either be used for self-study or as a text in an advanced graduate course. After the first four chapters, the remaining chapters are more or less independent of each other, so that a reader or instructor will be able to pick and choose according to their requirements.

It might be desirable to summarize how the subject of learning theory came to its current status. Even as the modern digital computer was being invented, the scientific community was beginning its attempts to formulate mathematical theories of how machines can be made to "learn" and to "generalize" on the basis of past experience. As early as 1943, J. C. McCulloch and W. Pitts [130] studied interconnections of switching elements that were simple approximations of the biological neurons found in the human brain, and proved that every finite-state automaton can be approximated by such a network, and vice versa. In the late 1950's Frank Rosenblatt introduced the perceptron and proved that, under suitable circumstances, a perceptron could learn to separate positive examples from negative examples. Very shortly thereafter, the training of perceptrons was given a statistical flavour by the Russian school, in a manner strongly reminiscent of subsequent developments more than twenty years later. The subject of "inductive" learning was sought to be formulated as a counterpart to deductive learning on the basis of mathematical logic. While the study of the original perceptron went into a hiatus following the publication of the book *Perceptrons* by Minsky and Papert [137], other models of learning systems such as learning automata continued to be invented and studied. The subject of neural networks, which can be thought of as an intellectual successor to perceptron theory, was revived in spectacular fashion following the publication of the multi-volume book *Parallel Distributed Processing* by Rumelhart and McClelland [168], [169]. At least part of the appeal of neural networks stems from their claimed ability to "generalize." A great deal of episodic evidence has been presented in the literature to support the claim that, once a neural network has been "trained" on a sufficient number of samples, it can then produce the correct output to a new, and previously unseen, input. It should be noted that, without the ability to generalize, much of the case for using neural networks would collapse – a simple table-lookup scheme would suffice if one were interested merely in constructing a network that could reproduce known input-output pairs. However, until recently there has not been a clear mathematical enunciation of just what "generalization" means, nor has there been any mathematical justification to back up the episodic evidence that neural networks seem to be able to generalize in specific situations.

While these developments were taking place in the neural networks community, the theoretical computer science community had its attention drawn to a novel formulation of the learning problem by the publication in 1984 of the paper "A Theory of the Learnable" by Leslie G. Valiant. This approach to learning has over the years come to be known as "probably approximately correct (PAC)" learning theory. In this paper, Valiant showed that Boolean functions in n variables are "learnable" in a very precise sense provided they can be expressed as a 3-CNF, that is, if they can be expressed as a conjunction of several clauses, each of which is a disjunction of no more than three variables. Several other classes of Boolean formulae were are also shown to be learnable in the same sense. Since the publication of Valiant's paper, many others have pursued the PAC formulation, refined and redefined it, derived necessary and sufficient conditions for learnability in Valiant's and related frameworks, developed several applications, and so on.

A parallel development in the theory of empirical processes was to have a profound impact on learning theory. More than two centuries ago, J. Bernoulli showed that, if a two-sided coin is tossed repeatedly, then the fraction of "heads" converges almost surely to the *true* probability of getting "heads" as the number of tosses approaches infinity. In more modern terminology and notation, the Glivenko-Cantelli lemma of the 1930's showed that, if one draws a sequence of random real numbers $x_1, \ldots, x_m$ in accordance with an unknown probability measure P, then the empirical distribution function converges uniformly and almost surely to the true distribution function. This result was subsequently generalized by Kolmogorov and Smirnov to vector-valued processes. Still more general problems were studied by various researchers, and culminated in the landmark 1971 paper "'On the Uniform Convergence of Relative Frequencies to Their Probabilities" by V.N. Vapnik and A.Ya. Chervonenkis [194]. This paper gave necessary and sufficient conditions for the empirical estimates of the probability measures of a family of sets to converge to their true values, as the number of samples approaches infinity. A combinatorial parameter, which has since come to be known as the Vapnik-Chervonenkis (VC-) dimension, plays a central role in these necessary and sufficient conditions.

The publication in 1989 of the paper "Learnability and the Vapnik-Chervonenkis Dimension" by Anselm Blumer *et al.* [32] represented another milestone in the development of PAC learning theory. This paper was apparently the first to make a connection between PAC learning theory and the theory of empirical processes. While there have been a few other papers that followed up this connection, my opinion is that by and large this connection remains unexplored, or perhaps merely unexplained to a wide audience. In particular, by reformulating the learning problem as a convergence problem for stochastic processes, it is possible to make a distinction between *information-theoretic* limitations to learning, and *complexity-theoretic* limitations to learning. Roughly speaking, information-based learning theory attempts to study what is learnable *in principle*, whereas complexity-based learning theory attempts to study what is learnable *in practice*.

By now it was widely appreciated that the PAC learning formulation presents a mathematically rigourous, as well as tractable, formulation of the intuitive idea that "neural networks can generalize." Moreover, by estimating the VC-dimension of a neural network architecture, it is possible to make *quantitatively precise*, albeit quite conservative, estimates of the "rates" at which a neural network can "learn" and "generalize." This naturally led several researchers to investigate ways of estimating the VC-dimension of various types of neural network architectures. The outcome of these researches is a rich theory, ranging from simple counting arguments to extremely sophisticated methods involving model theory of real numbers, algebraic geometry, and so on. It is therefore clear that the problem of estimating the VC-dimension is by now an important specialization in itself.

The issues of intelligence, learning, and generalization are also relevant in the context of control systems. In its broadest sense, a control system can be any object, natural or man-made, whose behaviour one wishes to modify into something more desirable. There are at least a few problems in control theory that can be viewed from the learning theory perspective. However, in contrast with neural networks, it is not yet clear whether the learning theory perspective offers any advantages over existing methods of control theory. Moreover, in contrast with neural networks where the "standard" PAC learning problem formulation nicely captures the notion of generalization, the problems in control theory may perhaps require a reformulation of the basic PAC learning problem.

The present monograph attempts to achieve several related objectives: (i) To present a treatment of (PAC) learning theory that brings out clearly the connection between this theory and some of the fundamental results in the theory of empirical processes; (ii) to present an application of learning theory to generalization by neural networks; (iii) to indicate how some problems in control theory might be viewed as problems in learning, and how learning theory needs to be modified in order to be applicable to such problems; and (iv) to discuss some open problems in statistical learning theory that merit the attention of the research community. The various objectives are rather disparate in nature. In the first case, the theory of empirical processes is by now a mature subject, and an exposition of the principal results as given here will have some lasting reference value. In the second case, many new discoveries continue to be made on the computation of the VC-dimension of neural networks, and it is possible that some of the results given here will be subsumed by newer developments. Nevertheless, some of the fundamental discoveries will stand the test of time. In the third and fourth cases, the aim is more to trigger further activity than anything else.

At present there are several excellent texts on *computational* learning theory, such as [147], [9], and [99]. A deliberate choice has been made here to focus on the *statistical* aspects of learning theory, though the computational aspects are touched upon in Chapters 9 and 10. Thus the present work is intended to complement the books listed above. It should be mentioned that there is also a great deal of work in the probability theory community on the problem of nonparametric density estimation, which is closely related to the problems studied here. This body of research is not discussed at all in the present book; the interested reader is referred to [55] and the references therein. Though for the most part the book is a compendium of known results, in many places the known results are refined and/or the proofs streamlined. Thus it is hoped that both the novice as well as the expert will "learn" something from reading this book.

Now it is my pleasure to express my sincere gratitude to several individuals who have assisted me in the writing of this book. Specifically, I would like to thank (in chronological order)

- Ravi Kannan and Sanjoy Mitter for first introducing me to the fascinating world of PAC learning theory, and for encouraging me along once I got into the subject.
- Sanjeev Kulkarni for initially teaching me everything I knew about learning theory, and also for carefully reading various drafts of the book.
- Eduardo Sontag for his infectious enthusiasm for the idea of such a book, which encouraged me to complete the project in record time by my standards, and also for his careful critique of the chapter on neural networks.
- Vivek Borkar for educating me in probability theory and for serving as my own personal "oracle" (which, unlike those studied in the book, never made a mistake!).
- Dan Ocone for class-testing an early draft of the book, and giving me valuable feedback.
- Lennart Ljung and Roberto Tempo for giving me an opportunity to present this material in condensed form at their respective institutions.

- Vijay Chandru, Girish Deodhare, Piyush Gupta and S.H. Srinivasan for enthusiastically participating in the lectures given by me based on this book.
- Vishwambhar Pati for aiding me to understand the material on algebraic topology in the chapter on neural networks.

In addition, I would like to thank Wolfgang Maass for sharing several of his papers on neural networks and for useful electronic discussions, and Pascal Koiran for useful comments on the chapter on neural networks.

Finally, as for my family, what can I say in respect of their encouragement and moral support that I have not already said on several previous occasions? Sometimes I feel that my attitude towards my family mirrors that of the male protagonist in the O'Henry short story "The Pendulum." Once again I can only say "Thank you" – it is my pleasure to dedicate this book to them.

1. Introduction

Ever since the advent of the "modern" digital computer about fifty years ago, many researchers have explored the possibility of using a machine to perform not merely numerical computations, but also more human-like tasks such as learning new concepts, solving problems, and so on. One of the objectives of the present monograph is to formulate one specific class of models of "learning" that seems natural for a machine, and to explore which types of "concepts" are amenable to machine learning, and which are not. It turns out that the type of "learning" discussed here can also be interpreted naturally as "generalization," especially in the context of neural networks.

There are two mathematical themes running in parallel in this monograph. The first theme is that of estimating unknown quantities on the basis of experimentation. This subject is studied using a well-established branch of probability theory known as the theory of empirical processes. The second theme is that of learning and generalization, as discussed above. Let us illustrate each of these themes by giving some very simple examples.

Take the first the notion of estimating unknown quantities on the basis of experimentation. Two examples are given to illustrate this application.

Example 1.1. (Finding the Probability of "Heads" of a Coin): As a first example, suppose we are given a coin with two sides, and that it is desired to determine the probability of getting "heads" when the coin is tossed. Of course, if the coin is unbiased, then the probability of getting heads equals 0.5. However, given some coin, there is no *a priori* reason to assume that the coin is unbiased. Let p denote the probability of getting "heads." Of course p is unknown at first. To make an estimate of p, it would be natural to toss the coin a few times and to measure the number of heads. Suppose the coin is tossed m times, out of which l tosses result in "heads." Then the ratio l/m is called the "empirical probability" of getting "heads." Let us denote this ratio by $\hat{p}_m$. Then we cannot hope that $\hat{p}_m$ is *exactly* equal to the unknown quantity p. For one thing, $\hat{p}_m$ is always a rational number, whose denominator is (a divisor of) m, whereas p could be irrational, or even if p is rational, its denominator could be relatively prime to m, and so on. However, it is not unreasonable to expect that, as $m \to \infty$, $\hat{p}_m$ begins to *approach* p. The question is: In what sense does $\hat{p}_m$ converge to p?

The quantity $\hat{p}_m$ is itself a random variable. If we were to toss the coin m times and determine $\hat{p}_m$, and then start afresh by tossing the coin another m times, there is no reason to suppose that the $\hat{p}_m$ obtained on the basis of the second set of m tosses will equal the answer obtained after the first m tosses. To analyze this problem, let $\{0, 1\}$ denote the set of possible outcomes of tossing the coin once, where 0 denotes "tails" and 1 denotes "heads." Then the set of all possible outcomes of tossing the coin m times can be identified with the set $\{0, 1\}^m$, the set of all strings of length m consisting of 0's and 1's. Let us specify some number ϵ, which corresponds to the accuracy to which we would like to determine the unknown quantity p. We can then divide the 2^m possible outcomes of the coin-tossing experiment into two sets, namely: (i) those outcomes (i.e., elements of $\{0, 1\}^m$) for which $|\hat{p}_m - p| \leq \epsilon$; these can be thought of as the set of "good" samples, and (ii) those outcomes for which $|\hat{p}_m - p| > \epsilon$; these can be thought of as the set of "bad" samples. Associated with each of the 2^m strings in $\{0, 1\}^m$ is the corresponding probability that tossing the coin m times will result in that particular string. Specifically, if a string in $\{0, 1\}^m$ consists of k 1's and $(m - k)$ 0's, then the probability that the coin-toss experiment will generate that particular string of 0's and 1's equals $p^k(1-p)^{m-k}$. Now one can ask: What is the probability that tossing the coin m times will generate a "bad" sample, i.e., generate a sample for which $|\hat{p}_m - p| > \epsilon$? It can be shown, using the so-called Chernoff bounds (see Chapter 2) that this probability is no larger than $2e^{-2m\epsilon^2}$. This means that, after tossing the coin m times, we can state with a confidence of at least $1 - 2e^{-2m\epsilon^2}$ that the empirical probability $\hat{p}_m$ is no more than ϵ different from the true but unknown probability p. Note that, in some sense, this is about all that can be said. Irrespective of how many times the coin toss experiment is repeated, there will always be *some* samples of length m that will give a totally misleading estimate of p; all that one can say is that, as m becomes larger, such samples become less likely.

One can turn the above estimate around and ask: If one wishes to estimate the unknown quantity p to an accuracy of ϵ with a confidence of $1 - \delta$, how many times should the coin be tossed? The answer is obtained by choosing m large enough that

$$2e^{-2m\epsilon^2} \leq \delta,$$

or equivalently,

$$m \geq \frac{1}{2\epsilon^2} \ln \frac{2}{\delta}.$$

This is only a *sufficient* condition, in the sense that this many coin tosses are *definitely enough* for us to assert with confidence $1 - \delta$ that $|\hat{p}_m - p| \leq \epsilon$. To be specific, in order to be 99% sure that the empirical probability is within 5% of the true value, it is enough to toss the coin 1,060 times.

Example 1.2. (Numerical Integration of a Function): Suppose f is a real-valued function of k variables that takes values in the interval $[0, 1]$. (In fact, there is nothing special about the interval $[0,1]$, and any *bounded* interval would do just as well.) Thus $f : \mathbb{R}^k \to [0,1]$. Suppose it is desired to find the integral of f over some set $X \subseteq \mathbb{R}^k$. To avoid technical difficulties related to integration over unbounded sets, let us suppose that $X = [0,1]^k$, and that it is desired to compute

$$E(f) := \int_X f(x)\, dx.$$

Unless f is known in closed form, it is not possible to compute the above integral exactly, and perhaps not even then. As an alternative, one could resort to *numerical* integration, by picking some points $x_1, \ldots, x_m \in X$ and forming the estimate

$$\hat{E}(f; \mathbf{x}) := \frac{1}{m} \sum_{i=1}^{m} f(x_i).$$

Such an estimate is not of much use unless we have some idea of the *error* $|E(f) - \hat{E}(f; \mathbf{x})|$. If one were to choose the various points $x_1, \ldots, x_m$ by dividing X into a uniform grid, then it is possible to give an estimate of this error, *provided* f satisfies some fairly stringent conditions such as being uniformly continuous or having a bounded gradient everywhere over X, and in addition, explicit upper bounds are available for the magnitude of the gradient vector of f. Such an approach suffers from many drawbacks, such as: (i) the requirement that f be uniformly continuous or have bounded gradient rules out interesting function classes such as those containing step discontinuities, and (ii) the number of grid points needed to achieve a prespecified accuracy $|E(f) - \hat{E}(f; \mathbf{x})|$ *increases exponentially* as a function of the integer k, which is the number of independent variables. On the other hand, suppose we try a probabilistic approach, as follows: Choose $x_1, \ldots, x_m$ *at random* from X, with the uniform density, and define $\hat{E}(f; \mathbf{x})$ as above. Then it can be shown that, as in the coin-tossing example above, $|E(f) - \hat{E}(f; \mathbf{x})| \leq \epsilon$ with a probability of at least $1 - 2e^{-2m\epsilon^2}$. This is the same bound as above, even though the interpretation and the application are quite different. This approach has the overwhelming advantage that the number of samples needed to achieve a prescribed accuracy and prescribed confidence level is *independent of the number k of independent variables*. The above bound is the basis of Monte Carlo simulation, or probabilistic estimation of various expected values. ∎

The above two examples are extremely simple-minded instances of the so-called "theory of empirical processes." The subject itself is much deeper than what one might gather from the above examples. In the present monograph, we shall examine issues such as the above when it is desired not merely to estimate a *single* probability or a *single* integral, but rather to estimate simultaneously an *infinite number* of probabilities or integrals. The precise statement of the type of problems studied here can be found in Chapter 3, and the principal results in this direction, based on some concepts introduced in Chapter 4, are stated in Chapter 5.

4   1. Introduction

Now let us turn to the problem of learning, which is the second major theme of the monograph. This problem is also illustrated by some simple examples.

Example 1.3. (The "Pick a Number" Game): Let us begin by studying the type of game that children sometimes play, the kind that often begins "Pick a number – any number – between 1 and 10." One child does so, and then the other child asks a series of questions about the unknown number (other than, of course, directly asking what the number is). After a series of carefully selected questions, the second child is usually able to guess the number exactly, much to the dismay (or delight) of the first child. In the type of learning theory studied here, allowance is made for the fact that it is often impossible to guess the unknown entity *exactly*. Instead, it is permissible for the learner to make a guess that is *approximately* correct after a finite number of questions have been answered, provided that the error between the guess and the unknown quantity can be made arbitrarily small by asking a sufficient number of questions. To illustrate this idea in the context of the "guess a number" game, let us suppose that the first child, called here as "O" (for "oracle") picks a *real number* between 0 and 1.[1] The second child, called here as "L" (for "learner") then asks a series of questions to try to pin down the unknown number. Let us denote the unknown number by t, and let us suppose that L is restricted to asking only questions that can be answered "yes" or "no." Then L could ask a series of questions such as the following: Is t between 0 and 0.5? (If the answer is "yes"): Is t between 0 and 0.25? (If the answer is "no"): Is t between 0.5 and 0.75? And so on. The idea is clear, as is the rate at which L can converge to the unknown t. After m questions, and assuming that each question is answered honestly, the learner is able to make a guess that is guaranteed to differ from t by no more than 2^{-m}. To put it the other way, suppose L is obliged to get within ϵ of the unknown t; then $\lceil \lg(1/\epsilon) \rceil$ appropriately selected questions will suffice, where $\lceil a \rceil$ denotes the smallest integer greater than or equal to a, and $\lg a$ denotes $\log_2 a$. It is not entirely coincidental that $\lceil 1/\epsilon \rceil$ is the minimum number of intervals of length ϵ that are needed to cover the unit interval $[0,1]$, which is where the unknown t resides. This number is a very particular example of what is known as the "metric entropy" of a set, which plays a central role in learning theory.

Now let us reformulate the above guessing game in another format, which fits more closely into the general format of the learning problems discussed in the book. Let us suppose that O picks an *interval* $[0,t]$, which may be called the "target" interval, and let us denote $[0,t]$ by T. The learner is permitted to select any $x \in [0,1]$, and O returns the value of the membership function $I_T(x)$; that is, O returns the value of 1 if $x \in T$ and 0 if $x \notin T$. It is clear that the reformulated game is entirely equivalent to the first formulation given above, because knowing whether or not $x \in T$ is the same as knowing

---

[1] Let us for the moment ignore the fact that a "child" that understands the difference between integers and real numbers is precocious indeed!

whether or not $x \leq t$. Let us suppose that, after asking m questions, L makes an estimate of T that is consistent with the information available from the answers. This estimate is of the form $[0, h]$, where $h \in [0, 1]$, and is called a "hypothesis" in learning theory. It does not particularly matter how exactly h is selected – perhaps h can be the *smallest* number that is consistent with all the answers, or perhaps it can be the *largest* number, or anything else in between. Nevertheless, it is true that $|h - t| \leq 2^{-m}$ after m questions and answers.

The type of learning described above is usually referred to as "active" learning, in that L is free to choose the next question to be asked, and would presumably do so in a manner calculated to make the hypothesis converge to the unknown target interval as rapidly as possible. But in many problems, the learner is *not* free to select the probing questions. One illustration of this is time-series analysis, or system identification. In such an application, all that is available is a partial sample path of several random variables, or in the context of system theory, a finite collection of input-output time histories. To model such a learning problem, we can amend the above description of the "guess a number" game as follows: There is a target interval $T = [0, t]$ where $t \in [0, 1]$. A random number generator outputs a sequence of numbers $x_1, x_2, x_3, \ldots$, all of which are independent and identically distributed (i.i.d.) according to some probability P, which may itself be imperfectly known. For each "training input" x_i, O gives out the information as to whether or not x_i belongs to T; that is, O gives out the value of the membership function $I_T(x_i)$ for each i. Based on this information, L is supposed to make an educated guess as to t. Such a model of learning may perhaps be called "passive" learning, in that L cannot exercise any control over the choices of the x_i's. In the active learning framework described in the preceding paragraph, L makes the choices: (i) $x_1 = 0.5$, (ii) $x_2 = 0.25$ if $I_T(x_1) = 0$, or $x_2 = 0.75$ if $I_T(x_1) = 1$, and so on. Now it is clear that, in the passive learning problem, the hypothesis H cannot ever be *guaranteed* to be a good approximation to T, irrespective of how many times the learning process is repeated. To illustrate this point, suppose that the training samples x_i are i.i.d. according to the uniform probability on $[0, 1]$. Then, irrespective of how many samples m are drawn, there is nevertheless a nonzero probability (namely, 2^{-m}) that all m samples will belong to $[0, 0.5]$. Hence, if $t > 0.5$, the experimentation does not help L at all in getting a better idea of the value of t. As a result, in the worst case $|h - t|$ could be as large as 0.5, with a probability of at least 2^{-m}. Thus learning in the passive case must reflect both an *accuracy* parameter, which is an estimate of the difference $|h - t|$, as well as a *confidence* parameter, which measures the probability that the estimate is incorrect. In this case, L can be said to "learn" the target interval T if, given any accuracy parameter ϵ and any confidence parameter δ, there exists an integer $m := m(\epsilon, \delta)$ such that, after m i.i.d. samples are drawn, L can assert that $|h - t| \leq \epsilon$ with a probability of at least $1 - \delta$. The reader may be interested to know that, in the present example, it is enough for L to draw

$$m \geq \frac{32}{\epsilon} \ln \frac{1}{\epsilon \delta}$$

samples. Results of this type are proved in Chapter 6.

Example 1.4. (Guessing an Unknown Convex Polygon): As another illustration of the learning problem, let us study a game that might be called "guess the unknown convex polygon." Let $X = [0,1]^2$, the unit square, and suppose the target set T is a fixed but unknown (nontrivial) convex polygon in X. Suppose random vectors $x_1, x_2, \ldots$ are drawn from X in accordance with the uniform density, and for each x_i, O returns the value of the membership function $I_T(x_i)$. (See Figure 1.1). After m samples are drawn, L can choose

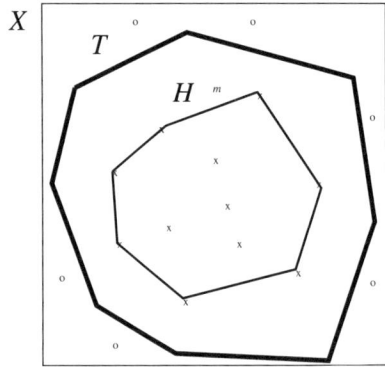

Fig. 1.1. Learning an Unknown Convex Polygon

the hypothesis H to be the convex hull of all the "positive" examples, that is, all the points x_i that are known to belong to T. It is clear that H is always a subset of T; also, as more and more points belonging to T are drawn, H becomes a better and better approximation to T, in the sense that the area of the difference $T - H$ becomes smaller and smaller. Moreover, as more and more samples are drawn, it is less and less likely that all of them will lie *outside* T. Thus, with this intuitive argument, the reader may perhaps be persuaded that "convex polygons in $[0,1]^2$ are learnable," even though we have not as yet defined precisely what learnability means. Another interesting point is that convex polygons are learnable in the present instance only because the random samples are being drawn with *uniform* probability density. If one were to ask: "Is the set of convex polygons learnable no matter *which* probability measure is used to generate the training inputs?" then the answer is "no." This follows from the results of Chapters 4 through 7.

Example 1.5. (A Nonlearnable Problem): Now let us study a *nonlearnable* example. Let $X = [0,1)$, and observe that every $x \in X$ has a binary expansion of the form

$$x = \sum_{i=1}^{\infty} b_i(x)\, 2^{-i}.$$

For example, if $x = 0.68$, then $x = 0.1010111\cdots$. Figure 1.2 shows the collection of sets

$$B_i = \{x \in X : b_i(x) = 1\}.$$

Now suppose O picks one of these sets B_n as the target; equivalently, O picks

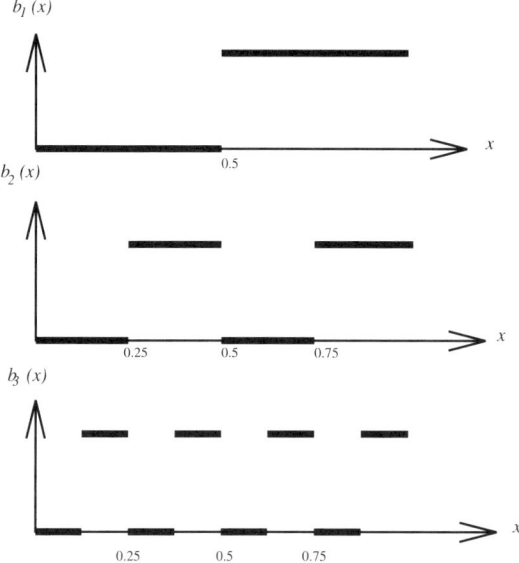

Fig. 1.2. A Nonlearnable Problem

some integer $n \geq 1$. This is a variation on the "pick a number – any number" theme, in that the set of numbers is *unbounded*. Let us suppose that learning is active, in that L is free to select any $x \in X$, and O returns the value of $b_n(x)$, which is the same as the membership function $I_{B_n}(x)$. After drawing some m numbers $x_1, \ldots, x_m$, L is expected to make a guess as to what integer n O is using to generate the output. To put it slightly differently, after m training inputs, L is supposed to make a prediction as to what the output of O would be on the *next* input, which is a randomly selected $x \in X$. This idea can be made quantitatively precise, as follows: Suppose that, after m training inputs, L makes a guess that n equals h_m, which is some integer. Thus, on a randomly selected x, the prediction of L is $b_{h_m}(x)$, whereas the actual output of O is of course $b_n(x)$. To avoid notational clutter, let us simply denote $b_{h_m}(\cdot)$ by $h_m(\cdot)$. Then the probability that L *fails* to predict the output of O correctly is equal to

$$\int_X |h_m(x) - b_n(x)|\, dx.$$

This can be thought of as a "figure of merit" for the hypothesis h_m. The objective of the learning exercise is to make the above number approach 0 as $m \to \infty$, if possible.

The functions $b_i(\cdot)$ have the following very interesting properties, as can be easily verified:

$$\int_X b_n(x)\, dx = 0.5,\ \forall n,\ \text{and}$$

$$\int_X |b_n(x) - b_l(x)|\, dx = 0.5,\ \forall n \neq l.$$

These two identities can be interpreted as follows: The first identity means that, for a variety of simple-minded strategies, L has a 50% probability of correctly predicting the output of O, even without performing any experiments. For example, if L were to *always* predict an output of 1 (or 0), or if L were to predict the output of O merely by tossing an unbiased coin, then L would have a 50% probability of guessing the output correctly, irrespective of what n is. Therefore, even without bothering to engage in any experimentation, L can be 50% successful in predicting the output of O on a randomly selected x. Thus the objective of learning must be, at the very least, to improve upon this figure of 50%. The second identity means that, unless the hypothesis h_m is *exactly* equal to b_n, the probability that L can correctly predict the output of O on a randomly selected input x *remains* at 50%. Thus the only way that L can improve upon the initially achievable performance of 50% is to *determine n exactly* after a finite number of experiments.

Now a heuristic argument is put forward to suggest that there does not exist any systematic procedure for choosing the values $x_1, \ldots, x_m$ that would *always* find the unknown value n exactly in a fixed number of steps. Given any one input x_i and the output of O to x_i, one can rule out some possible values for n; perhaps one can rule out *infinitely many* values for n. However, after a finite number of inputs x_i, the number of possible values of n that remain will always be infinite. To amplify this point further, suppose one were to try the sequence of inputs $x_1 = 0.5$, $x_2 = 0.25$, $x_3 = 0.125$, and so on. If the output of O to x_1 equals 1, then the game is over because obviously $n = 1$. (Note that the binary representation of x_1 contains a 1 in the first place and 0's elsewhere.) However, if the output of O to x_1 equals 0, all that L has gained is the knowledge that O is not looking at the first bit of x – but an infinite number of alternatives still remain. The same argument can be repeated to show that the above sequence of inputs will always succeed in finding the unknown n after a finite number of inputs; but unfortunately that "finite" number of inputs required is n itself. To put it another way, there is no *a priori* upper bound on the number of test inputs, *independent of the unknown n*.

Up to now, we have proceeded on the assumption that L tries to predict the output of O by making a guess as to the value of the unknown integer n; with this assumption, we have shown that no algorithm can succeed in reducing the prediction error below 0.5 for *every* possible value of n within a fixed number of inputs. Of course, this argument has not been very precise. In Chapter 6 it is shown that, no matter how L tries to predict the output of O (i.e., whether L tries to guess the value of n directly or uses some other method), it is not possible for the prediction error to be reduced below 0.5. ∎

The nonlearnability in the present example is a consequence of requiring that the algorithm should perform equally well for all unknown integers n. However, if it is assumed that O selects the integers according to a prior probability $\{p_n\}$, where p_n denotes the probability that O selects the integer n, then the problem becomes learnable; see Chapter 9 for details.

In the learning examples above, we have distinguished between passive learning and active learning. In active learning, the learner is at liberty to select the training inputs, whereas in passive learning, the training inputs are generated according to some underlying probability measure. If this underlying probability is known, then the (passive) learning problem is said to be of the "fixed-distribution" type; such learning problems are studied in Chapter 6. If on the other hand the random training inputs are generated by an underlying probability measure that is itself *completely* unknown, then the problem is said to be one of "distribution-free" learning; such problems are studied in Chapter 7. The intermediate case where the underlying probability measure is partially known is studied in Chapter 8.

Other distinctions are also possible. For example, it is possible that the oracle O does not answer each question honestly, or if one prefers, that the honest answer of O is communicated to L through a noisy channel that occasionally causes the answer to be misinterpreted at the receiving end, say with a probability of α. This type of learning is called by a variety of names, but "learning with imperfect oracles" seems to be as good a name as any. One of the interesting aspects of learning theory is that, in a wide variety of situations, learnability or the lack of it is "robust." This means that, for example, if a problem is learnable with perfect oracles, it remains so even with imperfect oracles. The only consequence of the oracle making a few mistakes now and then is that the number of training inputs required to achieve a specified level of accuracy and confidence increases as the probability of error by the oracle increases; but a learnable problem does not suddenly become nonlearnable merely because the oracle occasionally gives incorrect outputs. Similarly, if a problem is nonlearnable with passive choice of inputs, it remains so even if the learner has the option of choosing the inputs. The ability to choose inputs actively can only reduce the number of training inputs required (as in Example 1.3 above), but cannot turn a nonlearnable situation into a learnable one. Such results are reassuring, because they suggest that learnability is an intrinsic property and is not too sensitive to the particular "model" of learning used. For a discussion of these and related topics, see Chapter 9.

Once the theoretical development in the two areas of empirical processes and learning theory is completed, the monograph addresses the application of these ideas to neural networks. Recent years have witnessed a resurgence of research activity in this area, not least because of the claimed ability of neural networks to "generalize." A great deal of episodic evidence has been presented in the literature to support the claim that, once a neural network has been "trained" on a sufficient number of samples, it can then produce the correct output to a new, and previously unseen, input. It should be noted that, without the ability to generalize, much of the case for using neural networks would simply collapse – a simple table-lookup scheme would suffice if one were interested merely in constructing a network that could reproduce known input-output pairs. However, until recently there has not been a clear mathematical statement of just what "generalization" means, nor has there been any mathematical justification to back up the episodic evidence of the generalization ability of neural networks in specific situations. In this monograph, it is shown that learning theory provides a natural setting for discussing the generalization ability of neural networks, and also of quantifying the *rate* at which a neural network can learn. Chapter 3 contains several examples that support this viewpoint (see in particular Examples 3.6 and 3.7), while Chapter 10 gives quantitatively precise, albeit somewhat conservative, estimates of the rate at which a neural network can be made to learn. In particular, a combinatorial parameter known as the Vapnik-Chervonenkis (VC-) dimension of a neural network architecture provides a quick estimate of the complexity of training a neural network – the larger this parameter, the more difficult it is to train the network. The topic of estimating the VC-dimension of various neural network architectures is a topic of current research, and many of the known results are summarized in Chapter 10.

The issues of intelligence, learning, and generalization are also relevant in the context of control systems. At least two branches of control theory, namely system identification and adaptive control, deal with control systems that "learn." However, the mathematical formulations used in these areas are quite different from the problem formulations used here. It appears that the type of learning theory studied here cannot be *directly* applied to problems in control systems, except in special situations. What is needed is a reformulation of the basic problems so as to reflect more closely the goals of control theory. However, the *methodology* will perhaps prove useful even in a control-theoretic setting. Chapter 11 gives an indication of how such a theory might be expected to evolve.

The theory presented in this book is reasonably complete and self-contained. Nevertheless, there are several interesting open problems that merit the attention of the research community. Some such problems are discussed in Chapter 12.

Now some suggestions on how this book might be used. Learning theory uses the terminology of probability theory and stochastic processes, though in fact many of the deep concepts of probability theory are not used. Thus a reader who is not familiar with probability theory might be put off at first by the heavy probabilistic flavour of the subject. It is suggested that such a reader ought to read Chapter 1, skip Chapter 2, and go straight to Chapter 3, which gives the all-important *raison d'être* of the subject. He/she can always return to Chapter 2 as the need arises. On the other hand, a reader who is well-versed in probability theory can read the book sequentially. The first four chapters form the foundation for all that follows. Chapter 5 is at the heart of empirical process theory, as it contains the basic necessary and sufficient conditions that are used repeatedly in subsequent chapters. However, the proofs in this chapter are long and technical, and a reader who is uninterested in these details can accept the main results "on faith" and proceed to subsequent chapters. The introduction to Chapter 5 gives a more detailed "road map" on how to navigate through this chapter. The remaining chapters are more or less independent.

The book is more or less self-contained, and references to the literature are included where appropriate. Chapter 2 gives a summary of the background required of the reader. Throughout, **bold face** is used in definitions to delineate the notion being defined, while italics are used for *emphasis*. The end of proof symbol ■ is also used where necessary to indicate the end of an example. Throughout, the symbols $O(\cdot)$ and $\Omega(\cdot)$ are used in their conventional meaning: $x = O(y)$ if there exist constants y_0 and k such that $x \leq ky$ for all $y \in (0, y_0)$. Similarly, $x = \Omega(y)$ if there exist constants y_0 and k such that $x \geq ky$ for all $y \in (0, y_0)$.

2. Preliminaries

In this preliminary chapter, we review some of the basic concepts and results used in the remainder of the book. The main purpose of this chapter is to collect all of these results in a single place, so that a reader can become aware at once as to what background he/she needs in order to be able to understand the material that follows. Though some attempt is made to begin from first principles and to give a few proofs and examples, it must be emphasized that the chapter is *not* intented to be a complete treatment of the topics mentioned. In particular, a reader who is encountering a topic for the first time is encouraged to consult one of the standard references cited at the end of the chapter.

2.1 Pseudometric Spaces, Packing and Covering Numbers

2.1.1 Pseudometric Spaces

Suppose X is a set. A function $\rho : X \times X \to \mathbb{R}_+$ is said to be a **pseudometric** if

(i) $\rho(x,x) = 0$, $\forall x \in X$,
(ii) $\rho(x,y) = \rho(y,x)$, $\forall x, y \in X$, and
(iii) $\rho(x,z) \leq \rho(x,y) + \rho(y,z)$, $\forall x, y, z \in X$.

If, in addition, ρ satisfies

(iv) $\rho(x,y) = 0 \Rightarrow x = y$,

then ρ is said to be a **metric**.

If ρ is a pseudometric (metric) on X, we say that (X, ρ) is a **pseudometric space** (**metric space**).

Suppose (X, ρ) is a pseudometric space, and suppose a binary relation $\sim$ on X is defined as follows:

$$x \sim y \Leftrightarrow \rho(x,y) = 0.$$

Then it is easy to verify that $\sim$ is in fact an *equivalence* relation on X. The reflexivity of $\sim$ follows from (i) above, the symmetry from (ii), and the transitivity from (iii). Hence X can be partitioned into its equivalence classes under $\sim$. Let $\tilde{X}$ denote the resulting collection of equivalence classes; thus a typical element of $\tilde{X}$ is of the form $[x]$ where $x \in X$ and $[x]$ is the corresponding equivalence class under $\sim$. Now, given $[x], [y] \in \tilde{X}$, define

$$\tilde{\rho}([x],[y]) = \rho(x,y).$$

It can be easily verified that $\tilde{\rho}$ is well-defined; that is, $\tilde{\rho}([x],[y])$ is independent of the particular $x \in [x], y \in [y]$ used in the right side of the above equation. Also, $\tilde{\rho}$ is a *metric* on $\tilde{X}$. Thus the fact that a pseudometric ρ might not satisfy (iv) above need not cause much consternation, and in fact, all the familiar metric space concepts of neighbourhoods, open sets, closed sets, etc. can be readily adapted to pseudometric spaces.

Suppose (X, ρ) is a pseudometric space, and that $x \in X, \epsilon > 0$. Then we denote

$$\mathcal{B}(\epsilon, x, \rho) := \{y \in X : \rho(x,y) < \epsilon\}, \text{ and}$$

$$\bar{\mathcal{B}}(\epsilon, x, \rho) := \{y \in X : \rho(x,y) \leq \epsilon\}.$$

Thus $\mathcal{B}(\epsilon, x, \rho)$ and $\bar{\mathcal{B}}(\epsilon, x, \rho)$ are respectively the open and closed balls (with respect to the pseudometric ρ) of radius ϵ centered at x.

2.1.2 Packing and Covering Numbers

Suppose (X, ρ) is a pseudometric space, and that $S \subseteq X$. Given $\epsilon > 0$, a set $\{a_1, \ldots, a_n\} \subseteq S$ is said to be an ϵ-**cover** of S if, for each $x \in S$, there exists an index i such that $\rho(x, a_i) \leq \epsilon$. Equivalently, a set $\{a_1, \ldots, a_n\}$ is an ϵ-*cover* of S if $a_i \in S$ for all i, and in addition

$$\bigcup_{i=1}^{n} \bar{\mathcal{B}}(\epsilon, a_i, \rho) \supseteq S.$$

The ϵ-**covering number** of S (with respect to the pseudometric ρ) is defined as the smallest number n such that S has an ϵ-cover of cardinality n, and is denoted by $N(\epsilon, S, \rho)$. An ϵ-cover of this cardinality is said to be a **minimal ϵ-cover**. Note that a minimal ϵ-cover need not be unique, but the ϵ-covering number is well-defined (and could perhaps be infinite). Similarly, a set $\{b_1, \ldots, b_l\} \subseteq X$ is said to be an **external ϵ-cover** of S if

$$\bigcup_{i=1}^{l} \bar{\mathcal{B}}(\epsilon, b_i, \rho) \supseteq S.$$

The key point to note here is that the b_i's need not themselves belong to S. The **external ϵ-covering number** of S is defined as the smallest number l such that S has an external ϵ-cover of cardinality l, and is denoted by $L(\epsilon, S, \rho)$.

2.1 Pseudometric Spaces, Packing and Covering Numbers

The above definitions are not quite standard, in two ways. First, some authors define the (external) ϵ-covering number as the smallest number of *open* balls of radius ϵ needed to cover S, as opposed to the smallest number of *closed* balls of radius ϵ, as is done here. In the context of learning theory, the definition adopted here offers some advantages. Second, the term "covering number" is used with different meanings by different authors. For instance, Vapnik [190] uses "cover" to mean what we call here as an "external cover," while our "cover" is his "proper cover."

Lemma 2.1. *For each $S \subseteq X$ and each $\epsilon > 0$, it is true that*

$$N(2\epsilon, S, \rho) \leq L(\epsilon, S, \rho) \leq N(\epsilon, S, \rho).$$

In particular, the following statements are equivalent:

(i) The ϵ-covering number of S is finite for each ϵ.
(ii) The external ϵ-covering number of S is finite for each ϵ.

Proof. Obviously the right inequality is valid, because every ϵ-cover is also an external ϵ-cover. To prove the left inequality, suppose $\{b_1, \ldots, b_m\} \subseteq X$ (not S!) is an external ϵ-cover of S of minimal cardinality. Then each closed ball $\bar{B}(\epsilon, b_i, \rho)$ contains an element of S – if not, then b_i can be dropped from the ϵ-cover, thus contradicting the minimality of the cover. For each $i = 1, \ldots m$, choose an $a_i \in S \cap \bar{B}(\epsilon, b_i, \rho)$. Then, by the triangle inequality, it follows that $\{a_1, \ldots, a_m\}$ is a 2ϵ-cover of S, because every $x \in S$ is within a distance ϵ of some b_i, which in turn is within ϵ of a_i. ∎

A set $\{b_1, \ldots, b_m\} \subseteq S$ is said to be ϵ-**separated** if $\rho(b_i, b_j) > \epsilon \; \forall i \neq j$. The ϵ-**packing number** of S is defined as the *largest* number m such that S contains an ϵ-separated set of cardinality m, and is denoted by $M(\epsilon, S, \rho)$. An ϵ-separated set of this cardinality is called a **maximal ϵ-separated set**.

Note that some authors call a set $\{b_1, \ldots, b_m\}$ "ϵ-separated" if $\rho(b_i, b_j) \geq \epsilon$ for all $i \neq j$, as opposed to the present definition which requires that $\rho(b_i, b_j) > \epsilon \; \forall i \neq j$.

Lemma 2.2. *For each $S \subseteq X$ and $\epsilon > 0$, it is true that*

$$M(2\epsilon, S, \rho) \leq L(\epsilon, S, \rho) \leq N(\epsilon, S, \rho) \leq M(\epsilon, S, \rho).$$

Proof. Suppose $\{a_1, \ldots, a_k\}$ is a maximal 2ϵ-separated set in S. Then no closed ball of radius ϵ can contain more than one a_i (by the triangle inequality). This is true irrespective of whether the center of the ball belongs to S or not. This proves the left inequality. The middle inequality is already proved in Lemma 2.1.

Suppose $\{b_1, \ldots, b_m\} \subseteq S$ is a maximal ϵ-separated set. Then $\{b_1, \ldots, b_m\}$ must be an ϵ-cover of S – otherwise there would exist a b_{m+1} that is more than ϵ-far from each of $b_1, \ldots, b_m$, thus contradicting the maximality. This establishes the right inequality. ∎

16   2. Preliminaries

See [107] for an excellent discussion of packing and covering numbers, as well as a wealth of examples.

Lemma 2.3. *Suppose (X, ρ) is a pseudometric space, and let $S \subseteq X$. Then each of the three functions $L(\epsilon, S, \rho)$, $N(\epsilon, S, \rho)$, and $M(\epsilon, S, \rho)$ is a nondecreasing function of ϵ as ϵ decreases towards zero; that is*

$$L(\epsilon_1, S, \rho) \geq L(\epsilon_2, S, \rho) \text{ if } \epsilon_1 \leq \epsilon_2,$$

and similarly for the other two functions. Each function is continuous from the right; that is

$$L(\epsilon_0, S, \rho) = \lim_{\epsilon \to \epsilon_0^+} L(\epsilon, S, \rho), \forall \epsilon_0 > 0,$$

and similarly for the other two functions.

Proof. The fact that each function is nondecreasing is obvious. For a proof of the second part of the lemma, see [107], Theorem III. ∎

2.1.3 Compact and Totally Bounded Sets

Suppose (X, ρ) is a pseudometric space, and that $S \subseteq X$. Then we say that S is **compact** if every open cover of S has a finite subcover, and that S is **totally bounded** if

$$N(\epsilon, S, \rho) < \infty \; \forall \epsilon > 0,$$

i.e., if S has a finite ϵ-covering number for each $\epsilon > 0$. Note that, from Lemmas 2.3 and 2.1, the above condition is equivalent to

$$L(\epsilon, S, \rho) < \infty \; \forall \epsilon > 0,$$

and to

$$M(\epsilon, S, \rho) < \infty \; \forall \epsilon > 0.$$

Note that, instead of "totally bounded," one could also call such a set "precompact," because of the following result:

Lemma 2.4. *Suppose (X, ρ) is a pseudometric space, and that $S \subseteq X$. Then S is compact if and only if it is totally bounded and closed.*

Proof. See [100], p. 198, Theorem 32. ∎

Thus total boundedness is "almost" the same as compactness, the only difference being that a compact set is also closed, whereas a totally bounded set may or may not be closed.

2.2 Probability Measures

2.2.1 Definition of a Probability Space

Suppose X is a set. A (nonempty) collection $\mathcal{S}$ of subsets of X is said to be a σ-algebra if it satisfies the following:

(i) $\mathcal{S}$ is closed under complementation; i.e., $A \in \mathcal{S} \Rightarrow A^c \in \mathcal{S}$.
(ii) $\mathcal{S}$ is closed under countable union; i.e., if $A_i \in \mathcal{S}$ for $i = 1, 2, \ldots$, then $\cup_{i=1}^{\infty} A_i \in \mathcal{S}$.

It is an easy consequence of (i) and (ii) that $\mathcal{S}$ is also closed under countable intersection.

Suppose (X, ρ) is a pseudometric space. Then the smallest σ-algebra of subsets of X that contains every closed subset of X is called the **Borel σ-algebra** of (X, ρ). Note that, by Condition (i) above, the Borel σ-algebra also contains every open subset of X.

If $\mathcal{S}$ is a set and $\mathcal{S}$ is a σ-algebra of subsets of X, then the pair $(X, \mathcal{S})$ is called a **measurable space**. Suppose $(X, \mathcal{S})$, $(Y, \mathcal{T})$ are measurable spaces, and that $f : X \to Y$. Then f is said to be a **measurable function** if $f^{-1}(T) \in \mathcal{S}$ whenever $T \in \mathcal{T}$.

A function $\mu : \mathcal{S} \to \mathbb{R}_+$ is said to be a **measure** if $\mu(\emptyset) = 0$, and μ is countably additive; that is, if $A_i \in \mathcal{S}$, $i = 1, 2, \ldots$ is a finite or countable collection of pairwise disjoint sets, then

$$\mu\left(\bigcup_i A_i\right) = \sum_i \mu(A_i).$$

A ready consequence of the above property is the subadditivity property: Suppose $A_i \in \mathcal{S}$, $i = 1, 2, \ldots$ is a countable collection of sets (not necessarily pairwise disjoint). Then

$$\mu\left(\bigcup_{i=1}^{\infty} A_i\right) \leq \sum_{i=1}^{\infty} \mu(A_i).$$

A measure $P : \mathcal{S} \to \mathbb{R}_+$ is said to be a **probability measure** if $P(X) = 1$. We refer to $(X, \mathcal{S}, P)$ as a **probability space**.

We do not discuss at all the notion of *integrating* a measurable function, as such a discussion would take us too far afield. The reader is referred instead to [77] for a thorough treatment. It is not really necessary for the reader to master the various intricacies of of integration in the measure-theoretic sense in order to follow the contents of the book.

2.2.2 A Pseudometric Induced by a Probability Measure

Suppose $(X, \mathcal{S}, P)$ is a probability space. Then P induces a pseudometric on $\mathcal{S}$, as follows: For each $A, B \subseteq X$, define their **symmetric difference** $A \Delta B$ by
$$A \Delta B = (A^c \cap B) \cup (A \cap B^c),$$
where A^c denotes the complement of the set A. An equivalent definition is:
$$A \Delta B = (A \cup B) - (A \cap B).$$
Evidently, $A \Delta B$ is the set of points that belong to *exactly* one of the two sets A and B. It is easy to see that $A \Delta B = B \Delta A$, so that Δ is indeed symmetric. Also, it is tedious but routine to verify that, for three sets A, B, C, we have
$$A \Delta B = (A \Delta C) \Delta (B \Delta C).$$
Clearly, if $A, B \in \mathcal{S}$, then $A \Delta B \in \mathcal{S}$. Thus it is possible to define the function $d_P : \mathcal{S} \times \mathcal{S} \to [0, 1]$ by
$$d_P(A, B) = P(A \Delta B).$$
Now it is a routine matter to verify that d_P is a pseudometric on $\mathcal{S}$. Axioms (i) and (ii) of the definition follow readily. To prove the triangle inequality, one uses the fact that
$$A \Delta C \subseteq (A \Delta B) \cup (B \Delta C),$$
whence
$$P(A \Delta C) \leq P(A \Delta B) + P(B \Delta C).$$
However, d_P is in general *not* a metric because $d_P(A, B) = 0$ whenever $A \Delta B$ is a set of zero measure, even if $A \Delta B \neq \emptyset$.

More generally, let $[0,1]^X$ denote the set of *measurable* functions mapping X into $[0, 1]$, when $[0, 1]$ is equipped with the Borel σ-algebra.[1] Then one can define a pseudometric d_P on $[0, 1]^X$ by
$$d_P(f, g) = \int_X |f(x) - g(x)| \, P(dx), \ \forall f, g \in [0, 1]^X.$$
It is easy to verify that this d_P is also a pseudometric. In general it is not a metric, because $d_P(f, g) = 0$ whenever f and g differ on a set of measure zero, even if $f \neq g$. Actually, this d_P is a *generalization* of the earlier d_P defined on $\mathcal{S}$, which justifies the use of the same symbol for both. To see this, observe that there is a one-to-one correspondence between sets in $\mathcal{S}$ and (measurable) functions mapping X into $\{0, 1\}$. Specifically, if $A \in \mathcal{S}$, then its indicator function $I_A(\cdot)$ defined by

---

[1] This is a slight abuse of notation because, strictly speaking, $[0,1]^X$ should denote the set of *all* functions mapping X into $[0, 1]$, measurable or otherwise.

$$I_A(x) = \begin{cases} 1 & \text{if } x \in A, \\ 0 & \text{if } x \notin A, \end{cases}$$

is measurable and maps X into $\{0,1\}$. Conversely, suppose $f : X \to \{0,1\}$ is measurable. Then its support defined by

$$\mathrm{supp}(f) = \{x \in X : f(x) = 1\}$$

belongs to $\mathcal{S}$. Now, if $A, B \in \mathcal{S}$, then it is easy to see that

$$d_P(A, B) = d_P(I_A, I_B),$$

where the d_P on the left side is defined on $\mathcal{S}$ while the d_P on the right side is defined on $[0,1]^X$. This justifies the use of the same symbol for both quantities.

2.2.3 A Metric on the Set of Probability Measures

Suppose $(X, \mathcal{S})$ is a measurable space, and let $\mathcal{P}^*$ denote the set of all probability measures on $(X, \mathcal{S})$. It is possible to define a metric on $\mathcal{P}^*$ as follows: Given $P, Q \in \mathcal{P}^*$, let

$$\rho(P, Q; \mathcal{S}) := \sup_{A \in \mathcal{S}} |P(A) - Q(A)|.$$

The function ρ is indeed a metric (and not merely a pseudometric) because, if P, Q are probability measures on $(X, \mathcal{S})$ and $P \neq Q$, then there exists at least one set $A \in \mathcal{S}$ such that $P(A) \neq Q(A)$; hence $\rho(P, Q) > 0$. Note that ρ is called the **total variation metric** on $\mathcal{P}^*$. In the above definition, the underlying σ-algebra $\mathcal{S}$ is explicitly highlighted, since there will be cases when we will compute the total variation metric between the same pair of probability measures, but with respect to different σ-algebras. However, if $\mathcal{S}$ is obvious from the context, then it can be omitted, and we can simply write $\rho(P, Q)$.

Suppose $X \subseteq \mathbb{R}$, and that P, Q are probability measures with densities $p(\cdot)$ and $q(\cdot)$ respectively. Thus, if $A \subseteq X$ is measurable, then

$$P(A) = \int_A p(x)\, dx, \text{ and } Q(A) = \int_A q(x)\, dx.$$

Then the total variation metric between P and Q equals

$$\rho(P, Q) = \int_X |f(x) - g(x)|\, dx;$$

that is, $\rho(P, Q)$ is the L_1-distance between the densities $p(\cdot)$ and $q(\cdot)$.

Now it is shown that the total variation metric has a very useful property. In order to present it, we begin by discussing the notion of product σ-algebras and product probability measures.

Suppose $(X, \mathcal{S}, P)$ and $(Y, \mathcal{T}, R)$ are probability spaces. Then a set of the form $A \times B, A \in \mathcal{S}, B \in \mathcal{T}$ is called a "cylinder set." The smallest σ-algebra on $X \times Y$ that contains all such cylinder sets is called the "product σ-algebra" and is denoted by $\mathcal{S} \times \mathcal{T}$. By defining

$$R(A \times B) := P(A) \cdot Q(B), \ \forall A \in \mathcal{S}, B \in \mathcal{T},$$

we can define the measure of cylinder sets, which can then be extended, via the Kolmogorov extension theorem, to all of $\mathcal{S} \times \mathcal{T}$. The resulting (probability) measure R is called the "product measure" and is denoted by $P \times Q$. Clearly the idea can be extended to any finite number of probability spaces, and indeed, to even more general situations; see Section 2.4.

Lemma 2.5. *Suppose* $(X, \mathcal{S}, P), (Y, \mathcal{T}, Q)$ *and* $(Y, \mathcal{T}, R)$ *are probability spaces. Then*

$$\rho(P \times Q, P \times R) = \rho(Q, R).$$

Proof. Consider the collection of sets of the form $C = \cup_{i=1}^{\infty}(A_i \times B_i)$, where $A_i \in \mathcal{S}$, $B_i \in \mathcal{T}$, and the A_i are pairwise disjoint. Such a collection of sets forms an "algebra" in that it is closed under complementation and finite union; but it might not be a σ-algebra since it might not be closed under *countable* union. However, it can be shown that $\rho(P \times Q, P \times R)$ equals the supremum of the difference $|(P \times Q)(C) - (P \times R)(C)|$ over all such sets C.

Now

$$
\begin{aligned}
|(P \times Q)(C) - (P \times R)(C)| &\leq \sum_{i=1}^{\infty} |(P \times Q)(A_i \times B_i) - (P \times R)(A_i \times B_i)| \\
&= \sum_{i=1}^{\infty} P(A_i)|Q(B_i) - R(B_i)| \\
&\leq \left(\sum_{i=1}^{\infty} P(A_i)\right) \cdot \rho(Q, R) \\
&\leq \rho(Q, R).
\end{aligned}
$$

This shows that $\rho(P \times Q, P \times R) \leq \rho(Q, R)$. The opposite inequality can be proven by considering sets of the form $X \times B, B \in \mathcal{T}$. ∎

Lemma 2.6. *Suppose* $(X, \mathcal{S})$ *is a measurable space, and that* P, Q *are probability measures on this space. Then*

$$\rho(P^k, Q^k) \leq k\rho(P, Q).$$

Proof. By the triangle inequality, we have

$$\rho(P^k, Q^k) \leq \rho(P^k, P^{k-1}Q) + \ldots + \rho(PQ^{k-1}, Q^k).$$

By Lemma 2.5, each of the quantities on the right side equals $\rho(P, Q)$. ∎

One consequence of Lemma 2.6 is that if $\{P_i\}$ is a sequence of probability measures on $(X, \mathcal{S})$ converging to the probability measure Q, then $P_i^k \to Q^k$ for each *fixed* integer k. At the same time, it can also be shown that P, Q are distinct probability measures on $(X, \mathcal{S})$, then $\rho(P^k, Q^k) \to 1$ as $k \to \infty$. That is, as $k \to \infty$, the measures P^k, Q^k tend to become "mutually singular" in that they are supported on disjoint sets.

2.2.4 Random Variables

Suppose $(\Omega, \mathcal{T}, Q)$ is a probability space. Thus Ω is a set, $\mathcal{T}$ is a σ-algebra of subsets of Ω, and Q is a probability measure on $(\Omega, \mathcal{T})$. Suppose $(X, \mathcal{S})$ is a measurable space. Then an X-**valued random variable** is defined as a measurable map, call it X, from $(\Omega, \mathcal{T}, Q)$ to $(X, \mathcal{S})$. Note that in the probability literature, it is common to restrict the term "random variable" to the situation where $X = \mathbb{R}$ and $\mathcal{S}$ is the Borel σ-algebra. At best, X is taken to be $\mathbb{R}^k$ for some integer k and $\mathcal{S}$ is taken to be the associated Borel σ-algebra. However, for present purposes, it is desirable to adopt the more general usage stated above.

Suppose f is a measurable map from a probability space $(\Omega, \mathcal{T}, Q)$ into $\mathbb{R}$. Thus one can also think of f as a real-valued random variable. The **expected value** of the function f is defined as

$$E(f, Q) := \int_{\Omega} f(\omega)\, Q(d\omega),$$

assuming of course that the integral is well-defined. In particular, we can speak of the expected value of a random variable X with probability measure P, and denote it by $E(\mathrm{X}, P)$. If P is clear from the context (or if it does not matter what P is), then we simply write $E(\mathrm{X})$.

Suppose X is a real-valued random variable. Then the **distribution function** of X is the function P_X mapping $\mathbb{R}$ into $[0, 1]$ defined by

$$P_\mathrm{X}(a) := Q\{\omega \in \Omega : \mathrm{X}(\omega) \leq a\}.$$

It is obvious that $P_\mathrm{X}(a)$ is nondecreasing as a function of a. The distribution function has a property known as "cadlag," which is an abbreviation for the French expression "continuité a droite, limité a gauche." What it means is that the distribution function is continuous from the right, and has well-defined limits from the left. Thus

$$\lim_{a \to a_0^+} P_X(a) = P_\mathrm{X}(a_0),\ \forall a_0 \in \mathbb{R}, \text{ and } \lim_{a \to a_0^-} P_\mathrm{X}(a) \text{ exists}.$$

If $\mathrm{X}_1, \mathrm{X}_2$ are real-valued random variables defined on a common probability space $(\Omega, \mathcal{T}, Q)$, then one can define their **joint distribution** as follows:

$$P_{\mathrm{X}_1, \mathrm{X}_2}(a_1, a_2) := Q\{\omega \in \Omega : \mathrm{X}_1(\omega) \leq a_1,\ \mathrm{X}_2(\omega) \leq a_2\}.$$

It is obvious that the notion of a joint distribution can be readily extended to any finite number of real-valued random variables.

Suppose $(\Omega, \mathcal{T}, Q)$ and $(\Omega', \mathcal{T}', Q')$ are probability spaces, and that X, X' are random variables mapping $(\Omega, \mathcal{T}, Q)$ into $\mathbb{R}$ and $(\Omega', \mathcal{T}', Q')$ into $\mathbb{R}$ respectively. Then the random variables X and X' are said to have the same "law" if they have the same distribution function, that is, $P_X(\cdot) = P_{X'}(\cdot)$. The point is that the domain of a random variable is really not important.

Suppose $(\Omega, \mathcal{T}, Q)$ is a probability space, and that $A, B \in \mathcal{T}$. Thus A and B are deemed to be "events." The events A and B are said to be **independent** under the probability measure Q if $Q(A \cap B) = Q(A) \cdot Q(B)$. With a little bit of work, this notion can be extended to define the notion of independence for random variables.

Suppose X_1, X_2 are random variables defined on a common probability space $(\Omega, \mathcal{T}, Q)$. Thus X_i maps Ω into a measurable space $(X_i, \mathcal{S}_i)$ for $i = 1, 2$. Then the random variables X_1 and X_2 are independent if, for every $A \in \mathcal{S}_1, B \in \mathcal{S}_2$, the preimages $X_1^{-1}(A)$ and $X_2^{-1}(B)$ are independent. Equivalently, the random variables X_1 and X_2 are independent if

$$Q\{X_1(\omega) \in A,\ X_2(\omega) \in B\} = Q\{X_1(\omega) \in A\}\, Q\{X_2(\omega) \in B\}.$$

Note that in the above equation we employ the commonly used abbreviation, whereby $Q\{S\}$ denotes $Q\{\omega \in \Omega : S \text{ is true }\}$. In particular, if X_1 and X_2 are real-valued random variables, then X_1 and X_2 are independent if and only if the joint distribution function is a product of the individual distribution functions, that is, if and only if

$$P_{X_1, X_2}(a_1, a_2) = P_{X_1}(a_1) \cdot P_{X_2}(a_2).$$

More generally, suppose as above that X_1, X_2 are random variables defined on a common probability space. Define $X = X_1 \times X_2$, and let $\mathcal{S}$ denote the *product σ-algebra*, $\mathcal{S}_1 \times \mathcal{S}_2$, which consists of the smallest σ-algebra that contains all "cylinder sets" of the form $A \times B$, $A \in \mathcal{S}_1, B \in \mathcal{S}_2$. Then one can think of the map $\omega \mapsto (X_1(\omega), X_2(\omega))$ as a random variable taking values in the measurable space $(X, \mathcal{S})$. Then the collection $X_1^{-1}(A)$, $A \in \mathcal{S}_1$ is a σ-algebra on Ω, called the σ-algebra **generated** by X_1, and denoted by $\Sigma(X_1)$. Similarly, X_2 also generates a σ-algebra on Ω, denoted by $\Sigma(X_2)$. Now consider the product σ-algebra generated by $\Sigma(X_1)$ and $\Sigma(X_2)$. Denote it by $\Sigma(X_1, X_2)$, since it is the σ-algebra generated jointly by the two variables X_1 and X_2. Since both X_1 and X_2 are measurable maps, $\Sigma(X_1, X_2) \subseteq \mathcal{T}$. Now the probability measure Q, restricted to $\Sigma(X_i)$, is called the **one-dimensional marginal probability measure** corresponding to X_i, and denoted by $Q_{\Sigma(X_i)}$. Note that the original σ-algebra $\mathcal{T}$ could be strictly bigger than $\Sigma(X_1, X_2)$, but for the purposes of checking whether or not X_1, X_2 are independent, we do not need to work with all of $\mathcal{T}$, only $\Sigma(X_1, X_2)$. With these definitions, it is easy to see that X_1, X_2 are independent if and only if

$$Q_{\Sigma(X_1, X_2)} = Q_{\Sigma(X_1)} \times Q_{\Sigma(X_2)}.$$

2.2.5 Conditional Expectations

In this we give a strictly functional description of conditional expectations. The present treatment is much less general than is normally found in most texts on probability, but it is good enough for present purposes.

Suppose $(\Omega, \mathcal{T}, P)$ is a probability space, and that f is a measurable map from $(\Omega, \mathcal{T})$ into the real numbers. Let $\mathcal{B}$ denote the Borel σ-algebra on $\mathbb{R}$. Then the collection of preimages under f of Borel measurable sets is called the **σ-algebra generated by** f, and is denoted by $\Sigma(f)$, or sometimes by $\mathcal{F}$. Thus

$$\mathcal{F} = \Sigma(f) := \{f^{-1}(S), S \in \mathcal{B}\}.$$

Suppose Σ is a subalgebra of $\mathcal{T}$, and let $\mathcal{M}(\Sigma)$ denote the set of all functions from Ω into $\mathbb{R}$ that are measurable with respect to Σ. Clearly $\mathcal{M}(\Sigma)$ is a linear vector space. Suppose now that g is another measurable mapping from $(\Omega, \mathcal{T}, P)$ into $\mathbb{R}$ with finite variance. This means that

$$\int_\Omega g^2(\omega) \, P(d\omega) < \infty.$$

To put it another way, $g(\cdot)$ belongs to $L_2(\Omega, \mathcal{T}, P)$. Then the **conditional expectation of g with respect to the σ-algebra** Σ is defined as the unique function $h \in \mathcal{M}(\Sigma)$ that minimizes the mean-squared error

$$\int_X |h(\omega) - g(\omega)|^2 \, P(d\omega).$$

The conditional expectation of g with respect to Σ is denoted by g_Σ. Note that, whereas the expected value of a random variable is a real number, the conditional expectation of a random variable is itself a random variable. This is why, even though it is customary in the probability literature to denote the conditional expectation by $E(g|\Sigma)$, we prefer to use the notation $g_\mathcal{F}$. Moreover, suppose $X_1, \ldots, X_n$ are random variables on $(\Omega, \mathcal{T}, P)$, and let Σ denote the σ-algebra generated by these random variables. Then any function $h \in \mathcal{M}(\Sigma)$ can be written as $h = h(X_1, \ldots, X_n)$ where h is a measurable function from $\mathbb{R}^n$ into $\mathbb{R}$. Therefore, if $h = g_\Sigma$, then we can write $g_\Sigma = h(X_1, \ldots, X_n)$. This relationship expresses the "best estimate" of g very explicitly as a function of the n random variables, and is the motivation for the notation $E(g|X_1, \ldots, X_n)$.

It is possible to define the notion of conditional expectation in much more general settings. However, by restricting attention to random variables of finite variance, the technicalities are kept to a minimum. In particular, it is a ready consequence of the projection theorem that there exists a unique conditional expectation, and that the error term $g - g_\mathcal{F}$ is "orthogonal" to the space $\mathcal{M}(\mathcal{F})$. Therefore

$$\int_X h(\omega)[g(\omega) - g_\mathcal{F}(\omega)] \, P(d\omega) = 0, \ \forall h \in \mathcal{M}(\mathcal{F}),$$

provided only that the integral is well-defined. In particular, the above equality holds for every h with finite variance. The above relationship can be written very simply as

$$E[h(g - g_{\mathcal{F}})] = 0, \ E(hg) = E(hg_{\mathcal{F}}), \ \forall h \in \mathcal{M}(\mathcal{F}). \tag{2.2.1}$$

2.3 Large Deviation Type Inequalities

An important, and recurring, theme in these notes is the use of so-called "large deviation" type inequalities. These inequalities give an estimate of the probability that an "average" of independent random variables differs considerably from its mean value. In this section, several inequalities used in the sequel are summarized.

2.3.1 Chernoff Bounds

Suppose X is a random variable with only two possible values, namely 0 and 1, and suppose further that the probability that X = 1 is p. Then clearly $E(X) = p$. Let $x_1, \ldots, x_m$ denote independent samples of X; these are also known as **Bernoulli trials**. Now define

$$S_m = \sum_{i=1}^{m} x_i, \ A_m = \frac{1}{m} \sum_{i=1}^{m} x_i.$$

Then A_m can be thought of as the *empirical mean* of the random variable X. In other words, A_m is an *estimate* of the probability that X = 1 based on m trials. Note that A_m is itself a random variable. The probability that A_m exceeds a number r can be expressed as

$$\Pr\{A_m \geq r\} = \Pr\{S_m \geq mr\} = \sum_{k \geq mr}^{m} \binom{m}{k} p^k (1-p)^{m-k}.$$

Similarly, the probability that A_m is less than a given number r is given by

$$\Pr\{A_m \leq r\} = \sum_{k=0}^{k \leq mr} \binom{m}{k} p^k (1-p)^{m-k}.$$

The **Chernoff bounds** give upper bounds on the right side of the above inequalities. The bounds can be given in either additive form or multiplicative form. In the additive form, the bounds state that, for all $\epsilon \in [0, 1]$,

$$\Pr\{A_m \geq p + \epsilon\} \leq \exp(-2m\epsilon^2), \text{ and}$$

$$\Pr\{A_m \leq p - \epsilon\} \leq \exp(-2m\epsilon^2).$$

Combining these two inequalities gives
$$\Pr\{|A_m - p| \geq \epsilon\} \leq 2\exp(-2m\epsilon^2).$$
In the multiplicative form the bounds state that, for all $\gamma \in [0, 1]$,
$$\Pr\{A_m \geq (1+\gamma)p\} \leq \exp(-\gamma^2 mp/3), \text{ and}$$
$$\Pr\{A_m \leq (1-\gamma)p\} \leq \exp(-\gamma^2 mp/2).$$
Note that, unlike the additive bounds, the multiplicative bounds are not "symmetric." Also, the multiplicative bounds require that $\gamma \leq 1$, whereas the additive bounds do not place any restrictions on the size of ϵ relative to p.

It is worth noting that, in case p is itself "small," the multiplicative bounds are less conservative than the additive bounds. To illustrate, suppose $p = \epsilon$, a "small" number, and let us bound the probability that $A_m \geq 1.5\epsilon$. The additive form of the Chernoff bound leads to the estimate
$$\Pr\{A_m \geq 1.5\epsilon\} \leq \exp(-m\epsilon^2/2),$$
whereas the multiplicative form leads to the estimate
$$\Pr\{A_m \geq 1.5\epsilon\} \leq \exp(-m\epsilon/12)$$
after substituting $\gamma = 0.5, p = \epsilon$. It is important to note that the first bound contains an ϵ^2 in the exponent, whereas the second bound contains only ϵ. Hence, when ϵ is small, the second bound is considerably superior to the first.

In the multiplicative form of the Chernoff bound, the restriction that $\gamma \leq 1$ is not serious when it is desired to estimate $\Pr\{A_m \leq (1-\gamma)p\}$. However, it is a bit of a nuisance when it is desired to estimate $\Pr\{A_m \geq (1+\gamma)p\}$, since the quantity $(1+\gamma)p$ is effectively limited to the range $[p, 2p]$. Obviously situations can arise where one would like to estimate $\Pr\{A_m \geq (1+\gamma)p\}$ with $\gamma > 1$. In such cases, the Chernoff bound cannot be applied directly. However, with a little imagination, the range of applicability of the bound can be "stretched." Specifically, observe that the map $p \mapsto p^k(1-p)^{m-k}$ is nondecreasing whenever $p \leq k/m$. This observation leads to the following alternate form of the multiplicative Chernoff bound. Suppose X is a Bernoulli process, and that the probability $\Pr\{X = 1\}$ is *less than or equal to* μ. Let A_m denote the m-fold average of independent observations of X, as above. Then
$$\Pr\{A_m \geq (1+\gamma)\mu\} \leq \exp(-\gamma^2 m\mu/3), \text{ for } 0 \leq \gamma \leq 1, E(X) \leq \mu. \quad (2.3.1)$$
Note that it is permissible for $(1+\gamma)\mu$ to exceed $2p$, where $p = \Pr\{X = 1\}$. The above inequality follows from
$$\Pr\{A_m \geq (1+\gamma)\mu\} = \sum_{k \geq (1+\gamma)\mu m} \binom{m}{k} p^k (1-p)^{m-k}$$

$$\leq \sum_{k \geq (1+\gamma)\mu m} \binom{m}{k} \mu^k (1-\mu)^{m-k}$$

$$\leq \exp(-\gamma^2 m\mu/3),$$

where the last inequality follows from the "standard" Chernoff bound. Now suppose it is desired to use this alternate form to estimate $\Pr\{A_m \geq (1+\delta)p\}$ where $\delta > 1$. Then one can apply the above bound with $\mu = (1+\delta)p/2$ and $\gamma = 1$, and derive that

$$\Pr\{A_m \geq (1+\delta)p\} \leq \exp(-(1+\delta)mp/6), \; \forall \delta > 1.$$

In this way, the multiplicative form of the Chernoff bound can be extended to cover all values in the range $[p, 1]$.

2.3.2 Chernoff-Okamoto Bound

The **Chernoff-Okamoto bound** is less conservative than the Chernoff bounds, but applies only when $p \leq 0.5$. It states that, if $p \leq 0.5$ and $r \leq p$, then

$$\Pr\{A_m \leq r\} \leq \exp\left[-\frac{m(p-r)^2}{2p(1-p)}\right].$$

By applying the above bound with $r = p - \epsilon$ and $r = (1-\gamma)p$ respectively, and observing that $2p(1-p) \leq 0.5$ for all $p \in [0, 1]$, one can derive two of the four Chernoff bounds above as consequences of the Chernoff-Okamoto bound.

2.3.3 Hoeffding's Inequality

Hoeffding's inequality is a very general inequality that applies to the sum of independent random variables with bounded range.

Lemma 2.7. *Suppose* $Y_1, \ldots, Y_m$ *are independent random variables, and that* $a_i \leq Y_i \leq b_i$ *for each* i. *Suppose* $y_1, \ldots, y_m$ *are realizations of these random variables. Then*

$$\Pr\{\sum_{i=1}^{m}(y_i - E(Y_i)) \geq \alpha\} \leq \exp\left[-2\alpha^2 / \sum_{i=1}^{m}(b_i - a_i)^2\right].$$

Remark: Note that the additive form of the Chernoff bounds can be derived readily from Hoeffding's inequality (but not the multiplicative form).

The proof of Hoeffding's inequality uses the following auxiliary lemma.

Lemma 2.8. *Suppose* X *is a zero-mean random variable assuming values in the interval* $[a, b]$. *Then for any* $s > 0$, *we have*

$$E[\exp(sX)] \leq \exp(s^2(b-a)^2/8).$$

2.3 Large Deviation Type Inequalities

Proof. **(of the auxiliary lemma)**: Since the exponential is a convex function, the value of e^{sx} is bounded by the corresponding convex combination of its extreme values; that is,

$$\exp(sx) \leq \frac{x-a}{b-a}e^{sb} + \frac{b-x}{b-a}e^{sa}, \quad \forall x \in [a,b].$$

Now take the expectation of both sides, and use the fact that $E(X) = 0$. This gives

$$\begin{aligned} E[\exp(sX)] &\leq \frac{b}{b-a}e^{sa} - \frac{a}{b-a}e^{sb} \\ &= (1-p+pe^{s(b-a)})e^{-ps(b-a)} \\ &=: \exp(\phi(u)), \end{aligned}$$

where $p := -a/(b-a)$, $u := s(b-a)$, and $\phi(u) := -pu + \ln(1-p+pe^u)$. Clearly $\phi(u) = 0$. Moreover, a routine calculation shows that

$$\phi'(u) = -p + \frac{p}{p+(1-p)e^{-u}},$$

whence $\phi'(u) = 0$ as well. Moreover,

$$\phi''(u) = \frac{p(1-p)e^{-u}}{(p+(1-p)e^{-u})^2} \leq 0.25.$$

Hence by Taylor's theorem, there exists a $\theta \in [0, u]$ such that

$$\phi(u) = \phi''(\theta)u^2/2 \leq u^2/8 = \frac{s^2(b-a)^2}{8}.$$

This completes the proof. ∎

Proof. **(of Hoeffding's inequality)**: For any nonnegative random variable, we have

$$\Pr\{X \geq \epsilon\} \leq \frac{E(X)}{\epsilon},$$

which is known as **Markov's inequality**. Hence, for every $s > 0$, we have

$$\Pr\{X \geq \epsilon\} = \Pr\{e^{sX} \geq e^{s\epsilon}\} \leq \frac{E[\exp(sX)]}{\exp(s\epsilon)}i = e^{-s\epsilon}E[\exp(sX)].$$

Now apply this inequality to the random variable

$$Z_m := \sum_{i=1}^{m}(Y_i - E(Y_i)),$$

which has zero mean since the Y_i's are independent. Then

$$\Pr\{Z_m \geq \epsilon\} \leq e^{-s\epsilon} E\left[\exp\left(s \sum_{i=1}^{m}(Y_i - E(Y_i))\right)\right]$$

$$= e^{-s\epsilon} \prod_{i=1}^{m} E[e^{s(Y_i - E(Y_i))}] \text{ by independence}$$

$$\leq e^{-s\epsilon} \prod_{i=1}^{m} e^{s^2(b_i - a_i)^2/8} \text{ by Lemma 2.8}$$

$$= \exp\left[-s\epsilon + s^2 \sum_{i=1}^{m} \frac{(b_i - a_i)^2}{8}\right]$$

$$= \exp\left[\frac{-2\epsilon^2}{\sum_{i=1}^{m}(b_i - a_i)^2}\right], \qquad (2.3.2)$$

where the last step follows by choosing

$$s = \frac{4\epsilon}{\sum_{i=1}^{m}(b_i - a_i)^2}.$$

This completes the proof. ∎

Suppose $f : X \to [0,1]$ is measurable with respect to the σ-algebra $\mathcal{S}$, and that P is a probability measure on $(X, \mathcal{S})$. Then

$$E_P(f) := \int_X f(x) \, P(dx)$$

is the **expected value** or **mean** of the function f. Now suppose $x_1, \ldots, x_m$ are i.i.d. samples drawn from X in accordance with P, and define

$$\hat{E}(f; \mathbf{x}) := \frac{1}{m} \sum_{i=1}^{m} f(x_i),$$

where $\mathbf{x} = [x_1 \ldots x_m]^t \in X^m$. Then $\hat{E}(f; \mathbf{x})$ is called the **empirical mean** of the function f corresponding to the multisample $x_1, \ldots, x_m$. Now $f(x_1) - E_P(f), \ldots, f(x_m) - E_P(f)$ are all zero mean random variables, to which Hoeffding's inequality can be applied. This leads to the following very useful bounds:

$$P^m\{\mathbf{x} \in X^m : \hat{E}(f; \mathbf{x}) - E_P(f) \geq \epsilon\} \leq \exp(-2m\epsilon^2),$$

$$P^m\{\mathbf{x} \in X^m : \hat{E}(f; \mathbf{x}) - E_P(f) \leq -\epsilon\} \leq \exp(-2m\epsilon^2),$$

$$P^m\{\mathbf{x} \in X^m : |\hat{E}(f; \mathbf{x}) - E_P(f)| \geq \epsilon\} \leq 2\exp(-2m\epsilon^2).$$

2.4 Stochastic Processes, Almost Sure Convergence

In this section, we introduce the technical tools needed to conclude that various stochastic processes converge *almost surely*, as opposed to merely converging in probability. This section may be omitted by readers who are not very interested in such nuances. In that case, they should also skip through the references to almost sure convergence in subsequent chapters.

2.4.1 Probability Measures on Infinite Cartesian Products

Suppose $(X, \mathcal{S}, P)$ is a probability space. In the sequel, we shall often encounter situations where we would like to study the probability of (sets of) *infinite sequences* $\{x_1, x_2, \ldots\}$ where $x_i \in X$ for each i. The machinery in this subsection is intended to enable us to do so.

Let $(X, \mathcal{S})$ be a given measurable space, and let $\mathbf{N}$ denote the set $\{1, 2, \ldots\}$ of natural numbers. We begin by defining a measurable space whose underlying set is the (countably) infinite Cartesian product X^∞, consisting of all sequences of the form $\{x_i\}_{i \geq 1}$ where $x_i \in X$ for each i. A **cylinder set** $A \subseteq X^\infty$ is a set of the form $\prod_{i=1}^\infty A_i$ where $A_i \in \mathcal{S}$ for all i, and in addition, $A_i = X$ for all except a finite number of indices i. Let $\mathcal{S}^\infty$ denote the smallest σ-algebra on X^∞ that contains all the cylinder sets. Suppose now that P is a probability measure on $(X, \mathcal{S})$; it is possible to define a corresponding probability measure P^∞ on the space $(X^\infty, \mathcal{S}^\infty)$. Given any cylinder set $A = \prod_{i=1}^\infty A_i \subseteq X^\infty$, define

$$P^\infty(A) = \prod_{i=1}^\infty P(A_i).$$

Observe that $A_i = X$ for all but a finite number of i, and as a result $P(A_i) = 1$ for all but a finite number of indices i. By the Kolmogorov extension theorem, there exists a *unique* probability measure P^∞ on $(X^\infty, \mathcal{S}^\infty)$ that satisfies the above relationship.

2.4.2 Stochastic Processes

In this section, a brief introduction is given to the notion of stochastic processes. As is the case elsewhere, the treatment here is strictly minimalist, and the reader is encouraged to consult an authoritative source for a more thorough treatment.

Suppose $(\Omega, \mathcal{T}, Q)$ is a probability space, and that $(X, \mathcal{S})$ is a measurable space. Then an X-**valued stochastic process** is a sequence of X-valued random variables, of the form $\{\mathbf{X}\}_{i=-\infty}^\infty$. Note that it is customary in the stochastic process literature to work with two-sided infinite sequences, as opposed to one-sided sequences of the form $\{\mathbf{X}_i\}_{i=0}^\infty$.

Suppose $X = \mathbb{R}$ and $\mathcal{S}$ is the Borel σ-algebra, so that $\{X_i\}$ is a real-valued stochastic process. For such a process, one can define the multivariate distribution function as follows: Suppose k is an integer, $i_1, \ldots, i_k$ is an increasing set of k indices, and a_{i_1}, a_{i_k} are real numbers. Then

$$P_{X_{i_1}, \ldots, X_{i_k}}(a_{i_1}, \ldots, a_{i_k}) := Q\{\omega \in \Omega : X_{i_j}(\omega) \le a_{i_j},\ j = 1, \ldots, k\}.$$

As in the case of real-valued random variables, one can define two real-valued stochastic processes to be **equivalent** if they have the same multivariate distribution functions, for all multi-indices. Again as in the case of a single random variable, it is possible to change the underlying domain set to something that is a little more natural and easy to work with. This is called the **canonical representation**, and is defined as follows: Given a measurable space $(X, \mathcal{S})$, define the corresponding infinite cartesian product space and product σ-algebra $(X^\infty, \mathcal{S}^\infty)$ as above, and let $\tilde{P}$ denote a probability measure on $(X^\infty, \mathcal{S}^\infty)$. Then we define the canonical representation of an X-valued stochastic process as a measurable mapping $\mathbf{X}$ from the probability space $(X^\infty, \mathcal{S}^\infty, \tilde{P})$ into $(X^\infty, \mathcal{S}^\infty)$. Thus if ω denotes a typical element of X^∞, the image $\mathbf{X}(\omega)$ is a *sequence* $X_i(\omega)$ where each element belongs to X. We define the map $\omega \mapsto X_i(\omega)$ as the **coordinate random variable**.

The stochastic process $\{\mathbf{X}\}$ is said to be **stationary** if the probability measure is shift-invariant. This means that, for every finite set of indices $i_1, \ldots, i_l$, the marginal probability of the l-tuple $(X_{i_1}, \ldots, X_{i_l})$ is the same as that of $(X_{i_1+1}, \ldots, X_{i_l+1})$.

The stochastic process is said to be **i.i.d. (independent and identically distributed)** if the coordinate random variables are pairwise independent, and all have the same one-dimensional marginal probability measure.

2.4.3 The Borel-Cantelli Lemma and Almost Sure Convergence

Suppose $(\Omega, \mathcal{T}, Q)$ is a probability space, and let $\{f_m\}_{m \ge 1}$ be (the top half of) a stochastic process on $(\Omega, \mathcal{T}, Q)$. Thus, for each m, $f_m : \Omega \to \mathbb{R}$ and is measurable. Suppose $g : \Omega \to \mathbb{R}$ is measurable. We say that $\{f_m\}$ **converges to g in probability** if

$$Q\{\omega \in \Omega : |f_m(\omega) - g(\omega)| > \epsilon\} \to 0 \text{ as } m \to \infty,\ \forall \epsilon > 0.$$

We say that $\{f_m\}$ **converges almost surely to** g if

$$Q\{\omega \in \Omega : \lim_{m \to \infty} f_m(\omega) = g(\omega)\} = 1.$$

It is easy to see that almost sure convergence implies convergence in probability; however, the converse is not true in general. On the other hand, convergence in probability is often much easier to prove than almost sure convergence. A common method of deducing almost sure convergence from "sufficiently fast" convergence in probability is to appeal to the result below, which is known as the **Borel-Cantelli Lemma**.

2.4 Stochastic Processes, Almost Sure Convergence

Lemma 2.9. *Suppose $(\Omega, \mathcal{T}, Q)$ is a probability space, and let $\{A_m\}$ be a sequence of sets in $\mathcal{T}$. Define*

$$B = \bigcap_{m=1}^{\infty} \bigcup_{n=m}^{\infty} A_n.$$

Suppose

$$\sum_{m=1}^{\infty} Q(A_m) < \infty.$$

Then $Q(B) = 0$.

Remarks Note that a point $\omega \in \Omega$ belongs to B if and only if it belongs to *infinitely many* sets A_m, that is, for each $m \geq 1$ there exists an $n \geq m$ such that $\omega \in A_n$. The point of the lemma is that, if the measures of the sets A_m decrease sufficiently rapidly that the sequence $\{Q(A_m)\}$ is summable, then the set of points that belong to infinitely many A_m has measure zero.

Proof. Define

$$B_m = \bigcup_{n=m}^{\infty} A_n, \ m \geq 1.$$

Then, by the subadditivity property of Q, it follows that

$$Q(B_m) \leq \sum_{n=m}^{\infty} Q(A_n) \to 0 \text{ as } m \to \infty,$$

because of the assumption that $\sum_{m=1}^{\infty} Q(A_m) < \infty$. Now note that $B = \cap_{m=1}^{\infty} B_m$. Consequently we have

$$Q(B) \leq Q(B_m) \ \forall m \Rightarrow Q(B) \leq \inf_{m \geq 1} Q(B_m) = 0.$$

This completes the proof. ■

The Borel-Cantelli lemma leads at once to the following sufficient condition for almost sure convergence.

Lemma 2.10. *Suppose $(\Omega, \mathcal{T}, Q)$ is a probability space, that $\{f_m\}_{m \geq 1}$ is a stochastic process on $(\Omega, \mathcal{T}, Q)$, and that $g : \Omega \to \mathbb{R}$ is measurable. Finally, suppose*

$$\sum_{m=1}^{\infty} Q\{\omega \in \Omega : |f_m(\omega) - g(\omega)| > \epsilon\} < \infty, \ \forall \epsilon > 0.$$

Then $\{f_m\}$ converges almost surely to g.

Remarks: The hypothesis of the lemma states that the stochastic process $\{f_m(\cdot)\}$ converges in probability to $g(\cdot)$ *sufficiently rapidly* that the sequence $\{q_{m,\epsilon}\}$ is summable for each $\epsilon > 0$, where

$$q_{m,\epsilon} := Q\{\omega \in \Omega : |f_m(\omega) - g(\omega)| > \epsilon\}.$$

Proof. For each positive integer k, define the set

$$B_k = \bigcap_{m=1}^{\infty} \bigcup_{n=m}^{\infty} \{\omega \in \Omega : |f_n(\omega) - g(\omega)| > 1/k\}.$$

Applying the Borel-Cantelli lemma with

$$A_{m,k} = \{\omega \in \Omega : |f_m(\omega) - g(\omega)| > 1/k\},$$

we conclude that $Q(B_k) = 0$ for all k. Hence

$$Q\left(\bigcup_{k=1}^{\infty} B_k\right) = 0,$$

by the countable subadditivity property of Q. Now note that

$$\bigcup_{k=1}^{\infty} B_k = \{\omega \in \Omega : f_m(\omega) \not\to g(\omega)\}.$$

Therefore $\{f_m\}$ converges almost surely to g. ∎

Example 2.1. As an application of this lemma, let us return to the problem of empirically estimating the expected value (i.e., mean) of a function based upon i.i.d. samples. Suppose $(X, \mathcal{S}, P)$ is a probability space, and that $f : X \to [0, 1]$ is measurable. As before, let

$$E_P(f) := \int_X f(x)\, P(dx)$$

denote the expected value of f. Now suppose $x_1, \ldots, x_m$ are i.i.d. samples drawn from X in accordance with P, and define

$$\hat{E}(f; \mathbf{x}_m) := \frac{1}{m} \sum_{i=1}^{m} f(x_i).$$

It is now shown that $\hat{E}(f; \mathbf{x}_m)$ converges almost surely to $E_P(f)$ in a sense to be made precise next.

Let $\mathbf{x}^* \in X^\infty$; thus $\mathbf{x}^*$ is a sequence $\{x_i\}_{i \geq 1}$ where each $x_i \in X$. Now one can define $\hat{E}_m(f; \mathbf{x}^*)$ as the random variable mapping X^∞ into $[0, 1]$ according to

$$\hat{E}_m(f; \mathbf{x}^*) := \frac{1}{m} \sum_{i=1}^{m} f(x_i) = \hat{E}(f; \mathbf{x}_m).$$

Note that $\hat{E}_m(f; \mathbf{x}^*)$ depends only on the first m components of $\mathbf{x}^*$. Now Hoeffding's inequality states that

$$P^m\{\mathbf{x} \in X^m : |\hat{E}(f; \mathbf{x}_m) - E_P(f)| > \epsilon\} \leq 2\exp(-2m\epsilon^2).$$

One can recast this as

$$P^\infty\{\mathbf{x}^* \in X^\infty : |\hat{E}_m(f; \mathbf{x}^*) - E_P(f)| > \epsilon\} \leq 2\exp(-2m\epsilon^2).$$

Since the sequence $\{2\exp(-2m\epsilon^2)\}_{m \geq 1}$ is summable for each $\epsilon > 0$, it follows from Lemma 2.10 that the sequence of random variables $\{\hat{E}_m(f; \cdot)\}$ converges almost surely to $E_P(f)$ (or more precisely, to the "random" variable whose value equals $E_P(f)$ for all $\mathbf{x}^* \in X^\infty$). This means that

$$P^\infty\{\mathbf{x}^* \in X^\infty : \hat{E}_m(f; \mathbf{x}^*) \to E_P(f)\} = 1.$$

This property is known as the "strong law of large numbers."

As a very useful application of the above property, suppose $A \in \mathcal{S}$ is a measurable set, and let $f = I_A(\cdot)$, the indicator function of the set A. Then it is easy to see that $E_P(f)$ is the same as $P(A)$. Moreover, given an infinite sequence $\mathbf{x}^* \in X^\infty$, one can define the random variable $\hat{P}_m(A; \mathbf{x}^*)$ by

$$\hat{P}_m(A; \mathbf{x}^*) := \frac{1}{m} \sum_{i=1}^m I_A(x_i).$$

Note that $\hat{P}_m(A; \mathbf{x}^*)$ is just the fraction of the first m samples that belong to the set A. One can think of $\hat{P}_m(A; \mathbf{x}^*)$ as an empirical estimate of the probability of the set A, based on the first m elements of the sequence $\mathbf{x}^*$. By the preceding argument, it follows that

$$P^\infty\{\mathbf{x}^* \in X^\infty : \hat{P}_m(A; \mathbf{x}^*) \to P(A)\} = 1.$$

2.5 Mixing Properties of Stochastic Processes

This section is devoted to a discussion of an advanced notion called the "mixing" of stochastic processes. Up to now (as in Example 2.1 for instance), we have dealt with i.i.d. processes. Indeed, much of the "classical" form of statistical learning theory is couched in terms of i.i.d. processes. However, independence is a very restrictive concept, in several ways. First, it is often an assumption, rather than a deduction on the basis of observations. Second, it is an "all or nothing" property, in the sense that two random variables are either independent or they are not – the definition does not permit an intermediate notion of being "nearly" independent. As a result, many of the proofs based on the assumption that the underlying stochastic process is i.i.d. are rather "fragile." The notion of mixing allows one to put the notion of "near independence" on a firm mathematical foundation, and moreover, permits one to derive a "robust" rather than a "fragile" theory, by allowing one to prove that most of the desirable properties of i.i.d. stochastic processes are preserved when the underlying process is mixing.

2.5.1 Definitions of Various Kinds of Mixing Coefficients

There are several diverse notions of mixing used in the literature, but we shall be concerned with only two, namely α-mixing and β-mixing. In the interests of completeness, we also define one more notion called ϕ-mixing, but we also show why this is not a very useful concept, at least in learning theory.

To define these concepts, let us begin with a stationary stochastic process $\{X_i\}_{i=-\infty}^{\infty}$ defined on a probability space $(X^\infty, \mathcal{S}^\infty, \tilde{P})$. It is assumed that a canonical representation is used for the stochastic process, so that each X_i maps $(X^\infty, \mathcal{S}^\infty, \tilde{P})$ into X. For each index k, let $\Sigma_{-\infty}^{k}$ denote the σ-algebra generated by the coordinate random variables $X_i, i \leq k$, and similarly let Σ_k^∞ denote the σ-algebra generated by the coordinate random variables $X_i, i \geq k$. Let $\tilde{P}_{-\infty}^{k}$ and $\tilde{P}_k^\infty$ denote the corresponding marginal probability measures. Then, by the Kolmogorov extension theorem, there exists a unique probability measure on $(X^\infty, \mathcal{S}^\infty)$, denoted by $\tau_0(\tilde{P})$, such that

1. The laws of $\{X_i, i \leq 0\}$ under $\tilde{P}$ and under $\tau_0(\tilde{P})$ are the same.
2. The laws of $\{X_j, j \geq 1\}$ under $\tilde{P}$ and under $\tau_0(\tilde{P})$ are the same.
3. Under the measure $\tau_0(\tilde{P})$, the variables $\{X_i, i \leq 0\}$ are independent of $\{X_j, j \geq 1\}$. This means that each $X_i, i \leq 0$ is independent of each $X_j, j \geq 1$.

Some authors denote this new probability measure by the symbol $\tilde{P}_{-\infty}^0 \times \tilde{P}_1^\infty$. However, in the proofs it is more convenient to use the symbol $\tau_0(\tilde{P})$, where the subscript 0 serves to remind us of the place at which the two halves of the stochastic process are "split." To make the present theorem statements resemble those found in the literature, the two symbols $\tilde{P}_{-\infty}^0 \times \tilde{P}_1^\infty$ and $\tau_0(\tilde{P})$ are used interchangeably. For future use, let us also introduce the symbol $\bar{\Sigma}_1^{k-1}$ to denote the σ-algebra generated by the random variables $X_i, i \leq 0$ as well as $X_j, j \geq k$. Thus the bar over the Σ serves to remind us that the random variables between 1 and $k-1$ are missing from the list of variables that generate Σ.

With this notation, we can now define various mixing coefficients.

Definition 2.1. *The α-mixing coefficient of the stochastic process $\{X_i\}$ is defined as*

$$\alpha(k) := \sup_{A \in \Sigma_{-\infty}^0, B \in \Sigma_k^\infty} |\tilde{P}(A \cap B) - \tilde{P}(A) \cdot \tilde{P}(B)|. \quad (2.5.1)$$

The β-mixing coefficient of the stochastic process is defined as

$$\beta(k) := \sup_{C \in \bar{\Sigma}_1^{k-1}} |\tilde{P}(C) - (\tilde{P}_{-\infty}^0 \times \tilde{P}_1^\infty)(C)|$$
$$= \rho(\tilde{P}, \tau_0(\tilde{P}); \bar{\Sigma}_1^{k-1}). \quad (2.5.2)$$

The ϕ-mixing coefficient of the stochastic process is defined as

2.5 Mixing Properties of Stochastic Processes

$$\phi(k) := \sup_{A \in \Sigma_{-\infty}^0, B \in \Sigma_k^\infty} |\tilde{P}(B|A) - \tilde{P}(B)|. \qquad (2.5.3)$$

In the definition of the α-mixing coefficient, A is an event that depends only on the "past" random variables $\{X_i, i \leq 0\}$ while B is an event that depends only on the "future" random variables $\{X_i, i \geq k\}$. Thus if the future event B were to be truly independent of the past event A, then the probability $\tilde{P}(A \cap B)$ would exactly equal $\tilde{P}(A)\tilde{P}(B)$. Thus the α-mixing coefficient measures how near to independence future events are of the past events, by taking the supremum of the difference between the two quantities $\tilde{P}(A \cap B)$ and $\tilde{P}(A)\tilde{P}(B)$. Similarly, if the future event B were to be truly independent of the past event A, then the conditional probability $\tilde{P}(B|A)$ would exactly equal the unconditional probability $\tilde{P}(B)$. The ϕ-mixing coefficient measures how near to independence future events are of the past events, by taking the supremum of the difference between the two quantities $\tilde{P}(B|A)$ and $\tilde{P}(B)$. The β-mixing coefficient has a somewhat more involved interpretation. If the future events beyond time k were to be truly independent of the past events before time 0, then the probability measure $\tilde{P}$ would exactly equal the "split" measure $\tau_0(\tilde{P})$, or $\tilde{P}_{-\infty}^0 \times \tilde{P}_k^\infty$ as some authors write it. The β-mixing coefficient thus measures how nearly the product measure approximates the actual measure $\tilde{P}$.

Now a few properties of these coefficients are discussed.

1. Note that, in the definitions of $\alpha(k)$ and $\phi(k)$, we can write $\tilde{P}_{-\infty}^0(A)$ instead of $\tilde{P}(A)$; similarly we can write $\tilde{P}_1^\infty(B)$ instead of $\tilde{P}(B)$.
2. In the definition of the ϕ-mixing coefficient, the conditional probability $\tilde{P}(B|A)$ is taken as $\tilde{P}(B)$ if $\tilde{P}(A) = 0$.
3. Since $\Sigma_{k+1}^\infty \subseteq \Sigma_k^\infty$, it is obvious that the α-, β- and ϕ-mixing coefficients are all nonincreasing. Thus

$$\alpha(k+1) \leq \alpha(k),\ \beta(k+1) \leq \beta(k),\ \phi(k+1) \leq \phi(k),\ \forall k.$$

4. If we write $C = A \cap B$ where $A \in \Sigma_{-\infty}^0$ and $B \in \Sigma_k^\infty$, then we have

$$(P_{-\infty}^0 \times P_1^\infty)(A \cap B) = P_{-\infty}^0(A) \cdot P_1^\infty(B).$$

Thus, when C is restricted to intersections of the above type, the right sides of (2.5.1) and 2.5.2) coincide. However, in (2.5.2), the supremum is taken not only over sets C of the form $A \cap B$, but over the σ-algebra generated by all such intersections. Thus it follows that

$$\alpha(k) \leq \beta(k),\ \forall k \geq 1.$$

Similarly, since $\tilde{P}(B|A) = \tilde{P}(A \cap B)/\tilde{P}(A)$ if $\tilde{P}(A) \neq 0$, it is easy to see that

$$\alpha(k) \leq \phi(k),\ \forall k \geq 1.$$

It can also be shown that

$$\beta(k) \leq \phi(k),\ \forall k \geq 1.$$

36   2. Preliminaries

5. It is obvious that if the stochastic process $\{X_i\}$ consists of independent and identically distributed (i.i.d.) random variables, then $\tilde{P}$ equals the measure $(\tilde{P}_0)^\infty$, which denotes the measure on $(X^\infty, \mathcal{S}^\infty)$ under which each X_i has marginal probability $\tilde{P}_0$, and the X_i's are pairwise independent. In such a case, all the three mixing coefficients are zero, for each k.
6. It is somewhat ironic that some authors refer to α-mixing as "strong" mixing, even though it is the weakest of the various notions of mixing studied in the literature.

Definition 2.2. *The stochastic process* $\{X_i\}$ *is said to be* α**-mixing**, *or* **strongly regular** *if* $\alpha(k) \to 0$ *as* $k \to \infty$. *The stochastic process* $\{X_i\}$ *is said to be* β**-mixing** *or* **completely regular** *if* $\beta(k) \to 0$ *as* $k \to \infty$. *The stochastic process* $\{X_i\}$ *is said to be* ϕ**-mixing** *or* **uniformly regular** *if* $\phi(k) \to 0$ *as* $k \to \infty$.

2.5.2 Inequalities for Mixing Processes

In this subsection, a few consequences of mixing processes are derived. We begin with β-mixing processes.

Lemma 2.11. *Suppose* $\{X_i\}$ *is a* β-*mixing process on a probability space* $(X^\infty, \mathcal{S}^\infty, \tilde{P})$. *Suppose* $f : X^\infty \to \mathbb{R}$ *is essentially bounded and is measurable with respect to the* σ-*algebra* $\bar{\Sigma}_1^{k-1} = \Sigma(X_i, i \leq 0 \text{ or } i \geq k)$. *Then*

$$|E(f, \tilde{P}) - E(f, P_{-\infty}^0 \times P_1^\infty)| \leq \beta(k) \cdot \| f \|_\infty . \qquad (2.5.4)$$

Proof. Note that, by definition, the quantity $\beta(k)$ is precisely the total variation metric between the two probability measures $\tilde{P}$ and $P_{-\infty}^0 \times P_1^\infty$. The desired inequality now follows readily. ∎

Theorem 2.1. *Suppose* $\{X_i\}$ *is a* β-*mixing process on a probability space* $(X^\infty, \mathcal{S}^\infty, \tilde{P})$. *Suppose* $f : X^\infty \to \mathbb{R}$ *is essentially bounded and depends only on the variables* $x_{ik}, 0 \leq i \leq l$. *Let* $\tilde{P}_0$ *denote the one-dimensional marginal probability of each of the* X_i. *Then*

$$|E(f, \tilde{P}) - E(f, \tilde{P}_0^\infty)| \leq l\beta(k) \| f \|_\infty . \qquad (2.5.5)$$

The proof of theorem depends on the following auxiliary lemma.

Lemma 2.12. *Suppose* P, Q *are probability measures on* $(U, \mathcal{S})$, *that* X, Y *are measurable real-valued functions on* $(U, \mathcal{S})$. *Suppose further that (i)* X, Y *are independent random variables under each of* P, Q, *and (ii) the marginal probabilities of* X *under* P *and* Q *are equal. Then*

$$\rho(P, Q; \Sigma(X, Y)) = \rho(P, Q; \Sigma(Y)).$$

2.5 Mixing Properties of Stochastic Processes

Proof. **(of the lemma)**: Consider a set $S \in \Sigma(\mathsf{X},\mathsf{Y})$ of the form

$$S = \bigcup_{i=1}^{\infty}(A_i \cap B_i), \ A_i \in \Sigma(\mathsf{X}), \ B_i \in \Sigma(\mathsf{Y}), \ \forall i,$$

where the A_i are pairwise disjoint. As in the case of Lemma 2.5, $\rho(P,Q;\Sigma(\mathsf{X},\mathsf{Y}))$ equals the supremum of $|P(S) - Q(S)|$ as S varies over all such sets. Now

$$|P(S) - Q(S)| \le \sum_{i=1}^{\infty} |P(A_i \cap B_i) - Q(A_i \cap B_i)|.$$

Next, since X,Y are independent under both P and Q, we have

$$P(A_i \cap B_i) = P(A_i)P(B_i), \ Q(A_i \cap B_i) = Q(A_i)Q(B_i).$$

Moreover, since the marginal probabilities of X under P and Q are equal, we have

$$P(A_i) = Q(A_i).$$

Therefore it follows that

$$\begin{aligned}|P(S) - Q(S)| &\le \sum_{i=1}^{\infty} P(A_i)|P(B_i) - Q(B_i)| \\ &\le \left(\sum_{i=1}^{\infty} P(A_i)\right) \rho(P,Q;\Sigma(\mathsf{Y})) \\ &\le \rho(P,Q;\Sigma(\mathsf{Y})).\end{aligned}$$

This shows that $\rho(P,Q;\Sigma(\mathsf{X},\mathsf{Y})) \le \rho(P,Q;\Sigma(\mathsf{Y}))$. The opposite inequality follows readily since $\Sigma(\mathsf{X},\mathsf{Y})$ is a superset of $\Sigma(\mathsf{Y})$. ∎

Proof. **of the theorem**: Recall the notation $\tau_0(\tilde{P})$, which "splits" the original measure $\tilde{P}$ into a kind of product measure, in such a way that the variables $\mathsf{X}_i, i \le 0$ are independent of $\mathsf{X}_i, i \ge 1$, and both sets of variables have the same marginals as under $\tilde{P}$. Since we shall be "splitting" measures repeatedly, to make the notation less cumbersome let us define the symbol $\tilde{Q}$ recursively as follows:

$$\tilde{Q}_1 = \tau_0(\tilde{P}), \ \tilde{Q}_{i+1} = \tau_{ik}(\tilde{Q}_i), \ i = 1,\ldots, l-1.$$

Thus $\tilde{Q}_l$ consists of the original measure $\tilde{P}$ split at the time instants $0, k, 2k, \ldots, (l-1)k$. In the interests of brevity, let us use the symbol Σ to denote the σ-algebra $\Sigma(\mathsf{X}_{ik}, 0 \le i \le l)$. We shall display it in full form when needed. Now the claim is that

$$\rho(\tilde{P},\tilde{Q}_l;\Sigma) \le l\beta(k). \tag{2.5.6}$$

To establish this claim, note that by the triangle inequality we have

2. Preliminaries

$$\rho(\tilde{P}, \tilde{Q}_l; \Sigma) \leq \rho(\tilde{P}, \tilde{Q}_1; \Sigma) + \sum_{i=1}^{l-1} \rho(\tilde{Q}_i, \tilde{Q}_{i+1}; \Sigma). \tag{2.5.7}$$

Then (2.5.6) will follow if it can be shown that each of the above terms on the right side is less than $\beta(k)$. Note that, since the stochastic process is stationary, the β-mixing coefficient also equals

$$\beta(k) = \rho(\tilde{P}, \tau_j(\tilde{P}); \bar{\Sigma}_{j+1}^{j+k-1}).$$

In other words, the original measure $\tilde{P}$ can be "split" at *any* time instant j, and the total variation distance does not depend on j. With this background, note first that

$$\rho(\tilde{P}, \tilde{Q}_1; \Sigma) \leq \rho(\tilde{P}, \tau_0(\tilde{P}); \bar{\Sigma}_1^{k-1}) \leq \beta(k),$$

since $\tilde{Q}_1 = \tau_0(\tilde{P})$ and Σ is a subalgebra of $\bar{\Sigma}_1^{k-1}$. For the remaining terms, recall that $\tilde{Q}_{i+1}$ is obtained by splitting $\tilde{Q}_i$ at the time instant ik. Thus, under both $\tilde{Q}_i$ and $\tilde{Q}_{i+1}$, the variables $X_0, X_k, \ldots, X_{(i-1)k}$ are independent of $X_{(i+1)k}, \ldots, X_{lk}$. Moreover, the marginal probabilities of $\{X_0, \ldots, X_{(i-1)k}\}$ are the same under both $\tilde{Q}_i$ and $\tilde{Q}_{i+1}$. Hence, by Lemma 2.12, it follows that

$$\rho(\tilde{Q}_i, \tilde{Q}_{i+1}; \Sigma) \leq \rho(\tilde{Q}_i, \tilde{Q}_{i+1}; \Sigma(X_{ik}, \ldots, X_{lk})).$$

But on this algebra, $\tilde{Q}_i$ equals $\tilde{P}$, while $\tilde{Q}_{i+1}$ equals $\tau_{ik}(\tilde{P})$. Moreover, this algebra is a subalgebra of $\bar{\Sigma}_{ik+1}^{(i+1)k-1}$. Hence, by definition, it follows that

$$\rho(\tilde{Q}_i, \tilde{Q}_{i+1}; \Sigma(X_{ik}, \ldots, X_{lk})) \leq \rho(\tilde{P}, \tau_{ik}(\tilde{P}); \bar{\Sigma}_{ik+1}^{(i+1)k-1}) = \beta(k).$$

Hence each of the terms on the right side of (2.5.7) is less than $\beta(k)$. This establishes (2.5.6). Consequently, it follows that

$$|E(f, \tilde{P}) - E(f, \tilde{Q}_l)| \leq l\beta(k) \, \|f\|_\infty .$$

To complete the proof, note that under the measure $\tilde{Q}_l$, the $l+1$ variables X_0, X_k, X_{lk} are pairwise independent. Hence

$$E(f, \tilde{Q}_l) = E(f, \tilde{P}_0^\infty).$$

This completes the proof. ∎

Corollary 2.1. *Suppose* $i_0 < i_1 < \ldots < i_l$ *are integers, and define*

$$k := \min_{0 \leq j \leq l-1} i_{j+1} - i_j.$$

Suppose f is essentially bounded and depends only on $X_{i_0}, \ldots, X_{i_l}$. *Then*

$$|E(f, \tilde{P}) - E(f, \tilde{P}_0^\infty)| \leq \beta(k) \, \|f\|_\infty .$$

The proof is a routine modification of that of Theorem 2.1 and is left to the reader.

Now we present an inequality pertaining to α-mixing processes.

Theorem 2.2. *Suppose $\{X_i\}$ is an α-mixing process on a probability space $(X^\infty, \mathcal{S}^\infty, \tilde{P})$. Suppose $f, g : X^\infty \to \mathbb{R}$ are essentially bounded, that f is measurable with respect to $\Sigma(X_i, i \leq 0)$, and that g is measurable with respect to $\Sigma(X_i, i \geq k)$. Then*

$$|E(fg, \tilde{P}) - E(f, \tilde{P}) \, E(g, \tilde{P})| \leq 4\alpha(k) \, \|f\|_\infty \cdot \|g\|_\infty \, . \quad (2.5.8)$$

Theorem 2.2 is a ready consequence of the following more general result.

Lemma 2.13. *Suppose f, g are essentially bounded random variables on a probability space $(\Omega, \mathcal{T}, \mathcal{P})$, that $\mathcal{F}, \mathcal{G}$ are sub σ-algebras of $\mathcal{T}$, and that $f \in \mathcal{M}(\mathcal{F})$, $g \in \mathcal{M}(\mathcal{G})$. Define the coefficient*

$$\alpha(\mathcal{F}, \mathcal{G}) := \sup_{A \in \mathcal{F}, B \in \mathcal{G}} |P(A \cap B) - P(A)P(B)|.$$

Then

$$|E(fg) - E(f)E(g)| \leq 4 \, \|f\|_\infty \|g\|_\infty \, \alpha(\mathcal{F}, \mathcal{G}).$$

Proof. Define $\eta := \text{sign}(g_\mathcal{F} - E(g))$. Thus

$$\eta(\omega) := \begin{cases} 1 & \text{if } g_\mathcal{F}(\omega) - E(g) \geq 0, \\ 0 & \text{if } g_\mathcal{F}(\omega) - E(g) < 0. \end{cases}$$

Then we have $|g_\mathcal{F} - E(g)| = \eta(g_\mathcal{F} - E(g))$. Now by (2.2.1) we have that $E(fg) = E(fg_\mathcal{F})$. Therefore

$$\begin{aligned} |E(fg) - E(f)E(g)| &= |E(fg_\mathcal{F}) - E[fE(g)]| \\ &= |E[f(g_\mathcal{F} - E(g))]| \\ &\leq \|f\|_\infty \, E[|g_\mathcal{F} - E(g)|] \\ &= \|f\|_\infty \, E[\eta(g_\mathcal{F} - E(g))] \\ &= \|f\|_\infty \, E[\eta(g - E(g))], \end{aligned}$$

where in the last step we use the fact that $\eta \in \mathcal{M}(\mathcal{F})$ so that $E(\eta g) = E(\eta g_\mathcal{F})$ by (2.2.1). Now define $\xi = \text{sign}(g - E(g))$ and note that $\xi \in \mathcal{M}(\mathcal{G})$. Then it is possible to mimic the above argument and show that

$$|E[\eta(g - E(g))]| \leq \|g\|_\infty \, |E(\eta\xi) - E(\eta)E(\xi)|.$$

Combining this with the preceding inequality shows that

$$|E(fg) - E(f)E(g)| \leq \|f\|_\infty \|g\|_\infty \, |E(\eta\xi) - E(\eta)E(\xi)|.$$

The proof is therefore complete if it can be shown that

$$|E(\eta\xi) - E(\eta)E(\xi)| \le 4\alpha(\mathcal{F},\mathcal{G}). \tag{2.5.9}$$

Towards this end, define the four sets

$$A_+ := \{\omega : \eta(\omega) = 1\},\ A_- := \{\omega : \eta(\omega) = -1\},$$
$$B_+ := \{\omega : \xi(\omega) = 1\},\ B_- := \{\omega : \xi(\omega) = -1\},$$

and note that $A_+, A_- \in \mathcal{F}, B_+, B_- \in \mathcal{G}$. Now it is easy to see that

$$E(\eta\xi) = P(A_+ \cap B_+) + P(A_- \cap B_-) - P(A_+ \cap B_-) - P(A_- \cap B_+),$$
$$E(\eta) = P(A_+) - P(A_-),\ E(\xi) = P(B_+) - P(B_-).$$

Substituting all this shows that

$$\begin{aligned}|E(\eta\xi) - E(\eta)E(\xi)| = &\ P(A_+ \cap B_+) + P(A_- \cap B_-) \\ &- P(A_+ \cap B_-) - P(A_- \cap B_+) \\ &- P(A_+)P(B_+) - P(A_-)P(B_-) \\ &+ P(A_+)P(B_-) + P(A_-)P(B_+).\end{aligned}$$

Now by the definition of $\alpha(\mathcal{F},\mathcal{G})$ we have

$$|P(A_+ \cap B_+) - P(A_+)P(B_+)| \le \alpha(\mathcal{F},\mathcal{G}),$$

and similarly for the other three terms. This finally leads to the desired bound (2.5.9) and completes the proof. ∎

The next result is a multiplicative analog of Theorem 2.1.

Corollary 2.2. *Suppose $\{X_i\}$ is an α-mixing stochastic process. Suppose $f_0, \ldots, f_l$ are essentially bounded functions, where f_i depends only on X_{ik}. Then*

$$\left| E\left[\prod_{i=0}^{l} f_i\right] - \prod_{i=0}^{l} E(f_i) \right| \le 4l\alpha(k) \prod_{i=0}^{l} \|f_i\|_\infty. \tag{2.5.10}$$

The proof by induction on l is simple and is therefore omitted.

Notes and References The material in Sections 2.1 through 2.4 can be found in standard texts. For concepts from topology, see [100], See [107] for a survey of covering numbers, packing numbers, and their interrelationships, as well as explicit computations of these numbers for various specific sets. For concepts from probability, see [68], [74], or [119]. Hoeffding's inequality is proven in [86] and generalizes earlier work of Chernoff [45]. The material in Section 2.5 is more advanced. For definitions of mixing coefficients of stochastic processes, see [25]. For some reason, many texts on stochastic processes discuss only α-mixing and ϕ-mixing, but not β-mixing. As shown in subsequent chapters, ϕ-mixing is too strong an assumption, while α-mixing is too weak; but β-mixing is "just right." Theorem 2.1 is due to Yu [212], while Theorem 2.2 is due to Ibragimov [88], with the proof reproduced in [78], Theorem A.5. The importance of mixing processes originally arose from the fact that the simple law of large numbers established in Example 2.1 for i.i.d. processes can be readily extended even to α-mixing processes, which satisfy the weakest type of mixing condition. Several authors have studied conditions under which the output sequence of a Markov chain is mixing in any of the three senses discussed here. We shall return to this topic again in Chapter 3.

3. Problem Formulations

In this chapter, we present several problem formulations that form the "universe of discourse" for the remainder of the notes. In all, there are three "abstract" problem formulations, and these are presented in turn. Several specific applications are given throughout to illustrate the abstract formulations.

3.1 Uniform Convergence of Empirical Means

In this section, we begin by defining a notion called UCEM (Uniform Convergence of Empirical Means). Then this notion is extended to the case where there is a family of probability measures, to the so-called UCEMUP property. Finally, we extend both definitions to the case where the inputs are not necessarily independent.

3.1.1 The UCEM Property

Suppose $(X, \mathcal{S})$ is a measurable space, and that P is a probability measure on $(X, \mathcal{S})$. Suppose $A \in \mathcal{S}$, and that it is desired to compute the measure $P(A)$. One way to do this is as follows: Generate i.i.d. samples $x_1, \ldots, x_m$ from X, distributed according to P. For each j, check whether or not x_j belongs to A. After m samples, form the **empirical probability** of A according to

$$\hat{P}(A) = \frac{\text{No. of times that } x_j \in A}{m}.$$

One can now ask whether $\hat{P}(A)$ approaches the "true" probability $P(A)$, and if so, in what sense does $\hat{P}(A)$ converge to $P(A)$.

In order to make the question more precise, let us define

$$\mathbf{x}_m = [x_1 \ldots x_m]^t \in X^m,$$

and let $I_A(\cdot)$ denote the **indicator function** of A; that is,

$$I_A(x) = \begin{cases} 1 & \text{if } x \in A, \\ 0 & \text{if } x \notin A. \end{cases}$$

44   3. Problem Formulations

Then the empirical probability of A can be written as

$$\hat{P}(A;\mathbf{x}_m) = \frac{1}{m}\sum_{j=1}^{m} I_A(x_j).$$

The notation $\hat{P}(A;\mathbf{x}_m)$ is intended to highlight the explicit dependence of the empirical probability of A on the particular multisample $[x_1 \ldots x_m]^t$. Note that $\hat{P}(A;\mathbf{x}_m)$ is itself a random variable, because if the experiment is repeated by drawing another set of m i.i.d. samples, the samples thus drawn will in general be different, and will thus lead to a different value for $\hat{P}(A;\mathbf{x}_m)$. Given any $\epsilon > 0$, define

$$q(m,\epsilon,P) := P^m\{\mathbf{x}_m \in X^m : |\hat{P}(A;\mathbf{x}_m) - P(A)| > \epsilon\},$$

where P^m denotes the m-fold product measure on X^m induced by P. One can think of $q(m,\epsilon,P)$ as the measure of the set of "bad" multisamples $\mathbf{x}_m$, where a multisample $\mathbf{x}_m$ is considered to be "bad" if the empirical probability of A obtained from $\mathbf{x}_m$ deviates from the true value by more than ϵ. Let us say that $\hat{P}(A;\mathbf{x}_m)$ **converges in probability** to $P(A)$ if

$$q(m,\epsilon,P) \to 0 \text{ as } m \to \infty, \text{ for each fixed } \epsilon,$$

or equivalently,

$$\forall \epsilon, \delta > 0, \exists m_0 = m_0(\epsilon,\delta) \text{ s.t. } q(m,\epsilon,P) \leq \delta \ \forall m \geq m_0.$$

The terminology (convergence in probability) can be justified by thinking of $\hat{P}(A;\mathbf{x}_m)$ as a random variable mapping the *infinite Cartesian product* X^∞ into $[0,1]$, as follows[1]

$$\hat{P}_m(A;\mathbf{x}^*) := \frac{1}{m}\sum_{j=1}^{m} I_A(x_j) = \hat{P}(A;\mathbf{x}_m). \quad (3.1.1)$$

Of course, $\hat{P}(A;\mathbf{x}_m)$ depends only on the first m components of $\mathbf{x}^*$. In this way, *all* the empirical probabilities can be thought of as random variables defined on the *same* probability space, namely $(X^\infty, \mathcal{S}^\infty, P^\infty)$. Moreover, the sequence of random variables $\{\hat{P}_m(A;\cdot)\}$ converges in probability (with respect to P^∞) to the fixed value $P(A)$ if and only if $q(m,\epsilon,P) \to 0$ as $m \to \infty$ for each $\epsilon > 0$. In the same vein, one can also say that $\{\hat{P}_m(A;\cdot)\}$ converges almost surely to $P(A)$ if

$$P^\infty\{\mathbf{x}^* \in X^\infty : \hat{P}_m(A;\mathbf{x}^*) \to P(A) \text{ as } m \to \infty\} = 1.$$

Hereafter, we shall be dealing with sequences of random variables where the m-th random variable maps X^m into $[0,1]$, and we shall speak of such a

---
[1] See also Example 2.1 for a similar construction.

3.1 Uniform Convergence of Empirical Means

sequence converging to a constant value, usually zero. Whether the convergence is in probability or almost sure, such a statement should be interpreted as above, namely by embedding X^m into X^∞ for each m, and modifying the definition of the random variable accordingly.

It is easy to show, using the Chernoff bounds, that $\hat{P}(A; \mathbf{x}_m)$ does indeed converge almost surely to $P(A)$. One can think of $I_A(x)$ as a Bernoulli process, with
$$\Pr\{I_A(x) = 1\} = P(A).$$
Thus, from the additive form of the Chernoff bounds, it follows that
$$q(m, \epsilon, P) \leq 2 \exp(-2m\epsilon^2).$$
Note that the bound on the right side does not explicitly depend on the probability measure P. Hence $q(m, \epsilon, P) \to 0$ as $m \to \infty$ for each fixed ϵ. This shows that $\hat{P}(A; \mathbf{x}_m)$ converges in probability to $P(A)$. Moreover, since the sequence $\{2\exp(-2m\epsilon^2)\}$ is summable, it follows from Lemma 2.10 that $\hat{P}(A; \mathbf{x}_m)$ converges almost surely to $P(A)$.

Now suppose that we are given, not just *a single set* A, but *a collection of sets* $\mathcal{A} \subseteq \mathcal{S}$, where the collection need not be finite. As i.i.d. samples $x_1, \ldots, x_m$ are drawn from X in accordance with P, we can form empirical probabilities of *each* of the sets in $\mathcal{A}$ in the obvious manner, namely,[2]
$$\hat{P}(A; \mathbf{x}) = \frac{1}{m} \sum_{j=1}^{m} I_A(x_j), \ \forall A \in \mathcal{A}.$$

Now define
$$q(m, \epsilon, P) := P^m\{\mathbf{x} \in X^m : \exists A \in \mathcal{A} \text{ s.t. } |\hat{P}(A; \mathbf{x}) - P(A)| > \epsilon\}.$$
An equivalent definition of $q(m, \epsilon, P)$ is
$$q(m, \epsilon, P) = P^m\{\mathbf{x} \in X^m : \sup_{A \in \mathcal{A}} |\hat{P}(A; \mathbf{x}) - P(A)| > \epsilon\}.$$
Note that
$$1 - q(m, \epsilon, P) = P^m\{\mathbf{x} \in X^m : |\hat{P}(A; \mathbf{x}) - P(A)| \leq \epsilon \ \forall A \in \mathcal{A}\}.$$
In other words, it can be said with confidence $1 - q(m, \epsilon, P)$ that *every* empirical probability is within ϵ of its true value. With this rationale, we say that the **empirical probabilities converge uniformly** to their true values if $q(m, \epsilon, P) \to 0$ as $m \to \infty$, for each fixed ϵ. We also say that the collection $\mathcal{A}$ has the property of **uniform convergence of empirical probabilities**, or **UCEP** for short. Some authors refer to a collection of sets $\mathcal{A}$ that have the UCEP property as satisfying a "uniform law of large numbers."

---
[2] Hereafter we drop the subscript "m" on $\mathbf{x}$ so as to reduce clutter.

46   3. Problem Formulations

Given a probability measure P and a collection of sets $\mathcal{A}$, let us define a stochastic process $\{a_m(\cdot)\}$ on the space X^∞ as follows: For each integer m and each $\mathbf{x}^* \in X^\infty$, let

$$a_m(\mathbf{x}^*) := \sup_{A \in \mathcal{A}} |\hat{P}_m(A; \mathbf{x}^*) - P(A)|, \qquad (3.1.2)$$

where the symbol $\hat{P}_m(A; \mathbf{x}^*)$ denotes the empirical probability of the set A based on the first m components of $\mathbf{x}^*$, as defined in (3.1.1). Clearly

$$q(m, \epsilon, P) = P^\infty \{\mathbf{x}^* \in X^\infty : a_m(\mathbf{x}^*) > \epsilon\}.$$

Hence the collection of sets $\mathcal{A}$ has the UCEP property if and only if the stochastic process $\{a_m\}_{m \geq 1}$ converges to the zero function in probability (with respect to P^∞) as $m \to \infty$. Similarly, one can think of a stronger property whereby the empirical probabilities converge *almost surely* to their true values; this property corresponds to the stochastic process $\{a_m\}_{m \geq 1}$ converging almost surely to the zero function as $m \to \infty$. Precisely, we say that the collection $\mathcal{A}$ has the property of **almost sure convergence of empirical probabilities (ASCEP)** if

$$P^\infty \{\mathbf{x}^* \in X^\infty : \sup_{A \in \mathcal{A}} |\hat{P}_m(A; \mathbf{x}^*) - P(A)| \to 0 \text{ as } m \to \infty\} = 1.$$

It is easy to see that the ASCEP property implies the UCEP property, since convergence almost surely implies convergence in probability. A deep and surprising result states that the converse is also true (see Theorem 5.2).

Note that if $\mathcal{A}$ is a *finite* collection, then it follows from Hoeffding's inequality that

$$q(m, \epsilon, P) \leq 2|\mathcal{A}| \exp(-2m\epsilon^2).$$

Hence $q(m, \epsilon, P) \to 0$ as $m \to \infty$. In other words, *every* finite collection has the property that empirical probabilities converge uniformly. Moreover, since the sequence $\{2|\mathcal{A}| \exp(-2m\epsilon^2)\}$ is summable for each $\epsilon > 0$, it follows from Lemma 2.10 that $\{a_m\}$ converges to zero almost surely; in other words, for every finite collection of sets, empirical probabilities converge almost surely to the true values as the number of samples approaches infinity. However, if $\mathcal{A}$ is infinite, some additional conditions are required on the family $\mathcal{A}$ in order to ensure that it has the uniform convergence property – the property does not hold in general. This is illustrated by some examples.

Example 3.1. (Glivenko-Cantelli Lemma; see, e.g., [74], p. 448.) Let $X = \mathbb{R}$, $\mathcal{S} =$ the Borel σ-algebra on $\mathbb{R}$, and let $\mathcal{A} = \{A_t, t \in \mathbb{R}\}$ denote the collection of semi-infinite intervals of the form $(-\infty, t]$. Then by the Glivenko-Cantelli lemma, the empirical probability of each set A_t converges almost surely to the true probability of A_t, uniformly with respect to t. This means that the stochastic process $\{a_m\}$ defined in (3.1.2) above converges almost surely to the zero function, when all symbols are defined as above. Subsequently this classical lemma is obtained as a special case of the more general results presented in Chapter 7.

Example 3.2. (Benedek-Itai [22]) Let $X = [0,1]$, $\mathcal{S}$ = the Borel σ-algebra on X, and let P be the uniform probability on X. Suppose $\mathcal{A}$ is the collection of all *finite* subsets of X. Then $P(A) = 0$ for all $A \in \mathcal{A}$. On the other hand, let $\mathbf{x} = [x_1 \ldots x_m]^t \in X^m$ be arbitrary, and define

$$A(\mathbf{x}) = \{x_1, \ldots, x_m\}$$

(after deleting any repeated x_i's if necessary). Then $A(\mathbf{x}) \in \mathcal{A}$, and

$$\hat{P}[A(\mathbf{x}); \mathbf{x}] = 1,$$

because every x_i belongs to $A(\mathbf{x})$. Thus, whenever $\epsilon < 1$, we have that

$$\{\mathbf{x} \in X^m : \exists A \in \mathcal{A} \text{ s.t. } |\hat{P}(A; \mathbf{x}) - P(A)| > \epsilon\} = X^m,$$

because the condition is satisfied with $A = A(\mathbf{x})$ as defined above. Hence, whenever $\epsilon < 1$,

$$P^m\{\mathbf{x} \in X^m : \exists A \in \mathcal{A} \text{ s.t. } |\hat{P}(A; \mathbf{x}) - P(A)| > \epsilon\} = 1, \forall m.$$

Clearly the family $\mathcal{A}$ does *not* have the property of uniform convergence of empirical probabilities. ∎

The preceding example can be put down to the fact that the family $\mathcal{A}$ is in some sense too "rich" – in fact, $d_P(A, B) = 0 \ \forall A, B \in \mathcal{A}$. Hence, if $\mathcal{A}$ is replaced by the collection of equivalence classes $\mathcal{A}/\sim$, where $A \sim B$ if $d_P(A, B) = 0$, then the entire family $\mathcal{A}$ collapses into a *single* equivalence class. From the standpoint of computing probability measures, replacing $\mathcal{A}$ by $\mathcal{A}/\sim$ is quite natural and permissible. Let $B \in \mathcal{A}$ be a *fixed finite* set; then $\mathcal{A}/\sim = \{[B]\}$. To determine $P(B)$ empirically, pick a *single* $x \in X$ at random according to P. Then $I_B(x) = 0$ with probability one, and hence $\hat{P}(B; x) = 0$ with probability one. Therefore the singleton family $\{B\}$ *does* have the uniform convergence property.

Example 3.3. (Kulkarni) Once again let $X, \mathcal{S}, P$ be as in the previous example. Now let $\mathcal{A}$ consist of all sets of the form $A \cup B$, where $A \subseteq X$ is a finite set, say $\{a_1, \ldots, a_n\}$ with $a_1 > \ldots > a_n$, and $B = [0, b]$ where

$$b = \sum_{i=1}^{n} a_i (0.5)^{i+1}.$$

Clearly $b < 0.5$ for every finite A. Also, if $C = A \cup B \in \mathcal{A}$, then $P(C) = P(B) = b < 0.5$. Now look at the equivalence relation $\sim$ induced by the pseudometric d_P, as discussed above. Then each $C = A \cup B$ is equivalent to the corresponding $B = [0, b]$. Of course, more than one $C \in \mathcal{A}$ can be equivalent to the same B. For example, whether $A = \{0.98\}$ or $A = \{0.7, 0.56\}$, we get $b = 0.49$. On the other hand, every $b \in (0, 0.5)$ is of the above form for some finite set A. Hence the collection of equivalence classes is precisely the collection of intervals $[0, b], 0 < b < 0.5$. Thus the collection of equivalence classes is infinite, unlike in the previous example. Nevertheless, this family also *fails* to have the property of uniform convergence of empirical probabilities.

To see this, let $\mathbf{x} = [x_1 \ldots x_m]^t \in X^m$ be arbitrary. Let $\mathbf{a} = [a_1 \ldots a_n]^t$ denote the set of *distinct* elements in $x_1, \ldots, x_m$, and define b as above. Now let $A = \{a_1, \ldots, a_n\}$, $B = [0, b]$, and $C(\mathbf{x}) = A \cup B$. Then

$$\hat{P}[C(\mathbf{x}); \mathbf{x}] = 1$$

because every x_i belongs to C. On the other hand, $P[C(\mathbf{x})] = b < 0.5$. Hence, whenever $\epsilon \leq 0.5$, we have that

$$\{\mathbf{x} \in X^m : \exists C \in \mathcal{A} \text{ s.t. } |\hat{P}(C; \mathbf{x}) - P(C)| > \epsilon\} = X^m,$$

because the condition is satisfied with $C = C(\mathbf{x})$. Hence

$$P^m\{\mathbf{x} \in X^m : \exists C \in \mathcal{A} \text{ s.t. } |\hat{P}(C; \mathbf{x}) - P(C)| > \epsilon\} = 1, \ \forall m.$$

Therefore $\mathcal{A}$ does *not* have the property of uniform convergence of empirical probabilities.

Finally, suppose we define b as

$$b = \sum_{i=1}^{n} a_i (0.5)^{i+k},$$

where $k \geq 1$ is some integer; then $b \leq 0.5^k$ for every finite set A. Thus the relationship

$$P^m\{\mathbf{x} \in X^m : \exists C \in \mathcal{A} \text{ s.t. } |\hat{P}(C; \mathbf{x}) - P(C)| > \epsilon\} = 1, \ \forall m$$

holds whenever $\epsilon \leq 1 - (0.5)^k$. In this way it is possible to modify the family $\mathcal{A}$ in such a way that uniform convergence fails to hold for ϵ arbitrarily close to one; in other words, the "worst-case" disparity between the true probability and the empirical probability can be made as close to one as we wish. ∎

We shall see in Chapter 7 that the collection of intervals of the form $[0, b], 0 < b < 0.5$ *does* have the property that empirical means converge uniformly to their true values. Moreover, this property is "distribution-free," that is, it holds for *every* probability measure P on $[0, 1]$, and not merely when P is the uniform measure. Finally, the convergence of the empirical means is almost sure (and not merely in probability), in the sense that the stochastic process $\{a_m\}$ defined in (3.1.2) converges almost surely to the zero function. In contrast, in the present instance, we have restricted P to be a single known probability measure, namely the uniform measure. Under this probability measure, the collection of sets $\mathcal{A}$ is "functionally" equivalent to the collection of intervals $[0, b], 0 < b < 0.5$, which has all the nice properties alluded to above. And yet, the collection $\mathcal{A}$ *fails* to have the UCEP property. What goes wrong?

The explanation lies in the *representation* of the sets in $\mathcal{A}$. The issue of representation lies at the heart of learning theory, but is as yet poorly understood. However, a *specific* explanation can be attempted in the case of the present example. For each real number $b \in (0, 0.5)$, the collection $\mathcal{A}$ contains many sets that are equivalent[3] to the interval $(0, b)$ – too many in fact. As a result, given *any* multisample $\mathbf{x} \in X^m$, there always exists a set $A \in \mathcal{A}$ such that $\hat{P}(A; \mathbf{x}) = 1$. As a result, $\mathcal{A}$ fails to have the UCEP property. On the other hand, if we "prune" the collection $\mathcal{A}$ by including only a single representative of each equivalence class, leading to the identification $\mathcal{A}/\sim = \{[0, b], 0 < b < 0.5\}$, then the resulting collection of sets *does* have the UCEP property.

The issue of representation is studied further in the next example. As the example is a little involved, a reader who is uninterested in these intricacies may skip it in a first reading, without any loss of continuity.

Example 3.4. (Kulkarni) Let $X = [0, 1]$, $\mathcal{S}$ = the Borel σ-algebra on X, and let P be the uniform distribution on X. Now define $\mathcal{A}$ to be the collection of all unions of the form $[0, a] \cup G$, where $a \in [0, 0.5]$ and G is a finite set. Then it can be shown as in the preceding two examples that $\mathcal{A}$ *does not* have the UCEP property. More generally, *any* collection $\mathcal{A}$ fails to have the UCEP property, provided two conditions are satisfied: (i) There exists a number $\alpha < 1$ such that $P(A) \leq \alpha$ for all $A \in \mathcal{A}$. (ii) For *every* finite set G, there exists a corresponding set $A \in \mathcal{A}$ such that $G \subseteq A$. The proof of this statement is easy and is left to the reader. The collection of Example 3.2 satisfies these two conditions with $\alpha = 0$, while those of Example 3.3 and the present example satisfy these conditions with $\alpha = 0.5$. Hence $\mathcal{A}$ does not have the UCEP property.

Now let us define the equivalence relation $\sim$ on $\mathcal{A}$ as before, i.e., by defining $A \sim B$ if $d_P(A, B) = 0$. Let $\mathcal{A}/\sim$ denote the collection of equivalence classes under this relation. The objective of the example is to show that, if the representative elements of the family of equivalence classes $\mathcal{A}/\sim$ are selected on one way, then the resulting reduced collection *does* have the UCEP property, whereas if they are selected in another way, then the resulting reduced collection *fails* to have the UCEP property. Therefore, whether or not a collection has the UCEP property is very much dependent on the choice of the representative elements of the equivalence classes. In this sense, the UCEP property is rather "fragile."

Suppose we define the collection $\mathcal{A}_1$ as $\{[0, a], a \in [0, 0.5]\}$. Then each set in $\mathcal{A}_1$ belongs to a different equivalence class in $\mathcal{A}$; moreover, every set in $\mathcal{A}$ is equivalent to exactly one set in $\mathcal{A}_1$. Thus $\mathcal{A}_1$ is a "reduced" version of $\mathcal{A}$ consisting of one representative element from each equivalence class in

---

[3] under the relationship $\sim$ defined in the preceding example

$\mathcal{A}/\sim$. This choice might be considered as the natural choice. By the Glivenko-Cantelli lemma, it follows that $\mathcal{A}_1$ does indeed have the UCEP property.

Now consider another collection $\mathcal{A}_2$, defined next. Let $\mathcal{G}$ denote the collection of all finite subsets of X, and suppose $\tau : \mathcal{G} \to [0, 0.5)$ is a *one-to-one* (but not necessarily onto) map. For each number $a \in [0, 0.5)$, if a belongs to the range of the map τ, so that $a = \tau(F)$ for a *unique* finite set F, choose $[0, a] \cup F$ to be the representative of (the equivalence class of) all unions of the form $[0, a] \cup G$, $G \in \mathcal{G}$. If a does not belong to the range of τ, then choose $[0, a]$ to be the representative of all unions of the form $[0, a] \cup G$, $G \in \mathcal{G}$. As a varies over $[0, 0.5)$, this defines a collection $\mathcal{A}_2$ that also represents $\mathcal{A}/\sim$. Now the previous argument can be applied to show that $\mathcal{A}_2$ *does not* have the UCEP property: Clearly $P(A) \leq 0.5$ for all $A \in \mathcal{A}_2$, and every finite set F is contained in the set $[0, \tau(F)] \cup F$ which belongs to $\mathcal{A}_2$. Thus the example is complete once it is demonstrated that there exists a map τ with the desired properties, namely that $\tau : \mathcal{G} \to [0, 0.5)$ and is one-to-one.

Let $I_0, I_1, I_2, \ldots$ be a partition of the set $\{2, 3, 4, \ldots\}$ such that each set I_i is infinite. For example, let p_i denote the i-th prime number, and let I_i consist of all powers of p_i, that is, $I_i = \{p_i, p_i^2, p_i^3, \ldots\}$. Finally, let I_0 be the complement of the union $\cup_{i \geq 1} I_i$. Thus I_0 consists of all numbers that have at least two distinct prime divisors.[4] Given a finite set $F = \{x_1, \ldots, x_n\}$, arrange the x_i's such that $x_1 > x_2 > \ldots > x_n$. Now the number $\tau(F)$ is defined in terms of its binary expansion

$$\tau(F) = \sum_{i=1}^{\infty} b_i 2^{-i}.$$

Set $b_1 = 0$ always; this ensures that $\tau(F) \leq 0.5$. Next, set the first n bits in I_0 equal to 1, and the rest to 0; this encodes the value of the integer n, i.e., the cardinality of the set F. Finally, for each $i \leq n$, set the bits in I_i equal to the bits in the binary representation of x_i; and for $i > n$, set all bits in I_i equal to zero. Then it is easy to see that τ is one-to-one. Moreover, since only a finite number of bits in I_0 are nonzero, it follows that $\tau(F) < 0.5$. ∎

The preceding ideas can be extended in a natural way to the problem of empirically estimating the *mean value of a function*, as opposed to the *probability measure of a set*. Suppose $f : X \to [0, 1]$ is measurable with respect to the σ-algebra $\mathcal{S}$, and that P is a probability measure on $(X, \mathcal{S})$. Then the expected (or mean) value of f is given by

$$E_P(f) = \int_X f(x) \, P(dx).$$

It is possible to estimate $E_P(f)$ empirically by generating an i.i.d. sequence $x_1, \ldots, x_m \in X$ distributed according to P, and defining

---

[4] This particular definition of the sets I_i plays no role in the argument below, and is intended only for illustrative purposes.

$$\hat{E}(f;\mathbf{x}) := \frac{1}{m}\sum_{j=1}^{m} f(x_j)$$

as the **empirical mean** of the function f based on the multisample $\mathbf{x}$. Now it follows from Hoeffding's inequality that

$$P^m\{\mathbf{x} \in X^m : |\hat{E}(f;\mathbf{x}) - E_P(f)| > \epsilon\} \leq 2\exp(-2m\epsilon^2).$$

Now suppose that we are given a *family* of functions $\mathcal{F}$, where each $f \in \mathcal{F}$ maps X into $[0,1]$ and is measurable with respect to the σ-algebra $\mathcal{S}$.[5] For each $f \in \mathcal{F}$, we can define its empirical mean $\hat{E}(f;\mathbf{x})$ as above, and then define

$$q(m,\epsilon,P) := P^m\{\mathbf{x} \in X^m : \exists f \in \mathcal{F} \text{ s.t. } |\hat{E}(f;\mathbf{x}) - E_P(f)| > \epsilon\}.$$

An equivalent definition of $q(m,\epsilon,P)$ is

$$q(m,\epsilon,P) := P^m\{\mathbf{x} \in X^m : \sup_{f \in \mathcal{F}} |\hat{E}(f;\mathbf{x}) - E_P(f)| > \epsilon\}.$$

Then $1 - q(m,\epsilon,P)$ is the probability that *every* empirical mean is within ϵ of its true value. We say that the family of functions $\mathcal{F}$ has the property of **uniform convergence of empirical means**, or **UCEM** for short, if $q(m,\epsilon,P) \to 0$ as $m \to \infty$ for each $\epsilon > 0$.

Note that the problem of uniform convergence of empirical probabilities can be incorporated into this framework by identifying a set $A \in \mathcal{S}$ with its indicator function $I_A(\cdot)$. In this case, $E_P(I_A) = P(A)$, and the definition of $q(m,\epsilon,P)$ reduces to the earlier definition.

By Hoeffding's inequality, it follows that every *finite* family of functions has the property of the uniform convergence of empirical means. Specifically, if $\mathcal{F}$ is a finite family, then clearly

$$q(m,\epsilon,P) \leq 2|\mathcal{F}|\exp(-2m\epsilon^2).$$

Hence the study of uniform convergence is pertinent only when $\mathcal{F}$ is an *infinite* family.

In analogy with (3.1.2), let us define a stochastic process $\{a_m(\cdot)\}$ on X^∞ by

$$a_m(\mathbf{x}^*) := \sup_{f \in \mathcal{F}} |\hat{E}_m(f;\mathbf{x}^*) - E_P(f)|, \qquad (3.1.3)$$

where $\hat{E}_m(f;\mathbf{x}^*)$ denotes the empirical mean of f based on the first m components of $\mathbf{x}^*$, i.e.,

$$\hat{E}_m(f;\mathbf{x}^*) = \frac{1}{m}\sum_{j=1}^{m} f(x_j).$$

---

[5] Actually, there is nothing special about the interval $[0,1]$, and it can be replaced throughout by any *bounded* interval.

52   3. Problem Formulations

Notice that the above stochastic process is a generalization of that in (3.1.2), if one identifies sets in $\mathcal{S}$ with their indicator functions; this justifies the use of the same symbol in both cases. With this definition, the family $\mathcal{F}$ has the UCEM property if and only if the stochastic process $\{a_m(\cdot)\}$ converges to zero in probability. We say also that the family $\mathcal{F}$ has the property of **almost sure convergence of empirical means (ASCEM)** if the stochastic process $\{a_m(\cdot)\}$ converges almost surely to the zero function.

It is easy to see that the ASCEM property implies the UCEM property. However, the converse is also true, as shown in Theorem 5.2.

3.1.2 The UCEMUP Property

Up to now the problem formulations have focused on the case where there is a *single fixed* probability P, and it is desired to estimate the mean values of functions belonging to some family. It is natural to extend this problem formulation to encompass the case where it is desired simultaneously to estimate the mean values of a family of functions *with respect to each probability measure within a given family of probability measures*. To state this problem precisely, let $(X, \mathcal{S})$ be a given measurable space; let $\mathcal{F}$ be a family of measurable functions mapping X into $[0, 1]$; finally, let $\mathcal{P}$ be a family of probability measures on $(X, \mathcal{S})$. Let $P \in \mathcal{P}$ be fixed but unknown, and let $x_1, \ldots, x_m \in X$ be an i.i.d. sequence distributed according to P. Then one can define the "true" mean of f to be $E_P(f)$ as above. Now define

$$\bar{q}(m, \epsilon, \mathcal{P}) := \sup_{P \in \mathcal{P}} P^m\{\mathbf{x} \in X^m : \sup_{f \in \mathcal{F}} |\hat{E}(f; \mathbf{x}) - E_P(f)| > \epsilon\}. \qquad (3.1.4)$$

This is the same definition as above, with the obvious modification of taking the supremum with respect to $P \in \mathcal{P}$. We say that the pair $(\mathcal{F}, \mathcal{P})$ has the property of **uniform convergence of empirical means uniformly in probability**, or **UCEMUP** for short, if $\bar{q}(m, \epsilon, \mathcal{P}) \to 0$ as $m \to \infty$ for each $\epsilon > 0$. Note that there are *two* notions of uniformity here: one with respect to $f \in \mathcal{F}$ and another with respect to $P \in \mathcal{P}$.

The problem of estimating the probability $P(A)$ as A varies over a family of sets $\mathcal{A}$ and P varies over a family of probability measures $\mathcal{P}$ can be formulated as a special case of the above problem in an obvious manner. We say that the collection of sets $\mathcal{A}$ has the property of **uniform convergence of empirical probabilities uniformly in probability**, or **UCEPUP** for short, if the quantity

$$\bar{q}(m, \epsilon) := \sup_{P \in \mathcal{P}} P^m\{\mathbf{x} \in X^m : \sup_{A \in \mathcal{A}} |\hat{P}(A; \mathbf{x}) - P(A)| > \epsilon\}$$

approaches zero as $m \to \infty$ for each $\epsilon > 0$.

There is a subtle issue that is being glossed over in the above definition. Given a function $f \in \mathcal{F}$, a multisample $\mathbf{x} \in X^m$, and a probability measure

P, the *empirical* mean $\hat{E}(f;\mathbf{x})$ is independent of P, but the *true* mean $E_P(f)$ does depend on P. Thus, the set

$$S(m,\epsilon,P) := \{\mathbf{x} \in X^m : \sup_{f \in \mathcal{F}} |\hat{E}(f;\mathbf{x}) - E_P(f)| > \epsilon\}$$

representing the set of "bad" multisamples, also depends on P. The UCEMUP property demands only that, for each $\epsilon > 0$, the *measure* $P^m[S(m,\epsilon,P)]$ be bounded by a quantity that approaches zero as $m \to \infty$, even though the sets $S(m,\epsilon,P)$ themselves may vary with P.

The interpretation of the UCEMUP (or UCEPUP) property in terms of the convergence of stochastic processes is not very "clean." Let us define the stochastic process

$$a_{m,P}(\mathbf{x}^*) := \sup_{f \in \mathcal{F}} |\hat{E}_m(f;\mathbf{x}^*) - E_P(f)|,$$

which is indexed by the parameter P. This is the same as the stochastic process defined in (3.1.2), except that the dependence of a on P is explicitly identified. This is not mere pedantry – note that, while $\hat{E}_m(f;\mathbf{x}^*)$ depends only on $\mathbf{x}^*$ and not on P, $E_P(f)$ does depend on P. Thus the above stochastic process is indeed dependent on P. Now the family $\mathcal{F}$ has the UCEMUP property if the stochastic process $\{a_{m,P}(\cdot)\}$ converges to zero in probability with respect to the measure P^∞ for each $P \in \mathcal{P}$, and moreover, the convergence is somehow "uniform" also with respect to $P \in \mathcal{P}$.

Note that, while the notion of UCEM generalizes naturally to the case where the single fixed probability is replaced by a *family* of probability measures, the notion of ASCEM does not seem to have such a natural generalization. In particular, for a single fixed P, the almost sure convergence of $\{a_{m,P}(\cdot)\}$ to zero implies convergence in probability to zero; however, when there is a famiy of probability measures $\mathcal{P}$, even if $\{a_{m,P}(\cdot)\}$ converges to zero almost surely for each $P \in \mathcal{P}$, this is still not enough to imply the UCEMUP property.

Finally, let us discuss briefly what is meant by "almost sure convergence" of empirical means (or probabilities) to their true values in the case where P is itself variable. Define the stochastic process $\{a_{m,P}(\cdot)\}$ as above. Now we can say that the family $\mathcal{F}$ has the property that empirical means converge almost surely to their true values if it is true that the stochastic process $\{a_{m,P}(\cdot)\}$ converges almost surely to zero with respect to P^∞ for every $P \in \mathcal{P}$; that is

$$P^\infty\{\mathbf{x}^* \in X^\infty : a_{m,P}(\mathbf{x}^*) \to 0 \text{ as } m \to \infty\} = 1, \forall P \in \mathcal{P}.$$

However, there does not appear to be any natural notion of "uniformity" with respect to $P \in \mathcal{P}$ in this setting. Moreover, if we accept for a moment the result proved in Theorem 5.2 that the UCEM property and the ASCEM property are equivalent *for a fixed probability* P, then one can see that the above condition is equivalent to the UCEM property *for every fixed* $P \in \mathcal{P}$, *without* any notion of uniformity with respect to P. Hence it is possible that the above property is in fact weaker than the UCEMUP property, though this is an open question at the moment.

3.1.3 Extension to Dependent Input Sequences

Up to now the notions of UCEM and UCEMUP have been defined for the case where the random samples $\{x_i\}$ are i.i.d. However, it is relatively straightforward to extend these notions to the case where the inputs are not necessarily independent.

Suppose that $\{X_i\}$ is a stochastic process on the probability space $(X^\infty, \mathcal{S}^\infty, \tilde{P})$, and that $\mathbf{x}^* := \{x_m\}_{m=-\infty}^{\infty}$ is a realization of this stochastic process. Define the empirical mean of a measurable function $f : X \to [0,1]$ based on this realization as

$$\hat{E}_m(f;\mathbf{x}^*) := \frac{1}{m} \sum_{i=1}^{m} f(x_i).$$

Given a family of measurable functions $\mathcal{F}$, each of which maps X into $[0,1]$, define

$$q(m,\epsilon,\tilde{P}) := \tilde{P}\{\mathbf{x}^* \in X^\infty : \sup_{f \in \mathcal{F}} |\hat{E}_m(f;\mathbf{x}^*) - E(f,\tilde{P}_0)| > \epsilon\}, \quad (3.1.5)$$

where $\tilde{P}_0$ denotes the one-dimensional marginal probability of $\tilde{P}$ and $E(f,\tilde{P}_0)$ denotes the expectation of f with respect to $\tilde{P}_0$. Then the pair $(\mathcal{F},\tilde{P})$ is said to have the property of UCEM if $q(m,\epsilon,\tilde{P}) \to 0$ as $m \to \infty$. More generally, suppose $\tilde{\mathcal{P}}$ is a family of probability measures on $(X^\infty, \mathcal{S}^\infty)$, and define

$$\bar{q}(m,\epsilon,\tilde{\mathcal{P}}) := \sup_{\tilde{P} \in \tilde{\mathcal{P}}} q(m,\epsilon,\tilde{P}). \quad (3.1.6)$$

Then the pair $(\mathcal{F},\tilde{\mathcal{P}})$ is said to have the UCEMUP property if $\bar{q}(\mathcal{F},\tilde{\mathcal{P}}) \to 0$ as $m \to \infty$.

It is easy to see that each of these definitions reduces to the corresponding definition in the preceding subsections in case $\tilde{P} = (\tilde{P}_0)^\infty$, i.e., the input sequence is i.i.d.

In the general case too, the UCEM property can be interpreted in terms of the convergence of a stochastic process to zero. Specifically, consider once again the stochastic process $\{a_m(\cdot)\}$ of (3.1.3). Then the pair $(\mathcal{F},\tilde{P})$ has the UCEM property if and only if the stochastic process converges to zero in probability. Under very general conditions (e.g., the input sequence $\{X_i\}$ being ergodic), it can be shown that the convergence of this stochastic process to zero *in probability* also implies its convergence to zero *almost surely*. (The implication in the opposite direction is trivial.) But as in the i.i.d. case, the UCEMUP property does not extend naturally to almost sure convergence.

3.2 Learning Concepts and Functions

In this section we introduce the notion of "learning" an unknown concept, or more generally, an unknown function. It turns out that the same notion can also be used to give a precise meaning to the idea of "generalization." An abstract definition of concept learning is given first, and this abstract definition is illustrated by some examples. Then the definition is extended to the case of function learning.

The basic "ingredients" of learning theory are:

- A set X,
- A σ-algebra $\mathcal{S}$ of subsets of X,
- A family $\mathcal{P}$ of probability measures on $(X, \mathcal{S})$, and
- A subset $\mathcal{C} \subseteq \mathcal{S}$, called the *concept class*, or else a family $\mathcal{F}$ of measurable functions mapping X into $[0, 1]$, called the *function class*.

3.2.1 Concept Learning

Let us begin by discussing concept learning. The discussion is culminated with a formal definition of so-called "PAC" (Probably Approximately Correct) learning of concept classes. The definition is then illustrated through a few examples.

The basic premises in the formulation of the concept learning problem are as follows: There is a fixed but unknown concept $T \in \mathcal{C}$, called the **target** concept. The objective is to "learn" the target concept on the basis of observation, consisting of i.i.d. samples $x_1, \ldots, x_m \in X$ drawn in accordance with a fixed probability measure $P \in \mathcal{P}$. In contrast with the target concept T, the probability measure P may either be known or unknown. For each sample x_j, an "oracle" tells us whether or not $x_j \in T$; equivalently, the oracle returns the value of $I_T(x_j)$, where $I_T(\cdot)$ is the indicator function of T. Thus, after m samples have been drawn, the information available consists of the "labelled multisample"

$$[(x_1, I_T(x_1)), \ldots, (x_m, I_T(x_m))] \in [X \times \{0, 1\}]^m.$$

The objective is to construct a suitable approximation to the unknown target concept T on the basis of the labelled multisample, using an appropriate algorithm. For the purposes of the present discussion, an "algorithm" is merely an indexed family of maps $\{A_m\}_{m \geq 1}$, where

$$A_m : [X \times \{0, 1\}]^m \to \mathcal{C}.$$

If the probability measure P is known, then A_m may depend on P; otherwise A_m must be independent of P, though it can depend on $\mathcal{P}$. Similarly, A_m must be independent of T, but it can depend on $\mathcal{C}$.

Suppose m i.i.d. samples have been drawn, and define

56   3. Problem Formulations

$$H_m(T; \mathbf{x}) := A_m[(x_1, I_T(x_1)), \ldots, (x_m, I_T(x_m))].$$

Thus $H_m(T; \mathbf{x})$ is the output of the algorithm when the target concept is T and the multisample is $\mathbf{x} = [x_1 \ldots x_m]^t$. It is customary to refer to $H_m(T; \mathbf{x})$ as the **hypothesis** generated by the algorithm. When there is no danger of confusion, one can abbreviate $H_m(T; \mathbf{x})$ by H_m. Now the number $d_P(T, H_m) = P(T \Delta H_m)$ provides a quantitative measure of how well H_m approximates T. In particular, since T is the unknown target concept and H_m is an approximation to it, the number $d_P(T, H_m)$ corresponds to the probability that a randomly selected point $x \in X$ is *misclassified* by H_m. Roughly speaking, the algorithm $\{A_m\}$ can be said to "learn" the target concept T if $d_P(T, H_m)$ approaches zero as $m \to \infty$. However, $d_P(T, H_m)$ is itself a random number, because it depends on $H_m = H_m(T; \mathbf{x})$, which in turn depends on the random multisample $\mathbf{x} = [x_1 \ldots x_m]^t$. Thus the convergence of $d_P(T, H_m)$ to zero can only be in a probabilistic sense. Define

$$r(m, \epsilon, P) := \sup_{T \in \mathcal{C}} P^m \{\mathbf{x} \in X^m : d_P[T, H_m(T; \mathbf{x})] > \epsilon\}, \quad (3.2.1)$$

$$\bar{r}(m, \epsilon, \tilde{\mathcal{P}}) := \sup_{P \in \mathcal{P}} r(m, \epsilon, P). \quad (3.2.2)$$

Thus one can think of $r(m, \epsilon, P)$ as the measure of the set of "bad" samples, whereby a sample $\mathbf{x} \in X^m$ is considered to be "bad" if it generates a hypothesis H_m that differs from the true concept T by more than ϵ. The quantity $\bar{r}(m, \epsilon, \mathcal{P})$ is the supremum of $r(m, \epsilon, P)$ as the underlying probability measure P varies over $\mathcal{P}$. Now a precise definition of learning can be given in terms of the behaviour of the quantity $\bar{r}(m, \epsilon, \mathcal{P})$.

Before that, however, a brief caution is in order. Actually, there is a subtle measurability issue that is being glossed over here. For each m-tuple $\mathbf{x} \in X^m$, the number $d_P[T, H_m(T; \mathbf{x})]$ is well-defined because both T and $H_m(T, \mathbf{x})$ belong to $\mathcal{S}$. However, in the absence of some assumptions on the nature of the maps A_m, one cannot be sure that the function $\mathbf{x} \mapsto d_P[T, H_m(T; \mathbf{x})]$ is measurable. Blumer *et al.* [32] contains an example of a situation in which this function is in fact not measurable. Fortunately, however, very mild conditions on the A_m's are enough to ensure that this function is measurable for each m, so that $r(m, \epsilon, P)$ is well-defined. For this reason, such measurability issues are ignored throughout this book.

Definition 3.1. *The algorithm $\{A_m\}$ is said to be **probably approximately correct** (**PAC**) to accuracy ϵ if $\bar{r}(m, \epsilon, \mathcal{P}) \to 0$ as $m \to \infty$. The algorithm $\{A_m\}$ is said to be **probably approximately correct** (**PAC**) if $\bar{r}(m, \epsilon, \mathcal{P}) \to 0$ as $m \to \infty$ for each $\epsilon > 0$. The concept class $\mathcal{C}$ is said to be **PAC learnable to accuracy ϵ with respect to the family $\mathcal{P}$** if there exists an algorithm that is PAC to accuracy ϵ. Finally, the concept class $\mathcal{C}$ is said to be **PAC learnable with respect to the family $\mathcal{P}$** if there exists an algorithm that is PAC.*

Remarks

1. The somewhat unusual nomenclature "probably approximately correct" is motivated by the nature of the convergence of the hypothesis H_m to the target concept T. Now H_m is only an *approximation* to T in the sense that $d_P(T, H_m)$ is not required to be zero – only small. Also, one can only assert *with high probability* that $d_P[T, H_m(T; \mathbf{x})]$ is small – there is a small probability that the multisample $\mathbf{x}$ could lead to an unacceptably bad approximation H_m. It may be a good idea to think of a "probably approximately correct" algorithm as one that "works reasonably well most of the time."

2. One can recast the definition in the following equivalent form: The algorithm $\{A_m\}$ is PAC if, for every $\epsilon, \delta > 0$, there exists an integer $m_0 = m_0(\epsilon, \delta)$ such that
$$P^m\{\mathbf{x} \in X^m : d_P[T, H_m(T; \mathbf{x})] > \epsilon\} \leq \delta, \ \forall m \geq m_0, \ \forall T \in \mathcal{C}, \ \forall P \in \mathcal{P}. \tag{3.2.3}$$
Here ϵ is called the **accuracy** parameter, and δ is called the **confidence** parameter.[6] Thus, after drawing m i.i.d. samples at random and applying a PAC algorithm on the resulting labelled multisample, one can state with confidence of at least $1 - \delta$ that the resulting hypothesis H_m will correctly classify a randomly selected point $x \in X$ with a probability of at least $1 - \epsilon$. Thus the essence of (PAC) learning theory as studied here consists of (i) determining conditions under which a concept class is PAC learnable, and (ii) obtaining estimates (both upper and lower bounds) for m_0 as a function of ϵ and δ. Some authors refer to the *smallest* number $m_0(\epsilon, \delta)$ such that the above inequality holds as the **sample complexity** of the learning algorithm.

3. The definition of learnability given here does not place any restrictions on the nature of the "algorithm" $\{A_m\}$. In particular, there are no restrictions as to the effective computability of the function A_m, nor on its computational complexity (either in terms of time or storage). Thus the emphasis here is on what can be achieved *in principle* rather than on what can be achieved *in practice*. The brand of learning theory studied here can perhaps be called "statistical learning theory." This is in contrast to "computational learning theory," in which attention is paid also to the *nature* of the algorithm $\{A_m\}$, e.g., the effective computability of the function A_m, its computational complexity, and the like.

When there is no danger of confusion, we shall drop the prefix "PAC", and speak merely of "learnability." However, we shall continue to say that "an algorithm is PAC." If some other form of learnability is intended (e.g., PUAC learnability introduced in a later subsection), we shall use the complete name.

---

[6] Though this terminology has by now become standard, it is perhaps better to think of ϵ as the *inaccuracy* parameter, and of δ as the *lack of confidence* parameter.

58   3. Problem Formulations

Now the preceding abstract definition is illustrated through several examples.

Example 3.5. As a very simple illustration, suppose $X = [0,1]^2$, $\mathcal{S}$ = the Borel σ-algebra on X, and let $\mathcal{C}$ equal the collection of all "axis-parallel" rectangles of the form $[x_l, x_u] \times [y_l, y_u]$ in X. Finally, let $\mathcal{P} = \mathcal{P}^*$, the family of *all* probability measures on X. It is shown that this concept class is learnable by presenting an algorithm that is PAC.

The algorithm is very simple and intuitive: After m samples have been drawn at random, the hypothesis H_m is chosen as the *smallest* rectangle containing all the "positive" samples, i.e., all samples x_j for which $I_T(x_j)$ equals one. Figure 3.1 illustrates the idea. Note that the "negative" samples

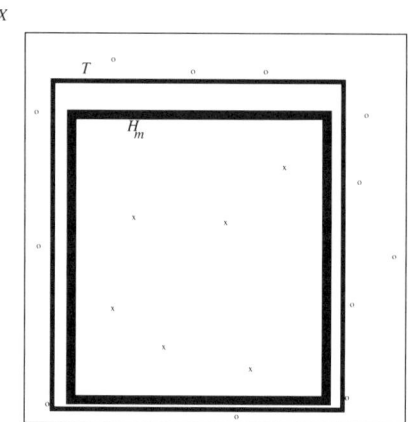

Fig. 3.1. Learning Axis-Parallel Rectangles

do not play any role in the construction of H_m, and that H_m is always a subset of the target concept T.

To estimate the error $d_P(T, H_m)$, let us define four smaller rectangles D_1 through D_4 as shown in Figure 3.2. Clearly

$$T \Delta H_m = T - H_m \subseteq \bigcup_{i=1}^{4} D_i.$$

Also, the four rectangles are not disjoint in general.

To show that the algorithm is PAC, let us estimate the quantity $\bar{r}(m, \epsilon, \mathcal{P}^*)$ defined in (3.2.1). Suppose $\epsilon > 0$ is specified. Now, if $P(T) < \epsilon$, then $d_P(T, H_m) = P(T \Delta H_m) < \epsilon$, because $T \Delta H_m$ is a subset of T. So it can be supposed without loss of generality that $P(T) \geq \epsilon$. Now choose a rectangle A_1 in such a way that the left, right, and top sides of A_1 are aligned with those of $[0,1]^2$, and A_1 is the smallest such rectangle with the property

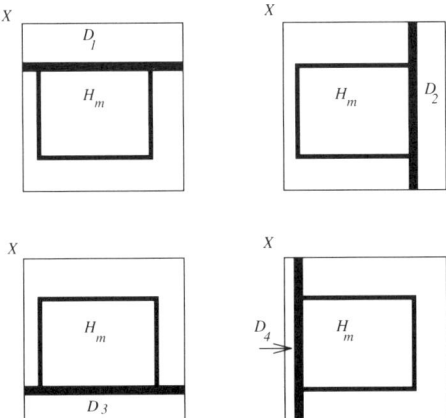

Fig. 3.2. The Rectangles D_1 through D_4

that $P(A_1) \geq \epsilon/4$. It is necessary to define A_1 in this tortuous manner because $P(A_1)$ need not in general be a *continuous* function the coordinate y_1 (see Figure 3.3). However, it follows from elementary probability theory (see,

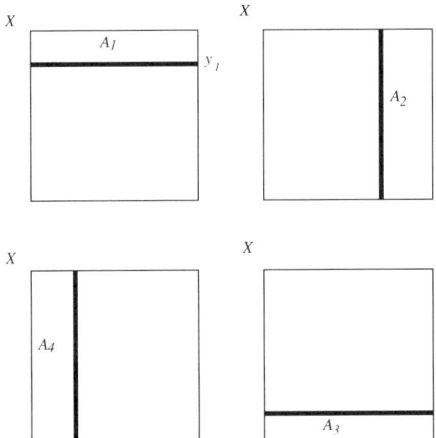

Fig. 3.3. The Rectangles A_1 through A_4

e.g., [74], p. 162) that $P(A_1)$ as a function of y_1 has the property referred to in [162] as "cadlag", that is, continuity from the right (for decreasing y_1) and existence of limits from the left (for increasing y_1). Hence there exists a *unique* smallest rectangle A_1 such that $P(A_1) \geq \epsilon/4$. Similarly, choose rectangles A_2 through A_4 as shown such that they too are the smallest rectangles with the property that $P(A_i) \geq \epsilon/4$. If m i.i.d. samples are chosen in accordance with P, then the probability that *none* of the samples falls within

A_1 equals $[1 - P(A_1)]^m \leq (1 - \epsilon/4)^m$. If none of the samples falls inside A_1, then the "top" edge of the hypothesis H_m is lower than the "bottom" edge of A_1, or equivalently, A_1 is a subset of D_1. Hence the probability that $P(D_1) \geq P(A_1) \geq \epsilon/4$ is at most $(1 - \epsilon/4)^m$. Similar remarks apply to D_2 through D_4. Hence, with probability at least $1 - 4(1 - \epsilon/4)^m$, it is true that $P(D_i) \leq \epsilon/4$ for *each* i, and hence that $P(\cup_i D_i) = P(T - H_m) \leq \epsilon$. This shows that

$$\bar{r}(m, \epsilon, \mathcal{P}^*) \leq 4(1 - \epsilon/4)^m, \; \forall m, \epsilon.$$

In proving the above bound, no use was made of the nature of T or P. Since the right side of the above inequality approaches zero as $m \to \infty$ for each $\epsilon > 0$, the algorithm is PAC.

Using the above upper bound for $\bar{r}(m, \epsilon, \mathcal{P}^*)$, it is possible to derive an upper bound for the "sample complexity," that is, the number of samples sufficient to estimate every unknown target rectangle to an error of ϵ with a confidence of at least $1 - \delta$. Suppose ϵ, δ are specified, and select m large enough that

$$4(1 - \epsilon/4)^m \leq \delta.$$

This bound on m can be made to appear more "explicit" if one observes that, from the Taylor series expansion of the ln function, one has

$$\ln(1 - x) \leq -x, \; \forall x \in (0, 1).$$

Hence the above inequality is satisfied if

$$m \geq \frac{4}{\epsilon} \ln \frac{4}{\delta}.$$

With this choice of m, it follows that

$$1 - 4(1 - \epsilon)^m \geq 1 - \delta,$$

i.e., that $d_P(T, H_m) \leq \epsilon$ with probability at least $1 - \delta$.

This example can be used to bring out one of the nuances in the definition of PAC learnability. For this purpose, suppose P is a *purely atomic* measure concentrated at points $\mathbf{z}_1, \ldots, \mathbf{z}_k$, where k is a finite number (see Figure 3.4). Suppose the target rectangle T *does not* contain any of these points on its boundary, and let H denote the smallest axis-parallel rectangle inside T such that $P(T) = P(H)$ (again, see Figure 3.4). If the samples $x_1, x_2, \ldots$ are drawn i.i.d. with respect to *this particular* probability measure P, then each sample x_j equals one of $\mathbf{z}_1, \ldots, \mathbf{z}_k$ (with probability one). Now the algorithm given here consists of choosing H_m to be the *smallest* axis-parallel rectangle consistent with the data. As a result, H_m can never equal T. However, H_m "converges in probability" (in a sense made precise later in this section) to the rectangle H. Now this is perfectly acceptable, because $P(T \Delta H) = 0$ with this particular probability measure, even though T and H are distinct rectangles in the normal sense.

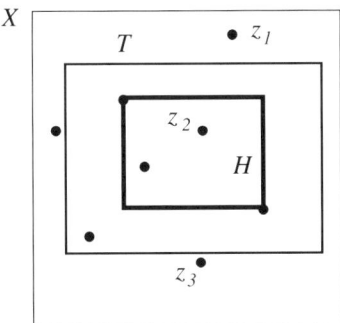

Fig. 3.4. Learning Axis-Parallel Rectangles Under a Purely Atomic Measure

Example 3.6. The point of this example is to show how the basic PAC inequality can be interpreted as "generalization with small error with high probability" in the context of neural network training. To keep the exposition simple, the discussion in this example is restricted to the problem of "identifying" a fixed but unknown perceptron. However, as shown in the next section, the theory also applies to a much more general class of problems known as model-free learning; note that some authors refer to this type of problem as "agnostic" learning. The problem in model-free learning is to find, with high probability, a neural network from within a given class that most closely approximates a given set of data; see Section 3.3 for details.

Recall that a **perceptron** is a threshold switching device described by the input-output relationship

$$y = \eta \left[\sum_{i=1}^{k} w_i u_i - \theta \right],$$

where $\mathbf{u} = [u_1 \ldots u_k]^t \in \mathbb{R}^k$ is the input, $y \in \{0, 1\}$ is the output, and the step function (also referred to as the Heaviside function) $\eta(\cdot)$ is defined by

$$\eta(x) = \begin{cases} 0, & \text{if } x < 0, \\ 1, & \text{if } x \geq 0. \end{cases}$$

The numbers $w_1, \ldots, w_k$ are called the "weights" of the perceptron, while θ is called the "threshold."

With each set of perceptron parameters $[w_1, \ldots, w_k, \theta]^t \in \mathbb{R}^{k+1}$, one can associate a unique subset of $\mathbb{R}^k$ representing all inputs in $\mathbb{R}^k$ that are mapped into the output 1. It is easy to see that this subset is a closed half-space. (See Figure 3.5 for a representation of the situation when $k = 2$.) However, the association of the perceptron parameters with the corresponding half-space is *not* one-to-one. In particular, if one replaces $[w_1, \ldots, w_k, \theta]^t$ by $\alpha[w_1, \ldots, w_k, \theta]^t$ where $\alpha > 0$, then the corresponding half-space remains the same. Thus, in order to set up the "perceptron identification" problem, we make the following definitions:

62   3. Problem Formulations

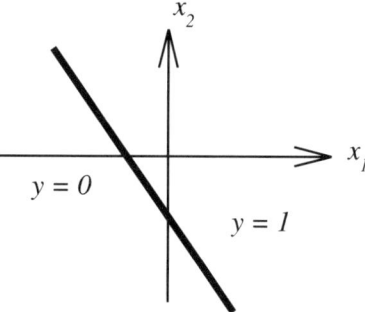

Fig. 3.5. A Perceptron with Two Inputs

Let $X = \mathbb{R}^k$, $\mathcal{S}$ = the Borel σ-algebra on $\mathbb{R}^k$, and let $\mathcal{C}$ = the collection of closed half-spaces in $\mathbb{R}^k$. With these definitions, the "learning" problem becomes the following: There is a fixed but unknown half-space $T \in \mathcal{C}$. (One can think of a fixed but unknown perceptron.) Given any input $\mathbf{u} \in \mathbb{R}^k$, an oracle tells us whether or not $\mathbf{u} \in T$. (In other words, though the unknown perceptron is a "black box," we are permitted to apply any input $\mathbf{u} \in \mathbb{R}^k$ to it and observe the output, 0 or 1.) To "learn" T, we draw inputs $\mathbf{u}_1, \ldots, \mathbf{u}_m \in \mathbb{R}^k$ that are generated i.i.d. in accordance with a probability measure P that may itself be unknown. After m samples are drawn, an estimate H_m of T is formed of T. (See Figure 3.6.) We can think of H_m as a "candidate" percep-

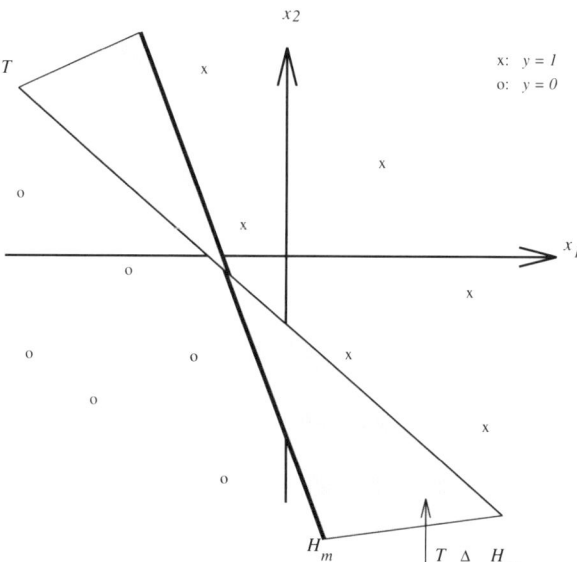

Fig. 3.6. The Error Region for a Perceptron

tron that has been "trained" on the inputs $\mathbf{u}_1, \ldots, \mathbf{u}_m$. Now the symmetric difference $T \Delta H_m$ gives an indication of how well the candidate perceptron is able to "generalize." *Specifically, $1 - P(T \Delta H_m)$ is the probability that, if an input $\mathbf{u} \in \mathbb{R}^k$ is selected at random, the candidate perceptron H_m gives the correct output.* Thus the aim of the learning algorithm is to drive $P(T \Delta H_m)$ towards zero; if the "training inputs" $\mathbf{u}_i \in \mathbb{R}^k$ are selected at random, then the convergence of $P(T \Delta H_m)$ to zero is also in probability.

Thus far we have seen that the PAC learning problem formulation is one way to formalize the notion that a perceptron (or more generally, a neural network) can "generalize." It is perhaps possible to come up with alternate formulations of this notion, besides the PAC formulation. However, it is important to realize that "perfect" generalization is *impossible*, even in the simplest possible case of a single perceptron. To clarify this point, suppose that a "candidate" perceptron has been "trained" on a *finite* number of inputs $\mathbf{u}_1, \ldots, \mathbf{u}_m$, as depicted in Figure 3.7. Then *any* perceptron whose boundary

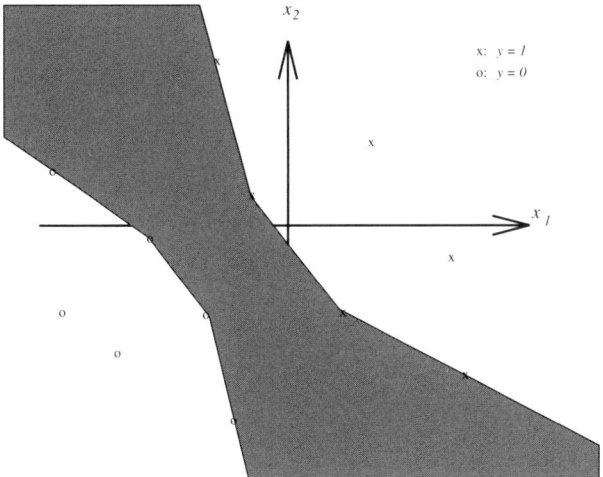

Fig. 3.7. Impossibility of "Perfect" Generalization

passes through the shaded region in Figure 3.7 could have produced this data. In other words, there are *infinitely many* perceptrons that are consistent with the observed data. Note that this statement is true whenever the observed data consists of a *finite* number of inputs $\mathbf{u}_1, \ldots, \mathbf{u}_m$, irrespective of what the actual data is. Now, if the training algorithm makes use of *only* the training inputs and the corresponding correct outputs (and any reasonable training algorithm can make use of only this information and nothing else), it follows that the "trained" perceptron must be exactly the same for *every* true perceptron whose boundary passes through the shaded region. As there are infinitely many such perceptrons whenever the data set is finite, it follows that the trained perceptron can never *exactly* equal the "true" perceptron after a *finite* number of training inputs. Hence *perfect* generalization by a neural network is an impossibility, even in the simplest case of a single perceptron. In the case of more general neural networks of the classification type (i.e., where the output is either 0 or 1), similar arguments apply. Now the PAC learning problem formulation gets around this difficulty by asking, not for *perfect* generalization, but only for generalization with probability *close to* (but not necessarily equal to) 1.

3.2.2 Function Learning

The preceding problem formulation can be extended in a natural way to encompass *function* learning. Suppose (as before) that $(X, \mathcal{S})$ is a measurable space, $\mathcal{P}$ is a family of probability measures on $(X, \mathcal{S})$, and that $\mathcal{F}$ is a family of functions mapping X into $[0, 1]$, each of which is measurable with respect to $\mathcal{S}$. By a slight abuse of notation, let us agree to use the symbol $[0, 1]^X$ to denote the set of all *measurable* functions mapping X into $[0, 1]$, as opposed to the set of *all* functions mapping X into $[0, 1]$, which is the normal meaning. Given two measurable functions $a, b : X \to [0, 1]$ and a probability measure P on $(X, \mathcal{S})$, one can define

$$d_P(a, b) = \int_X |a(x) - b(x)|\, P(dx).$$

This is a pseudometric on $[0, 1]^X$, and equals the expected value of the difference $|a(x) - b(x)|$. Moreover, this $d_P(\cdot)$ is a natural generalization of the pseudometric d_P on $\mathcal{S}$ defined earlier. Given two sets $A, B \in \mathcal{S}$, let a, b denote their indicator functions respectively. Then

$$d_P(A, B) = d_P(a, b).$$

This identity justifies the dual usage of the symbol d_P.

In function learning, there is a fixed but unknown target function $f \in \mathcal{F}$, and a probability measure $P \in \mathcal{P}$ that may or may not be known. In the present context, an "algorithm" is an indexed family of maps $\{A_m\}$ where

$$A_m : (X \times [0, 1])^m \to \mathcal{F}.$$

Learning proceeds as follows: i.i.d. samples $x_1, \ldots, x_m \in X$ are drawn in accordance with P, and for each sample x_i, an oracle returns the value $f(x_i)$. Define

$$h_m(f; \mathbf{x}) := A_m[(x_1, f(x_1)), \ldots, (x_m, f(x_m))], \text{ and}$$

$$\bar{r}(m, \epsilon, \mathcal{P}) := \sup_{P \in \mathcal{P}, f \in \mathcal{F}} P^m\{\mathbf{x} \in X^m : d_P[f, h_m(f; \mathbf{x})] > \epsilon\}. \quad (3.2.4)$$

3.2 Learning Concepts and Functions

Definition 3.2. *The algorithm $\{A_m\}$ is* **probably approximately correct (PAC) to accuracy** ϵ *if $\bar{r}(m, \epsilon, \mathcal{P}) \to 0$ as $m \to \infty$, and is* **probably approximately correct (PAC)** *if $\bar{r}(m, \epsilon, \mathcal{P}) \to 0$ as $m \to \infty$ for each $\epsilon > 0$. The function class $\mathcal{F}$ is* **PAC learnable to accuracy ϵ with respect to the family** $\mathcal{P}$ *if there exists an algorithm that is PAC to accuracy ϵ, and is* **PAC learnable with respect to the family** $\mathcal{P}$ *if there exists an algorithm that is PAC.*

Example 3.7. Consider a so-called "standard sigmoidal" neuron, mapping $\mathbb{R}^k$ into $[0, 1]$, whose input-output relationship is

$$y = \frac{1}{1 + e^{-z}},$$

where

$$z = \sum_{i=1}^{k} w_i u_i - \theta.$$

Let $\mathcal{F}$ denote the set of all functions of this form, as $[w_1, \ldots, w_k, \theta]^t$ varies over $\mathbb{R}^{k+1}$. The learning problem is the same as in Example 3.6, with the obvious modifications.

The interpretation of "generalization" in the present instance is slightly different from that in the case of perceptrons. In the case where the objective is to learn a fixed but unknown perceptron, the error measure $d_P(T, H_m)$ is the probability that a randomly chosen $\mathbf{u} \in \mathbb{R}^k$ is incorrectly classified by the hypothesized perceptron H_m. In the present instance, the error measure is the *expected value* of the (magnitude of the) difference between the correct output $f(\mathbf{u})$ and the predicted output $h_m(\mathbf{u})$. Notice that the error measure says nothing at all about the size of the set $\{\mathbf{u} \in \mathbb{R}^k : f(\mathbf{u}) \neq h_m(\mathbf{u})\}$. The measure of this set can be thought of as the probability of getting an incorrect output. This particular error measure is somewhat misleading in that it does not take into account *by how much* the correct output $f(\mathbf{u})$ differs from the predicted output $h_m(\mathbf{u})$. In fact, it is easy to see that, with any reasonable learning algorithm, $f(\mathbf{u})$ does *not* equal $h_m(\mathbf{u})$ for almost all $\mathbf{u} \in \mathbb{R}^k$, i.e., that $P\{\mathbf{u} \in \mathbb{R}^k : f(\mathbf{u}) \neq h_m(\mathbf{u})\} = 1$. This observation once again underlines the fact that "perfect" generalization is an illusion – all that one can hope to achieve is to make the expected value of the difference $|f(x) - h_m(x)|$ approach zero in probability. ■

It can be seen that the function learning problem has a natural interpretation in terms of *interpolation*, in the sense that one is given the value of an unknown function f at various points $x_1, \ldots, x_m$, and the objective is to deduce the function itself from its values at these sample points.

3.2.3 Extension to Dependent Input Sequences

It is straight-forward to extend the above definition of PAC learning to the case where the input sequence is not necessarily i.i.d. As in Section 3.1, suppose that $\{\mathbf{X}_i\}$ is a stochastic process on the probability space $(X^\infty, S^\infty, \tilde{P})$, and that $\mathbf{x}^* = \{x_m\}_{m=-\infty}^{\infty}$ is a realization of this stochastic process. Let $h_m(f; \mathbf{x}^*)$ denote the output of the learning algorithm after m inputs when the target function is f and the input sequence is $\mathbf{x}^*$. Define

$$r(m, \epsilon, \tilde{P}) := \sup_{f \in \mathcal{F}} \tilde{P}\{\mathbf{x}^* \in X^\infty : d_{\tilde{P}_0}[f, h_m(f; \mathbf{x}^*)] > \epsilon\}. \qquad (3.2.5)$$

Here, as elsewhere, $\tilde{P}_0$ denotes the one-dimensional marginal probability associated with $\tilde{P}$. Then the algorithm is PAC if $r(m, \epsilon, \tilde{P}) \to 0$ as $m \to \infty$. In case the single probability measure $\tilde{P}$ is replaced by a family of probability measures $\tilde{\mathcal{P}}$, we need to take the supremum with respect to $\tilde{P} \in \tilde{\mathcal{P}}$ as well.

The function learning problem can also be interpreted in terms of the convergence of *an indexed family* of stochastic processes to zero, in a uniform sense. This is in contrast to the UCEM property, in which *a single stochastic process* is required to converge to zero in probability. Specifically, with the above notation, define the stochastic process $\{b_{m,f}(\cdot)\}$ as follows: Given a sample sequence $\mathbf{x}^*$, let

$$b_{m,f}(\mathbf{x}^*) := d_P[f; h_m(f; \mathbf{x}^*)]. \qquad (3.2.6)$$

In other words, $b_{m,f}(\mathbf{x}^*)$ equals the generalization error after m samples when the target function is f and the multisample is $\mathbf{x}^*$. Then it is clear from the definition that the algorithm is PAC if and only if the stochastic process $\{b_{m,f}\}$ converges to zero in probability with respect to $\tilde{P}$, uniformly with respect to $f \in \mathcal{F}$. This reinterpretation of PAC algorithms in terms of the convergence of a family of stochastic processes is very useful when we study the behaviour of learning algorithms when the input sequence is not necessarily i.i.d.

The above definition can be extended in an obvious manner to the case of a *family* of probability measures, call it $\tilde{\mathcal{P}}$, on the measurable space (X^∞, S^∞). Then define

$$\bar{r}(m, \epsilon, \tilde{\mathcal{P}}) := \sup_{\tilde{P} \in \tilde{\mathcal{P}}} r(m, \epsilon, \tilde{P}). \qquad (3.2.7)$$

Then the algorithm is PAC if $\bar{r}(m, \epsilon, \tilde{\mathcal{P}}) \to 0$ as $m \to \infty$, for each $\epsilon > 0$. The pair $(\mathcal{F}, \tilde{\mathcal{P}})$ is PAC learnable if there exists an algorithm that is PAC.

3.2.4 Assumptions Underlying the Model of Learning

Having introduced the basic model of PAC learning of concept and function classes, we digress briefly to highlight a few fundamental assumptions underlying the definition of PAC learning. Three points are discussed here, namely: (i) the distinction between the "training" probability and the "testing" probability, (ii) the possibility of using an algorithm that makes explicit use of the accuracy and the confidence parameters, and (iii) the possibility that the hypothesis class is distinct from the concept (or function) class. Each of these points is discussed in turn.

Let us first consider the distinction between the "testing" probability and the "training" probability. This distinction is most meaningful in concept learning problems, though such a distinction could also be made in function learning problems. To bring out this distinction clearly, let us for a moment fix the probability P, and rewrite the PAC inequality (3.1.2) in the form

$$P^m\{\mathbf{x} \in X^m : P(T \Delta H_m(T; \mathbf{x})) > \epsilon\} \leq \delta, \; \forall m \geq m_0, \; \forall T \in \mathcal{C}. \quad (3.2.8)$$

Now one can see that the probability P plays *two distinct* roles in the above inequality. On the one hand, the number $P(T \Delta H_m(T; \mathbf{x}))$ measures the "goodness of fit" between the hypothesis H_m and the unknown target concept T. As pointed out elsewhere, this number is the probability that, if a "testing input" $x \in X$ is selected at random (according to P), then x is misclassified by the hypothesis. In this role, P is the probability that generates the *testing inputs*. Now let us examine the set

$$S_m := \{\mathbf{x} \in X^m : P(T \Delta H_m(T; \mathbf{x})) > \epsilon\}.$$

One can think of S_m as the set of bad "training inputs," where a multisample $\mathbf{x} \in X^m$ is considered to be "bad" if it leads to a hypothesis that misclassifies a randomly selected testing input with a probability greater than ϵ. If the training inputs are generated i.i.d. according to P, then it is natural to use the product measure P^m on X^m to measure the "size" of the set S_m of bad training inputs. In this role, P is the probability that generates the *training inputs*.

In view of the dual role being played by P, suppose we replace (3.2.8) by the more general inequality

$$P^m\{\mathbf{x} \in X^m : Q(T \Delta H_m(T; \mathbf{x})) > \epsilon\} \leq \delta, \; \forall m \geq m_0, \; \forall T \in \mathcal{C}. \quad (3.2.9)$$

In this inequality, one can think of P as the probability measure that generates the training inputs, whereas Q is the probability measure that generates the testing input. One could then define an algorithm to be (P, Q)-PAC if, for every $\epsilon, \delta > 0$, there exists an $m_0 = m_0(\epsilon, \delta)$ such that (3.2.9) holds. Now it is argued that, in order for such a definition to be meaningful, the probabilities P and Q must be "compatible" in the sense that

$$P(A) = 0 \Rightarrow Q(A) = 0. \quad (3.2.10)$$

Suppose to the contrary that the above condition is violated, and that there exists an $A \in \mathcal{S}$ such that $P(A) = 0$ and $Q(A) > 0$. Then, since the training inputs are drawn i.i.d. according to P, *no* training input will belong to the set A (with probability one). On the other hand, since $Q(A) > 0$, there is a positive probability that the *testing* input belongs to the set A. Thus the learning algorithm would be penalized for not being able to produce the correct output $I_T(x)$ when $x \in A$, even though the algorithm has no opportunity to learn how the target concept looks when $x \in A$ (since $P(A) = 0$). Therefore, in order for the more general notion of PAC learning[7] to be meaningful, it is necessary that the condition (3.2.10) be satisfied.

Now suppose we replace (3.2.10) by the stronger condition: There exists a constant μ such that

$$Q(A) \leq \mu P(A), \ \forall A \in \mathcal{S}.$$

Then it is easily shown that (P,Q)-PAC learnability is implied by the standard PAC learnability. This is the rationale for assuming that $P = Q$.

In Chapter 9, we study a somewhat different form of learning known as "active" learning, in which the learner has the freedom to *choose* the training inputs. In this situation, the number $Q(T \Delta H_m(T; \mathbf{x}))$ can be thought of as a "cost function" that should be minimized by choosing $\mathbf{x}$ suitably. Even in this case, the interpretation of Q as a "testing" probability still holds good. After the algorithm is run m times (with the training inputs chosen by the learner), a testing input $x \in X$ is selected according to Q, and the algorithm is evaluated according to the probability that the hypothesis produced by the algorithm misclassifies this randomly selected x.

All of the preceding comments apply *mutatis mutandis* to function learning as well.

The next issue to be discussed is the nature of the "algorithm" employed by the learner. As stated here, an algorithm is an indexed family of mappings from $(X \times [0,1])^m$ to $\mathcal{F}$. One could generalize the notion of an algorithm by permitting it to make use of not only the labelled samples $(x_j, f(x_j))$, but also the accuracy parameter ϵ and the confidence parameter δ. Such an algorithm can be thought of as an indexed family of mappings $\{B_m\}$ where B_m maps $(X \times [0,1])^m \times (0,1)^2$ into $\mathcal{F}$. (Note that the term $(0,1)^2$ arises from the dependence of the algorithm on ϵ and δ.) In other words, the algorithm is specifically tailored to learn to a *prescribed level* of accuracy and confidence; if it is desired to reduce either ϵ or δ, then in principle one would have to change the manner in which the labelled samples (belonging to $(X \times [0,1])^m$) are processed. It turns out, however, that such a distinction is not meaningful, as shown next. Since there is no particular advantage to assuming that the learning inputs are independent, we treat the case of (possibly) dependent inputs.

Theorem 3.1. *Suppose $X, \mathcal{S}, \tilde{\mathcal{P}}, \mathcal{F}$ are as above, and suppose that there exists a family of maps $\{B_m\}_{m \geq 1}$ where*

$$B_m : (X \times [0,1])^m \times (0,1)^2 \to \mathcal{F} \qquad (3.2.11)$$

with the following property: For each $(\epsilon, \delta) \in (0,1)^2$, there exists an integer $s = s(\epsilon, \delta)$ such that

---

[7] That is, when the testing probability need not be the same as the training probability

3.2 Learning Concepts and Functions 69

$$\sup_{\tilde{P} \in \tilde{\mathcal{P}}} \sup_{f \in \mathcal{F}} \tilde{P}\{\mathbf{x}^* \in X^\infty : d_P[f, h_s(f; \mathbf{x}^*)] > \epsilon\} \leq \delta, \quad (3.2.12)$$

where

$$h_s(f; \mathbf{x}^*) := B_s[(x_1, f(x_1)), \ldots, (x_s, f(x_s)); \epsilon, \delta].$$

Then there exists an algorithm that PAC learns $\mathcal{F}$.

Remarks In words, the lemma states that, if there exists an "ϵ, δ-dependent algorithm" that is "PAC" for each ϵ, δ, then in fact there exists an algorithm that is *independent* of ϵ, δ that is also PAC in the conventional sense.

Proof. The objective is to construct an algorithm that does not depend on ϵ and δ, but nevertheless is PAC in the conventional sense. Towards this end, let $s : (0, 1)^2 \to \mathbf{N}$ be the "sample complexity" map defined in the hypothesis; that is, $s(\epsilon, \delta)$ is the integer such that (3.2.12) above holds. Now define the map $\bar{s} : (0, 1) \to \mathbf{N}$ by

$$\bar{s}(\alpha) := \max_{\epsilon \geq \alpha, \delta \geq \alpha} s(\epsilon, \delta), \ \forall \alpha \in (0, 1).$$

It is easy to see that $\bar{s}$ is a nonincreasing function of α; that is, $\alpha_1 < \alpha_2$ implies that $\bar{s}(\alpha_1) \geq \bar{s}(\alpha_2)$. Next, choose a continuous and strictly decreasing function $\phi : (0, 1) \to \mathbb{R}$ such that $\phi(\alpha) \geq \bar{s}(\alpha)$ for all α.

Now an algorithm $\{A_m\}_{m \geq 1}$, where

$$A_m : (X \times [0, 1])^m \to \mathcal{F}$$

is defined as follows: Given a multisample $\mathbf{x}^* \in X^\infty$, an integer m, and a vector $\mathbf{v} \in [0, 1]^m$, let $\alpha = \phi^{-1}(m)$. Define $s = s(\alpha, \alpha)$, and notice that $s \leq m$ by the manner in which the function ϕ is defined. Now define

$$A_m(\mathbf{x}^*, \mathbf{v}) := B_s(\mathbf{x}_s, \mathbf{v}_s, \alpha, \alpha),$$

where $\mathbf{x}_s \in X^s, \mathbf{v}_s \in [0, 1]^s$ consist of the first s components of $\mathbf{x}^*$ and $\mathbf{v}$ respectively. To put into words, apply the map B_s to the first s components of $\mathbf{x}^*$ and $\mathbf{v}$ with $\epsilon = \delta = \alpha$, and define the resulting element of $\mathcal{F}$ to be the output of the algorithm A_m. Note that all the components of $\mathbf{x}^*$ and $\mathbf{v}$ may not be used by the algorithm. Moreover, by the hypothesis on the map B_s, it follows that if the vector $\mathbf{v}$ equals $[f(x_1) \ldots f(x_m)]^t$ for some target function $f \in \mathcal{F}$, then we have that $d_P(f, h_m) \leq \alpha$ with a probability of at least $1 - \alpha$, where h_m is the output of the algorithm.

It is claimed that the algorithm is PAC. To establish this claim, suppose $\epsilon, \delta > 0$ are specified. It is shown that there exists an $m_0 = m_0(\epsilon, \delta)$ such that

$$\tilde{P}d_P[f, h_m(f; \mathbf{x}^*)] > \epsilon\} \leq \delta, \ \forall f \in \mathcal{F}, \forall \tilde{P} \in \tilde{\mathcal{P}}.$$

Let $\mu = \min\{\epsilon, \delta\}$, and let $m_0 = \lceil \phi(\mu) \rceil$. Suppose $m \geq m_0$. Then $\alpha = \phi^{-1}(m) \leq \phi^{-1}(m_0) \leq \mu$, since $\phi(\cdot)$ is a decreasing function. Now the manner in which the algorithm $\{A_m\}$ is defined ensures that

$$\tilde{P}d_P[f, h_m(f; \mathbf{x}^*)] > \alpha\} \leq \alpha.$$

However, since $\alpha \leq \mu \leq \min\{\epsilon, \delta\}$, the above inequality implies the desired conclusion. ∎

The above proof follows along the lines of a corresponding proof in [81]. Though the notation is a bit elaborate, the basic idea behind the proof is simple enough. For each integer m, one can think of an "achievable accuracy and confidence" with m samples; this is the number $\alpha = \phi(m)$. By applying an "ϵ, δ-dependent algorithm" to these m samples (or less, if $s(\alpha) < m$) with the accuracy and confidence both set equal to α, one can ensure that the resulting hypothesis is accurate to within α with a probability of at least $1 - \alpha$. As the number of samples approaches infinity, the achievable accuracy and confidence both approach zero. Hence the algorithm is PAC. Thus, if a function class is learnable using an ϵ, δ-dependent algorithm, then it is also learnable by an algorithm that does not explicitly depend on ϵ and δ.

The last issue to be discussed is the possibility of identifying a separate **hypothesis class** $\mathcal{H} \subseteq \mathcal{S}$, which may or may not equal the concept class $\mathcal{C}$. In other words, the algorithm consists of a family of indexed maps $\{A_m\}$, where A_m maps $[X \times \{0, 1\}]^m$ into $\mathcal{H}$, not $\mathcal{C}$. The distinction between the hypothesis class and the concept class is common in the computer science literature. The main reason for making such a distinction is that, in some cases, it is an NP-complete or NP-hard problem to find a hypothesis in $\mathcal{C}$ is "consistent" with data (i.e., reproduces the labelled multisample exactly). However, if $\mathcal{C}$ is replaced by a larger set $\mathcal{H}$, then the problem of finding a consistent hypothesis can become tractable; see for instance Example 9.5. Therefore, if one is interested only in the *statistical* aspects of learning, then there is no advantage to replacing the concept class $\mathcal{C}$ by a strictly larger class $\mathcal{H}$.

3.2.5 Alternate Notions of Learnability

The notion of PAC learnability defined in the preceding subsection can be thought of as the "standard" notion of learnability. In the present subsection, we introduce a few other types of learnability that are stronger than PAC learnability. The reason for introducing these stronger notions is that in many situations, the assumptions permit us to deduce these stronger types of learnability.

Now that the reader is presumably comfortable with the notion of learnability even when the inputs to the learning algorithm are not necessarily independent, in the remainder of this subsection we study the general case where the input sequence is a realization of a stochastic process, not necessarily i.i.d.

As before, let $(X, \mathcal{S})$ denote a measurable space, $\tilde{\mathcal{P}}$ a family of probability measures on $(X^\infty, \mathcal{S}^\infty)$, and suppose $\mathcal{F} \subseteq [0,1]^X$ is a family of measurable functions. Let $\{A_m\}$ be an algorithm, i.e., let $A_m : (X \times [0,1])^m \to \mathcal{F}$ for each $m \geq 1$. Finally, let $h_m(f; \mathbf{x}^*) \in \mathcal{F}$ denote the hypothesis produced by the algorithm when the target function is f and the multisample is $\mathbf{x}^*$.

Definition 3.3. *The family $\{A_m\}$ is said to be* **probably uniformly approximately correct (PUAC)** *if the quantity*

$$s(m, \epsilon) := \sup_{\tilde{P} \in \tilde{\mathcal{P}}} \tilde{P}\{\mathbf{x}^* \in X^\infty : \sup_{f \in \mathcal{F}} d_{\tilde{P}_0}[f, h_m(f; \mathbf{x}^*)] > \epsilon\} \qquad (3.2.13)$$

approaches zero as $m \to \infty$, for each fixed $\epsilon > 0$. The function class $\mathcal{F}$ is said to be **PUAC learnable** *if there exists an algorithm that is PUAC.*

To contrast the definition of a PUAC algorithm with that of a PAC algorithm, let us recall the quantity $r(m, \epsilon)$ defined in (3.2.5), namely

$$r(m, \epsilon) := \sup_{\tilde{P} \in \tilde{\mathcal{P}}} \sup_{f \in \mathcal{F}} \tilde{P}\{\mathbf{x}^* \in X^\infty : d_{\tilde{P}_0}[f, h_m(f; \mathbf{x}^*)] > \epsilon\}.$$

Let us for the moment fix the probability measure $\tilde{P}$ in the interests of clarity. Suppose an accuracy parameter $\epsilon > 0$ and an integer m are specified. Given a target function $f \in \mathcal{F}$, let us refer to a multisample $\mathbf{x}^* \in X^\infty$ as "bad" if the hypothesis $h_m(f; \mathbf{x}^*)$ produced by the algorithm differs from f by more than ϵ, i.e., if $d_{\tilde{P}_0}[f, h_m(f; \mathbf{x}^*)] > \epsilon$. Let

$$B_m(f, \epsilon) := \{\mathbf{x}^* \in X^\infty : d_{\tilde{P}_0}[f, h_m(f; \mathbf{x}^*)] > \epsilon\}$$

denote the set of bad samples for the target function f. Then

$$B_m(\epsilon) := \bigcup_{f \in \mathcal{F}} B_m(f, \epsilon) = \{\mathbf{x}^* \in X^\infty : \sup_{f \in \mathcal{F}} d_{\tilde{P}_0}[f, h_m(f; \mathbf{x}^*)] > \epsilon\}$$

consists of the set of multisamples that are bad for *even a single* target function $f \in \mathcal{F}$. Therefore

$$r(m, \epsilon) = \sup_{f \in \mathcal{F}} \tilde{P}[B_m(f, \epsilon)],$$

whereas

$$s(m, \epsilon) = \tilde{P}\left[\bigcup_{f \in \mathcal{F}} B_m(f, \epsilon)\right].$$

Thus an algorithm is PAC if the measure of *each single* set $B_m(f, \epsilon)$ approaches zero as $m \to \infty$, uniformly with respect to $f \in \mathcal{F}$; an algorithm is

PUAC if the measure of *their union* approaches zero as $m \to \infty$. With this interpretation, it is obvious that every PUAC algorithm is also PAC. If the function class is finite, then every PAC algorithm is also PUAC, because

$$\tilde{P}\left[\bigcup_{f \in \mathcal{F}} B_m(f, \epsilon)\right] \leq |\mathcal{F}| \sup_{f \in \mathcal{F}} \tilde{P}[B_m(f, \epsilon)].$$

In the case where the probability measure $\tilde{P}$ itself varies over a set $\tilde{\mathcal{P}}$, one should take into account the fact that the set of "bad" multisamples $B_m(f, \epsilon)$ defined above is now a function of $\tilde{P}$ as well as f and ϵ, since the disparity $d_{\tilde{P}_0}[f, h_m(f; \mathbf{x}^*)]$ between the target function f and the hypothesis $h_m(f; \mathbf{x}^*)$ now depends on $\tilde{P}$.[8] Let us denote the $B_m(f, \epsilon)$ defined above as $B_m(f, \epsilon; \tilde{P})$ to reflect this dependence. Then

$$r(m, \epsilon) = \sup_{f \in \mathcal{F}} \tilde{P}[B_m(f, \epsilon; \tilde{P})],$$

whereas

$$s(m, \epsilon) = \tilde{P}\left[\bigcup_{f \in \mathcal{F}} B_m(f, \epsilon; \tilde{P})\right].$$

In this case also, it is easy to see that every PUAC algorithm is PAC, and that if $\mathcal{F}$ is a finite function class, then every PAC algorithm is also PUAC.

One can try to understand the distinction between a PAC algorithm and a PUAC algorithm as follows: Suppose a random sample generator outputs an i.i.d. sequence $x_1, \ldots, x_m$, and suppose a *single* target function is "loaded" into the oracle at a time. The resulting labelled multisample $(x_1, f(x_1)), \ldots, (x_m, f(x_m))$ is fed into the algorithm. If the algorithm is PAC, then as $m \to \infty$, the distance $d_{\tilde{P}_0}[f, h_m(f; \mathbf{x}^*)]$ between the target function f and the hypothesis h_m approaches zero in a probabilistic sense, as discussed previously. This is similar to the problem of empirically estimating the probability measure of a *single* set, as discussed in Section 3.1. Now suppose the *same* multisample $x_1, \ldots, x_m$ is fed into a "bank" of oracles, each of which is loaded with one target function in $\mathcal{F}$. Thus there are as many oracles as there are target functions (finite, countable, or even uncountable). The collection of labelled samples is then fed into a corresponding bank of replicas of the algorithm, and each replica produces a hypothesis approximating the corresponding target function. If the algorithm is PUAC, then the distance $d_{\tilde{P}_0}[f, h_m(f; \mathbf{x}^*)]$ converges to zero *uniformly* with respect to $f \in \mathcal{F}$. This is similar to the problem of *simultaneously* estimating the probability measure of a *collection* of sets, as discussed in Section 3.1.

The distinction between an algorithm being PAC and being PUAC can also be discussed in terms of the convergence of stochastic processes, when the

---

[8] Note that the hypothesis $h_m(f; \mathbf{x}^*)$ itself is independent of $\tilde{P}$.

underlying family of probabilities $\tilde{\mathcal{P}}$ is a singleton set $\{\tilde{P}\}$. In this connection, recall the indexed family of stochastic processes

$$b_{m,f}(\mathbf{x}^*) := d_{\tilde{P}_0}[f, h_m(f; \mathbf{x}^*)]$$

defined previously. Now define a new stochastic process

$$\bar{b}_m(\mathbf{x}^*) := \sup_{f \in \mathcal{F}} b_{m,f}(\mathbf{x}^*).$$

Then the algorithm is PAC if the family $\{b_{m,f}\}$ converges to zero in probability, uniformly with respect to $f \in \mathcal{F}$, whereas the algorithm is PUAC if $\bar{b}_m$ converges to zero in probability. In case $\tilde{\mathcal{P}}$ is not a singleton, the convergence has to be uniform with respect to $\tilde{P} \in \tilde{\mathcal{P}}$ as well.

Example 3.8. Consider again the learning problem of Example 3.5, whereby $X = \mathbb{R}^2$, $\mathcal{S}$ = the Borel σ-algebra on X, and $\mathcal{C}$ consists of all axis-parallel rectangles. It can be shown that the algorithm of choosing H_m as the smallest rectangle containing all the positive examples is PUAC, and not merely PAC. This claim follows from the results of Chapter 5. Clearly this is a much stronger property than the claim proven in Example 3.5 that the algorithm is PAC.

Example 3.9. The objective of this example is to present an algorithm that is PAC but not PUAC. Let $X = [0, 1]$, $\mathcal{S}$ = the Borel σ-algebra on X. and let P denote the uniform probability measure on X. Let $\mathcal{G}$ denote the collection of all *finite* subsets of X, and let $\tau : \mathcal{G} \to [0, 0.5)$ be a one-to-one (but not necessarily onto) mapping. Such a mapping is previously defined in Example 3.4, but the exact nature of the mapping is not important for the present example. If $a \in [0, 0.5)$ belongs to the range of the map τ, let $\tau^{-1}(a)$ denote the unique finite set G such that $\tau(G) = a$. If a does not belong to the range of τ, define $\tau^{-1}(a)$ to equal the emptyset $\emptyset$. With this notation, let the concept class $\mathcal{C}$ consist of all unions of the form $[0, a] \cup \tau^{-1}(a)$ as a varies over $[0, 0.5)$, together with X itself. In symbols,

$$\mathcal{C} = \{[0, a] \cup \tau^{-1}(a) : 0 \leq a < 0.5\} \cup \{X\}.$$

The algorithm is defined next. Suppose $\mathbf{x} \in X^m$, and let $\mathbf{L}(\mathbf{x}) = \lfloor I_T(x_1) \ldots I_T(x_m) \rfloor^t \in \{0, 1\}^n$ denote the set of labels of the components of $\mathbf{x}$ generated by the unknown target concept T. The algorithm is as follows: If the label vector $\mathbf{L}(\mathbf{x})$ consists of all 1's, then define $H_m(T; \mathbf{x}) := X$. If the label vector $\mathbf{L}(\mathbf{x})$ does not equal the vector of all 1's, then define $H_m(T; \mathbf{x}) := [0, h] \cup \tau^{-1}(h)$, where

$$h := \min\{0.5, \max\{x_i : I_T(x_i) = 1\}\}.$$

The algorithm is intuitively as follows: If each component of the multisample $\mathbf{x}$ is labelled with a 1, then the algorithm declares that the unknown target concept is the entire set X. If at least one component of the multisample $\mathbf{x}$ *fails* to belong to the unknown target concept, then the algorithm declares that T is the largest interval of the form $[0, h]$ that contains all the positive examples (i.e., all x_i that belong to T), together with the finite "tail" $\tau^{-1}(h)$. The "min" is introduced to ensure that h is never larger than 0.5.

It is shown first that the algorithm is PAC. Observe that $P(G) = 0$ for every finite set G. Hence, given an $x \in X$ selected at random according to P, we have

$$I_{[0,a] \cup \tau^{-1}(a)}(x) = I_{[0,a]}(x) \text{ w.p. } 1,$$

where "w.p. 1" is an abbreviation for "with probability one." Hence, if the target concept T is of the form $[0, a] \cup \tau^{-1}(a)$, then the label $I_T(x)$ is the same as the indicator function $I_{[0,a]}(x)$, with probability one. Hence, after a multisample $\mathbf{x}$ is drawn, if the target concept is of the above form, then it can be assumed with probability one that none of the samples x_i belongs to the "tail" $\tau^{-1}(a)$, and that

$$h = \max\{x_i : x_i \leq a\}.$$

Now it is easy to see that, w.p. 1, the hypothesis interval $[0, h]$ is a subset of the target interval $[0, a]$, so that $d_P(T, H_m) = a - h$. Suppose $\epsilon > 0$ is specified. Then $d_P(T, H_m) > \epsilon$ only if every one of the samples x_i *fails* to belong to the interval $[a-\epsilon, a]$. For a fixed i, the probability of this happening is $1 - \epsilon$, whence it follows that the probability of this happening m times in a row is $(1 - \epsilon)^m$. In other words, it has been shown that

$$P^m\{\mathbf{x} \in X^m : d_P(T, H_m) > \epsilon\} \leq (1 - \epsilon)^m.$$

The above analysis applies whenever the target concept T is of the form $[0, a] \cup \tau^{-1}(a)$. If on the other hand the target concept T equals X, then the labels $I_T(x)$ will all equal one, in which case the algorithm will output $H_m = X$, which happens to be correct; thus $d_P(T, H_m) = 0$ for all $\mathbf{x}$ and all m in this case. Combining these steps, we conclude that the quantity $r(m, \epsilon)$ defined in (3.2.1) is bounded by

$$r(m, \epsilon) \leq (1 - \epsilon)^m.$$

Since the right side of this inequality approaches zero as $m \to \infty$ for every fixed ϵ, we conclude that the algorithm is PAC.

It is shown next that the algorithm is *not* PUAC. To establish this claim, let m and $\mathbf{x} \in X^m$ be arbitrary, and define $G(\mathbf{x}) := \{x_1, \ldots, x_m\}$ after deleting repeated components if any. Now suppose the target concept T is $[0, \tau(G(\mathbf{x}))] \cup G(\mathbf{x})$. Then the label vector $I_T(\mathbf{x}_i)$ equals the vector of all 1's; as a result, the algorithm returns the hypothesis $H_m = X$. Since the measure of T equals $\tau(G(\mathbf{x})) < 0.5$, it follows that, *for this particular choice of target concept,*

$$d_P[T; H_m(T; \mathbf{x})] > 0.5.$$

This reasoning can be applied to *every* multisample **x**. Hence we conclude that
$$\sup_{T \in \mathcal{C}} d_P[T; H_m(T; \mathbf{x})] > 0.5, \; \forall \mathbf{x} \in X^m.$$
In other words, whenever $\epsilon \leq 0.5$, we have
$$\{\mathbf{x} \in X^m : \sup_{T \in \mathcal{C}} d_P[T; H_m(T; \mathbf{x})] > \epsilon\} = X^m.$$
Hence the quantity $s(m, \epsilon)$ defined in (3.2.13) is given by
$$s(m, \epsilon) = P^m(X^m) = 1, \; \forall m.$$
This shows that the algorithm is *not PUAC*. ∎

Though the details of the above example are a little messy, the idea is simple: For each fixed target concept, the set of multisamples **x** that lead to a poor hypothesis have small measure. But *every* multisample is "bad" for *some* target concept, so that the union of the multisamples that are "bad" for at least one target concept is in fact the entire space X^m.

The next definition is applicable only to the case where $\mathcal{P}$ is a *singleton set*, i.e., the so-called fixed-distribution learning problem.

Definition 3.4. *With all symbols as before, an algorithm $\{A_m\}$ is said to be* **almost surely eventually correct (ASEC)** *if*
$$P^\infty \{\mathbf{x}^* \in X^\infty : \lim_{m \to \infty} \sup_{f \in \mathcal{F}} d_P[f, h_m(f; \mathbf{x})] = 0\} = 1.$$

The distinction between an ASEC algorithm and a PUAC algorithm can be demonstrated by once again defining an appropriate stochastic process on the countable Cartesian product X^∞. Define the stochastic process $\{b_m(\cdot)\}$ on X^∞ by
$$b_m(\mathbf{x}^*) := \sup_{f \in \mathcal{F}} d_P[f, h_m(f; \mathbf{x})].$$
Then $b_m(\mathbf{x}^*)$ (which depends only on the first m terms of the sequence $\mathbf{x}^*$) is the *worst-case error* between a target function and the corresponding hypothesis, when the multisample is $\mathbf{x}^*$. Now it is easy to see that the algorithm is PUAC if and only if the stochastic process $\{b_m(\cdot)\}$ converges to zero *in probability*, whereas the algorithm is ASEC if the same stochastic process converges to zero *almost surely*. Thus the distinction between the PUAC and ASEC properties is quite similar to that between the UCEM and AS-CEM properties introduced in Section 3.1. Clearly an ASEC algorithm is also PUAC. However, it is not known whether the converse is true. (It is shown in Theorem 5.2 that the UCEM and the ASECM properties are equivalent.)

As in Section 3.1, the ASEC property does not extend naturally to learning problems where the underlying probability P is itself not known precisely.

3.3 Model-Free Learning

In this section we introduce a more general type of learning problem, known as "model-free" learning, which includes the previous notion of PAC learning as a special case. Note that some authors refer to this type of learning as "agnostic" learning. Then it is shown that the model-free learning problem can in effect be partitioned into two subproblems: one of establishing that a particular family of functions has the UCEM property, and another of choosing an algorithm that minimizes the "empirical" estimate of the so-called risk function.

3.3.1 Problem Formulation

The PAC learning problem formulation of the preceding section comprises several assumptions that may be questioned.

1. The assumption that the target concept T belongs to $\mathcal{C}$ (or that $f \in \mathcal{F}$) means, roughly, that the data to which we are trying to fit a hypothesis is assumed, *a priori*, to be generated by a member of the family of models $\mathcal{C}$ (or $\mathcal{F}$). However, there are many situations where the data is *not* generated by a member of the family of models, and the objective is to achieve the best possible fit to the data using an element of the model class.
2. The assumption that the oracle returns $I_T(x)$ [or $f(x)$] rules out the possibility of noisy measurements, or an imperfect oracle.
3. The assumption that the range of A_m equals $\mathcal{C}$ (or $\mathcal{F}$) rules out the case where the "hypothesis class" is different from the concept class or the function class. However, there are several plausible learning problems in which the aim is to *find the best approximation* to the unknown target (concept or function) *within another class of hypotheses*, which need not coincide exactly with the concept class or the function class.
4. Implicit in the PAC learning problem formulation is the assumption that the "hypothesis class" is the same as (or is a superset of) the concept class or the function class. This assumption means that the problem at hand is assumed to be one of identifying the source of generating the data, rather than one of deciding what to do next. This assumption is not always valid. For example, in so-called *direct* adaptive control (as opposed to indirect adaptive control), there is an unknown plant for which it is desired to design an appropriate controller. The plant can be "identified" by exciting it with various inputs, but the ultimate objective is not to identify the plant as such, but to modify the controller on the basis of observation.

The learning problem formulated in this section is a generalization of function learning, and is intended to address the above concerns. The ingredients of the present learning problem are:

3.3 Model-Free Learning

- Sets X, Y, and U.
- A σ-algebra $\bar{\mathcal{S}}$ on $X \times Y$, and a family of probability measures $\bar{\mathcal{P}}$ on $(X \times Y, \bar{\mathcal{S}})$.
- A family of functions $\mathcal{H}$ mapping X into U, called the set of *hypotheses*.
- A function ℓ mapping $Y \times U$ into $[0, 1]$, called the "loss" function.

Learning takes place as follows: An unknown probability measure $\bar{P} \in \bar{\mathcal{P}}$ is fixed, and i.i.d. samples $(x_1, y_1), \ldots, (x_m, y_m)$ are drawn from $X \times Y$ in accordance with $\bar{P}$. In this setting, an "algorithm" is an indexed family of maps $\{A_m\}_{m \geq 1}$, where

$$A_m : (X \times Y)^m \to \mathcal{H}.$$

As before, define
$$h_m := A_m[(x_1, y_1), \ldots, (x_m, y_m)].$$

Associated with h_m and $\bar{P}$ is an error measure

$$J(h_m, \bar{P}) := \int_{X \times Y} \ell[y, h_m(x)] \, \bar{P}(dx, dy).$$

Also, associated with $\bar{P}$ alone is a number

$$J^*(\bar{P}) := \inf_{h \in \mathcal{H}} \int_{X \times Y} \ell[y, h(x)] \, \bar{P}(dx, dy).$$

One can think of $J^*(\bar{P})$ as the *best possible* performance by any hypothesis function $h \in \mathcal{H}$, when the samples are drawn in accordance with the probability measure $\bar{P}$. Similarly, $J(h_m, \bar{P})$ can be thought of as the *actual* performance of the algorithm after m samples are drawn in accordance with $\bar{P}$. The quantity $J(h_m, \bar{P})$ is sometimes referred to as the **risk** associated with the hypothesis h_m, when the underlying probability measure is $\bar{P}$. Clearly

$$0 \leq J^*(\bar{P}) \leq J(h_m, \bar{P}) \leq 1.$$

Also, while $J^*(\bar{P})$ is a deterministic number, $J(h_m, \bar{P})$ is a random number, since it depends on the random samples $(x_1, y_1), \ldots, (x_m, y_m)$. Now define

$$r_{\mathrm{mf}}(m, \epsilon) := \sup_{\bar{P} \in \bar{\mathcal{P}}} \bar{P}^m \{ (\mathbf{x}, \mathbf{y}) \in X^m \times Y^m : J(h_m, \bar{P}) > J^*(\bar{P}) + \epsilon \}, \quad (3.3.1)$$

where
$$\mathbf{x} := [x_1 \ldots x_m]^t \in X^m, \ \mathbf{y} := [y_1, \ldots, y_m]^t \in Y^m,$$

and by a slight abuse of notation we write the m-fold sample $[(x_1, y_1), \ldots, (x_m, y_m)]$ as an element of $X^m \times Y^m$ (rather than as an element of $(X \times Y)^m$, which it is). Thus, for a fixed ϵ, one can think of $r_{\mathrm{mf}}(m, \epsilon)$ as the measure of the set of "bad" samples, where a sample is deemed to be "bad" if it leads to a hypothesis that performs more than ϵ-worse compared to the optimum achievable performance. The quantity $r_{\mathrm{mf}}(m, \epsilon)$ is analogous to the quantity $r(m, \epsilon)$ defined in (3.2.1), except that the subscript "mf" is used to remind us that the problem under study is one of model-free learning.

78   3. Problem Formulations

Definition 3.5. *The algorithm* $\{A_m\}$ *is* **probably approximately correct (PAC) to accuracy** ϵ *if* $r_{\mathrm{mf}}(m,\epsilon) \to 0$ *as* $m \to \infty$, *and is* **probably approximately correct (PAC)** *if* $r_{\mathrm{mf}}(m,\epsilon) \to 0$ *as* $m \to \infty$ *for each* $\epsilon > 0$. *The triplet* $(\mathcal{H}, \bar{\mathcal{P}}, \ell)$ *is* **model-free learnable to accuracy** ϵ *if there exists an algorithm that is PAC to accuracy* ϵ, *and is* **model-free learnable** *if there exists an algorithm that is PAC.*

Example 3.10. In this example, it is shown how the function learning problem discussed earlier can be viewed as a special case of the present problem of learning nearly optimum decision rules. Let $X, \mathcal{S}, \mathcal{P}, \mathcal{F}$ be as in Section 3.2.2, let $Y = U = [0,1]$, and let $\mathcal{H} = \mathcal{F}$. For each probability measure $P \in \mathcal{P}$ and each function $f \in \mathcal{F}$, define a corresponding probability measure $\bar{P}_f$ on $X \times Y$ as follows: Suppose $A \in \mathcal{S}$ and that B is a Borel subset of $[0,1]$. Define

$$\bar{P}_f(A \times B) := \int_X I_A(x)\, I_B[f(x)]\, P(dx),$$

or equivalently,

$$\bar{P}_f(A \times B) := \int_A I_B[f(x)]\, P(dx).$$

This defines the measure of "rectangular" sets of the form $A \times B$, which can then be extended to all measurable subsets of $X \times Y$ using standard techniques.[9] Now the family $\bar{\mathcal{P}}$ of probability measures on $X \times Y$ is defined as

$$\bar{\mathcal{P}} = \{\bar{P}_f : P \in \mathcal{P}, f \in \mathcal{F}\}.$$

The loss function $\ell : Y \times U \to [0,1]$ is defined as

$$\ell(y,z) = |y - z|.$$

Let us see what the general learning problem becomes with these definitions. Suppose a probability measure $\bar{P}_f \in \bar{\mathcal{P}}$ is fixed. This means that a probability measure $P \subset \mathcal{P}$ and a function $f \in \mathcal{F}$ are fixed. In view of the definition of $\bar{P}_f$, a randomly selected element of $X \times Y$ has the form $(x, f(x))$ where x is randomly distributed according to P. Thus an i.i.d. sequence $[(x_1, y_1), \ldots, (x_m, y_m)]$ has the form

$$[(x_1, f(x_1)), \ldots, (x_m, f(x_m))].$$

Next, observe that, since $f \in \mathcal{H} = \mathcal{F}$, we have that

$$J^*(\bar{P}_f) = 0,$$

which can be achieved by choosing $h = f$. Similarly,

---
[9] This definition can be made less opaque in case P has a density function, say $p(\cdot)$. In this case $\bar{P}_f$ also has a density function, namely $p(x)\,\delta(y - f(x))$.

$$J(h_m, f) = \int_X |f(x) - h_m(x)|\, P(dx) = d_P(f, h_m).$$

Finally,

$$r_{\mathrm{mf}}(m, \epsilon) = \sup_{P \in \mathcal{P}, f \in \mathcal{F}} P^m\{\mathbf{x} \in X^m : d_P(f, h_m) > \epsilon\} = r(m, \epsilon).$$

Thus it is clear that the problem at hand reduces to the function learning problem defined earlier.

Example 3.11. In this example we study the problem of learning concept classes in the case where the oracle outputs are "noisy." That is, the oracle sometimes outputs a "1" when the correct output is 0, and vice versa. The problem formulation is as follows: There is a measurable space $(X, \mathcal{S})$, and a concept class $\mathcal{C} \subseteq \mathcal{S}$. There is also a family of probability measures $\mathcal{P}$ on $(X, \mathcal{S})$. Learning takes place as follows: A target concept $T \in \mathcal{C}$ is loaded into the oracle, and i.i.d. samples $x_1, x_2, \ldots \in X$ are generated according to some probability $P \in \mathcal{P}$. For each x_i, the oracle returns $I_T(x_i)$ with probability $1 - \alpha$, and $1 - I_T(x_i)$ with probability α, where $\alpha < 0.5$ is the probability that the oracle makes a mistake. It is assumed that the process that generates the output of the oracle is independent of the x_i. Moreover, it is assumed that the probability of an incorrect output from the oracle is the same for all i. However, it is not necessary that α be known, other than that α is known to be less than 0.5.[10]

The above problem can be put into the present framework as follows: Let $Y = U = \{0, 1\}$, and $\ell(y, z) := |y - z|$. Define a family of probability measures $\{\bar{P}_T, T \in \mathcal{C}\}$ on $X \times \{0, 1\}$ as follows: If $A \subseteq \mathcal{S}$, the probability $\bar{P}_T(A \times \{1\})$ is given by

$$\bar{P}_T(A \times \{1\}) := \int_A [(1 - \alpha) I_T(x) + \alpha(1 - I_T(x))]\, P(dx)$$

$$= \int_A [\alpha + (1 - 2\alpha) I_T(x)]\, P(dx) = \alpha P(A) + (1 - 2\alpha) P(A \cap T).$$

Similarly,

$$\bar{P}_T(A \times \{0\}) := \int_A [\alpha I_T(x) + (1 - \alpha)(1 - I_T(x))]\, P(dx)$$

$$= (1 - \alpha) P(A) - (1 - 2\alpha) P(A \cap T).$$

Given a concept (hypothesis) $H \in \mathcal{C}$, we have

$$J(H, \bar{P}_T) = \int_{X \times \{0,1\}} |y - I_H(x)|\, \bar{P}_T(dx, dy).$$

---

[10] Note that if $\alpha > 0.5$, then one could simply "flip" the output of the oracle. If $\alpha = 0.5$, then the output of the oracle is pure noise, and learning is clearly impossible.

80   3. Problem Formulations

To evaluate this integral, note that if $I_H(x) = I_T(x)$, then $y = 0$ with probability $1 - \alpha$, and $y = 1$ with probability α. The situation is reversed if $I_H(x) \neq I_T(x)$. Therefore, after a little manipulation, we get

$$J(H, \bar{P}_T) = \alpha + (1 - 2\alpha)P(H\Delta T) = \alpha + (1 - 2\alpha)\, d_P(H, T).$$

Finally,
$$J^*(\bar{P}_T) = \inf_{H \in \mathcal{C}} J(H, \bar{P}_T) = \alpha,$$

which can be achieved by choosing $H = T$. Therefore

$$J(H, \bar{P}_T) > J^*(\bar{P}_T) + \epsilon \iff d_P(H, T) > \frac{\epsilon}{1 - 2\alpha}.$$

Hence the quantity $r_{\mathrm{mf}}(m, \epsilon)$ defined in (3.3.1) is equal to

$$r_{\mathrm{mf}}(m, \epsilon) = P^m\{\mathbf{x} \in X^m : d_P(H_m, T) > \epsilon/(1 - 2\alpha)\} = r(m, \epsilon/(1 - 2\alpha)),$$

where H_m is the hypothesis produced by the algorithm. Thus learning a concept class with a noisy oracle with a noise rate of α to an accuracy of ϵ is the same as learning the same concept class to an accuracy of $\epsilon/(1 - 2\alpha)$ with a noise-free oracle. In subsequent chapters we will be deriving learning algorithms for both the noise-free as well as the noisy case. One can gauge the deleterious effects of a noisy oracle by comparing the sample complexity estimates in the noise-free case for an accuracy parameter of $\epsilon/(1 - 2\alpha)$ with the sample complexity estimates in the noisy case for an accuracy parameter of ϵ.

Example 3.12. In this example, it is shown how the problem of function learning with noisy measurements can be formulated in the present set-up. Let $X, \mathcal{S}, \mathcal{F}, \mathcal{P}$ be as before, and let η be a real-valued zero-mean random variable representing measurement noise, with a probability density function $\phi(\cdot)$. Thus, if an input $x \in X$ is presented to the oracle, it returns the value $f(x) + \eta$, where f is the target function and η is distributed according to the density $\phi(\cdot)$. The learning problem in this case is as follows: A probability measure $P \in \mathcal{P}$ and a target function $f \in \mathcal{F}$ are fixed; i.i.d. samples $x_1, \ldots, x_m \in X$ are drawn in accordance with P, and the labelled sample that forms the input to the algorithm consists of each x_j together with a noisy measurement of $f(x_j)$, i.e., the number $f(x_j) + \eta_j$ where each η_j has the density function $\phi(\cdot)$ and is independent of all other η_i's. With this information, one would like to construct an estimate of the target function $f \in \mathcal{F}$.

To define the problem in the present set-up, define $Y = \mathbb{R}$, $U = [0, 1]$, and $\mathcal{H} = \mathcal{F}$. Given $P \in \mathcal{P}$ and $f \in \mathcal{F}$, define a probability measure $\bar{P}_f$ on $X \times Y$ as follows: If P has a density function $p(\cdot)$, then the density function of $\bar{P}_f$ is defined as

$$\bar{p}_f(x, y) = [p(x), \phi(y - f(x))].$$

More generally, if $A \in \mathcal{S}$ and $B \subseteq Y$ is a Borel set, then

$$\bar{P}_f(A \times B) := \int_A \int_B \phi(y - f(x)) \, dy \, P(dx).$$

Now $\bar{P}_f$ can be extended to all measurable subsets of $X \times Y$ using standard methods. Next, the loss function $\ell : Y \times U \to [0,1]$ can be defined as

$$\ell(y,z) := \frac{|y-z|}{1+|y|+|z|}.$$

3.3.2 Relationship to the Uniform Convergence of Empirical Means

In this subsection, we explore the relationship between the model-free learning problem formulated in the preceding subsection, and the property of uniform convergence of empirical means uniformly with respect to probability (UCEMUP) introduced in Section 3.1. In particular, the following useful result is established: A family of functions $\mathcal{L}_{\mathcal{H}}$ is associated with the model-free learning problem. It is shown that if (i) the family $\mathcal{L}_{\mathcal{H}}$ has the UCEMUP property with respect to the family of probability measures $\bar{\mathcal{P}}$, and (ii) the algorithm generates, with high probability, a hypothesis that "nearly" minimizes an empirical estimate of the risk function, then such an algorithm is PAC. This subsection is somewhat abstract, and the reader might not be able to appreciate the full significance of the contents of this subsection at a first reading. Thus it is suggested that the reader quickly peruse the subsection, and return to it while reading Chapters 5 and 10.

Let all symbols be as in the preceding subsection. Some additional notation is now introduced. Given a hypothesis function $h \in \mathcal{H}$, define the associated function $\ell_h : X \times Y \to [0,1]$ by

$$\ell_h(x,y) := \ell(y, h(x)), \ \forall x, y.$$

Let $\mathcal{L}_{\mathcal{H}}$ denote the collection of functions ℓ_h as h varies over $\mathcal{H}$. In what follows, the UCEMUP property of the family $\mathcal{L}_{\mathcal{H}}$ is related to the PAC learnability of the pair $(\mathcal{H}, \bar{\mathcal{P}})$.

For notational convenience, let z denote the pair (x,y), and let Z denote the Cartesian product space $X \times Y$. Thus the multisample $\{(x_1, y_1), \ldots, (x_m, y_m)\}$ available to the algorithm can be denoted by $\mathbf{z} = [z_1 \ldots z_m]^t \in Z^m$. Suppose $h \in \mathcal{H}, \bar{P} \in \bar{\mathcal{P}}$. Then, as before, the risk function $J(h, \bar{P})$ is defined by

$$J(h, \bar{P}) := \int_{X \times Y} \ell(y, h(x)) \, \bar{P}(dx, dy) = \int_Z \ell(y, h(x)) \, \bar{P}(dz).$$

Now define

$$\hat{J}(h; \mathbf{z}) := \frac{1}{m} \sum_{i=1}^{m} \ell[y_i, h(x_i)].$$

Observe that $\hat{J}(h; \mathbf{z})$ is just the *empirical estimate* of the risk $J(h, \bar{P})$ based on the multisample $\mathbf{z}$. Therefore one can refer to $\hat{J}(h; \mathbf{z})$ as the "empirical risk" based on $\mathbf{z}$. Now it is natural for the algorithm to try to choose a hypothesis h_m that makes the empirical risk $\hat{J}(h; \mathbf{z})$ as close as possible to the optimum performance measure $J^*(\bar{P})$, defined as before as

$$J^*(\bar{P}) := \inf_{h \in \mathcal{H}} J(h, \bar{P}).$$

The difficulty is that $J^*(\bar{P})$ is itself unknown. As a compromise, one can compute the quantity

$$\hat{J}^*(\mathbf{z}) := \inf_{h \in \mathcal{H}} \hat{J}(h; \mathbf{z}),$$

which is the minimum achievable *empirical* risk based on the multisample $\mathbf{z}$. Observe that $\hat{J}^*(\mathbf{z})$ is independent of the underlying probability measure $\bar{P}$, but does depend on the multisample $\mathbf{z}$. Moreover, unlike $J^*(\bar{P})$, the quantity $\hat{J}^*(\mathbf{z})$ can be computed on the basis of available information. Now, once the multisample $\mathbf{z}$ is drawn, one can think of choosing the hypothesis h_m such that the empirical risk estimate $\hat{J}(h; \mathbf{z})$ equals the infimum $\hat{J}^*(\mathbf{z})$. This turns out to be a bit of "overkill," and it is enough that the empirical risk $\hat{J}(h; \mathbf{z})$ is "nearly" equal to $\hat{J}^*(\mathbf{z})$ with high probability. To make this notion precise, let $h_m(\mathbf{z})$ denote the hypothesis generated by the algorithm based on the multisample $\mathbf{z}$, and define

$$t(m, \epsilon) := \bar{P}^m \{ \mathbf{z} \in Z^m : \hat{J}[h_m(\mathbf{z}); \mathbf{z}] > \hat{J}^*(\mathbf{z}) + \epsilon \}.$$

Thus $t(m, \epsilon)$ is the probability that, after m random samples are drawn, the empirical risk $\hat{J}[h_m(\mathbf{z}); \mathbf{z}]$ is more than ϵ-worse compared to the minimum achievable value $\hat{J}^*(\mathbf{z})$. Then the algorithm is said to **nearly minimize empirical risk with high probability**, or to be **NMER**, if $t(m, \epsilon) \to 0$ as $m \to \infty$.

Now we come to the main result of this subsection.

Theorem 3.2. *Suppose*

1. *The family of functions $\mathcal{L}_\mathcal{H}$ has the UCEMUP property with respect to $\bar{\mathcal{P}}$, and*
2. *The algorithm $\{A_m\}$ has the NMER property; that is, $t(m, \epsilon) \to 0$ as $m \to \infty$.*

Then the algorithm $\{A_m\}$ is PAC.

Proof. Let $\epsilon, \delta > 0$ be specified. We now construct an $m_0(\epsilon, \delta)$ such that

$$r_{\mathrm{mf}}(m, \epsilon) \leq \delta \; \forall m \geq m_0(\epsilon, \delta),$$

where $r(m, \epsilon)$ is defined in (3.3.1). Choose $m_0 = m_0(\epsilon, \delta)$ such that

$$\bar{P}^m \{ \mathbf{z} \in Z^m : \sup_{h \in \mathcal{H}} |\hat{J}(h; \mathbf{z}) - J(h; \bar{P})| > \epsilon/4 \} \leq \delta/2, \; \forall m \geq m_0, \; \forall \bar{P} \in \bar{\mathcal{P}}, \text{ and}$$

$$\bar{P}^m\{\mathbf{z} \in Z^m : \hat{J}[h_m(\mathbf{z});\mathbf{z}] > \hat{J}^*(\mathbf{z}) + \epsilon/4\} \leq \delta/2, \ \forall m \geq m_0, \ \forall \bar{P} \in \bar{\mathcal{P}}.$$

Such an m_0 can always be found. Observe that $\hat{J}(h;\mathbf{z})$ is just the empirical mean of the function L_h based on the multisample $\mathbf{z}$, whereas $J(h;\bar{P})$ is the true mean of the same function. Since it is assumed that the family of functions $\mathcal{L}_{\mathcal{H}}$ has the UCEMUP property with respect to $\bar{\mathcal{P}}$, the first inequality can be satisfied by choosing m_0 large enough. Similarly, since the algorithm is assumed to have the NMER property, it follows that $t(m,\epsilon) \to 0$ as $m \to \infty$, whence the second inequality can also be satisfied by choosing m_0 large enough.

Now let $\mathbf{z} \in Z^m$ be chosen at random according to some $\bar{P} \in \bar{\mathcal{P}}$. Then, with probability at least $1 - \delta$, it can be assumed that *both* the following inequalities are satisfied:

$$|\hat{J}(h;\mathbf{z}) - J(h;\bar{P})| \leq \epsilon/4 \ \forall h \in \mathcal{H}, \text{ and}$$

$$\hat{J}[h_m(\mathbf{z});\mathbf{z}] \leq \hat{J}^*(\mathbf{z}) + \epsilon/4.$$

Select a hypothesis $h_\epsilon \in \mathcal{H}$ such that

$$J(h_\epsilon, \bar{P}) \leq J^*(\bar{P}) + \epsilon/4.$$

Such an h_ϵ exists, by the definition of $J^*(\bar{P})$. Then it follows from the first inequality above that

$$\hat{J}(h_\epsilon;\mathbf{z}) \leq J(h_\epsilon, \bar{P}) + \epsilon/4.$$

It also follows from the same inequality that

$$\hat{J}[h_m(\mathbf{z});\mathbf{z}] \geq J(h_m, \bar{P}) - \epsilon/4.$$

On the other hand, the property of the algorithm implies that

$$\begin{aligned}\hat{J}[h_m(\mathbf{z});\mathbf{z}] &\leq \hat{J}^*(\mathbf{z}) + \epsilon/4 \leq \hat{J}(h_\epsilon;\mathbf{z}) + \epsilon/4 &&\text{by the definition of } \hat{J}^*(\mathbf{z}), \\ &\leq J(h_\epsilon, \bar{P}) + \epsilon/4 + \epsilon/4 &&\text{by the UCEMUP property,} \\ &\leq J^*(\bar{P}) + \epsilon/4 + \epsilon/4 + \epsilon/4 &&\text{by the definition of } h_\epsilon.\end{aligned}$$

Combining these two inequalities shows that

$$J(h_m, \bar{P}) \leq J^*(\bar{P}) + \epsilon, \text{ with probability } \geq 1 - \delta.$$

In other words, $r_{\mathrm{mf}}(m,\epsilon) \leq \delta$ whenever $m \geq m_0$. ∎

3.4 Preservation of UCEMUP and PAC Properties

This section is addressed to the resolution of two rather dissimilar questions. In the first case, we study the situation where the sequence of samples is

84   3. Problem Formulations

not i.i.d. but satisfies a mixing condition. It is shown that if the pair $(\mathcal{F}, \mathcal{P})$ satisfies the UCEMUP property, then the property is preserved if the i.i.d. input sequence is replaced by a β-mixing sequence. Moreover, it is possible to relate the rates of convergence in the two cases. In the PAC learning problem, two different kinds of results are proved. First, if there exists a PAC algorithm for a pair $(\mathcal{F}, \mathcal{P})$ when the inputs are i.i.d., then *there exists* a corresponding PAC algorithm when the inputs are β-mixing. Second, if the learning algorithm satisfies a kind of "quasi-subadditivity" condition, then *the same algorithm* continues to be PAC even if the input sequence is β-mixing. In the second case, we restrict our attention to the case of i.i.d. inputs, and ask: What happens if the underlying family of probability measures $\mathcal{P}$ is replaced by its closure $\bar{\mathcal{P}}$, where the closure is taken with respect to the total variation metric? It is shown that in such a case, both the UCEMUP property and the PAC property are preserved. More precisely, if the pair $(\mathcal{F}, \mathcal{P})$ has the UCEMUP property, then so does the pair $(\mathcal{F}, \bar{\mathcal{P}})$. Similarly, if an algorithm is PAC for the pair $(\mathcal{F}, \mathcal{P})$, then it continues to be PAC for the pair $(\mathcal{F}, \bar{\mathcal{P}})$. Thus both the UCEMUP property and the PAC property are "preserved" when $\mathcal{P}$ is replaced by its closure $\bar{\mathcal{P}}$.

3.4.1 Preservation of UCEMUP Property with Beta-Mixing Inputs

Consider, as always, a family of functions $\mathcal{F}$ mapping X into $[0, 1]$, and a family of probability measures $\tilde{\mathcal{P}}$ on $(X^\infty, \mathcal{S}^\infty)$. As in (3.1.5), let us define the quantity

$$q(m, \epsilon, \tilde{P}) := \tilde{P}\{\mathbf{x}^* \in X^\infty : \sup_{f \in \mathcal{F}} |\hat{E}_m(f; \mathbf{x}^*) - E(f, \tilde{P}_0)| > \epsilon\},$$

and as in (3.1.6), let us define

$$\bar{q}(m, \epsilon, \tilde{\mathcal{P}}) := \sup_{\tilde{P} \in \tilde{\mathcal{P}}} q(m, \epsilon, \tilde{P}).$$

The idea is to compare the behaviour of these quantities with their counterparts when the underlying samples are i.i.d. with the law $\tilde{P}_0^\infty$. Thus we wish to compare $q(m, \epsilon, \tilde{P})$ with $q(m, \epsilon, \tilde{P}_0^\infty)$. Similarly we wish to compare $\bar{q}(m, \epsilon, \tilde{\mathcal{P}})$ with $\bar{q}(m, \epsilon, \mathcal{P})$, where in the interests of simplicity we use the symbol $\mathcal{P}$ to denote the set $\{\tilde{P}_0^\infty : \tilde{P} \in \tilde{\mathcal{P}}\}$, that is, the set of i.i.d. laws corresponding to the probability measures in $\tilde{\mathcal{P}}$. As per the notational conventions introduced previously, the pair $(\mathcal{F}, \tilde{P})$ has the UCEM property if $q(m, \epsilon, \tilde{P}) \to 0$ as $m \to \infty$. Similarly, the pair $(\mathcal{F}, \tilde{\mathcal{P}})$ has the UCEMUP property if $\bar{q}(m, \epsilon, \tilde{\mathcal{P}}) \to 0$ as $m \to \infty$.

In this subsection, two distinct results are proven. In the first, it is shown that whenever a pair $(\mathcal{F}, \tilde{P}_0^\infty)$ has the UCEM property, the pair $(\mathcal{F}, \tilde{P})$ also has the UCEM property provided the underlying sample process is β-mixing.

3.4 Preservation of UCEMUP and PAC Properties

In the second, *explicit estimates* are given relating the rate at which $q(m, \epsilon, \tilde{P})$ approaches zero in terms of the rate of convergence of $q(m, \epsilon, \tilde{P}_0^\infty)$ and the β-mixing coefficients of the sample process. However, these estimates *do not* imply the result in the preceding sentence. Naturally, both types of results are extended to the case where there is a family of probability measures $\tilde{\mathcal{P}}$ instead of a single probability measure $\tilde{P}$.

A technical lemma is presented first that facilitates the proofs of the main theorems.

Lemma 3.1. *Suppose $\beta(k) \downarrow 0$ as $k \to \infty$, and $g : Z_+ \to \mathbb{R}$ is strictly increasing. Then it is possible to choose a sequence $\{k_m\}$ such that $k_m \leq m$, and with $l_m = \lfloor m/k_m \rfloor$ we have*

$$l_m \to \infty, \ \beta(k_m)g(l_m) \to 0 \ as \ m \to \infty.$$

Proof. Though the function β is defined only for integer-valued arguments, it is convenient to replace it by another function defined for all real-valued arguments. Moreover, it can be assumed that $\beta(\cdot)$ is continuous and monotonically decreasing, so that β^{-1} is well-defined, by replacing the given function by a larger function if necessary. With this convention, choose any sequence $\{a_i\}$ such that $a_i \downarrow 0$ as $i \to \infty$. Define

$$m_i := i \lceil \beta^{-1}(a_i/g(i)) \rceil.$$

Clearly $a_i/g(i) \downarrow 0$, so $\beta^{-1}(a_i/g(i)) \uparrow \infty$. Therefore $i\beta^{-1}(a_i/g(i)) \uparrow \infty$. Thus $\{m_i\}$ is a monotonically increasing sequence. Given an integer m, choose a unique integer $i = i(m)$ such that $m_i \leq m < m_{i+1}$. Define $l_m = i(m)$, and choose k_m as the largest integer such that $l_m = \lfloor m/k_m \rfloor$. Note that $i(m) \to \infty$ as $m \to \infty$, so that $l_m \to \infty$. Next, since $i\lceil \beta^{-1}(a_i/g(i))\rceil = m_i \leq m$, it follows that

$$k_m \geq \lceil \beta^{-1}(a_i/g(i)) \rceil.$$

So

$$\beta(k_m) \leq \beta(\lceil \beta^{-1}(a_i/g(i)) \rceil) \leq \beta[\beta^{-1}(a_i/g(i))] = a_i/g(i).$$

Since $l_m = i$, we have $g(l_m) = g(i)$. Finally

$$\beta(k_m)g(l_m) \leq a_i.$$

Since $a_i \to 0$ as $i \to \infty$, the result follows. ■

Theorem 3.3. *Suppose $\mathcal{F}$ is a given family of measurable functions mapping X into $[0, 1]$. Suppose $\{X_i\}$ is a β-mixing stochastic process on a probability space $(X^\infty, \mathcal{S}^\infty, \tilde{P})$, and that $q(m, \epsilon, \tilde{P}_0^\infty) \to 0$ as $m \to \infty$, for each $\epsilon > 0$. Then $q(m, \epsilon, \tilde{P}) \to 0$ as $m \to \infty$, for each $\epsilon > 0$.*

Proof. Given an integer m, choose any integer $k_m \leq m$, and define $l_m = \lfloor m/k_m \rfloor$ to be the integer part of m/k_m. For the time being, k_m and l_m

are denoted respectively by k and l, so as to reduce notational clutter. The dependence of k and l on m is restored near the end of the proof. As in (3.1.3), define

$$a_m(\mathbf{x}^*) := \sup_{f \in \mathcal{F}} \left| \frac{1}{m} \sum_{j=1}^{m} [f(x_j) - E(f, \tilde{P}_0)] \right|.$$

As we have already seen, the pair $(\mathcal{F}, \tilde{P})$ has the UCEM property if and only if the stochastic process $\{a_m(\cdot)\}$ converges to zero in probability with respect to the measure $\tilde{P}$. Since a_m assumes values in $[0, 1]$, it is easy to see that for every $\epsilon > 0$ and m, we have

$$\epsilon q(m, \epsilon, \tilde{P}) \le E(a_m, \tilde{P}) \le \epsilon + q(m, \epsilon, \tilde{P}). \tag{3.4.1}$$

Therefore, the UCEM property is equivalent to the requirement that

$$E(a_m, \tilde{P}) \to 0 \text{ as } m \to \infty. \tag{3.4.2}$$

Next, let $r = m - kl$, and define the index sets $I_i, i = 1, \ldots, k$ as follows.

$$I_i = \{i, i+k, \ldots, i+lk\},\ i = 1, \ldots, r,$$

$$I_i = \{i, i+k, \ldots, i+(l-1)k\},\ i = r+1, \ldots, k.$$

Note that $\cup_i I_i$ equals the index set $\{1, \ldots, m\}$ and that, within each set I_i, the elements are pairwise separated by at least k. Now define

$$\alpha_i(\mathbf{x}^*) := \sup_{f \in \mathcal{F}} \left| \frac{1}{|I_i|} \sum_{j \in I_i} [f(x_j) - E(f, \tilde{P}_0)] \right|,\ i = 1, \ldots, k.$$

Note that, for each fixed function $f \in \mathcal{F}$, we have

$$\frac{1}{m} \sum_{j=1}^{m} [f(x_j) - E(f, \tilde{P}_0)] = \frac{1}{k} \sum_{i=1}^{k} \frac{1}{|I_i|} \sum_{j \in I_i} [f(x_j) - E(f, \tilde{P}_0)].$$

Taking absolute values of both sides, applying the triangle inequality, and taking suprema with respect to $f \in \mathcal{F}$ leads to

$$\sup_{f \in \mathcal{F}} \left| \frac{1}{m} \sum_{j=1}^{m} [f(x_j) - E(f, \tilde{P}_0)] \right| \le \frac{1}{k} \sum_{i=1}^{k} \sup_{f \in \mathcal{F}} \left| \frac{1}{|I_i|} \sum_{j \in I_i} [f(x_j) - E(f, \tilde{P}_0)] \right|.$$

Hence it follows that

$$a_m(\mathbf{x}^*) \le \frac{1}{k} \sum_{i=1}^{k} \alpha_i(\mathbf{x}^*). \tag{3.4.3}$$

Therefore

$$E(a_m, \tilde{P}) \le \frac{1}{k} \sum_{i=1}^{k} E(\alpha_i, \tilde{P}). \qquad (3.4.4)$$

Now note that, for $i = 1, \ldots, r$, the quantities $E(\alpha_i, \tilde{P})$ are all the same, since the stochastic process is stationary. Moreover, since the components in the index set I_i are separated by at least k, it follows from Theorem 2.1 and (2.5.5) that

$$E(\alpha_i, \tilde{P}) \le E(\alpha_i, \tilde{P}_0^\infty) + l\beta(k),$$

where $\beta(k)$ is the β-mixing coefficient of the underlying stochastic process. Similarly, for $i = r+1, \ldots, k$, $E(\alpha_i, \tilde{P})$ is the same due to the stationarity of the stochastic process. Moreover, it follows from the same theorem as above that

$$E(\alpha_i, \tilde{P}) \le E(\alpha_i, \tilde{P}_0^\infty) + (l-1)\beta(k) \le E(\alpha_i, \tilde{P}_0^\infty) + l\beta(k).$$

Next, note that, since the pair $(\mathcal{F}, \tilde{P}_0^\infty)$ has the UCEM property, the quantity $q(m, \epsilon, \tilde{P}_0^\infty) \to 0$ as $m \to \infty$, for each $\epsilon > 0$. In turn this implies, by the analog of (3.4.2) with $\tilde{P}$ replaced by $\tilde{P}_0^\infty$ that

$$c_m := E(a_m, \tilde{P}_0^\infty) \to 0 \text{ as } m \to \infty.$$

Now we come to the last steps in the proof. Define $\bar{c}_m = \max\{c_m, c_{m+1}\}$. Note that, under the probability measure $\tilde{P}_0^\infty$, the process $\{X_i\}$ is i.i.d. Therefore

$$E(\alpha_i, \tilde{P}_0^\infty) = c_{l+1} \le \bar{c}_l, \ i = 1, \ldots, r,$$

$$E(\alpha_i, \tilde{P}_0^\infty) = c_l \le \bar{c}_l, \ i = r+1, \ldots, k.$$

Substituting these estimates into (3.4.4) shows that

$$E(a_m, \tilde{P}) \le \bar{c}_{l_m} + l_m \beta(k_m). \qquad (3.4.5)$$

Finally, as $m \to \infty$, choose a corresponding sequence of integers $\{k_m\}$ in such a way that $l_m \to \infty$, and $l_m \beta(k_m) \to 0$. This is possible irrespective of how slowly the quantity $\beta(k)$ approaches zero as k increases; see Lemma 3.1. With such a choice, the right side of the above inequality approaches zero. In turn this establishes the desired conclusion (3.4.2). ∎

If we know how to construct the sequence $\{k_m\}$, (3.4.5) gives a relationship between the rates of convergence to zero of the quantities $E(a_m, \tilde{P}_0^\infty)$ and $E(a_m, \tilde{P})$. Moreover, in case the original sequence $\{X_i\}$ is i.i.d., we can "almost" recover the relationship $E(a_m, \tilde{P}) = E(a_m, \tilde{P}_0^\infty)$. In the i.i.d. case, we can choose $k_m = 1$ for all m, in which case (3.4.5) reduces to

$$E(a_m, \tilde{P}) \le \max\{E(a_m, \tilde{P}_0^\infty), E(a_{m+1}, \tilde{P}_0^\infty)\}.$$

Thus the inequality (3.4.5) is not very conservative.

To extend the above theorem to the case of a family of probability measures $\tilde{\mathcal{P}}$, let us refer to a stochastic process $\{X_i\}$ as "uniformly β-mixing" if $\{X_i\}$ is β-mixing for each $\tilde{P} \in \tilde{\mathcal{P}}$, and moreover,

$$\bar{\beta}(k) := \sup_{\tilde{P} \in \tilde{\mathcal{P}}} \beta(k) \to 0 \text{ as } k \to \infty.$$

Corollary 3.1. *With all symbols as above, suppose $\bar{q}(m, \epsilon, \tilde{P}) \to 0$ as $m \to \infty$, and that the sample process is uniformly β-mixing. Then $\bar{q}(m, \epsilon, \tilde{\mathcal{P}}) \to 0$ as $m \to \infty$.*

The proof consists of nothing more complicated than simply inserting $\sup_{\tilde{P} \in \tilde{\mathcal{P}}}$ at several places in the proof of Theorem 3.3.

While the above proof is very elegant and direct, it is not very useful when it comes to deriving explicit quantitative estimates relating the quantities $q(m, \epsilon, \tilde{P})$ and $q(m, \epsilon, \tilde{P}_0^\infty)$. Ideally one would expect to prove a bound involving these two quantities plus the β-mixing coefficient $\beta(k)$, in such a way that if $\beta(k) = 0$ for all k, then the bound for $q(m, \epsilon, \tilde{P})$ becomes simply $q(m, \epsilon, \tilde{P}_0^\infty)$. Though the estimate (3.4.5) is fairly tight, it involves expected values and not the "tail" probabilities $q(m, \epsilon, \cdot)$. The next result gives an explicit estimate that comes quite close to achieving this.

Theorem 3.4. *With all symbols as above, we have*

$$q(m, \epsilon, \tilde{P}) \leq k_m q'(l_m, \epsilon, \tilde{P}_0^\infty) + m\beta(k_m), \qquad (3.4.6)$$

where

$$q'(m, \epsilon, \tilde{P}_0^\infty) = \max\{q(m, \epsilon, \tilde{P}_0^\infty), q(m+1, \epsilon, \tilde{P}_0^\infty)\}.$$

Proof. It follows from (3.4.3) that if $a_m(\mathbf{x}^*) > \epsilon$, then at least one quantity among $\alpha_1(\mathbf{x}^*)$ through $\alpha_k(\mathbf{x}^*)$ must exceed ϵ. Therefore

$$\tilde{P}\{a_m > \epsilon\} \leq \sum_{i=1}^{k} \tilde{P}\{\alpha_i > \epsilon\}. \qquad (3.4.7)$$

Now a probability measure of a set is simply the expected value of its indicator function. Thus it follows from Theorem 2.1 that

$$\tilde{P}\{\alpha_i > \epsilon\} \leq \tilde{P}_0^\infty\{\alpha_i > \epsilon\} + l\beta(k) \leq q(l+1, \epsilon, \tilde{P}_0^\infty) + l\beta(k), \ i = 1, \ldots, r,$$

$$\tilde{P}\{\alpha_i > \epsilon\} \leq q(l, \epsilon, \tilde{P}_0^\infty) + (l-1)\beta(k), \ i = r+1, \ldots, k.$$

Substituting these two bounds into (3.4.7) establishes (3.4.6). ∎

In the case of i.i.d. inputs, it is possible to take $k_m = 1$ for all m, since $\beta(k) = 0$ for all $k \geq 1$. With this choice, the estimate of (3.4.6) becomes $q'(m, \epsilon, \tilde{P}_0^\infty)$, which is almost the same as $q(m, \epsilon, \tilde{P}_0^\infty)$. Thus, in this restricted sense, the estimate (3.4.6) is reasonable. Another situation of practical interest arises when the stochastic process $\{X_i\}$ exhibits "finite" dependence, that is, there exists a constant k such that $\mathtt{X}_i$ and $\mathtt{X}_j$ are independent whenever $|i-j| \geq k$. In such a case $\beta(k) = 0$ and one can therefore choose $k_m = k$ for all m in (3.4.6). This leads to the estimate

$$q(m, \epsilon, \tilde{P}) \leq k \, \max\{q(\lfloor m/k \rfloor, \epsilon, (\tilde{P}_0)^\infty), q(\lfloor m/k \rfloor + 1, \epsilon, (\tilde{P}_0)^\infty)\}.$$

On the other hand, Theorem 3.3 does *not* follow from this estimate, since in general the product term $m\beta(k_m)$ cannot be made to go to zero if, for instance, $\beta(k)$ approaches zero more slowly than $1/k$. Nevertheless, this estimate is useful to have. In particular, suppose the stochastic process $\{\mathtt{X}_i\}$ is "geometrically" β-mixing in the sense that there exist constants μ and $\lambda < 1$ such that

$$\beta(k) \leq \mu \lambda^k, \; \forall k.$$

Then the bound (3.4.6) *does* imply that $q(m, \epsilon, \tilde{P}) \to 0$ as $m \to \infty$. The details are easy and are left to the reader.

Corollary 3.2. *With all symbols as above, suppose the sample process $\{\mathtt{X}_i\}$ is uniformly β-mixing as $\tilde{P}$ varies over $\tilde{\mathcal{P}}$. Then*

$$\bar{q}(m, \epsilon, \tilde{\mathcal{P}}) \leq k_m \bar{q}'(l_m, \epsilon, \mathcal{P}) + m \bar{\beta}(k_m),$$

where

$$\bar{q}'(m, \epsilon, \mathcal{P}) = \max\{\bar{q}(m, \epsilon, \mathcal{P}), \bar{q}(m+1, \epsilon, \mathcal{P})\}.$$

The proof is obvious and is therefore omitted.

3.4.2 Law of Large Numbers Under Alpha-Mixing Inputs

In the preceding subsection, it is shown that the UCEMUP property is preserved if the i.i.d. input sequence $\{\mathtt{X}_i\}$ is replaced by a β-mixing process. The reader might wonder why two other notions were introduced in Chapter 2, namely α-mixing and ϕ-mixing, if they are not to be studied further. Actually, these two notions are also useful in their own way, but β-mixing processes are more meaningful in the context of both UCEM and PAC learning. Specifically, the situation can be summarized as follows: If $\{\mathtt{X}_i\}$ is α-mixing, then every *finite* set $\mathcal{F}$ has the UCEM property. However, it is not yet known whether infinite families of functions have the UCEM property, in case the UCEM property holds with i.i.d. inputs. At the other end of the spectrum, since every ϕ-mixing process is also β-mixing, all of the results in the preceding section hold also for ϕ-mixing input sequences. However, as shown below, the assumption of ϕ-mixing is very restrictive. In contrast, it is shown in the next section that a very large class of Markov chains and hidden Markov models produce β-mixing sequences. Thus, while α-mixing is "too weak" an assumption and ϕ-mixing is "too strong" an assumption, β-mixing is "just right."

90   3. Problem Formulations

Let us begin with α-mixing sequences. We shall take up ϕ-mixing sequences at the end of the subsection. The set-up is as follows: As before, $\{X_i\}$ is an α-mixing process defined on a probability space $(X^\infty, \mathcal{S}^\infty, \tilde{P})$, where we use the canonical representation for the stochastic process. Suppose now that $f : X \to [0, F]$ is a measurable function. Given a realization $\{x_i\}$ of the stochastic process, define the empirical mean

$$\hat{E}_m(f; \mathbf{x}) := \frac{1}{m} \sum_{i=1}^{m} f(x_i).$$

Let $\tilde{P}_0$ denote the one-dimensional marginal probability of $\tilde{P}$, and let $E(f, \tilde{P}_0)$ denote the expectation of f with respect to $\tilde{P}_0$. With this notation, we have the following result.

Theorem 3.5. *Define*

$$q_\alpha(m, \epsilon) := \tilde{P}\{\mathbf{x} \in X^\infty : |\hat{E}_m(f; \mathbf{x}) - E(f, \tilde{P}_0)| > \epsilon\}. \tag{3.4.8}$$

Given an integer m, choose $k \leq m$, and define $l := \lfloor m/k \rfloor$. Then

$$q_\alpha(m, \epsilon) \leq 2 \exp[-2\epsilon^2(l+1)/F^2] + 8\alpha(k) l \exp[4\epsilon(l+1)/F]. \tag{3.4.9}$$

In particular, if $\alpha(k) \to 0$ as $k \to \infty$, then $q_\alpha(m, \epsilon) \to 0$ as $m \to \infty$.

Proof. By replacing f by $f - E(f, \tilde{P}_0)$, it can be assumed that f has zero mean. Note that if f assumes values in $[0, F]$, then $f - E(f, \tilde{P}_0)$ assumes values in $[-E(f, \tilde{P}_0), F - E(f, \tilde{P}_0)]$, which is an interval of width F. Moreover, since $E(f, \tilde{P}_0) \in [0, F]$, it follows that $\| f \|_\infty \leq F$. Hence it is assumed hereafter that f has zero mean, assumes values in an interval of width F, and in addition $\| f \|_\infty \leq F$.

With k and l defined as above, let $r := m - lk$, and define

$$I_i = \begin{cases} \{i, i+k, \ldots, i+lk\} & \text{for } 1 \leq i \leq r, \\ \{i, i+k, \ldots, i+(l-1)k\} & \text{for } r+1 \leq i \leq k. \end{cases}$$

Let $p_i = |I_i|/m$ for $i = 1, \ldots, k$. Define the stochastic processes $\{a_m\}$ and b_i as follows:

$$a_m(\mathbf{x}) := \frac{1}{m} \sum_{i=0}^{m} f(x_i),$$

$$b_i(\mathbf{x}) := \frac{1}{|I_i|} \sum_{j \in I_i} f(x_j).$$

Then

$$a_m(\mathbf{x}) = \sum_{i=1}^{k} p_i b_i(\mathbf{x}).$$

Step 1. It is claimed that

3.4 Preservation of UCEMUP and PAC Properties

$$E[\exp(\gamma a_m), \tilde{P}] \leq \sum_{i=1}^{k} p_i E[\exp(\gamma b_i), \tilde{P}], \quad \forall \gamma > 0.$$

Since $\exp(\cdot)$ is convex, we have

$$\exp(\gamma a_m) = \exp\left[\sum_{i=1}^{k} p_i \gamma b_i\right] \leq \sum_{i=1}^{k} p_i \exp(\gamma b_i).$$

Now take the expectation of both sides with respect to $\tilde{P}$.

Step 2. It is claimed that

$$E[\exp(\gamma b_i), \tilde{P}] \leq \{E[\exp(\gamma f/|I_i|), \tilde{P}_0]\}^{|I_i|} + 4(|I_i|-1)\alpha(k)e^{\gamma F}, \quad \forall i.$$

To show this, note that

$$\exp(\gamma b_i) = \exp\left[\frac{\gamma}{|I_i|}\sum_{j \in I_i} f(x_j)\right] = \prod_{j \in I_i} \exp[\gamma f(x_j)/|I_i|].$$

By Corollary 2.2, it now follows that

$$\left|E[e^{\gamma b_i}, \tilde{P}] - E[e^{\gamma b_i}, \tilde{P}_0^\infty]\right| \leq 4(|I_i|-1)\alpha(k)[e^{\gamma F/|I_i|}]^{|I_i|}$$
$$= 4(|I_i|-1)\alpha(k)e^{\gamma F}.$$

Now

$$E[e^{\gamma b_i}, \tilde{P}_0^\infty] = E\left[\prod_{j \in I_i} \exp(\gamma f(x_i)/|I_i|), \tilde{P}_0^\infty\right]$$
$$= \{E[\exp(\gamma f/|I_i|), \tilde{P}_0]\}^{|I_i|},$$

since under the measure $\tilde{P}_0^\infty$ the various x_i are independent.

Step 3. Apply Lemma 2.8 to the function f. Since f has zero mean and assumes values in an interval of width F, it follows from the above lemma that

$$E[\exp(\gamma f/|I_i|), \tilde{P}_0] \leq \exp(\gamma^2 F^2/8|I_i|^2).$$

So

$$E[\exp(\gamma b_i), \tilde{P}] \leq \exp(\gamma^2 F^2/8|I_1|) + 4\alpha(k)(|I_i|-1)e^{\gamma F},$$

$$E[\exp(\gamma a_m), \tilde{P}] \leq \sum_{i=1}^{k} p_i \left[\exp(\gamma^2 F^2/8|I_1|) + 4\alpha(k)(|I_i|-1)e^{\gamma F}\right].$$

Step 4. By Markov's inequality,

$$\tilde{P}\{a_m > \epsilon\} \leq e^{-\gamma\epsilon} E[\exp(\gamma a_m), \tilde{P}]$$
$$\leq \sum_{i=1}^{k} p_i \left[\exp(-\gamma(\epsilon - \gamma F^2/8|I_1|)) + 4\alpha(k)(|I_i| - 1)e^{-\gamma\epsilon+\gamma F}\right]$$

Let us replace $|I_i|$ by its upper bound $l + 1$. This leads to

$$\tilde{P}\{a_m > \epsilon\} \leq \exp[-\gamma(\epsilon - \gamma F^2/8(l+1))] + 4\alpha(k)le^{-\gamma\epsilon+\gamma F},$$

where we have used the fact that $\sum_{i=1}^{k} p_i = 1$. The above inequality holds for *any* $\gamma > 0$. Now choose

$$\gamma = \frac{4\epsilon(l+1)}{F^2}, \text{ so that } \frac{\gamma F^2}{8(l+1)} = \frac{\epsilon}{2}.$$

Then it follows (after noting that $e^{-\gamma\epsilon+\gamma F} \leq e^{\gamma F}$) that

$$\tilde{P}\{a_m > \epsilon\} \leq \exp[-2\epsilon^2(l+1)/F^2] + 4\alpha(k)l\exp[4\epsilon(l+1)/F].$$

By symmetry (i.e., replacing f by $-f$), we also have

$$\tilde{P}\{a_m < -\epsilon\} \leq \exp[-2\epsilon^2(l+1)/F^2] + 4\alpha(k)l\exp[4\epsilon(l+1)/F].$$

Combining these two bounds leads to the desired inequality (3.4.9).

Step 5. By assumption, the stochastic process is α-mixing. Moreover, the map $l \mapsto l\exp(4\epsilon(l+1)/F)$ is strictly increasing. Hence, by Lemma 3.1, it is possible to choose a sequence $\{k_m\}$ such that, with $l_m = \lfloor m/k_m \rfloor$, we have

$$l_m \to \infty, \quad \alpha(k_m)l_m \exp(4\epsilon(l_m+1)/F) \to 0, \text{ as } m \to \infty.$$

With such a choice, it follows that $\tilde{P}\{a_m > \epsilon\} \to 0$ as $m \to \infty$. ∎

The preceding argument is a little conservative because of the possibility that m might not be exactly divisible by k. If we restrict ourselves to the case where $m = lk$, then the inequality becomes a little less conservative.

Corollary 3.3. *With all notation as in Theorem 3.5, suppose $m = lk$. Then*

$$q_\alpha(lk, \epsilon) \leq 2\exp(-2\epsilon^2 l/F^2) + 4\alpha(k)l\exp(4\epsilon l/F). \qquad (3.4.10)$$

Proof. If $m = lk$, then $p_i = 1/k$ for all i, and $|I_i| = l$ for all i. The desired inequality now follows. ∎

An important special case arises when the stochastic process $\{X_i\}$ has **finite dependence**, that is, there exists an integer k such that X_i and X_j are independent whenever $|i - j| \geq k$. In such a case, it is clear that $\alpha(k) = 0$. So we can apply the above inequality, which shows that whenever $m = lk$, we have

$$q_\alpha(lk, \epsilon) \leq 2\exp(-2\epsilon^2 l/F^2).$$

3.4 Preservation of UCEMUP and PAC Properties

In particular, if the samples are i.i.d., then we can take $k = 1$ for all m and recover the Hoeffding inequality as a special case.

Theorem 3.5 shows that a law of large numbers holds when we attempt to compute the expected value of *a single function* f by averaging $f(x_i)$, provided the x_i come from an α-mixing process. Clearly the result extends to *a finite number* of bounded functions. However, it is not known at present whether the result holds for *an infinite family* of functions.

Corollary 3.4. *Suppose $\mathcal{F}$ is a finite family of measurable functions taking values in $[0, F]$, and that $\{X_i\}$ is an α-mixing process. For each $f \in \mathcal{F}$, define its empirical mean $\hat{E}_m(f; \mathbf{x})$ based on a sample path $\mathbf{x} \in X^\infty$ as before, and define*

$$q_\alpha(m, e, \mathcal{F}) := \tilde{P}\{\mathbf{x} \in X^\infty : \max_{f \in \mathcal{F}} |\hat{E}(f; \mathbf{x}) - E(f, P)| > \epsilon\}.$$

Then

$$q_\alpha(m, \epsilon; \mathcal{F}) \leq 2|\mathcal{F}| \sum_{i=1}^{k} p_i [\exp(-2\epsilon^2 |I_i|/F^2) + 4\alpha(k)(|I_i| - 1) \exp(4\epsilon |I_i|/F)].$$

Next, let us turn to ϕ-mixing sequences. Since every ϕ-mixing sequence is also β-mixing, Theorem 3.3 applies. However, the following result due to Athreya and Pantula shows that ϕ-mixing is an extremely restrictive concept.

Lemma 3.2. ([12], Theorem 2) *Consider the first-order recursion*

$$X_{t+1} = \lambda X_t + e_t,$$

where $\lambda \in [0, 1)$ is some constant, and $\{e_t\}$ is an i.i.d. sequence independent of X_0. Suppose

1. $E[\{\log(e_1)\}_+] < \infty$, *where $(\cdot)_+$ denotes the positive part.*
2. *For some $n \geq 1$, the random variable $\sum_{i=1}^{n} \lambda^i e_i$ has a nontrivial absolutely continuous component. (This assumption is satisfied if $\lambda > 0$ and e_1 has a nontrivial absolutely continuous component.)*

Then $\{X_t\}$ is ϕ-mixing if and only if the noise sequence $\{e_t\}$ is essentially bounded, that is, there exists a constant M such that

$$|e_t| \leq M \text{ a.s.}$$

The interesting part of the above lemma is the "only if" part. This lemma implies that even the simple situation of a stable recursion driven by Gaussian noise is not ϕ-mixing, since Gaussian noise is unbounded. In contrast, it is shown in the next section that such a sequence is indeed β-mixing. Thus it appears that β-mixing is a more natural and useful notion than ϕ-mixing.

94   3. Problem Formulations

3.4.3 Preservation of PAC Learning Property with Beta-Mixing Inputs

In this subsection we state and prove a "universal" result which states that PAC learnability is preserved if the i.i.d. input sequence is replaced by a uniformly β-mixing input sequence. There is less to this theorem than one might suppose, because what it says is that *there exists an algorithm that is PAC* when the input sequence is uniformly β-mixing. This is a much less strong statement than saying that *the same algorithm* continues to be PAC even if an i.i.d. input sequence is replaced by a uniformly β-mixing input sequence. Strong results of the latter type are proved later in the next subsection. The purpose of the rather weak result presented here is to provide some completeness to the theory.

Theorem 3.6. *Suppose the pair $(\mathcal{F}, \tilde{\mathcal{P}})$ is PAC learnable when the learning inputs are i.i.d. Then the pair $(\mathcal{F}, \tilde{\mathcal{P}})$ continues to be PAC learnable if the learning input sequence is uniformly β-mixing.*

Proof. Suppose a set $A \in \Sigma\{X_i, X_{i+k}, \ldots, X_{i+lk}\}$. Apply Theorem 2.1 to the indicator function $I_A(\cdot)$. Then it follows from (2.5.5) that

$$|\tilde{P}(A) - (\tilde{P}_0)^\infty(A)| \le l\beta(k, \tilde{P}).$$

Define

$$\bar{\beta}(k) := \sup_{\tilde{P} \in \tilde{\mathcal{P}}} \beta(k, \tilde{P}).$$

Then by the assumption that the inputs are uniformly β-mixing, it follows that $\bar{\beta}(k) \to 0$ as $k \to \infty$.

As before, in the interests of simplicity define

$$\mathcal{P} := \{(\tilde{P}_0)^\infty, \tilde{P} \in \tilde{\mathcal{P}}\}. \tag{3.4.11}$$

Suppose $\{A_m\} : (X \times [0,1])^m \to \mathcal{F}$ is an algorithm that is PAC when the inputs are i.i.d. By assumption, this means that the quantity $\bar{r}(m, \epsilon, \mathcal{P})$ approaches zero as $m \to \infty$ for each $\epsilon > 0$. Now let us define the following modified algorithm for the case of mixing inputs. Given an input sequence **x** of length m, choose an integer $k_m \le m$, and define $l_m := \lfloor m/k_m \rfloor$. Run the algorithm A_{l_m} on the inputs $\{x_{ik_m}, i = 1, \ldots, l_m\}$. Call the resulting hypothesis h. By definition,

$$P^\infty\{d_{\tilde{P}_0}(f, h) > \epsilon\} \le \bar{r}(l_m, \epsilon, \mathcal{P}).$$

Note that the event $\{d_{\tilde{P}_0}(f, h) > \epsilon\}$ belongs to the σ-algebra $\Sigma\{X_{k_m}, \ldots, X_{l_m k_m}\}$. Hence by Theorem 2.1, we have

$$\tilde{P}\{d_{\tilde{P}_0}(f, h) > \epsilon\} \le \bar{r}(l_m, \epsilon, \mathcal{P}) + (l_m - 1)\bar{\beta}(k_m).$$

3.4 Preservation of UCEMUP and PAC Properties

Since the right side is independent of $\tilde{P}$, it can serve as an upper bound for the quantity $\bar{r}(m, \epsilon, \tilde{P})$ when the inputs are mixing. Now as $m \to \infty$, choose the integer sequence $\{k_m\}$ in such a way that $l_m \to \infty$ and $l_m \bar{\beta}(k_m) \to 0$ as $m \to \infty$. This is possible in view of Lemma 3.1. With such a choice, the right side of the above inequality approaches zero as $m \to \infty$, which shows that the algorithm is PAC. ∎

3.4.4 Preservation of PAC Learning Property with Beta-Mixing Inputs: Continued

Theorem 3.6 shows that, if a pair $(\mathcal{F}, \mathcal{P})$ is PAC learnable via a particular algorithm $\{A_m\}$, then it is possible to adjust the algorithm so as to be PAC even when the sample sequence is β-mixing. The modified algorithm consists of nothing more than resampling the inputs, keeping every k_m-th input and throwing away the rest. Thus the algorithm requires a knowledge of the nature of the input sequence and its mixing coefficients. It would be much more natural to prove a result whereby *the original algorithm* that is PAC with i.i.d. inputs continues to be PAC even when the input sequence is β-mixing, even if the learning rate is a little slower. Such a result is proved in this subsection. In order to present this result, we introduce the notion of a "quasi-subadditive" learning algorithm. A given algorithm $\{A_m\}$ said to be **quasi-subadditive** for a pair $(\mathcal{F}, \mathcal{P})$ (with i.i.d. inputs) if the following condition holds: Let $\mathbf{x}^* \in X^\infty$ be any learning sample, and suppose m is an integer. Let $h_m(f; \mathbf{x}^*)$ denote the hypothesis generated by applying the algorithm to the input sequence $\mathbf{x}^*$ up to time m, that is,

$$h_m(f; \mathbf{x}^*) := A_m[(x_1, f(x_1)), \ldots, (x_m, f(x_m))].$$

For every $k_m \leq m$, define $l_m = \lfloor m/k_m \rfloor$, and let $r_m = m - k_m l_m$. For $i = 1, \ldots, k_m$, let $g_i(f; \mathbf{x}^*)$ denote the hypothesis generated by applying the same algorithm to a sampled version of $\mathbf{x}^*$; specifically[11],

$$g_i(f; \mathbf{x}^*) := A_{l_m+1}[(x_i, f(x_i)), (x_{i+k_m}, f(x_{i+k_m})), \ldots, (x_{i+l_m k_m}, f(x_{i+l_m k_m}))],$$

for $1 \leq i \leq r_m$ and

$$g_i(f; \mathbf{x}^*) := A_{l_m}[(x_i, f(x_i)), \ldots, (x_{i+(l_m-1)k_m}, f(x_{i+(l_m-1)k_m}))]$$

for $r_m + 1 \leq i \leq k_m$. Then the algorithm is said to be **quasi-subadditive** if

$$d_P[f, h_m(f; \mathbf{x}^*)] \leq \frac{1}{k_m} \sum_{i=1}^{k_m} d_P[f; g_i(f; \mathbf{x}^*)], \ \forall \mathbf{x}^*, f. \tag{3.4.12}$$

---

[11] Actually the hypothesis g_i depends on a few other arguments as well, but these are suppressed in order to minimize notational clutter.

Thus an algorithm is quasi-subadditive if (roughly speaking) the generalization error that results when all m inputs are used to generate the hypothesis is less than the average of the generalization errors when the hypotheses are generated using sub-sampled versions of the learning inputs. Since we would expect that, for a "reasonable" learning algorithm, the generalization error decreases as the number of samples increases, the assumption of quasi-subadditivity is quite moderate. Moreover, we shall see in subsequent chapters that many of the widely used learning algorithms do indeed possess this property.

Theorem 3.7. *Suppose an algorithm $\{A_m\}$ PAC-learns a pair $(\mathcal{F}, \mathcal{P})$ with i.i.d. inputs. Suppose further that the algorithm has the quasi-subadditivity property. Then the algorithm $\{A_m\}$ continues to be PAC for the pair $(\mathcal{F}, \tilde{\mathcal{P}})$. Moreover, for every m and every $k_m \leq m$, define $l_m = \lfloor m/k_m \rfloor$. Then*

$$\bar{r}(m, \epsilon, \tilde{\mathcal{P}}) \leq k_m \max\{\bar{r}(l_m, \epsilon, \mathcal{P}), \bar{r}(l_m + 1, \epsilon, \mathcal{P})\} + m\beta(k_m). \quad (3.4.13)$$

Proof. The proof uses arguments that are already very familiar from those of Theorems 3.3 and 3.4. We begin by replacing the problem of estimating $\bar{r}(m, \epsilon, \tilde{\mathcal{P}})$ by that of estimating the expected value. Define

$$b_m := \sup_{\tilde{P} \in \tilde{\mathcal{P}}} E(d_{\tilde{P}_0}[f, h_m(f, \mathbf{x}^*)], \tilde{P}),$$

$$c_m := \sup_{P \in \mathcal{P}} E(d_P[f, h_m(f, \mathbf{x}^*)], P^\infty),$$

where as before $\mathcal{P}$ denotes $\{(\tilde{P}_0)^\infty, \tilde{P} \in \tilde{\mathcal{P}}\}$. Then, by a relationship analogous to (3.4.1), it follows that the algorithm $\{A_m\}$ is PAC for the pair $(\mathcal{F}, \tilde{\mathcal{P}})$ if and only if $b_m \to 0$ as $m \to \infty$. Similarly, the algorithm $\{A_m\}$ is PAC for the pair $(\mathcal{F}, \mathcal{P})$ if and only if $c_m \to 0$ as $m \to \infty$. Define $\bar{c}_m = \max\{c_m, c_{m+1}\}$. Since the algorithm is PAC for $(\mathcal{F}, \mathcal{P})$, it follows that $\bar{c}_m \to 0$ as $m \to \infty$. Now we use the quasi-subadditivity of the algorithm. By (3.4.12) it follows that

$$b_m \leq \frac{1}{k_m} \sum_{i=1}^{k_m} \sup_{\tilde{P} \in \tilde{\mathcal{P}}} E(d_{\tilde{P}_0}[f, g_i(f, \mathbf{x}^*)], \tilde{P}).$$

Now, since the hypothesis g_i depends on only $x_i, \ldots x_{i+l_m k_m}$ or $x_i, \ldots, x_{i+(l_m-1)k_m}$, we can apply Theorem 2.1. This shows that

$$E(d_{\tilde{P}_0}[f, g_i(f, \mathbf{x}^*)], \tilde{P}) \leq E(d_{\tilde{P}_0}[f, g_i(f, \mathbf{x}^*)], P^\infty) + l_m \beta(k_m), \; \forall i.$$

Moreover, we have

$$E(d_{\tilde{P}_0}[f, g_i(f, \mathbf{x}^*)], P^\infty) = \max\{c_{l_m+1}, c_{l_m}\} = \bar{c}_{l_m}, \; \forall i.$$

Therefore

$$b_m \leq l_m \beta(k_m) + \bar{c}_{l_m}, \; \forall m.$$

Now, as $m \to \infty$, choose a subsequence k_m such that $l_m \to \infty$ and $l_m \beta(k_m) \to 0$. This is possible in view of Lemma 3.1. Then it follows that $b_m \to 0$. This shows that the algorithm $\{A_m\}$ is PAC for the pair $(\mathcal{F}, \tilde{\mathcal{P}})$.

To establish the estimate (3.4.13), recall that for each $\tilde{P} \in \tilde{\mathcal{P}}$, we have

$$r(m, \epsilon, \tilde{P}) = \tilde{P}\{\mathbf{x}^* : d_P[f, h_m(f; \mathbf{x}^*)] > \epsilon\}.$$

However, by the quasi-subadditivity of the algorithm, and in particular (3.4.12), it follows that the following containment of events holds:

$$\{\mathbf{x}^* : d_P[f, h_m(f; \mathbf{x}^*)] > \epsilon\} \subseteq \bigcup_{i=1}^{k_m} \{\mathbf{x}^* : d_P[f, g_i(f; \mathbf{x}^*)] > \epsilon\}.$$

Thus

$$r(m, \epsilon, \tilde{P}) \leq \sum_{i=1}^{k_m} \tilde{P}\{d_P[f, g_i(f; \mathbf{x}^*)] > \epsilon\}.$$

Now the probabilities inside the summation can be estimated using Theorem 2.1. It follows that

$$\begin{aligned}\tilde{P}\{\mathbf{x}^* : d_P[f, g_i(f; \mathbf{x}^*)] > \epsilon\} &\leq l_m \beta(k_m) + P^\infty\{\mathbf{x}^* : d_P[f, g_i(f; \mathbf{x}^*)] > \epsilon\} \\ &= l_m \beta(k_m) + \max\{\bar{r}(l_m, \epsilon, \tilde{\mathcal{P}}), \bar{r}(l_m + 1, \epsilon, \tilde{\mathcal{P}})\}.\end{aligned}$$

The desired estimate (3.4.13) now follows upon substituting in the preceding inequality and noting that $l_m k_m \leq m$. ∎

3.4.5 Replacing $\mathcal{P}$ by its Closure

In this section it is shown that both the UCEMUP property and the PAC learnability property are preserved when the underlying family of probability measures $\mathcal{P}$ is replaced by its closure $\bar{\mathcal{P}}$ under the total variation metric.

Theorem 3.8. *Given a pair $(\mathcal{F}, \mathcal{P})$, we have*

$$\bar{q}(m, \epsilon, \mathcal{P}) \leq \bar{q}(m, \epsilon, \bar{\mathcal{P}}) \leq \lim_{\epsilon' \to \epsilon^-} \bar{q}(m, \epsilon', \mathcal{P}), \; \forall m, \epsilon > 0. \qquad (3.4.14)$$

Consequently, suppose a pair $(\mathcal{F}, \mathcal{P})$ has the UCEMUP property. Then the pair $(\mathcal{F}, \bar{\mathcal{P}})$ also has the UCEMUP property.

Proof. Recall the definition

$$\bar{q}(m, \epsilon, \mathcal{P}) := \sup_{P \in \mathcal{P}} P^m\{\mathbf{x} \in X^m : \sup_{f \in \mathcal{F}} |\hat{E}(f; \mathbf{x}) - E_P(f)| > \epsilon\}.$$

By analogy,

$$\bar{q}(m, \epsilon, \bar{\mathcal{P}}) := \sup_{P \in \bar{\mathcal{P}}} P^m\{\mathbf{x} \in X^m : \sup_{f \in \mathcal{F}} |\hat{E}(f; \mathbf{x}) - E_P(f)| > \epsilon\}.$$

Thus the left inequality in (3.4.14) is obvious, since $\mathcal{P} \subseteq \bar{\mathcal{P}}$.

Next, observe that $\bar{q}(m, \epsilon, \mathcal{P})$ is a monotonic, nonincreasing function of ϵ for fixed $m, \mathcal{P}$. In other words,

$$\epsilon > \epsilon' \;\Rightarrow\; \bar{q}(m, \epsilon, \mathcal{P}) \leq \bar{q}(m, \epsilon', \mathcal{P}).$$

To prove the right inequality in (3.4.14), fix $m, \epsilon > 0$, and let

$$\delta := \bar{q}(m, \epsilon, \bar{\mathcal{P}}).$$

Choose a "large" integer $n \geq 1$; eventually we will let $n \to \infty$. Then by the definition of δ, there exists a $P_n \in \bar{\mathcal{P}}$ such that

$$P_n^m \{ \mathbf{x} \in X^m : \sup_{f \in \mathcal{F}} |\hat{E}(f; \mathbf{x}) - E_P(f)| > \epsilon \} > (1 - 1/n)\delta.$$

Since $P_n \in \bar{\mathcal{P}}$, there exists a $Q_n \in \mathcal{P}$ such that $\rho(P_n, Q_n) \leq \min\{\epsilon/n, \delta/mn\}$. Then it follows from Lemma 2.6 that $\rho(P_n^m, Q_n^m) \leq \delta/n$. Since each $f \in \mathcal{F}$ is bounded by one, it follows that

$$|E(f, P_n) - E(f, Q_n)| \leq \epsilon/n.$$

As a result,

$$|E(f, P_n) - \hat{E}(f; \mathbf{x})| \leq |E(f, Q_n) - \hat{E}(f; \mathbf{x})| + \epsilon/n.$$

Consequently, for a given $\mathbf{x} \in X^m$,

$$\sup_{f \in \mathcal{F}} |E(f, P_n) - \hat{E}(f; \mathbf{x})| > \epsilon \;\Rightarrow\; \sup_{f \in \mathcal{F}} |E(f, Q_n) - \hat{E}(f; \mathbf{x})| > (1 - 1/n)\epsilon.$$

So if we define

$$K_n := \{ \mathbf{x} \in X^m : \sup_{f \in \mathcal{F}} |E(f, P_n) - \hat{E}(f; \mathbf{x})| > \epsilon \},$$

$$S_n := \{ \mathbf{x} \in X^m : \sup_{f \in \mathcal{F}} |E(f, Q_n) - \hat{E}(f; \mathbf{x})| > (1 - 1/n)\epsilon \},$$

then $K_n \subseteq S_n$. By assumption $P_n^m(K_n) > (1 - 1/n)\delta$. Since $\rho(P_n^m, Q_n^m) \leq \delta/n$, it follows that

$$Q_n^m(K_n) \geq P_n^m(K_n) - \delta/n = (1 - 2/n)\delta.$$

Since $K_n \subseteq S_n$, we have $Q_n^m(S_n) \geq (1 - 2/n)\delta$, that is,

$$Q_n^m \{ \mathbf{x} \in X^m : \sup_{f \in \mathcal{F}} |E(f, Q_n) - \hat{E}(f; \mathbf{x})| > (1 - 1/n)\epsilon \} \geq (1 - 2/n)\delta.$$

Since $Q_n \in \mathcal{P}$, this in turn implies that

$$\bar{q}(m, (1-1/n)\epsilon, \mathcal{P}) \geq (1-2/n)\delta.$$

Now let $n \to \infty$. The left side approaches the right side of (3.4.14), while the right side approaches $\delta = \bar{q}(m, \epsilon, \bar{\mathcal{P}})$. This establishes (3.4.14).

To establish the second sentence, suppose $(\mathcal{F}, \mathcal{P})$ has the UCEMUP property, and fix $\epsilon > 0$. Then $\bar{q}(m, \epsilon, \mathcal{P}) \to 0$ as $m \to \infty$. Now choose any $\epsilon' < \epsilon$. Then (3.4.14) implies that $\bar{q}(m, \epsilon, \bar{\mathcal{P}}) \leq \bar{q}(m, \epsilon', \mathcal{P}) \to 0$ as $m \to \infty$. ∎

Theorem 3.9. *Suppose an algorithm $\{A_m\}$ PAC-learns a pair $(\mathcal{F}, \mathcal{P})$. Then the same algorithm also PAC-learns the pair $(\mathcal{F}, \bar{\mathcal{P}})$, though with a different sample complexity.*

Proof. The proof is based on Theorem 3.1. Given $\epsilon, \delta > 0$ we show how to construct a sample complexity $s(\epsilon, \delta)$ such that (3.2.12) holds. Since the algorithm $\{A_m\}$ PAC-learns the pair $(\mathcal{F}, \mathcal{P})$, there exists an integer $m_0(\epsilon, \delta)$ such that

$$\sup_{P \in \mathcal{P}} \sup_{f \in \mathcal{F}} P^m\{\mathbf{x} \in X^m : d_P[f; h_m(f; x)] > \epsilon\} < \delta, \ \forall m \geq m_0(\epsilon, \delta). \quad (3.4.15)$$

It is claimed that the choice $s(\epsilon, \delta) = m_0(\epsilon/2, \delta/2)$ will satisfy (3.2.12). Suppose $P \in \bar{\mathcal{P}}, f \in \mathcal{F}$ are arbitrary. Since $P \in \bar{\mathcal{P}}$, there exists a sequence $\{P_i\}$ in $\mathcal{P}$ converging to P. Moreover, for any *fixed* integer k, Lemma 2.6 implies that $\rho(P^k, P_i^k) \to 0$ as $i \to \infty$. Now choose the integer i sufficiently large that (i) $\rho(P, P_i) < \epsilon/2$, and (ii) $\rho(P^s, P_i^s) \leq \delta/2$, where $s = m_0(\epsilon/2, \delta/2)$. Look at the set of "bad" samples

$$L := \{\mathbf{x} \in X^s : d_P[f, h_s(f; x)] > \epsilon\}.$$

Since $\rho(P, P_i) < \epsilon/2$, it follows that

$$|d_P[f, h_s(f; x)] - d_{P_i}[f, h_s(f; x)]| \leq \epsilon/2.$$

Consequently

$$d_P[f, h_s(f; x)] > \epsilon \Rightarrow d_{P_i}[f, h_s(f; x)] > \epsilon/2.$$

So if we define

$$M := \{\mathbf{x} \in X^s : d_{P_i}[f, h_s(f; x)] > \epsilon/2\},$$

then $L \subseteq M$. Now it follows from (3.4.15) that

$$P_i^s(M) \leq \delta/2.$$

However, since $\rho(P^s, P_i^s) \leq \delta/2$, in turn this implies that $P^s(M) \leq P_i^s(M) + \delta/2 \leq \delta$. This is precisely the desired conclusion. ∎

3.5 Markov Chains and Beta-Mixing

The previous section serves to establish that β-mixing is in some sense a very "natural" assumption on non-i.i.d. stochastic processes, because both the UCEMUP property and PAC learning are preserved if i.i.d. input sequences are replaced by β-mixing input sequences. At the same time, Lemma 3.2 serves to show that ϕ-mixing is a very strong assumption, which is not satisfied even by the "natural" stochastic process generated when a stable first-order recursion is driven by Gaussian noise. Against this background, the aim of the present section is to show that β-mixing sequences result when a stable *nonlinear* recursion is driven by noise with bounded *variance*, or in the case of "hidden" Markov models wherein the state variable is further processed through a probability transition function.

The section is organized as follows. We begin by introducing some terminology from the Markov chain literature, and define a notion called V-geometric ergodicity, which has been widely studied in the literature. It is shown that V-geometric ergodicity implies geometric β-mixing, and it is shown that the β-mixing coefficient can be expressed as an abstract integral of a V-geometrically ergodic process. Then it is shown that a large class of Markov chains generated by noise inputs with bounded variance are β-mixing. The mixing property is then extended to hidden Markov models.

3.5.1 Geometric Ergodicity and Beta-Mixing

Suppose $(X, \mathcal{S})$ is a measurable space. For the purposes of the present discussion, a **Markov chain** is a sequence of random variables $\{X_m\}_{m \geq 0}$ together with a set of probability measures $P^n(x, A), x \in X, A \in \mathcal{S}$ denoting the "transition probabilities." It is assumed that

$$\Pr\{X_{n+m} \in A | X_j, j \leq m, X_m = x\} = P^n(x, A).$$

Thus $P^n(x, A)$ denotes the probability that the state X will belong to the set A after n time steps, starting from the initial state x at time m. It is common to denote the "one-step" transition probability by $P(x, A)$, so that $P^1(x, A) = P(x, A)$. The fact that the transition probability does not depend on the values of X prior to time m is the Markov property, and the fact that the transition probability does not depend on the "initial time" m means that the Markov chain is stationary.

Suppose the Markov chain is set in motion with the initial state at time $t = 0$ distributed according to the probability measure Q_0. Then the definition of $P(\cdot, \cdot)$ implies that

$$Q_1(A) := \Pr\{x_1 \in A\} = \int_X P(x, A) \, Q_0(dx).$$

Under suitable conditions (see [135] for a detailed treatment), a stationary Markov chain has an **invariant measure** or a **stationary distribution** π on $(X, \mathcal{S})$ with the property that

$$\pi(A) = \int_X P(x, A) \, \pi(dx).$$

Thus, if the Markov chain is started off with the initial state distributed according to the the stationary distribution π, then at all subsequent times the state continues to be distributed according to π.

A (stationary) Markov chain is said to be **geometrically ergodic** if there exist constants μ and $\lambda < 1$ such that

$$\rho[P^n(x, \cdot), \pi] \leq \mu \lambda^n, \ \forall x \in X.$$

Note that here ρ denotes the total variation metric between two probability measures. Thus in a geometrically ergodic Markov chain, the total variation metric distance between the n-step transition probability $P^n(x, \cdot)$ and the stationary distribution π decays to zero at a geometric rate; moreover, this rate is *independent of the initial state* x. If the state space X is not compact, it is not reasonable to expect such a strong type of convergence to hold. To cater to the general situation, a more liberal notion called "V-geometric ergodicity" is introduced. A stationary Markov chain is said to be V-**geometrically ergodic** with respect to the measurable function $V : X \to [1, \infty)$ if there exist constants μ and $\lambda < 1$ such that

$$\rho[P^n(x, \cdot), \pi] \leq \mu \lambda^n V(x), \ \forall x \in X, \tag{3.5.1}$$

and in addition,

$$E[V, \pi] = \int_X V(x) \, \pi(dx) < \infty.$$

Actually, the notion of V-geometric ergodicity as defined in [135] is more restrictive than the above. Specifically, in [135] the total variation metric $\rho[P^n(x, \cdot), \pi]$ is replaced by a larger quantity that can be thought of as the total variation *with respect to all functions bounded by* V. Since V is bounded below by one, this latter quantity is no smaller than $\rho[P^n(x, \cdot), \pi]$. Consequently V-geometric ergodicity in the sense of [135] implies the above inequality.

Thus a Markov chain is V-geometrically ergodic if two conditions hold. First, there is a nonnegative-valued function V such that the total variation distance between the n-step transition probability $P^n(x, \cdot)$ and the invariant measure π approaches zero at a geometric rate *multiplied by* $V(x)$. Thus the rate of geometric convergence is independent of x, but the multiplicative constant is allowed to depend on x. To ensure that the property is meaningful, the second condition is imposed, namely that the "growth function" $V(\cdot)$ has finite expectation with respect to the invariant measure π. Thus, "on average" the total variation metric distance between the n-step transition probability and the stationary distribution decays to zero at a geometric rate.

The main result of this subsection is given next.[12]

Theorem 3.10. *Suppose a Markov chain is V-geometrically ergodic. Then the state sequence $\{X_t\}$ is geometrically β-mixing, i.e., there exist constants B and $\lambda < 1$ such that $\beta(m) \leq B\lambda^m$ for all m. Moreover, the β-mixing coefficient is given by*

$$\beta(m) = E\{\rho[P^m(x, \cdot), \pi], \pi\} = \int_X \rho[P^m(x, \cdot), \pi] \, \pi(dx). \tag{3.5.2}$$

In order to prove the theorem we require two preliminary lemmas. The first is a kind of "selection lemma," while the second lemma shows that two distinct ways of defining the β-mixing coefficient are in fact equivalent.

Let us begin with a little notation. Suppose X_1, X_2 are complete separable metric spaces, and that $\mathcal{S}_1, \mathcal{S}_2$ are the corresponding Borel σ-algebras of subsets of X_1 and X_2 respectively. Define $X = X_1 \times X_2$ and let $\mathcal{S} = \mathcal{S}_1 \times \mathcal{S}_2$ be the corresponding product algebra. Define $\mathcal{G}_1 = \mathcal{S}_1 \times \{\emptyset, X_2\}$, and similarly $\mathcal{G}_2 = \{\emptyset, X_1\} \times \mathcal{S}_2$. Suppose P is a probability measure on $(X, \mathcal{S})$, and let P_1, P_2 denote the marginal probability measures of P on X_1 and X_2 respectively. Thus for $A_1 \in \mathcal{S}_1, A_2 \in \mathcal{S}_2$ we have

$$P_1(A_1) = P\{(x_1, x_2) : x_1 \in A_1\},$$

and similarly for P_2. Now we are ready to state the selection lemma.

Lemma 3.3. *With the above notation, there exists a probability transition function $Q : X_1 \times \mathcal{S}_2 \to [0, 1]$, that is, $Q(x_1, \cdot)$ is a probability measure on $(X_2, \mathcal{S}_2)$ for all $x_1 \in X_1$, and $Q(\cdot, A_2) \in \mathcal{S}_1$ for all $A_2 \in \mathcal{S}_2$, such that for all $A \in \mathcal{S}$ we have*

$$P(A) = \int_{X_1} Q(x_1, A(x_1)) \, P_1(dx_1),$$

where

$$A(x_1) := \{x_2 : (x_1, x_2) \in A\}.$$

Further,

$$E_P(I_A | \mathcal{G}_1) = Q(\cdot, A_{(\cdot)}),$$

where as before $E_P(I_A | \mathcal{G}_1)$ denotes the best approximation to the indicator function $I_A(\cdot)$ among functions measurable with respect to $\mathcal{G}_1$, and the error measure is the L_2-norm with respect to the measure P. In other words, $f(x_1) = Q(x_1, A(x_1))$ satisfies

$$E_P(I_A | \mathcal{G}_1) = f(\cdot).$$

---

[12] The theorem and proof are due to R. L. Karandikar.

The proof can be found in, for example, [35].

Next, it is shown that two distinct-looking definitions of the β-mixing coefficient that are widely used in the literature are in fact equivalent.

Lemma 3.4. *With the notation as above, let $\mathcal{H}_2 \subseteq \mathcal{S}_2$ be a sub-σ-algebra on X_2 such that $(X_1, \mathcal{S}_1)$, $(X_2, \mathcal{H}_2)$ are standard-Borel. Let*

$$\beta := \sup_{A \in \mathcal{S}_1 \times \mathcal{H}_2} |P(A) - (P_1 \times P_2)(A)|,$$

$$\theta := E[\sup_{A_2 \in \mathcal{H}_2} |Q(x_1, A_2) - P_2(A_2)|, P_1]$$

$$= \int_{X_1} \sup_{A_2 \in \mathcal{H}_2} |Q(x_1, A_2) - P_2(A_2)| \, P_1(dx_1).$$

Then $\beta = \theta$.

Proof. **(of the lemma)** To show that $\beta \leq \theta$, suppose that $A \in \mathcal{S}_1 \times \mathcal{H}_2$. Then

$$P(A) = \int_{X_1} Q(x_1, A(x_1)) \, P_1(dx_1), \quad (P_1 \times P_2)(A) = \int_{X_1} P_2(A(x_1)) \, P_1(dx_1).$$

Hence

$$|P(A) - (P_1 \times P_2)(A)| \leq \int_{X_1} |Q(x_1, A(x_1)) - P_2(A(x_1))| \, P_1(dx_1)$$

$$\leq \int_{X_1} \sup_{A_2 \in \mathcal{H}_2} |Q(x_1, A_2) - P_2(A_2)| \, P_1(dx_1) = \theta.$$

Here we use the fact that $A(x_1) \in \mathcal{H}_2$ since $A \in \mathcal{S}_1 \times \mathcal{H}_2$. Since the above argument holds for every A, it follows that $\beta \leq \theta$.

To prove that $\theta \leq \beta$, we proceed as follows. Fix an $\epsilon > 0$, $x_1 \in X_1$, and choose $A_2(x_1) \in \mathcal{H}_2$ such that

$$|Q(x_1, A_2(x_1)) - P_2(A_2(x_1))| \geq \sup_{A_2 \in \mathcal{H}_2} |Q(x_1, A_2) - P_2(A_2)| - \epsilon.$$

This can be done for each $x_1 \in X_1$. Now by appealing to the principle of measurable selection (which applies since $(X_1, \mathcal{S}_1)$, $(X_2, \mathcal{H}_2)$ are standard Borel), it is possible to select $A_2(x_1)$ in such a way that the set

$$A := \{(x_1, x_2) : x_2 \in A_2(x_1)\}$$

belongs to $\mathcal{S}_1 \times \mathcal{H}_2$. Now

$$P(A) - (P_1 \times P_2)(A) = \int_{X_1} [Q(x_1, A_2(x_1)) - P_2(A_2(x_1))] \, P_1(dx_1)$$

$$\geq \int_{X_1} [\sup_{A_2 \in \mathcal{H}_2} |Q(x_1, A_2) - P_2(A_2)| - \epsilon] \, P_1(dx_1)$$

$$\geq \theta - \epsilon.$$

Therefore
$$\beta = \sup_{A \in \mathcal{S}_1 \times \mathcal{H}_2} |P(A) - (P_1 \times P_2)(A)| \geq \theta - \epsilon.$$
Since this argument can be repeated for every $\epsilon > 0$, it follows that $\beta \geq \theta$. ∎

Proof. **(of the theorem)** Let $X^\infty, \mathcal{S}^\infty$ be as before. Let $Y = \prod_{i=1}^{\infty} X$ and let $\mathcal{T}$ be the corresponding product σ-algebra generated by $\mathcal{S}$. (The difference between X^∞ and Y is that X^∞ is a doubly infinite Cartesian product whereas Y is a singly infinite Cartesian product; similarly for $\mathcal{S}^\infty$ versus $\mathcal{T}$.) Let P be a probability measure on $(X^\infty, \mathcal{S}^\infty)$ such that $\{X_n\}$ is a stationary Markov chain with the transition probability $P(x, A)$ and stationary distribution π. Similarly, let Q_x be a probability measure on $(Y, \mathcal{T})$ such that $\{Y_n\}$ is a stationary Markov chain with transition probability $P(x, A)$ and initial distribution δ_x. To apply Lemma 3.4, we identify

$$\mathcal{X}_1 = \prod_{i=-\infty}^{0} X, \ \mathcal{X}_2 = \prod_{i=1}^{\infty} X.$$

Let P_1, P_2 be the marginal measures of P on $\mathcal{X}_1$ and $\mathcal{X}_2$ respectively. Finally, we have
$$P_2(A_2) = \int_X Q_{x_0}(A_2) \, \pi(dx_0).$$
Here we make use of the Markov property which implies that the probability $P_2\{(x_1, x_2, \ldots) \in A_2 \subseteq \mathcal{X}_2\}$ depends only on x_0. Hence we identify $Q(x, A_2)$ with $Q_x(A_2)$.

For each subset $D \subseteq Y$, we have
$$P\{(x_m, x_{m+1}, \ldots) \in D | \Sigma\{X_i, i \leq m\}\} = Q_{x_m}(D).$$

Now define $\mathcal{H}_2 := \Sigma\{\mathsf{X}_i, i \geq m\}$. Then the β-mixing coefficient is given, using the result of Lemma 3.4, by

$$\beta(m) = \int_{X^\infty} \sup_D |P\{(x_m, x_{m+1}, \ldots) \in D | x_i, i \leq 0\}$$
$$- P\{(x_m, x_{m+1}, \ldots) \in D\}| \, dP.$$

Note that
$$P\{(x_m, x_{m+1}, \ldots) \in D\} = \int_{X^\infty} P\{(x_m, x_{m+1}, \ldots) \in D | \Sigma(X_m)\} dP$$
$$= \int_{X^\infty} Q_{x_m}(D) dP = \int_X Q_y(D) \, \pi(dy),$$
since the only independent variable under the integral sign is x_m. Similarly

$$P\{(x_m, x_{m+1}, \ldots) \in D | x_i, i \leq 0\} = E[P\{(x_m, x_{m+1}, \ldots) \in D$$
$$| \Sigma\{X_i, i \leq m\} | \Sigma\{X_i, i \leq 0\}]$$
$$= E[Q_{x_m}(D) | \Sigma\{X_i, i \leq 0\}]$$
$$= \int_X Q_y(D)\, P^m(x_0, dy)$$

Thus

$$\beta(m) = \int_X \sup_D \left| \int_X Q_y(D)\, P^m(x_0, dy) - \int_X Q_y(D)\, \pi(dy) \right| \pi(dx_0). \quad (3.5.3)$$

Since $Q_y(D) \leq 1$, it is clear that

$$\sup_D \left| \int_X Q_y(D)\, P^m(x_0, dy) - \int_X Q_y(D)\, \pi(dy) \right| \leq \rho[P^m(x_0, \cdot), \pi(\cdot)],$$

where ρ is the total variation metric. If in (3.5.3) we take D to be of the form $B \times X \times X \times \ldots$ where $B \in S$, it follows that the left side is in fact no smaller than the right side. Therefore we finally have

$$\beta(m) = \int_X \rho[P^m(x_0, \cdot), \pi(\cdot)]\, \pi(dx_0).$$

Now, since the Markov chain is V-geometrically ergodic, it follows that

$$\rho[P^m(x_0, \cdot), \pi(\cdot)] \leq V(x_0) \mu \lambda^m$$

for some function $V : X \to [1, \infty)$ such that $E[V, \pi] < \infty$, and some constants μ and $\lambda < 1$. Consequently

$$\beta(m) \leq \mu E[V, \pi] \lambda^m, \; \forall m,$$

which shows that $\{X_t\}$ is geometrically β-mixing. ■

3.5.2 Beta-Mixing Properties of Markov Sequences

In this subsection, it is shown that a very significant class of Markov chains produces output sequences that are β-mixing. Consequently, the results of the preceding subsections have wide applicability.

In order to state the main theorem of this subsection, we introduce a little notation. In the remainder of this subsection, we shall be dealing with vector-valued random variables. Suppose $\mathbf{v}_t(\omega)$ is a random vector defined on the probability space (Ω, Σ, Q), assuming values in $\mathbb{R}^k$. Then it is necessary to make a distinction between the vector norm $\| \mathbf{v}_t(\omega) \|_2$, which is the Euclidean norm on $\mathbb{R}^k$, and the function space norm $\| \mathbf{v}_t(\cdot) \|_2$, which is a norm on $L_2(\Omega)$. To avoid confusion, in the remainder of this subsection we will use $|\cdot|$ to denote the Euclidean norm of a vector, and $\| \cdot \|_2$ to denote the L_2-norm on (Ω, Σ, Q). For more details on Markov chains, see [135], and for more details on stability theory, see [199].

We consider Markov chains described by the recursion relation

$$x_{t+1} = f(x_t) + e_t, \qquad (3.5.4)$$

where $x_t \in \mathbb{R}^k$ for some integer k, subject to the following assumptions:

1. The function f is globally Lipschitz continuous. Thus there exists a constant L such that

$$|f(x) - f(y)| \leq L(|x - y|), \ \forall x, y \in \mathbb{R}^k,$$

where $|\cdot|$ denotes the Euclidean norm.
2. When $e_t = 0 \ \forall t$, the "unforced" system

$$x_{t+1} = f(x_t)$$

has a globally exponentially stable equilibrium at $x = 0$. This means that there exist constants M and $\lambda < 1$ such that

$$|x_t| \leq M|x_0|\lambda^t, \ \forall t \geq 1, \ \forall x_0.$$

More generally, define the function $f^n : \mathbb{R}^k \to \mathbb{R}^k$ recursively by

$$f^1(x) := f(x), \ f^n(x) := f[f^{n-1}(x)]. \qquad (3.5.5)$$

Then it is assumed that f^n is globally Lipschitz continuous with the Lipschitz constant $M\lambda^n$ for some constants M and $\lambda < 1$. Therefore

$$|f^n(x) - f^n(y)| \leq M\lambda^n |x - y|, \ \forall x, y \in \mathbb{R}^k.$$

3. The noise sequence $\{e_t\}$ is i.i.d., is assumed to be defined on a probability space (Ω, Σ, Q), has finite variance, and has a continuous density function that is everywhere positive.

Since the unforced system is globally exponentially stable, it is possible to define the **Lyapunov function** $V : \mathbb{R}^k \to \mathbb{R}_+$ as follows:[13]

$$V(x) := 1 + \sum_{i=0}^{\infty} |f^i(x)|^2. \qquad (3.5.6)$$

Lemma 3.5. *The function V satisfies the following properties:*

---

[13] Traditionally, in stability theory, the constant 1 in front of the summation is absent, since in stability theory the Lyapunov function V is required to satisfy $V(0) = 0$. However, in the present application, the stochastic Lyapunov function V is required to bounded *below* by 1. This is the reason for introducing this constant.

3.5 Markov Chains and Beta-Mixing

1. $V[f(x)] - V(x) = -|x|^2$, $\forall x$.
2. $V(x) \geq 1 + |x|^2$, $\forall x$.
3. We have
$$V(x) \leq 1 + \frac{M^2}{1-\lambda^2}|x|^2.$$
4. We have
$$V(x+y) - V(x) \leq \frac{2M^2}{1-\lambda^2}|x|\cdot|y| + \frac{M^2}{1-\lambda^2}|y|^2.$$

Remark: In other words, the Lyapunov function V behaves essentially like a quadratic function of the form $1 + x'Px$.

Proof. All of the above inequalities are direct consequences of the properties of the iterated map f^n.

To prove (1), note that

$$V[f(x)] = 1 + \sum_{i=0}^{\infty}|f^n[f(x)]|^2 = 1 + \sum_{i=0}^{\infty}|f^{n+1}(x)|^2 = V(x) - |x|^2.$$

The relation (2) is obvious. To prove (3), note that since $f^n(0) = 0$, the Lipschitz condition (3.5.5) implies that

$$|f^n(x)|^2 \leq M^2\lambda^{2n}|x|^2.$$

Thus by the triangle inequality

$$V(x) \leq 1 + \sum_{i=0}^{\infty} M^2\lambda^{2n}|x|^2.$$

This is the same as property (3).

Finally, to prove property (4), note the elementary identity

$$|a|^2 - |b|^2 = (a+b)'(a-b) \leq |a+b|\cdot|a-b|$$

by Schwartz' inequality. Consequently

$$\begin{aligned}|f^n(x+y)|^2 - |f^n(x)|^2 &\leq |f^n(x+y) + f^n(x)|\cdot|f^n(x+y) - f^n(x)| \\ &\leq M\lambda^n[|x+y| + |x|]\cdot M\lambda^n|y| \\ &\leq M^2\lambda^{2n}(2|x| + |y|)\cdot|y|.\end{aligned}$$

Finally this leads to

$$V(x+y) - V(x) \leq \sum_{i=0}^{\infty} M^2\lambda^{2n}(2|x|+|y|)\cdot|y|,$$

from which the desired inequality follows.

3. Problem Formulations

Now we are ready to state the main result.

Theorem 3.11. *Suppose the above assumptions are satisfied. Then the sequence $\{x_t\}$ is geometrically β-mixing.*

Proof. The proof is based on a result found in [135], p. 354, which provides a sufficient condition for the process $\{x_t\}$ to be V-geometrically ergodic, which in turn implies geometric β-mixing. Recall ([135], p. 174) that the "drift" of the stochastic Lyapunov function V for the system (3.5.4) is defined as

$$\Delta V(x) = \int_X V(y)\, P_x(dy) - V(x) = \int_X [V(y) - V(x)] P_x(dy).$$

The fact that the noise term e_t enters additively into the dynamics of the Markov chain and the fact that e_t has a continuous density which is everywhere positive implies two things: First, the Markov chain is aperiodic and ψ-irreducible, and second, every compact set is "petite" in the sense of [135]. Thus, by the above-cited theorem, it follows that the Markov chain is V-geometrically ergodic (with the function V defined above) if there exists a constant ν such that

$$\Delta V(x) \leq -\gamma V(x) + I_{B(\nu)}\ \forall x, \qquad (3.5.7)$$

where $B(\nu)$ denotes the closed ball of radius ν centered at the origin, and $I.$ denotes the indicator function. Thus the proof consists of obtaining an upper bound for $\Delta V(x)$ and showing that (3.5.7) is satisfied. This will establish V-geometric ergodicity of the Markov chain, which in turn implies β-mixing by Theorem 3.10.

For notational convenience, define

$$\mu := \frac{M^2}{1 - \lambda^2},$$

and note that

$$V(x+y) - V(x) \leq 2\mu |x||y| + \mu |y|^2$$

by Property (4) of Lemma 3.5. Now let $y = f(x) + e_t$, where e_t is the noise term. Computing directly, we have

$$\begin{aligned} V(y) - V(x) &= V[f(x) + e_t] - V(x) \\ &= V[f(x)] - V(x) + (V[f(x) + e_t] - V[f(x)]) \\ &\leq -|x|^2 + 2\mu |f(x)| \cdot |e_t| + \mu |e_t|^2 \\ &\leq -|x|^2 + 2\mu L |x| \cdot |e_t| + \mu |e_t|^2. \end{aligned}$$

Thus

$$\begin{aligned}
\Delta V(x) &= \int_X [V(y) - V(x)]\, P_x(dy) \\
&= E[V(f(x) + e_t) - V(x)] \\
&\leq -|x|^2 + 2\mu L|x|\, \|e\|_1 + \mu\, \|e\|_2^2 \\
&\leq -|x|^2 + 2\mu L|x|(\|e\|_2^2)^{1/2} + \mu\, \|e\|_2^2,
\end{aligned}$$

where E refers to the expectation with respect to the laws of e_t. Now, since e_t has finite variance, it follows that $\|e\|_2 < \infty$. Let

$$a := 2\mu L(\|e\|_2^2)^{1/2}, \quad b := \mu\, \|e\|_2^2.$$

Then

$$\Delta V(x) \leq -|x|^2 + a|x| + b.$$

Suppose

$$|x| \geq \nu := \max\{4a, 2b^{1/2}\}.$$

Then

$$a|x| \leq |x|^2/4, \quad b \leq |x|^2/4,$$

and finally

$$\Delta V(x) \leq -|x|^2/2.$$

The rest is pure algebra. Define

$$\theta := \frac{M^2}{M^2 + \nu^2(1-\lambda^2)} < 1.$$

Then, whenever $|x| \geq \nu$, we can write

$$\Delta V(x) \leq -|x|^2/2 \leq -\frac{1}{2}[\theta \nu^2 + (1-\theta)|x|^2].$$

However, the way in which θ is chosen guarantees that

$$\frac{\theta}{1-\theta} = \frac{M^2}{\nu^2(1-\lambda^2)}.$$

Therefore

$$\Delta V(x) \leq -\frac{\theta \nu^2}{2}\left(1 + \frac{M^2}{1-\lambda^2}|x|^2\right) \leq -\frac{\theta \nu^2}{2} V(x).$$

Hence (3.5.7) is satisfied with $\gamma = \theta \nu^2/2$ and ν as above. ■

3.5.3 Mixing Properties of Hidden Markov Models

In some situations, it is not possible to observe the state X_t of the Markov process directly. Rather, one observes only some measured value Y_t, which is randomly generated according to a probability distribution which is itself a function of the current true state X_t. Such a model is called a "hidden Markov model." It is therefore of interest to ascertain what kind of mixing properties, if any, are possessed by the output sequences of hidden Markov models.

The main result of this subsection is Theorem 3.12, which shows that if the underlying Markov chain has a particular mixing property, then so does a corresponding hidden Markov model. Two preliminary results are first proved, so as to facilitate the proof of this main theorem.

Lemma 3.6. *Suppose a real-valued stochastic process $\{X_t\}$ is α-, β-, or ϕ-mixing, and that $Y_t = f(X_t)$ where $f : X \to \mathbb{R}$. Then $\{Y_t\}$ is also α-, β-, or ϕ-mixing, as appropriate.*

Proof. Note that mixing is really a property of the σ-algebras generated by the stochastic process. Since Y_t is a measurable function of X_t, we see that the σ-algebra generated by any collection of the Y_t is a subset of (and perhaps equal to) the σ-algebra generated by the corresponding collection of X_t. Hence the Y_t stochastic process inherits the mixing properties of the $\{X_t\}$ sequence. ∎

Lemma 3.7. *Suppose $\{X_t\}$ is β-mixing, and that $\{U_t\}$ is i.i.d. and also independent of $\{X_t\}$. Suppose $Y_t = f(X_t, U_t)$, where f is a fixed measurable function. Then $\{Y_t\}$ is also β-mixing.*

Proof. Note that under the hypotheses, it follows that the joint process $\{(X_t, U_t)\}$ is β-mixing. Now the desired conclusion follows from Lemma 3.6. ∎

Similar results apply to both α- and ϕ-mixing as well.
Now we come to the main result of this subsection.

Theorem 3.12. *Suppose $\{X_t\}_{t \geq 0}$ is a stationary Markovian stochastic process assuming values in a set X with associated σ-algebra $\mathcal{S}$. Suppose Y is a complete separable metric space, and let $\mathcal{B}(Y)$ denote the Borel σ-algebra on Y. Suppose $\mu : X \times \mathcal{B}(Y) \to [0,1]$ is a transition probability function. Thus for each $x \in X$, $\mu(x, \cdot)$ is a probability measure on Y, and for each $A \in \mathcal{B}(Y)$, $\mu(\cdot, A)$ is a measurable function on $(X, \mathcal{S})$. Finally, suppose $\{Y_t\}_{t \geq 0}$ is a Y-valued stochastic process such that*

$$\Pr\{y_t \in A | Y_i, i \leq t-1, X_j, j \leq t\} = \mu(X_t, A).$$

Under these assumptions, if $\{X_t\}$ is β-mixing, so is $\{Y_t\}$.

Proof. The theorem is proved by constructing a representation of Y_t as a deterministic function of X_t and another random variable U_t that is i.i.d. and also independent of X_t. The conclusion then follows from Lemma 3.7. Specifically, it is shown that there exists a measurable mapping $\psi : X \times [0,1] \to Y$ such that the process $\{Z_t\}_{t \geq 0}$ defined by

$$Z_t = \psi(X_t, U_t)$$

has the same distribution as $\{Y_t\}$, where $\{U_t\}_{t \geq 0}$ is a sequence of i.i.d. random variables whose common distribution is the uniform distribution on $[0,1]$.

Recall (see e.g., [159]) that if Y is a complete separable metric space, then there exists a Borel subset E of $[0,1]$ and a one-to-one onto mapping ϕ from Y into E such that both ϕ and ϕ^{-1} are measurable. With ϕ as above, define the transition function $\nu : X \times \mathcal{B}(E) \to [0,1]$ as follows:

$$\nu(x, B) := \mu(x, \phi^{-1}(B)), \ \forall x \in X, B \in \mathcal{B}(E).$$

Here $\mathcal{B}(E)$ denotes the σ-algebra of Borel subsets of E. Now define the map $\psi_0 : X \times [0,1] \to [0,1]$ as follows:

$$\psi_0(x, s) := \lim_{m \to \infty} \frac{1}{2^m} \inf\{k \geq 0 : \nu(x, (-\infty, k/2^m)) \geq s\}.$$

It readily follows from the above definition that the function ψ_0 is jointly measurable. Moreover, it is easy to see that

$$\psi_0(x, s) = \inf\{u \geq 0 : \nu(x, (-\infty, u)) \geq s\}.$$

However, the above equation is not used as a definition of ψ_0 since it involves an infimum over an uncountable set, and it is therefore not clear that the resulting function is jointly measurable.

From the above equation it can be seen that

$$\psi_0(x,s) \leq u \text{ if and only if } \nu(x, (-\infty, u)) \geq s.$$

Hence, if λ denotes the Lebesgue measure on $[0,1]$, it follows that

$$\lambda\{s : \psi_0(x,s) \leq u\} = \nu(x, (-\infty, u)).$$

Now define

$$\psi(x,s) := \phi^{-1}(\psi_0(x,s)).$$

Then for each $A \in \mathcal{B}(Y)$ we have

$$\lambda\{s : \psi(x,s) \in A\} = \mu(x, A).$$

Therefore, if $\{U_t\}_{t \geq 0}$ is a sequence of i.i.d. random variables whose common distribution is the uniform distribution on $[0,1]$, then the process $\{Z_t\}_{t \geq 0}$ defined by

$$Z_t = \psi(X_t, U_t)$$

has the same distribution as $\{Y_t\}$.

Finally, by Lemma 3.7, if $\{X_t\}$ is β-mixing, so is $\{Y_t\}$. ∎

112   3. Problem Formulations

Similar results can be proven for α- and ϕ-mixing. Moreover, notice that we have really not used the Markovian property of the process $\{X_t\}$. However, in the absence of a result like Theorem 3.11 that guarantees the mixing properties of the sequence $\{X_t\}$, Theorem 3.12 might not be very useful by itself.

Notes and References The problem of the uniform convergence of empirical probabilities of a collection of sets to their true values has its origin in attempts to generalize the classical Glivenko-Cantelli lemma. Very early on, this lemma was extended from the case where $X = \mathbb{R}$ to the case where $X = \mathbb{R}^k$ for some integer k and the collection of sets $\mathcal{A}$ is of the form $\prod_{i=1}^{k}(-\infty, t_i]$, where $t_i \in \mathbb{R}$ for each i. Apparently the first person to study this problem in a more general setting is Ranga Rao [164]. He studied this problem in the case where $\mathcal{A}$ consists of convex sets, and derived some *topological* conditions for the problem to have a solution. It is interesting to note that subsequent approaches to the problem have been based mostly on *combinatorial* conditions rather than topological conditions; see, e.g., Chapters 4 and 5. The related problem of the uniform convergence of empirical *means* (as opposed to empirical probabilities) is a very natural generalization. Seminal contributions to the solution of both problems have been made by Vapnik and Chervonenkis [193], [194], [196], [190]. These contributions are discussed in Chapter 5. The problem of the uniform convergence of empirical means *uniformly in probability* is not explicitly stated in Vapnik's book, but is implicit in the literature.

In this book, attention is restricted to what might be called the "two-sided convergence" of empirical means to their true values. Recall the definition of the quantity $q(m, \epsilon, P)$, namely:

$$q(m, \epsilon, P) := P^m\{\mathbf{x} \in X^m : \exists f \in \mathcal{F} \text{ s.t. } |\hat{E}(f; \mathbf{x}) - E_P(f)| > \epsilon\}.$$

We have defined a family of functions $\mathcal{F}$ as having the UCEM property if $q(m, \epsilon, P) \to 0$ as $m \to \infty$. However, a perusal of the proof of Theorem 3.2 shows that weaker form of convergence is good enough to apply that theorem. Specifically, if J denotes the actual risk function and $\hat{J}$ denotes the empirical risk, then it is enough if the *one-sided quantity*

$$\bar{P}^m\{\mathbf{z} \in Z^m : \sup_{h \in \mathcal{H}} \hat{J}(h; \mathbf{z}) - J(h; \bar{P}) > \epsilon/4\}$$

approaches zero as $m \to \infty$. In other words, we do not care if $\hat{J}(h; \mathbf{z})$ *under-estimates* $J(h; \bar{P})$. This suggests that we should study a one-sided version of $q(m, \epsilon, P)$, namely

$$q_u(m, \epsilon) := P^m\{\mathbf{x} \in X^m : \exists f \in \mathcal{F} \text{ s.t. } \hat{E}(f; \mathbf{x}) - E_P(f) > \epsilon\}.$$

Necessary and sufficient conditions for this one-sided quantity to converge to zero as $m \to \infty$ for each $\epsilon > 0$ are given in [197].

The definition of PAC learnability as given here is essentially the same as that given by Valiant [187], with one important difference. Valiant defines learnability in the context of Boolean formulae, in which case the set X has cardinality 2^n for some integer n. Thus learnability in the sense defined here is not an issue, because (as shown in Chapter 7), on a finite set *every* concept class is learnable. Valiant adds one more crucial requirement, namely that the resources used by the learning algorithm, both in terms of computational time and storage, must be *polynomial in* n. The imposition of this additional requirement is what distinguishes *computational* learning theory from *statistical* learning theory. Subsequent researchers, in extending the Valiant formulation to infinite sets, have not always insisted on the latter requirement. Thus the definition given here has its genesis in Valiant's work, but is in reality the outcome of further refinement. The terminology "probably approximately correct" learning seems to be due to Angluin [4]. It should be pointed out that the paper [1], in discussing the number of training steps required by a perceptron, comes very close to the "modern" formulation of PAC learning. The notions of probably *uniformly* approximately correct (PUAC) learning, and almost surely eventually correct (ASEC) learning appear to be new. The formulation of the problem of model-free learning (also referred to by other researchers as "agnostic" learning) is taken from Haussler [80].

Though the classical definition of PAC learning assumes independent input sequences, the generalization to dependent inputs is natural. Several authors have attempted to extend the known results to the more general case. See for example [71, 134, 143].

The results in Sections 3.4 and 3.5 are new or quite recent. Theorem 3.3 is proved in [153]; however, the proof given here is different. Theorem 3.4 giving explicit estimates of the rates of convergence of empirical means is taken from [93]. Theorem 3.5 showing that a law of large numbers holds with α-mixing inputs is new. In [138] a similar result is proved under the more restrictive assumption that the input sequence is *geometrically* α-mixing, that is, the coefficient $\alpha(k)$ approaches zero at a geometrical rate. The results of Theorems 3.6 and 3.7 are presented here for the first time. The results of Theorems 3.8 and 3.9 are taken from [204]. However, the present Theorem 3.8 is more general than the corresponding result in [204]. Finally, the contents of Section 3.5 are presented here for the first time.

4. Vapnik-Chervonenkis, Pseudo- and Fat-Shattering Dimensions

In this chapter, we introduce three distinct notions of "dimension" that play an important role in the subsequent development. The phrase "dimension" is rather unfortunate, as the three "dimensions" have nothing at all to do with the dimension of a vector space, except in very special situations. Rather, these "dimensions" are combinatorial parameters that measure the "richness" of concept classes or function classes. The Vapnik-Chervonenkis dimension, often referred to as the VC-dimension, is historically the first dimension to be introduced into the subject, and is defined for concept classes, or equivalently, binary-valued functions. The Pseudo-dimension, also referred to by some authors as the Pollard dimension, is a generalization of the VC-dimension to real-valued functions. The fat-shattering dimension, unlike the Pseudo-dimension, is a "scale-sensitive" measure of richness. All three of these dimensions are used in deriving conditions for the uniform convergence of empirical means and for PAC learnability.

In the first section, the definitions of these dimensions are given. The next section contains some useful inequalities in respect of some so-called growth functions. Finally, in the last section, it is shown how to bound the VC-dimension of collections of sets obtained by performing Boolean operations on other collections of sets. These inequalities enable us to derive simple upper bounds for the VC-dimension of complicated collections of sets.

4.1 Definitions

4.1.1 The Vapnik-Chervonenkis Dimension

Definition 4.1. *Let $(X, \mathcal{S})$ be a given measurable space, and let $\mathcal{A} \subseteq \mathcal{S}$. A set $S = \{x_1, \ldots, x_n\} \subseteq X$ is said to be* **shattered** *by $\mathcal{A}$ if, for every subset $B \subseteq S$, there exists a set $A \in \mathcal{A}$ such that $S \cap A = B$. The* **Vapnik-Chervonenkis dimension** *of $\mathcal{A}$, denoted by VC-dim($\mathcal{A}$), equals the largest integer n such that there exists a set of cardinality n that is shattered by $\mathcal{A}$.*

By identifying a set with its indicator function, it is easy to see that the notion of VC-dimension can also be defined for a collection $\mathcal{F}$ of *binary-valued* functions on X. Where convenient, we shall switch back and forth between the two interpretations of the VC-dimension (that is, as a property of collections of sets, and as a property of families of binary-valued functions).

Note that if a set S has n elements, then there are exactly 2^n distinct subsets of S, including the empty set and S itself. Also, if B_1 and B_2 are distinct subsets of S, then any sets $A_1, A_2 \in \mathcal{A}$ that satisfy

$$S \cap A_1 = B_1, \ S \cap A_2 = B_2$$

must themselves be distinct. Hence, in order to shatter a set of cardinality n, the collection $\mathcal{A}$ must contain at least 2^n distinct sets. Conversely, if the collection $\mathcal{A}$ is finite, then any set that is shattered by $\mathcal{A}$ can contain at most $\lfloor \lg |\mathcal{A}| \rfloor$ elements. This shows that every *finite* collection $\mathcal{A}$ has finite VC-dimension, and that

$$\text{VC-dim}(\mathcal{A}) \leq \lfloor \lg |\mathcal{A}| \rfloor \text{ if } |\mathcal{A}| < \infty.$$

On the other hand, if $\mathcal{A}$ is an *infinite* collection of sets, then the VC-dimension of $\mathcal{A}$ could be either finite or infinite.

One can think of the VC-dimension of $\mathcal{A}$ as a measure of the "richness" of the collection $\mathcal{A}$. A set S is "shattered" by $\mathcal{A}$ if $\mathcal{A}$ is rich enough to distinguish between all possible subsets of S. Note that S is shattered by $\mathcal{A}$ if (and only if) one can "pick off" every possible subset of S by intersecting S with an appropriately chosen set $A \in \mathcal{A}$. In this respect, perhaps "completely distinguished" or "completely discriminated" would be a better term than "shattered." However, the latter term has by now become standard in the literature. Note that, if $\mathcal{A}$ has finite VC-dimension, say d, then it is not rich enough to distinguish all subsets of *any* set containing $d + 1$ elements or more; but it is rich enough to distinguish all subsets of *some* set containing d elements (but not necessarily *all* sets of cardinality d).

The abstract definition above is illustrated through several examples.

Example 4.1. Let $X = \mathbb{R}$, $\mathcal{S} =$ the Borel σ-algebra on X, and let $\mathcal{A}$ denote the collection of semi-infinite intervals of the form $(-\infty, t]$ as t varies over $\mathbb{R}$. It is shown that the VC-dimension of $\mathcal{A}$ equals one. In order to show this, it is necessary to establish two things: (i) There exists a set of cardinality one that is shattered by $\mathcal{A}$, and (ii) *No* set of cardinality two is shattered by $\mathcal{A}$. To prove the first statement, let $S = \{a\}$, where $a \in \mathbb{R}$. Then there are two subsets of S, namely the empty set and S itself. Choose $A_1 = (-\infty, t_1], A_2 = (-\infty, t_2]$ such that $t_1 < a$ and $t_2 \geq a$. Then clearly

$$S \cap A_1 = \emptyset, \ S \cap A_2 = S.$$

Hence S is shattered by $\mathcal{A}$.[1] To prove the second statement, suppose $S = \{a, b\}$, and suppose without loss of generality that $a < b$. Now let $B = \{b\} \subseteq$

---

[1] Actually, what has been shown is that *every* set of cardinality one is shattered by $\mathcal{A}$; however, there is no extra advantage gained by this.

S. Then there does not exist any set $A \in \mathcal{A}$ such that $S \cap A = B$. This is because every $A \in \mathcal{A}$ is of the form $(-\infty, t]$ for some $t \in \mathbb{R}$, and if A contains b, then perforce A also contains $a < b$. Hence S is not shattered by $\mathcal{A}$. Since S is arbitrary, it follows that VC-dim$(\mathcal{A}) = 1$.

Example 4.2. Along the same lines as the preceding example, let $\mathcal{A}$ denote the collection of all closed intervals of the form $[\alpha, \beta]$ where $\alpha, \beta \in \mathbb{R}$. Then VC-dim$(\mathcal{A}) = 2$. To show that *no* set of cardinality three is shattered by $\mathcal{A}$, let $S = \{a, b, c\}$, and suppose without loss of generality that $a < b < c$. Let $B = \{a, c\} \subseteq S$. Then there does not exist any set $A \in \mathcal{A}$ such that $S \cap A = B$. This is because any closed interval that contains both a and c perforce contains b as well. The remainder of the details are left to the reader.

Example 4.3. ([210]) Let $X = \mathbb{R}^2$, $\mathcal{S}$ = the Borel σ-algebra on X, and let $\mathcal{A}$ denote the collection of closed half-planes in X, i.e., sets of the form

$$\{\mathbf{x} \in \mathbb{R}^2 : w_1 x_1 + w_2 x_2 - \theta \geq 0\}$$

where $(w_1, w_2) \neq (0, 0)$. One can also think of $\mathcal{A}$ as the collection of all input sets that are mapped into the output 1 by some perceptron (see Example 3.6). It is claimed that the VC-dimension of $\mathcal{A}$ equals three. First, let $\mathbf{a}, \mathbf{b}, \mathbf{c}$ denote any set of three *non-collinear* points in $\mathbb{R}^2$ (see Figure 4.1). Then it

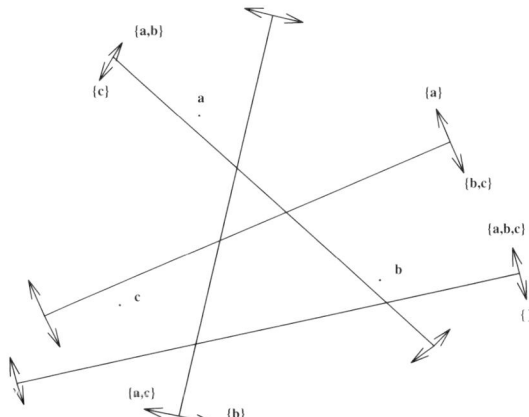

Fig. 4.1. Shattering Three Non-collinear Points

is possible to select $2^3 = 8$ closed half-planes that "pick off" each of the eight possible subsets of $\{\mathbf{a}, \mathbf{b}, \mathbf{c}\}$ (again, see Figure 4.1). Next, let $S = \{\mathbf{a}, \mathbf{b}, \mathbf{c}, \mathbf{d}\}$ be arbitrary; it is shown that S is *not* shattered by $\mathcal{A}$. There are two cases to consider, namely: (i) One of the four points, say $\mathbf{d}$, belongs to the convex hull of the remaining three points (see Figure 4.2a), and (ii) None of the four points belongs to the convex hull of the remaining three points (see Figure 4.2b). In the first case, let $B = \{\mathbf{a}, \mathbf{b}, \mathbf{c}\}$. Then there does not exist

Fig. 4.2. No Set of Four Points can be Shattered

any set $A \in \mathcal{A}$ such that $S \cap A = B$. The reason is that every $A \in \mathcal{A}$ is a convex set; consequently, if $A \in \mathcal{A}$ contains $\mathbf{a}, \mathbf{b}, \mathbf{c}$, then perforce A contains $\mathbf{d}$ as well. In the second case, let B consist of two "opposite" points, e.g., $B = \{\mathbf{a}, \mathbf{c}\}$ or $B = \{\mathbf{b}, \mathbf{d}\}$ in Figure 4.2b. In this case, the well-known XOR counterexample shows that B cannot be separated from its complement (in S) by any straight line. Thus there does not exist any set $A \in \mathcal{A}$ such that $S \cap A = B$. In either case, S is not shattered by $\mathcal{A}$. Since S is arbitrary, it follows that VC-dim$(\mathcal{A}) = 3$.

More generally, it can be shown that the collection of closed half-spaces in $\mathbb{R}^l$ has VC-dimension $l + 1$. For complete details, see [210].

Example 4.4. ([32]) Let $X = \mathbb{R}^2$, $\mathcal{S} = $ the Borel σ-algebra on X, and let $\mathcal{A}$ denote the collection of "axis-parallel" rectangles in X of the form $[x_l, x_u] \times [y_l, y_u]$, as first introduced in Example 3.5. It is shown that the VC-dimension of $\mathcal{A}$ equals four.

Given any finite set $S \subseteq X$ and any point $\mathbf{a} \in S$, let us say that $\mathbf{a}$ is the "left-most" point in S if
$$a_1 = \min_{\mathbf{b} \in S} b_1,$$
i.e., the first coordinate of $\mathbf{a}$ is the smallest among the first coordinates of all points in S. Similarly, let us say that $\mathbf{a}$ is the "unique leftmost" point in S if
$$a_1 < b_1 \; \forall \mathbf{b} \in S, \; \mathbf{b} \neq \mathbf{a}.$$

The phrases "(unique) rightmost," "(unique) topmost" and "(unique) bottommost" are defined analogously. Finally, let us say that $\mathbf{a}$ is a "(unique) sidemost" point of S if it has one of the above four properties.

First we construct a set S of cardinality four that is shattered by $\mathcal{A}$. Select $S = \{\mathbf{a}, \mathbf{b}, \mathbf{c}, \mathbf{d}\}$ such that $\mathbf{a}$ is the unique leftmost point of S, $\mathbf{b}$ is the unique topmost, $\mathbf{c}$ is the unique rightmost, and $\mathbf{d}$ is the unique bottommost point (see Figure 4.3). To show that S is shattered by $\mathcal{A}$, let B be any nonempty subset of S, and define A to be the smallest rectangle in $\mathcal{A}$ that contains B. In other words,
$$A = [x_l, x_u] \times [y_l, y_u], \text{ where}$$

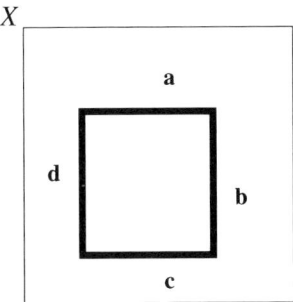

Fig. 4.3. Shattering Four Points by Axis-Parallel Rectangles

$$x_l = \min_{\mathbf{z} \in B} z_1, \quad x_u = \max_{\mathbf{z} \in B} z_1, \quad y_l = \min_{\mathbf{z} \in B} z_2, \quad y_u = \max_{\mathbf{z} \in B} z_2.$$

Then clearly $S \cap A$ *contains* B. To show that $S \cap A$ in fact *equals* A, it is enough to show that no point in $S - B$ belongs to $S \cap A$. Accordingly, suppose $\mathbf{z} \in S - B$. Then, by the assumption that every point in S is a unique sidemost point, it follows that either

$$z_1 < x_l, \text{ or } z_1 > x_u, \text{ or } z_2 < y_l, \text{ or } z_2 > y_u.$$

In any case, $\mathbf{z} \notin S \cap A$. Hence, for A constructed as above, we have that $S \cap A = B$. Such an A can be constructed for every nonempty subset B of S. Finally, it is easy to construct an $A \in \mathcal{A}$ such that $S \cap A = \emptyset$. Hence S is shattered by $\mathcal{A}$. This shows that VC-dim$(\mathcal{A}) \geq 4$.

To show that VC-dim$(\mathcal{A})$ in fact *equals* four, let $S = \{\mathbf{a}, \mathbf{b}, \mathbf{c}, \mathbf{d}, \mathbf{e}\}$ be arbitrary. Then not all of the five points of S can be *unique* sidemost points of S. Suppose without loss of generality that $\mathbf{e}$ is not a unique sidemost point. Then it is easy to see that any axis-parallel rectangle that contains $\mathbf{a}, \mathbf{b}, \mathbf{c}, \mathbf{d}$ must perforce contain $\mathbf{e}$ as well. Thus, if we define $B = \{\mathbf{a}, \mathbf{b}, \mathbf{c}, \mathbf{d}\}$, then there does not exist a set $A \in \mathcal{A}$ that satisfies $S \cap A = B$. Hence S is not shattered by $\mathcal{A}$. Since S is arbitrary, it follows that VC-dim$(\mathcal{A}) = 4$.

Similar arguments apply in a higher-dimensional space as well. Let $X = \mathbb{R}^n$ and let $\mathcal{A}$ equal the collection of axis-parallel hypercubes in X. Then VC-dim$(\mathcal{A}) = 2n$.

Example 4.5. Let $X = [0,1]^2$, $\mathcal{S} =$ the Borel σ-algebra on X, and let $\mathcal{A}$ equal the collection of convex polygons in X. It is shown that $\mathcal{A}$ has *infinite* VC-dimension.

Let S be the boundary of any strictly convex set in X, e.g., a circle (see Figure 4.4). Then no point in S can be expressed as a convex combination of any other set of points in S. Now let $B = \{x_1, \ldots, x_n\} \subseteq S$ be arbitrary. Define $A \in \mathcal{A}$ to be the convex hull of B. Then $S \cap A = B$. Since this can be done for *every* finite subset B in S, it follows that S is shattered by $\mathcal{A}$. Hence $\mathcal{A}$ has infinite VC-dimension.

120   4. Vapnik-Chervonenkis, Pseudo- and Fat-Shattering Dimensions

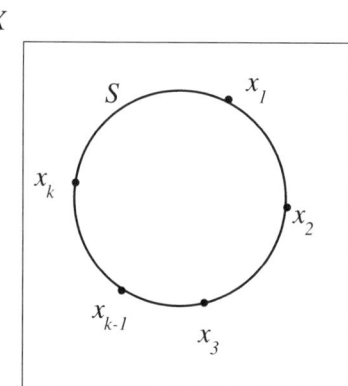

Fig. 4.4. Shattering an Infinite Set

4.1.2 The Pseudo-Dimension

The pseudo-dimension, also referred to by some authors as the Pollard dimension, is a generalization of the VC-dimension to function classes. To define the pseudo-dimension, it is convenient first to introduce the so-called "step function" $\eta(\cdot) : \mathbb{R} \to \mathbb{R}$ defined by

$$\eta(z) = \begin{cases} 1, & \text{if } z \geq 0, \\ 0, & \text{if } z < 0. \end{cases} \quad (4.1.1)$$

Some authors refer to the step function as the "Heaviside" function, after the British mathematician O. Heaviside.

Definition 4.2. *Let $(X, \mathcal{S})$ be a measurable space, and let $\mathcal{F} \subseteq [0, R]^X$ consist of measurable functions. A set $S = \{x_1, \ldots, x_n\} \subseteq X$ is said to be **P-shattered** by $\mathcal{F}$ if there exists a real vector $\mathbf{c} \in [0, R]^n$ such that, for every binary vector $\mathbf{e} \in \{0, 1\}^n$, there exists a corresponding function $f_\mathbf{e} \in \mathcal{F}$ such that*

$$f_\mathbf{e}(x_i) \geq c_i \text{ if } e_i = 1, \text{ and } f_\mathbf{e}(x_i) < c_i \text{ if } e_i = 0.$$

The above condition can be expressed equivalently as

$$\eta[f_\mathbf{e}(x_i) - c_i] = e_i, \ \forall i, \ \forall \mathbf{e}.$$

*In such a case, we say that $\mathbf{c}$ is a **witness** to the P-shattering. The **P-dimension** of $\mathcal{F}$, denoted by P-dim($\mathcal{F}$), is defined as the largest integer n such that there exists a set of cardinality n that is P-shattered by $\mathcal{F}$.*

The concept of P-shattering by a function class can be understood with reference to Figure 4.5. Fix a real vector $\mathbf{c} \in [0, R]^n$. At each point $x_i \in S$ and for each $f \in \mathcal{F}$, the graph of $f(x_i)$ can either pass *above* (or through) c_i, or else *below* c_i. Thus there are 2^n different possible behaviours as f varies

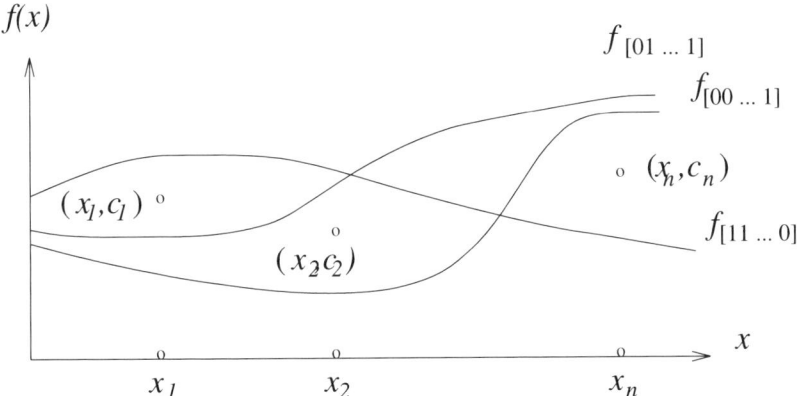

Fig. 4.5. Illustration of P-Shattering

over $\mathcal{F}$. Now the set $\{x_1, \ldots, x_n\}$ is said to be "P-shattered" by $\mathcal{F}$ if *each* of the 2^n possible behaviours is realized by some $f \in \mathcal{F}$.

It is easy to see that the P-dimension is a generalization of the VC-dimension, if the latter concept is recast in terms of binary-valued functions (as opposed to collections of sets). Suppose $F \subseteq \{0,1\}^X$ (not $[0,R]^X$!) is a family of measurable functions. Then there is a natural association between $\mathcal{F}$ and a collection $\mathcal{A} \subseteq \mathcal{S}$. Thus it is possible to speak of the VC-dimension of $\mathcal{F}$, using the earlier definition. Now a set $S = \{x_1, \ldots, x_n\} \subseteq X$ is shattered by $\mathcal{F}$ if, for each binary vector $\mathbf{e} \in \{0,1\}^n$, there exists a function $f \in \mathcal{F}$ such that $f(x_i) = e_i$ for $i = 1, \ldots, n$. Equivalently, a set $S = \{x_1, \ldots, x_n\} \subseteq X$ is shattered by $\mathcal{F}$ if all the 2^n possible maps from S into $\{0,1\}$ can be realized as restrictions to S of some function in $\mathcal{F}$. Thus, if $\mathcal{F}$ consists solely of binary-valued functions, then its VC-dimension is the same as its P-dimension. Now suppose $\mathcal{F} \subseteq [0,R]^X$ (not $\{0,1\}^X$!). A set $S = \{x_1, \ldots, x_n\} \subseteq X$ is P-shattered by $\mathcal{F}$ if there exists a vector $\mathbf{c} \in [0,R]^n$ such that, for every binary vector $\mathbf{e} \in \{0,1\}^n$, there exists a corresponding function $f \in \mathcal{F}$ such that

$$\eta[f(x_i) - c_i] = e_i, \ i = 1, \ldots, n.$$

Thus the only extra feature of the P-dimension is the possibility of introducing the "off-set" vector $\mathbf{c} \in [0,R]^n$. Equivalently, the set $S = \{x_1, \ldots, x_n\} \subseteq X$ is P-shattered by $\mathcal{F}$ if there exists a vector $\mathbf{c} \in [0,R]^n$ such that the class of *binary-valued* functions $x_i \mapsto \eta[f(x_i) - c_i]$ shatters $\{x_1, \ldots, x_n\} \subseteq X$. It is also easy to see that, for binary-valued functions, the VC-dimension and the P-dimension coincide.

The discussion above is captured in the following very useful lemma.

Lemma 4.1. *[126] Given a collection of functions $\mathcal{F}$ mapping X into $[0,R]$, define an associated collection of functions $\bar{\mathcal{F}}$ as follows: For each $f : X \to [0,R]$, define a corresponding $\bar{f} : X \times [0,R] \to \{0,1\}$ by*

4. Vapnik-Chervonenkis, Pseudo- and Fat-Shattering Dimensions

$$\bar{f}(x,c) = \eta[f(x) - c].$$

Let $\bar{\mathcal{F}} = \{\bar{f} : f \in \mathcal{F}\}$. Then

$$\text{P-dim}(\mathcal{F}) = \text{VC-dim}(\bar{\mathcal{F}}). \qquad (4.1.2)$$

Proof. Suppose a set $S = \{x_1, \ldots, x_n\} \subseteq X$ is P-shattered by $\mathcal{F}$. Then, by definition, there exists a vector $\mathbf{c} \in [0, R]^n$ such that the set

$$S' := \{(x_1, c_1), \ldots, (x_n, c_n)\}$$

is shattered by $\bar{\mathcal{F}}$. Thus P-dim($\mathcal{F}$) $\leq$ VC-dim($\bar{\mathcal{F}}$). To prove the opposite inequality, suppose a set $S' := \{\mathbf{z}_1, \ldots, \mathbf{z}_n\} \subseteq \mathbb{R}^k \times [0, R]$ is shattered by $\bar{\mathcal{F}}$, and partition each $\mathbf{z}_i$ as (x_i, c_i) where $x_i \in X$ and $c_i \in [0, R]$. Then it is clear that each of the x_i's must be distinct, and that the set $S = \{x_1, \ldots, x_n\} \subseteq \mathsf{X}$ is P-shattered by $\mathcal{F}$, with the vector $(c_1, \ldots, c_n)$ as the witness. Thus VC-dim($\mathcal{F}'$) $\leq$ P-dim($\mathcal{F}$). ■

4.1.3 The Fat-Shattering Dimension

In this subsection we introduce a notion called the fat-shattering dimension, which is usually referred to as a "scale-sensitive" version of the P-dimension introduced in the preceding subsection.

Definition 4.3. *Let $(X, \mathcal{S})$ be a measurable space, and let $\mathcal{F} \subseteq [0, R]^X$ consist of measurable functions. A set $S = \{x_1, \ldots, x_n\}$ is said to be* **fat-shattered to width** γ **with witness c** *if, for each binary vector $\mathbf{e} \in \{0, 1\}^n$, there exists a corresponding function $f_\mathbf{e} \in \mathcal{F}$ such that*

$$f_\mathbf{e}(x_i) \begin{cases} \geq c_i + \gamma & \text{if } e_i = 1, \\ \leq c_i - \gamma & \text{if } e_i = 0. \end{cases}$$

The **fat-shattering dimension of** $\mathcal{F}$ **to width** γ *is denoted by F-dim($\mathcal{F}, \gamma$), and is defined as the largest integer n such that there exists a set of cardinality n that is fat-shattered to width γ.*

The notion of fat-shattering can be compared and contrasted with that of P-shattering, with reference to Figure 4.5. In order for a set $S = \{x_1, \ldots, x_n\}$ to be merely P-shattered by the family $\mathcal{F}$, it is sufficient (and necessary) to be able to find functions $f_\mathbf{e}$ in $\mathcal{F}$ such that if $e_i = 1$ then the graph of $f_\mathbf{e}(x_i)$ passes above or through c_i, while if $e_i = 0$ the graph of $f_\mathbf{e}(x_i)$ passes below c_i. But in order for the set to be fat-shattered to width γ, the graph of $f_\mathbf{e}(x_i)$ must pass above $c_i + \gamma$ if $e_i = 1$, and below $c_i - \gamma$ if $e_i = 0$. Thus, in the case of P-shattering, the graph of $f_\mathbf{e}(x_i)$ is allowed to pass arbitrarily close to the value c_i (either above or below, as the case might be). But in the case of fat-shattering to width γ, the graph must steer clear of the value c_i by a width of γ in the appropriate direction.

It is clear that if a set S is fat-shattered by $\mathcal{F}$ to *any* finite width γ, then it is also P-shattered. Thus it is obvious that if $\mathcal{F}$ has finite P-dimension, then
$$\text{F-dim}(\mathcal{F}, \gamma) \leq \text{P-dim}(\mathcal{F}), \ \forall \gamma > 0.$$
It is also clear that
$$\gamma_1 < \gamma_2 \Rightarrow \text{F-dim}(\mathcal{F}, \gamma_1) \geq \text{F-dim}(\mathcal{F}, \gamma_2).$$
On the other hand, it is possible for a family $\mathcal{F}$ to have infinite P-dimension, and yet have finite fat-shattering dimension for each finite width γ. In such a case, it is clear that $\text{F-dim}(\mathcal{F}, \gamma) \to \infty$ as $\gamma \to 0$. Then the *rate* at which the quantity $\text{F-dim}(\mathcal{F}, \gamma)$ increases gives us valuable information about the rate at which empirical means converge to their true values. This connection will become clear in the sequel.

4.2 Bounds on Growth Functions

In this section, we prove several bounds that play a central role in subsequent proofs. These bounds depend on each of the three dimensions introduced in the previous section, namely, the VC-dimension, the pseudo-dimension, and the fat-shattering dimension.

4.2.1 Growth Functions of Collections of Sets

To motivate the bounds proved in this subsection, suppose $\mathcal{A}$ is a collection of sets with a finite VC-dimension, say d, and suppose that S is a set of cardinality $n > d$. By the definition of the VC-dimension, it follows that S is *not* shattered by $\mathcal{A}$. Hence, not every one of the 2^n subsets of S can be expressed in the form $S \cap A$ for some $A \in \mathcal{A}$. But let us ask: How many subsets of S can be expressed in this form? In this section, an explicit upper bound is given for this number as a function of n and d. It turns out that this bound is *polynomial* in n. Thus, as the cardinality of the set S increases, the fraction of subsets of S that can be written in the form $S \cap A$ for some $A \in \mathcal{A}$ approaches zero. It is this property that gives the VC-dimension its importance.

Now let us state the problem under study precisely. Let $(X, \mathcal{S})$ be a given measurable space, and suppose $\mathcal{A} \subseteq \mathcal{S}$ is a given collection of subsets of X. For a fixed finite set $S \subseteq X$, define $\pi(S; \mathcal{A})$ to be the number of distinct subsets of S of the form $S \cap A$ for some $A \in \mathcal{A}$. If S is shattered by $\mathcal{A}$, then of course $\pi(S; \mathcal{A}) = 2^{|S|}$, because *every* subset of S can be expressed in the form $S \cap A$ for some $A \in \mathcal{A}$. Otherwise, if S is *not* shattered by $\mathcal{A}$, then $\pi(S; \mathcal{A}) < 2^{|S|}$. Now define, for each integer $n \geq 1$,
$$\pi(n; \mathcal{A}) := \max_{|S|=n} \pi(S; \mathcal{A}).$$

124   4. Vapnik-Chervonenkis, Pseudo- and Fat-Shattering Dimensions

Some authors refer to $\pi(n;\mathcal{A})$ as the "growth function" of the collection of sets $\mathcal{A}$. If VC-dim$(\mathcal{A}) = n$, then $\pi(n;\mathcal{A}) = 2^n$ for every $n \leq d$. This follows from the fact that there exists a set S of cardinality d that is shattered by $\mathcal{A}$, whence every subset of S is also shattered by $\mathcal{A}$. Note that one could also define the VC-dimension of $\mathcal{A}$ as $n - 1$ where n is the smallest integer such that $\pi(n;\mathcal{A}) < 2^n$. But the object of interest here is the behaviour of the function $\pi(n;\mathcal{A})$ when $n > d$. This is the main result of the section, and is given next.

Theorem 4.1. ([194], [171], [32]) *Suppose $\mathcal{A}$ has finite VC-dimension, say d. Then*

$$\pi(n;\mathcal{A}) \leq 2\frac{n^d}{d!} \leq \left(\frac{en}{d}\right)^d, \; \forall n \geq d \geq 1. \qquad (4.2.1)$$

The proof of the theorem is given below. But the key point to note is that $\pi(n;\mathcal{A})$ increases no faster than a *polynomial* in n of degree d. Now suppose S is a set of cardinality n. Then the fraction of subsets of S that can be expressed in the form $S \cap A$ for some $A \in \mathcal{A}$ is no larger than $2^{-n}\pi(n;\mathcal{A})$ and this approaches zero as $n \to \infty$.

In order to prove the theorem, we introduce an auxiliary function $\phi(n,d)$ defined as follows: Given integers $d, n \geq 1$, define

$$\phi(n,d) := \sum_{i=0}^{d} \binom{n}{i} \text{ if } n > d, \; 2^n \text{ if } n \leq d, \qquad (4.2.2)$$

where

$$\binom{n}{i} = \frac{n!}{(n-i)! \, i!}$$

is the binomial coefficient. By adopting the convention that

$$\binom{n}{i} = 0 \text{ if } n < i,$$

we can simplify the definition of $\phi(n,d)$ to

$$\phi(n,d) := \sum_{i=0}^{d} \binom{n}{i}.$$

The function $\phi(n,d)$ has a simple intuitive interpretation. Suppose S is a set of cardinality n; then $\phi(n,d)$ is the number of subsets of S containing d elements or fewer.

Lemma 4.2. *Suppose $\mathcal{A}$ is a collection with finite VC-dimension, say d. Then*

$$\pi(n;\mathcal{A}) \leq \phi(n,d) \; \forall n \geq 1. \qquad (4.2.3)$$

Proof. If $n \leq d$, the above inequality holds (in fact, with equality) since both sides are equal to 2^n. Thus it can be supposed without loss of generality that $n > d$. In proving the lemma, it is useful to observe that $\phi(n,d)$ satisfies the recursion-type relationship

$$\phi(n,d) = \phi(n-1,d) + \phi(n-1,d-1). \tag{4.2.4}$$

To establish the above relationship, we begin with the simpler relationship

$$\binom{n}{i} = \binom{n-1}{i} + \binom{n-1}{i-1}, \tag{4.2.5}$$

which can be established readily by writing out both sides as fractions and then multiplying both sides by $(n-i)!\,i!$. Substituting (4.2.5) into the definition of $\phi(n,d)$ gives

$$\begin{aligned}
\phi(n,d) &= 1 + \sum_{i=1}^{d} \binom{n}{i} \\
&= 1 + \sum_{i=1}^{d} \left[\binom{n-1}{i} + \binom{n-1}{i-1} \right] \\
&= 1 + \sum_{i=1}^{d} \binom{n-1}{i} + \sum_{i=0}^{d-1} \binom{n-1}{i} \\
&= \phi(n-1,d) + \phi(n-1,d-1).
\end{aligned} \tag{4.2.6}$$

This establishes (4.2.4).

Now the proof of the lemma is given on the basis (4.2.4). Suppose S is fixed and that $|S| = n$. For the purposes of counting $\pi(S;\mathcal{A})$, one can identify two subsets A_1 and A_2 of X if $S \cap A_1 = S \cap A_2$. With this identification, $\mathcal{A}$ reduces to a *collection of subsets of S*, call it $\mathcal{B}$, with VC-dimension d or less. Also, it is clear that $|\mathcal{B}| = \pi(S;\mathcal{A})$. Thus, in order to establish (4.2.3), it is enough to prove the following claim: Suppose $|S| = n > d$, and suppose $\mathcal{B}$ is a collection of subsets of S with VC-dimension d or less. Then $|\mathcal{B}| \leq \phi(n,d)$.

The proof is by "double induction" on n and d. To start the inductive process, note that (4.2.3) is true for all n if $d = 0$, because $d = 0$ implies that $\mathcal{A}$ contains only one set[2] and $\pi(n;\mathcal{A}) = 1 \ \forall n$. Also, (4.2.3) is true for $n = 1$ for all $d \geq 1$. Now suppose for the purposes of the induction that (4.2.3) is true for all integers n and all $\mathcal{A}$ with VC-dimension $d-1$ or less, and for all integers up to $n-1$ and all $\mathcal{A}$ with VC-dimension d or less. Let S be a finite set of cardinality n, and let $\mathcal{A}$ be a collection of sets with a VC-dimension d where $d < n$. We show that $\pi(S;\mathcal{A}) \leq \phi(n,d)$. This is enough to complete the proof of the inductive step and thus establish (4.2.3).

---

[2] Note that if $\mathcal{A}$ contains two distinct subsets of X, then $d \geq 1$.

To prove this claim, choose $x \in S$ arbitrarily, and define

$$\mathcal{B} - x = \{A - \{x\} : A \in \mathcal{B}\}, \text{ and}$$

$$\mathcal{B}_x = \{A \in \mathcal{B} : x \notin A, A \cup \{x\} \in \mathcal{B}\}.$$

Then both $\mathcal{B} - x$ and $\mathcal{B}_x$ are families of subsets of $S - \{x\}$. Moreover,

$$|\mathcal{B}| = |\mathcal{B} - x| + |\mathcal{B}_x|. \tag{4.2.7}$$

This is because, under the association $A \mapsto A - \{x\}$ of subsets in $\mathcal{B}$, the images of the sets A and $A \cup \{x\}$ are the same; and $|\mathcal{B}_x|$ is precisely the number of pairs of sets in $\mathcal{B}$ that map into the same set in $\mathcal{B} - x$. Now VC-dim$(\mathcal{B} - x) \le d$, since $\mathcal{B} - x$ is a subcollection of $\mathcal{B}$. Also, every set in $\mathcal{B} - x$ is a subset of $S - \{x\}$, which is a set of cardinality $n - 1$. Hence, by the inductive hypothesis, it follows that

$$|\mathcal{B} - x| \le \phi(n - 1, d).$$

Next, it is shown that VC-dim$(\mathcal{B}_x) \le d - 1$. Suppose to the contrary that VC-dim$(\mathcal{B}_x) = d$, and let $V \subseteq S - \{x\}$ be a set of cardinality d that is shattered by $\mathcal{B}_x$. Then it is easy to see that $V \cup \{x\}$ is shattered by $\mathcal{B}$: Given any subset B of V, choose $A \in \mathcal{B}_x$ such that $V \cap A = B$. Then $A \cup \{x\} \in \mathcal{B}_x$, and

$$(V \cup \{x\}) \cap (A \cup \{x\}) = B \cup \{x\}.$$

Since this can be done for every subset B of V, it follows that $V \cup \{x\}$ is shattered by $\mathcal{B}$. Now $x \notin V$ since $V \subseteq S - \{x\}$. Hence $|V \cup \{x\}| = d + 1$, which contradicts the assumption that VC-dim$(\mathcal{B}) \le d$. Thus it must be the case that VC-dim$(\mathcal{B}_x) \le d - 1$. Hence, by the inductive hypothesis,

$$|\mathcal{B}_x| \le \phi(n - 1, d - 1).$$

Substituting the bounds on $|\mathcal{B} - x|$ and $\mathcal{B}_x$ into (4.2.7) leads to

$$|\mathcal{B}| \le \phi(n-1, d) + \phi(n-1, d-1) = \phi(n, d),$$

where the last step comes from (4.2.4). ∎

Example 4.6. ([11]) The purpose of this example is to show that the bound given in Lemma 4.2 is the best possible. Let X be any infinite set, let d be a fixed integer, and let $\mathcal{A}$ consist of all subsets of X of cardinality d or less. Then it is easy to see that the VC-dimension of $\mathcal{A}$ is d; in fact, $\mathcal{A}$ shatters *every* set of cardinality d. Now suppose S is a set of cardinality $n > d$. Then $\pi(S; \mathcal{A})$ equals the number of subsets of S of cardinality d or less, which is precisely $\phi(n, d)$.

Lemma 4.3. *The function $\phi(n, d)$ satisfies the inequality*

$$\phi(n, d) \le 2 \frac{n^d}{d!} \le \left(\frac{en}{d}\right)^d, \forall n \ge d \ge 1. \tag{4.2.8}$$

Proof. Let us first establish the left inequality. The proof is by double induction on n and d. If $d = 1$, then $\phi(n,d) = n + 1 \le 2n$ for all $n \ge 1$; hence the inequality holds for all n if $d = 1$. If $n = d$, then $\phi(n,d) = 2^d$. By the binomial expansion,

$$2 \le \left(1 + \frac{1}{d-1}\right)^{d-1} = \left(\frac{d}{d-1}\right)^{d-1}.$$

Now suppose by way of induction that

$$2^{d-1} \le 2\frac{(d-1)^{d-1}}{(d-1)!}.$$

Combining these two inequalities leads to

$$2^d \le 2\left(\frac{d}{d-1}\right)^{d-1} \frac{(d-1)^{d-1}}{(d-1)!} = 2\frac{d^{d-1}}{(d-1)!} = 2\frac{d^d}{d!}.$$

This establishes the inductive step when $n = d > 1$. Finally, suppose $n > d > 1$. Since $\phi(n,d) = \phi(n-1,d) + \phi(n-1,d-1)$ and since

$$\phi(n-1,d) \le 2\frac{(n-1)^d}{d!}, \quad \phi(n-1,d-1) \le 2\frac{(n-1)^{d-1}}{(d-1)!}$$

by the inductive hypothesis, it is enough to show that

$$2\frac{(n-1)^d}{d!} + 2\frac{(n-1)^{d-1}}{(d-1)!} \le 2\frac{n^d}{d!}. \tag{4.2.9}$$

After multiplication of both sides by $d!/2$, the above inequality is equivalent to

$$(n-1)^d + d(n-1)^{d-1} \le n^d,$$

which in turn is equivalent to

$$(d+n-1)(n-1)^{d-1} \le n^d, \text{ or}$$

$$\frac{d+n-1}{n-1} \le \left(\frac{n}{n-1}\right)^d, \text{ or}$$

$$1 + \frac{d}{n-1} \le \left(1 + \frac{1}{n-1}\right)^d.$$

But this last inequality follows readily from the binomial expansion. This completes the proof of the inductive step (4.2.9) for $n > d > 1$, and completes the proof of the left inequality in (4.2.8).

To establish the right inequality in (4.2.8), we use Stirling's approximation. This inequality clearly holds when $d = 1$, because $2n \le en$. For $d \ge 2$, we have

$$d! \geq \sqrt{2\pi d}\, d^d e^{-d}.$$

Hence

$$2\frac{n^d}{d!} \leq \sqrt{\frac{2}{\pi d}} \left(\frac{en}{d}\right)^d \leq \left(\frac{en}{d}\right)^d \text{ if } d \geq 2.$$

This completes the proof of the lemma. ∎

Proof. **of Theorem 4.1** The theorem is a ready consequence of Lemmas 4.2 and 4.3. ∎

4.2.2 Bounds on Covering Numbers Based on the Pseudo-Dimension

In the previous subsection, we have derived bounds on the number of different maps that can be realized by a collection of sets having finite VC-dimension. In the present subsection, we derive similar results based on the pseudo-dimension.

To present the main results, we begin by introducing some notation. Suppose $\mathcal{F} \subseteq [0, R]^X$ is a collection of measurable functions mapping a set X into the interval $[0, R]$. Suppose m is a given integer, and that $\mathbf{x}_m = (x_1, \ldots, x_m) \in X^m$ is an m-tuple. Then, for each $f \in \mathcal{F}$, we define

$$\mathbf{f}(\mathbf{x}_m) := (f(x_1) \ldots f(x_m)) \in [0, R]^m,$$

$$\mathcal{F}|_{\mathbf{x}_m} := \{\mathbf{f}(\mathbf{x}_m) : x_m \in X^m\} \subseteq [0, R]^m.$$

Recall that the symbol $\|\cdot\|_\infty$ denotes the ℓ_∞-norm on $\mathbb{R}^m$. Our objective in this subsection is to obtain an estimate of the *covering number* $N(\epsilon, \mathcal{F}|_{\mathbf{x}_m}, \|\cdot\|_\infty)$ in terms of ϵ, m and the pseudo-dimension of $\mathcal{F}$. We shall do this by actually obtaining an upper bound for the *packing number* $M(\epsilon, \mathcal{F}|_{\mathbf{x}_m}, \|\cdot\|_\infty)$.

Theorem 4.2. *Suppose $\mathcal{F} \subseteq [0, R]^X$ consists of measurable functions and that P-dim($\mathcal{F}$) = d. Then for each ϵ, m we have*

$$\begin{aligned} N(\epsilon, \mathcal{F}|_{\mathbf{x}_m}, \|\cdot\|_\infty) &\leq M(\epsilon, \mathcal{F}|_{\mathbf{x}_m}, \|\cdot\|_\infty) \\ &\leq \sum_{i=0}^{d} \binom{m}{i} \left(\frac{R}{\epsilon}\right)^i \\ &\leq \left(\frac{emR}{\epsilon d}\right)^d, \quad \forall m \geq d. \end{aligned} \quad (4.2.10)$$

In order to prove the theorem, let us observe that the left-most inequality is already proven in Lemma 2.2, while the right-most inequality follows from Lemma 4.3 after replacing $(R/\epsilon)^i$ by $(R/\epsilon)^d$. Thus in effect all we have to do is to establish the middle inequality. This is achieved by the following bound.

4.2 Bounds on Growth Functions

Lemma 4.4. *Suppose $|X| = n$, $Y = \{0, \ldots, N-1\}$, $\mathcal{F}$ is a collection of functions mapping X into Y, and that P-dim$(\mathcal{F}) = d < n$. Then*

$$|\mathcal{F}| \leq \psi(n, d, N) := \sum_{i=0}^{d} \binom{n}{i} (N-1)^i. \qquad (4.2.11)$$

Remark: Note that the function ψ defined above is a generalization of the function ϕ defined in (4.2.2). Specifically, we have

$$\psi(n, d, 2) = \phi(n, d).$$

Thus Lemma 4.4 contains Lemma 4.2 as a special case.

Proof. **of the Lemma**. Recall the identity

$$\binom{n}{i} = \binom{n-1}{i} + \binom{n-1}{i-1},$$

which is the same as (4.2.5). Substituting this identity into the definition of the function ψ shows that ψ satisfies the following recursion:

$$\psi(n, d, N) = \psi(n-1, d, N) + (N-1)\,\psi(n-1, d-1, N).$$

If $d \geq n$, then $\psi(n, d, N) = N^n$, from the binomial theorem. Moreover, since $|X| = n$ and $|Y| = N$, it is obvious that $|\mathcal{F}| \leq N^n$ in any case. Thus the bound is trivially true if $d \geq n$. Now let us examine the general case. As in the proof of Lemma 4.2, we use double induction on n and d. If $d = 0$, then $|\mathcal{F}| = 1$, so the bound is true for all n. Now suppose the statement is true for all function families with pseudo-dimension less than d for all n, and for all function families when $|X| < n$ for all $d \leq n - 1$. We shall now show the statement is true for $|X| = n$ and P-dim$(\mathcal{F}) = d$.

Given the set $X = \{x_1, \ldots, x_n\}$, define an equivalence relation $\sim$ on $\mathcal{F}$ as follows: $f \sim g$ if $f(x_i) = g(x_i)$ for $i = 1, \ldots, n-1$. Let $\mathcal{G}$ denote the collection of equivalence classes under $\sim$. If $\bar{X} := \{x_1, \ldots, x_{n-1}\} \subseteq X$, then we can identify $\mathcal{G}$ with the class $\mathcal{F}|_{\bar{X}}$ of the functions in $\mathcal{F}$ restricted to $\bar{X}$. For each equivalence class in $\mathcal{G}$, choose a representative function $f \in \mathcal{F}$ such that $f(x_n)$ is as small as possible. In this way, we can think of $\mathcal{G}$ as a subset of $\mathcal{F}$, whence P-dim$(\mathcal{G}) \leq$ P-dim$(\mathcal{F}) = d$. By the inductive assumption, we have $|\mathcal{G}| \leq \psi(n-1, d, N)$.

Next, look at the collection $\mathcal{H} := \mathcal{F} \setminus \mathcal{G}$. Then for each function $h \in \mathcal{H}$ there is a corresponding function $g \in \mathcal{G}$ such that $h \sim g$. Since the functions in $\mathcal{G}$ have been chosen such that $g(x_n)$ is as small as possible within each equivalence class, we see that it is *not* possible for $h(x_n)$ to equal 0 for any $h \in \mathcal{H}$. Thus $h(x_n) \in \{1, \ldots, N-1\}$ for every $h \in \mathcal{H}$. Now fix $j \in \{1, \ldots, N-1\}$, and define

$$\mathcal{H}_j := \{h \in \mathcal{H} : h(x_n) = j\}.$$

130   4. Vapnik-Chervonenkis, Pseudo- and Fat-Shattering Dimensions

Then $\mathcal{H} = \cup_{j=1}^{N-1} \mathcal{H}_j$. Now it is claimed that P-dim($\mathcal{H}_j$) $\leq d-1$ for each j. To see this, fix j and suppose to the contrary that P-dim($\mathcal{H}_j$) $= d$. (Since $\mathcal{H}_j \subseteq \mathcal{F}$, its pseudo-dimension cannot be larger than d.) Suppose a set $Y \subseteq X$ of cardinality d is P-shattered by $\mathcal{H}_j$. Since every function in $\mathcal{H}_j$ assumes exactly the same value at x_n, we see that in fact Y must be a subset of $\bar{X} = \{x_1, \ldots, x_{n-1}\}$. Denote $Y = \{y_1, \ldots, y_d\}$. Suppose $\mathbf{c} \in [0, R]^d$ witnesses the P-shattering. Thus, for each binary vector $\mathbf{e} \in \{0,1\}^d$, there exists a corresponding function $h_\mathbf{e} \in \mathcal{H}_j$ such that $\eta[h_\mathbf{e}(y_i) - c_i] = e_i$ for all i. Now it is claimed that the set $Y \cup \{x_n\}$ of cardinality $d+1$ is P-shattered by $\mathcal{F}$, with the witness $(\mathbf{c}, j) \in [0, R]^{d+1}$. To see this, recall that for each $h \in \mathcal{H}_j$ there exists another function $g \in \mathcal{G}$ such that $h \sim g$. Moreover, we have $g(x_n) < h(x_n)$ by the manner in which the representative elements of the equivalence class have been chosen. In particular, if we let $g_\mathbf{e}$ denote the function in $\mathcal{G}$ such that $g_\mathbf{e} \sim h_\mathbf{e}$ for each binary vector $\mathbf{e}$, then it follows that $g_\mathbf{e}(y_i) = h_\mathbf{e}(y_i)$ for each i, and in addition $g_\mathbf{e}(x_n) < j$, for each binary vector $\mathbf{e}$. This establishes the claim and shows that P-dim($\mathcal{G} \cup \mathcal{H}_j$) $= d+1$. But this is a contradiction since $\mathcal{G} \cup \mathcal{H}_j \subseteq \mathcal{F}$ and P-dim($\mathcal{F}$) $= d$. This contradiction shows that in fact we must have P-dim($\mathcal{H}_j$) $\leq d-1$ for each j. Moreover, since every function in $\mathcal{H}_j$ assumes exactly the same value at x_n, we can in fact think of $\mathcal{H}_j$ as a family mapping the set $\bar{X}$ of cardinality $n-1$ into Y, and having pseudo-dimension no larger than $d-1$. By the inductive assumption, this implies that

$$|\mathcal{H}_j| \leq \psi(n-1, d-1, N), \ j = 1, \ldots, N-1.$$

Since $\mathcal{H} = \cup_{j=1}^{N-1} \mathcal{H}_j$, it follows that

$$|\mathcal{H}| \leq \sum_{j=1}^{N-1} |\mathcal{H}_j| \leq (N-1)\,\psi(n-1, d-1, N).$$

Finally, since $\mathcal{F} = \mathcal{G} \cup \mathcal{H}$, we conclude that

$$|\mathcal{F}| = |\mathcal{G}| + |\mathcal{H}| \leq \psi(n, d-1, N) + (N-1)\psi(n-1, d-1, N) = \psi(n, d, N),$$

which is precisely the desired conclusion. ∎

A much more involved proof of the above lemma is given in [3, 7].

Proof. of the Theorem. The proof of the theorem makes use of the idea of quantization. Suppose α is a real number, and define the quantization function $Q_\alpha : \mathbb{R} \to \mathbb{R}$ as follows:

$$Q_\alpha(u) := \alpha \lfloor \frac{u}{\alpha} \rfloor.$$

If $f : X \to \mathbb{R}$, define

$$Q_\alpha(f) := Q_\alpha \circ f, \text{ i.e. } [Q_\alpha(f)](u) = Q_\alpha[f(u)].$$

Finally, let
$$Q_\alpha(\mathcal{F}) := \{Q_\alpha(f) : f \in \mathcal{F}\}.$$

Now the proof proceeds via a series of claims.

Claim 1. P-dim$(Q_\alpha(\mathcal{F})) \leq$ P-dim$(\mathcal{F})$ $\forall \alpha$.

To prove this claim, suppose a set S with $|S| = n$ is P-shattered by $Q_\alpha(\mathcal{F})$ with witness vector $\mathbf{c} \in [0,R]^n$. Then by definition for each binary vector $\mathbf{e} \in \{0,1\}^n$, there exists a function $f_{\mathbf{e}} \in \mathcal{F}$ such that

$$\eta[(Q_\alpha \circ f_{\mathbf{e}})(x_i) - c_i] = e_i, \ \forall x_i \in S.$$

Now $Q_\alpha \circ f_{\mathbf{e}}$ assumes only values that are multiples of α. So if $(Q_\alpha \circ f_{\mathbf{e}})(x_i) \geq c_i$ then it will be the case that $(Q_\alpha \circ f_{\mathbf{e}})(x_i) \geq \alpha \lceil c_i/\alpha \rceil$, that is, c_i rounded up to the next multiple of α. On the other hand if $(Q_\alpha \circ f_{\mathbf{e}})(x_i) < c_i$ then surely $(Q_\alpha \circ f_{\mathbf{e}})(x_i) < \alpha \lceil c_i/\alpha \rceil$. Thus without loss of generality we can replace c_i by $\alpha \lceil c_i/\alpha \rceil$. Let $\mathbf{v}$ denote this new vector whose components are $\alpha \lceil c_i/\alpha \rceil$. Then $\mathbf{v}$ also witnesses the P-shattering of S by $Q_\alpha(\mathcal{F})$. Moreover, again since $Q_\alpha \circ f$ assumes only values that are multiples of α, it follows that if $(Q_\alpha \circ f)(x) < v_i$, then in fact $(Q_\alpha \circ f)(x) \leq v_i - \alpha$. It is now shown that the same vector $\mathbf{v}$ also witnesses the P-shattering of the set S by the unquantized function family $\mathcal{F}$. Note that $Q_\alpha(y) \leq y < Q_\alpha(y) + \alpha$ $\forall y$, so that $(Q_\alpha \circ f_{\mathbf{e}})(x) \leq f(x) < (Q_\alpha \circ f_{\mathbf{e}})(x) + \alpha$ for all $f \in \mathcal{F}, x \in S$. So

if $e_i = 1$, then $f_{\mathbf{e}}(x_i) \geq (Q_\alpha \circ f_{\mathbf{e}})(x_i) \geq v_i$,

if $e_i = 0$, then $f_{\mathbf{e}}(x_i) < (Q_\alpha \circ f_{\mathbf{e}})(x_i) + \alpha = v_i$.

Thus $\mathbf{v}$ witnesses the P-shattering of S by $\mathcal{F}$. It has been shown that any set that is P-shattered by $Q_\alpha(\mathcal{F})$ is also P-shattered by $\mathcal{F}$. The claim now follows.

Claim 2. We have

$$\text{P-dim}(Q_\epsilon(\mathcal{F})|_{\mathbf{x}_m}) \leq d, \ \forall \mathbf{x}_m \in X^m.$$

This follows from Claim 1 with $\alpha = \epsilon$ and observing that P-dim$(\mathcal{F}) = d$.

Claim 3. We have

$$|Q_\alpha(a) - Q_\alpha(b)| \geq Q_\alpha(|a-b|), \ \forall a, b, \alpha.$$

To establish the above inequality, assume without loss of generality that $a > b$. (The inequality is trivial if $a = b$.) To be specific, suppose $a = l\alpha + r$ and $b = m\alpha + s$, where l, m are integers and $r, s \in [0, \alpha)$. Then $Q_\alpha(a) = l$, $Q_\alpha(b) = m$. Moreover, we have $a - b = (l-m)\alpha + (r-s)$. Now since $r, s \in [0, \alpha)$, definitely $r - s < \alpha$ (it could be negative as well). So $a - b < (l-m)\alpha + \alpha$, which means that

$$Q_\alpha(a-b) \leq (l-m)\alpha = Q_\alpha(a) - Q_\alpha(b),$$

which is the desired inequality.

Claim 4. Suppose α, ϵ are arbitrary. Then

$$M(\epsilon, \mathcal{F}|_{\mathbf{x}_m}, \|\cdot\|_\infty) \leq M(Q_\alpha(\epsilon), Q_\alpha(\mathcal{F})_{\mathbf{x}_m}, \|\cdot\|_\infty).$$

Fix α, ϵ and let $\mathbf{x}_m \in X^m$. Suppose two functions $f, g \in \mathcal{F}$ are ϵ-separated when restricted to $\mathbf{x}_m$. Thus there exists an index i such that $|f(x_i) - g(x_i)| > \epsilon$. Now it follows from Claim 3 that

$$|(Q_\alpha \circ f)(x_i) - (Q_\alpha \circ g)(x_i)| > Q_\alpha(\epsilon).$$

Thus the two functions $Q_\alpha(f)$ and $Q_\alpha(g)$ are $Q_\alpha(\epsilon)$-separated when restricted to $\mathbf{x}_m$. This establishes the claim.

Now we can complete the proof. Suppose $\epsilon, m, \mathbf{x}_m \in X^m$ are fixed, and consider the quantized family $Q_\epsilon(\mathcal{F})|_{\mathbf{x}_m}$ as a family of functions mapping the set $S = \{x_1, \ldots, x_m\}$ into the set $Y = \{0, \ldots, Q_\epsilon(R)\}$. Apply Claim 4 with $\alpha = \epsilon$ which incidentally results in $Q_\alpha(\epsilon) = \epsilon$. The ϵ-packing number of this set cannot be larger than the cardinality of the family $Q_\epsilon(\mathcal{F})_{\mathbf{x}_m}$. Now apply Lemma 4.4 and in particular (4.2.11) with X replaced by S, $m = n$, and $N = Q_\epsilon(R)$. This shows that

$$M(\epsilon, Q_\epsilon(\mathcal{F})_{\mathbf{x}_m}, \|\cdot\|_\infty) \leq \sum_{i=0}^{d} \binom{m}{i} \left(\frac{R}{\epsilon}\right)^i,$$

which is the middle inequality in (4.2.10).

To prove the right inequality in (4.2.10), simply replace $(R/\epsilon)^i$ by $(R/\epsilon)^d$ in the above summation. This leads to

$$\sum_{i=0}^{d} \binom{m}{i} \left(\frac{R}{\epsilon}\right)^i \leq \phi(m, d)(R/\epsilon)^d$$

$$\leq \left(\frac{cm}{d}\right)^d \cdot \left(\frac{R}{\epsilon}\right)^d = \left(\frac{emR}{d\epsilon}\right)^d,$$

where the last conclusion follows from Lemma 4.3. ■

4.2.3 Metric Entropy Bounds for Families of Functions

In this subsection we derive some upper bounds that complement those of the preceding subsection. In general, the bounds given here are tighter than those given in the previous subsection. On the other hand, the bounds derived earlier achieve a kind of unification between concept classes and function classes by being very similar-looking in the two cases. In this subsection, one is interested in the so-called *metric entropy* of various sets generated from these function classes, that is, the covering numbers of these sets.

4.2 Bounds on Growth Functions

Let $\mathcal{F} \subseteq [0,1]^X$ consist of measurable functions. One can define an integer $\pi(n; \mathcal{F})$ in analogy with $\pi(n; \mathcal{A})$ as defined above. However, the definition of $\pi(n; \mathcal{F})$ is a little more complicated, so as to take into account the fact that elements of $\mathcal{F}$ assume values in $[0,1]$ and not just $\{0,1\}$.

Recall the definition of the step function given in (4.1.1). Given a vector $\mathbf{v} \in \mathbb{R}^n$, define $\eta(\mathbf{v}) \in \{0,1\}^n$ by applying the step function componentwise to $\mathbf{v}$; that is,

$$[\eta(\mathbf{v})]_i = \eta(v_i) = \begin{cases} 1, & \text{if } v_i \geq 0, \text{ and} \\ 0, & \text{if } v_i < 0. \end{cases}$$

Suppose $S = \{x_1, \ldots, x_n\}$ has cardinality n. For a fixed vector $\mathbf{c} \in [0,1]^n$, define $\pi_{\mathbf{c}}(S; \mathcal{F})$ to equal the number of distinct binary vectors of the form

$$\eta\{[f(x_1) - c_1 \ldots f(x_n) - c_n]^t\}$$

generated by varying f over $\mathcal{F}$. Now define

$$\pi(S; \mathcal{F}) := \max_{\mathbf{c} \in [0,1]^n} \pi_{\mathbf{c}}(S; \mathcal{F}), \text{ and}$$

$$\pi(n; \mathcal{F}) := \max_{|S|=n} \pi(S; \mathcal{F}).$$

Lemma 4.5. *Suppose $\mathcal{F}$ has finite P-dimension, say d. Then*

$$\pi(n; \mathcal{F}) \leq \left(\frac{en}{d}\right)^d, \forall n \geq d.$$

Proof. As one might expect, the proof is based on Theorem 4.1. Fix a set S of cardinality n and a vector $\mathbf{c} \in [0,1]^n$. Now the collection of maps

$$x_i \mapsto \eta[f(x_i) - c_i], \ i = 1, \ldots, n; \ f \in \mathcal{F}$$

is a collection of *binary-valued* maps on S, and by assumption, this collection has a *VC-dimension* no larger than d. Hence, from Theorem 4.1, it follows that

$$\pi_{\mathbf{c}}(S; \mathcal{F}) \leq \left(\frac{en}{d}\right)^d, \forall n \geq d.$$

Now taking maxima, first with respect to $\mathbf{c} \in [0,1]^n$ and then with respect to S, establishes the theorem. ■

Aside from this, it is possible to derive an upper bound on the ϵ-*packing number* of a family of measurable functions $\mathcal{F}$ in terms of its P-dimension. The noteworthy aspect of this bound is that it is *independent* of the underlying probability measure.

We begin with several variations of a basic inequality that prove useful not only in this section but elsewhere as well.

134   4. Vapnik-Chervonenkis, Pseudo- and Fat-Shattering Dimensions

Lemma 4.6.   1. *Suppose $q > 4, m \geq 1$. Then*
$$m \geq 2q \lg q \;\Rightarrow\; m > q \lg m.$$
Equivalently,
$$m \leq q \lg m \;\Rightarrow\; m < 2q \lg q.$$
2. *Suppose $\alpha, \beta > 0, \alpha\beta > 4, m \geq 1$. Then*
$$m \geq 2\alpha \lg \alpha\beta \;\Rightarrow\; m > \alpha \lg \beta m.$$
Equivalently,
$$m \leq \alpha \lg \beta m \;\Rightarrow\; m < 2\alpha \lg \alpha\beta.$$
3. *Suppose $\alpha, \beta, \gamma > 0$, $\alpha\beta 2^{\gamma/\alpha} > 4$, and $m \geq 1$. Then*
$$m \geq 2\gamma + 2\alpha \lg \alpha\beta \;\Rightarrow\; m > \gamma + \alpha \lg \beta m.$$
Equivalently,
$$m \leq \gamma + \alpha \lg \beta m \;\Rightarrow\; m < 2\gamma + 2\alpha \lg \alpha\beta.$$

Proof. To prove the first part of the lemma, define $\phi(m) := m - q \lg m$, and let $m_0 = 2q \lg q$. Note that $m_0 > 2q$, in view of the assumption that $q > 4$. Then
$$\begin{aligned}\phi(m_0) &= 2q \lg q - q(\lg 2q + \lg \lg q) \\ &= 2q \lg q - q(1 + \lg q + \lg \lg q) \\ &= q(\lg q - 1 - \lg \lg q) \\ &> 0,\end{aligned}$$
provided
$$\lg q > 1 + \lg \lg q, \text{ or } q > 2 \lg q, \text{ or } 2^q > q^2,$$
after exponentiation twice. But the last inequality holds whenever $q > 4$. Hence $\phi(m_0) > 0$ if $q > 4$. Since $\phi'(m) = 1 - q/m > 0$ for all $m \geq m_0$, it is easy to see that $\phi(m) > 0 \;\forall m \geq m_0$. This establishes the first inequality. Taking the contrapositive establishes the second inequality.

To prove the second part of the lemma, observe that
$$m \geq 2\alpha \lg \alpha\beta \Leftrightarrow \beta m \geq 2\alpha\beta \lg \alpha\beta.$$
Applying the first two inequalities with q replaced by $\alpha\beta$ and m replaced by βm establishes the last two inequalities.

To prove the third part of the lemma, note that
$$m \leq \gamma + \alpha \lg \beta m \Leftrightarrow m \leq \alpha \left(\frac{\gamma}{\alpha} + \lg \beta m\right) = \alpha \lg(\beta 2^{\gamma/\alpha} m).$$
Now apply Part 2. This leads to
$$\begin{aligned}m \leq \gamma + \alpha \lg \beta m &\Rightarrow m < 2\alpha \lg(\alpha\beta 2^{\gamma/\alpha}) \\ &= 2\alpha \left(\frac{\gamma}{\alpha} + \lg \alpha\beta\right) = 2\gamma + 2\alpha \lg \alpha\beta.\end{aligned}$$
This completes the proof. ∎

Sometimes it is desirable to have a form of this inequality that involves the natural logarithm rather than the binary logarithm.

Corollary 4.1. *Suppose $\alpha, \beta, \gamma > 0$, $\alpha\beta e^{\gamma/\alpha} > 4\lg e$, and $m \geq 1$. Then*
$$m \leq \gamma + \alpha \ln \beta m \;\Rightarrow\; m < 2\gamma + 2\alpha \ln \alpha\beta.$$

Equivalently,
$$m \geq 2\gamma + 2\alpha \ln \alpha\beta \;\Rightarrow\; m > \gamma + \alpha \ln \beta m.$$

Proof. Rewrite the first inequality as follows:
$$m \leq \gamma + \alpha \ln \beta m \;\Leftrightarrow\; m \leq \gamma + \frac{\alpha}{\lg e} \lg \beta m.$$

The condition $\alpha\beta e^{\gamma/\alpha} > 2\lg e$ means that Part 3 of Lemma 4.6 can be applied with α replaced by $\alpha/\lg e$. Hence it follows that
$$m \leq \gamma + \alpha \ln \beta m \;\Rightarrow\; m < 2\gamma + \frac{2\alpha}{\lg e} \cdot \lg \frac{\alpha\beta}{\lg e}$$
$$= 2\gamma + 2\alpha \ln \frac{\alpha\beta}{\lg e} \leq 2\gamma + 2\alpha \ln \alpha\beta.$$

Taking the contrapositive establishes the other claim. ∎

Theorem 4.3. *Suppose $\mathcal{F} \subseteq [0,1]^X$ consists of measurable functions, and that $\mathcal{F}$ has finite P-dimension $d \geq 2$. Let P be any probability measure on the measurable space $(X, \mathcal{S})$, and define the pseudometric d_P on $[0,1]^X$ by*
$$d_P(f,g) := \int_X |f(x) - g(x)|\, P(dx).$$

Finally, let $0 < \epsilon < e/(4\lg e) \approx 0.47$. Then the ϵ-packing number $M(\epsilon, \mathcal{F}, d_P)$ is bounded above by
$$M(\epsilon, \mathcal{F}, d_P) \leq 2 \left(\frac{2e}{\epsilon} \ln \frac{2e}{\epsilon}\right)^d. \tag{4.2.12}$$

Corollary 4.2. *With all symbols as in Theorem 4.3, the ϵ-covering number $N(\epsilon, \mathcal{F}, d_P)$ and the external ϵ-covering number $L(\epsilon, \mathcal{F}, d_P)$ are bounded above by*
$$L(\epsilon, \mathcal{F}, d_P) \leq N(\epsilon, \mathcal{F}, d_P) \leq 2 \left(\frac{2e}{\epsilon} \ln \frac{2e}{\epsilon}\right)^d. \tag{4.2.13}$$

The proof of Theorem 4.3 is based on the following auxiliary lemma.

Lemma 4.7. *Suppose $\mathcal{F} \subseteq [0,1]^X$ consists of measurable functions, and let P be any probability measure on the measurable space $(X, \mathcal{S})$. Let $\mathbf{c} \in [0,1]^n$ be a random vector generated by the uniform distribution on $[0,1]^n$, and let $\mathbf{x} = [x_1 \ldots x_n]^t \in X^n$ be a random vector generated according to the product probability P^n on X^n. In analogy with the above, let $\pi_{\mathbf{c}}(\mathbf{x}; \mathcal{F})$ denote the number of distinct binary vectors of the form*

$$\eta\{[f(x_1) - c_1 \ldots f(x_n) - c_n]\}^t$$

generated by varying f over $\mathcal{F}$. Then, for each ϵ in $(0,1)$, we have

$$E[\pi_{\mathbf{c}}(\mathbf{x}; \mathcal{F})] \geq M(\epsilon, \mathcal{F}, d_P)[1 - M(\epsilon, \mathcal{F}, d_P) \, e^{-n\epsilon}], \quad (4.2.14)$$

where the expectation is taken jointly with respect to $\mathbf{c}$ and $\mathbf{x}$.

Proof. of the lemma For a fixed $f \in \mathcal{F}$ and $\mathbf{x} \in X^n$, let $\mathbf{f}(\mathbf{x}) \in [0,1]^n$ denote the vector

$$\mathbf{f}(\mathbf{x}) := [f(x_1) \ldots f(x_n)]^t.$$

Then

$$\pi_{\mathbf{c}}(\mathbf{x}; \mathcal{F}) = |\{\eta[\mathbf{f}(\mathbf{x}) - \mathbf{c}] : f \in \mathcal{F}\}|.$$

Let $\mathcal{G}$ be a maximal ϵ-separated set in $\mathcal{F}$; thus $|\mathcal{G}| = M(\epsilon, \mathcal{F}, d_P)$. Now

$$E[\pi_{\mathbf{c}}(\mathbf{x}; \mathcal{F})] \geq E[\pi_{\mathbf{c}}(\mathbf{x}; \mathcal{G})],$$

since $\mathcal{G}$ is a subset of $\mathcal{F}$. Further,

$$\begin{aligned}
E[\pi_{\mathbf{c}}(\mathbf{x}; \mathcal{G})] &\geq E[|\{f \in \mathcal{G} : \eta[\mathbf{f}(\mathbf{x}) - \mathbf{c}] \neq \eta[\mathbf{g}(\mathbf{x}) - \mathbf{c}] \, \forall \mathbf{g} \in \mathcal{G} - f\}|] \\
&= \sum_{f \in \mathcal{G}} \Pr\{\eta[\mathbf{f}(\mathbf{x}) - \mathbf{c}] \neq \eta[\mathbf{g}(\mathbf{x}) - \mathbf{c}] \, \forall \mathbf{g} \in \mathcal{G} - f\} \\
&= \sum_{f \in \mathcal{G}} 1 - \Pr\{\exists g \in \mathcal{G} - f : \eta[\mathbf{f}(\mathbf{x}) - \mathbf{c}] = \eta[\mathbf{g}(\mathbf{x}) - \mathbf{c}]\} \\
&\geq \sum_{f \in \mathcal{G}} 1 - |\mathcal{G}| \max_{g \in \mathcal{G} - f} \Pr\{\eta[\mathbf{f}(\mathbf{x}) - \mathbf{c}] = \eta[\mathbf{g}(\mathbf{x}) - \mathbf{c}]\}, (4.2.15)
\end{aligned}$$

where $\mathcal{G} - f$ is a shorthand for $\mathcal{G} - \{f\}$. Next, suppose $f, g \in \mathcal{G}$ are distinct. Since $\mathcal{G}$ is ϵ-separated, it follows that $d_P(f,g) > \epsilon$. Hence, if $x_i \in X$ is drawn at random according to P, then the probability that c_i lies between $f(x_i)$ and $g(x_i)$ is at least ϵ. Also, the condition

$$\eta[\mathbf{f}(\mathbf{x}) - \mathbf{c}] = \eta[\mathbf{g}(\mathbf{x}) - \mathbf{c}]$$

holds only if this fails to happen for each i. Thus

$$\Pr\{\eta[\mathbf{f}(\mathbf{x}) - \mathbf{c}] = \eta[\mathbf{g}(\mathbf{x}) - \mathbf{c}]\} \leq (1 - \epsilon)^n \leq e^{-n\epsilon}.$$

Since this holds for every distinct pair $f, g \in \mathcal{G}$, it follows from (4.2.15) that

$$E[\pi_{\mathbf{c}}(\mathbf{x}; \mathcal{F})] \geq |\mathcal{G}| \cdot (1 - |\mathcal{G}| e^{-n\epsilon}),$$

which is the desired conclusion. ∎

4.2 Bounds on Growth Functions

Proof. **of the Theorem** Suppose P-dim($\mathcal{F}$) = d. Then, from Lemma 4.5, it follows that

$$\pi_{\mathbf{c}}(\mathbf{x}; \mathcal{F}) \leq \left(\frac{en}{d}\right)^d, \quad \forall \mathbf{c} \in [0,1]^n, \; \forall \mathbf{x} \in X^n, \; \forall n \geq d.$$

As a consequence,

$$E[\pi_{\mathbf{c}}(\mathbf{x}; \mathcal{F})] \leq \left(\frac{en}{d}\right)^d, \quad \forall n \geq d.$$

Substituting this into (4.2.14) leads to

$$M(1 - Me^{-n\epsilon}) \leq \left(\frac{en}{d}\right)^d, \quad \forall n \geq d, \tag{4.2.16}$$

where M is a shorthand for $M(\epsilon, \mathcal{F}, d_P)$. If $M < 0.5e^{d\epsilon}$, then the bound (4.2.12) is satisfied; so it can be assumed without loss of generality that $M \geq 0.5e^{d\epsilon}$, i.e., that $d \leq \ln(2M)/\epsilon$. Now suppose $n = \ln(2M)/\epsilon$.[3] Then $n \geq d$ so that the bound (4.2.16) applies. Also,

$$n = \frac{\ln 2M}{\epsilon} \Rightarrow Me^{-n\epsilon} = 0.5 \Rightarrow 1 - Me^{-n\epsilon} = 0.5.$$

Thus (4.2.16) becomes

$$\left(\frac{e \ln 2M}{\epsilon d}\right)^d \geq \frac{M}{2}.$$

To derive the bound (4.2.12) from the above inequality, define $x = (M/2)^{1/d}$, and raise both sides to the power $1/d$. This gives

$$\frac{e \ln(4x^d)}{\epsilon d} \geq x,$$

or equivalently,

$$x \leq \frac{e}{\epsilon d} \ln 4 + \frac{e}{\epsilon} \ln x.$$

Now apply Corollary 4.1 with $\alpha = e/\epsilon$, $\beta = 1$, and $\gamma = e \ln 4/(\epsilon d)$. Then the condition $\alpha \beta c^{\gamma/\alpha} > 4 \lg e$ becomes

$$\frac{e}{\epsilon} \exp(\ln 4/d) > 4 \lg e,$$

or, after routine algebra,

$$\epsilon < \frac{e}{4 \lg e} 4^{1/d}.$$

Since $4^{1/d} > 1$ for all $d \geq 1$, Corollary 4.1 can be applied whenever $e < e/(4 \lg e)$. This leads to the estimate

---
[3] Strictly speaking, one should write that n equals the *integer part* of $\ln(2M)/\epsilon$. However, for small enough ϵ, such niceties are not essential.

$$x < \frac{2e \ln 4}{\epsilon d} + \frac{2e}{\epsilon} \ln \frac{e}{\epsilon}$$

$$= \frac{2e}{\epsilon} \ln \left(\frac{e}{\epsilon} \cdot 4^{1/d} \right)$$

$$\leq \frac{2e}{\epsilon} \ln \frac{2e}{\epsilon} \text{ if } d \geq 2,$$

since $4^{1/d} \leq 2$ if $d \geq 2$. Hence

$$\left(\frac{M}{2} \right)^{1/d} < \frac{2e}{\epsilon} \ln \frac{2e}{\epsilon},$$

which in turn implies (4.2.12). ∎

Proof. **of the corollary** This follows readily from Lemma 2.2. ∎

Now let us apply Corollary 4.2 to obtain a result analogous to Theorem 4.1 for families of functions. A little notation is introduced first to facilitate the statement of the result. Let $\| \cdot \|_{a1}$ denote the "averaged" l_1-norm on the space $\mathbb{R}^n$; that is, given $\mathbf{v} \in \mathbb{R}^n$, we define

$$\| \mathbf{v} \|_{a1} := \frac{1}{n} \sum_{i=1}^{n} |v_i|.$$

The extra factor of $1/n$ distinguishes the averaged l_1-norm from the usual l_1-norm. Given $\mathbf{x} \in X^n$ and $f \in \mathcal{F}$, let us define the vector $\mathbf{f}(\mathbf{x}) \in [0,1]^n$ by

$$\mathbf{f}(\mathbf{x}) := [f(x_1) \ldots f(x_n)]^t,$$

and the set $\mathcal{F}|_{\mathbf{x}} \subseteq [0,1]^n$ by

$$\mathcal{F}|_{\mathbf{x}} := \{[f(x_1) \ldots f(x_n)]^t, f \in \mathcal{F}\}. \tag{4.2.17}$$

Thus $\mathcal{F}|_{\mathbf{x}}$ is the set of vectors of the form $[f(x_1) \ldots f(x_n)]^t$ generated by varying f over the family $\mathcal{F}$. A measure of the richness of the family $\mathcal{F}$, localized at $\mathbf{x}$, is the ϵ-covering number $L(\epsilon, \mathcal{F}|_{\mathbf{x}}, \| \cdot \|_{a1})$. The next result gives an upper bound for this covering number using Corollary 4.2.

Lemma 4.8. *Let $L(\epsilon, \mathcal{F}|_{\mathbf{x}}, \| \cdot \|_{a1})$ be defined as above, and suppose the family $\mathcal{F}$ has finite P-dimension, call it d. Suppose $\epsilon < e/(4 \lg e) \approx 0.47$. Then*

$$L(\epsilon, \mathcal{F}|_{\mathbf{x}}, \| \cdot \|_{a1}) \leq 2 \left(\frac{2e}{\epsilon} \ln \frac{2e}{\epsilon} \right)^d.$$

Proof. The inequality (4.2.13) is valid for *any* probability measure P on $(X, \mathcal{S})$. In particular, let $P_{\mathbf{x}}$ denote the purely atomic measure concentrated uniformly on the components of the vector $\mathbf{x}$. (If $\mathbf{x}$ has repeated components, then the repeated component has a weight that is a multiple of $1/n$.)

With respect to this probability measure, the distance between two functions $f, g \in \mathcal{F}$ is given by

$$d_{P_\mathbf{x}}(f, g) = \frac{1}{n} \sum_{i=1}^{n} |f(x_i) - g(x_i)|.$$

But this is precisely the distance $\| \mathbf{f}(\mathbf{x}) - \mathbf{g}(\mathbf{x}) \|_{a1}$. As a result, the covering number $L(\epsilon, \mathcal{F}, d_{P_\mathbf{x}})$ is precisely the same as the covering number $L(\epsilon, \mathcal{F}|_\mathbf{x}, \| \cdot \|_{a1})$. The bound now follows from Corollary 4.2. ∎

Now let us see how "tight" the above bound is by specializing it to the case where $\mathcal{F}$ consists only of *binary-value* functions, i.e., where the problem is essentially one of estimating the growth function of a collection of sets. Suppose every function in $\mathcal{F}$ maps X into $\{0, 1\}$. Then it is clear that the set $\mathcal{F}|_\mathbf{x}$ is in fact a subset of $\{0, 1\}^n$, and not merely $[0, 1]^n$. Now, two distinct vectors in $\{0, 1\}^n$ are at a distance of at least $1/n$ (with respect to $\| \cdot \|_{a1}$). Hence, if $\epsilon < 1/n$, then the covering number $L(\epsilon, \mathcal{F}|_\mathbf{x}, \| \cdot \|_{a1})$ is the same as the cardinality of the set $\mathcal{F}|_\mathbf{x}$. This leads to the bound

$$|\mathcal{F}|_\mathbf{x}| \leq 2(2en \ln 2en)^d.$$

On the other hand, by definition the growth function $\pi(n; \mathcal{F})$ is the maximum value of $\mathcal{F}|_\mathbf{x}$ as $\mathbf{x}$ varies over all vectors $\mathbf{x} \in X^n$. Hence Lemma 4.8 leads to a bound for the growth function

$$\pi(n; \mathcal{F}) \leq 2(2en \ln 2en)^d.$$

This bound is somewhat worse than the bound $(en/d)^d$ derived in Theorem 4.1. In particular, the right side of the above bound grows as $(n \ln n)^d$, whereas the bound in (4.2.1) grows as n^d. Nevertheless, considering that the bound in Lemma 4.8 applies to real-valued functions as opposed to binary-valued functions, this small difference may perhaps be considered to be insignificant.

4.2.4 Bounds on Covering Numbers Based on the Fat-Shattering Dimension

In the preceding subsections, we have derived bounds on covering numbers based on the pseudo-dimension. In the present section, we derive bounds based on the fat-shattering dimension. Since it is possible for a family of functions to have infinite pseudo-dimension and yet have finite fat-shattering dimension for each nonzero width, in principle the bounds derived here are more general.

The main result of this subsection is now presented.

Theorem 4.4. *Suppose $\mathcal{F} \subseteq [0, R]^X$ consists of measurable functions, and suppose $\epsilon > 0$ is a given number. Let $d = $ F-dim$(\mathcal{F}, \epsilon/4)$, $r = \lfloor 2R/\epsilon \rfloor$, and define*

$$y = \sum_{i=1}^{d} \binom{m}{i} r^i.$$

Then

$$N(\epsilon, \mathcal{F}|_{\mathbf{x}_m}, \|\cdot\|_\infty) \leq M(\epsilon, \mathcal{F}|_{\mathbf{x}_m}, \|\cdot\|_\infty) \leq 2(mr^2)^{\lceil \lg y \rceil}. \tag{4.2.18}$$

The proof of the theorem makes use of the following combinatorial lemma.

Lemma 4.9. *Let $Y = \{0, 1, \ldots, r\}$, $|X| = m$, and suppose $\mathcal{H} \subseteq Y^X$ has F-dim$(\mathcal{H}, 1) = d$. Then*

$$M(2, \mathcal{H}, \|\cdot\|_\infty) \leq 2(mr^2)^{\lceil \lg y \rceil}.$$

The proof of this technical lemma can be found in [3] or [7], p. 176.

Proof. **of the Theorem.** The left inequality in (4.2.18) is standard, and it remains only to prove the right inequality.

By Claim 4 in the proof of Theorem 4.2, we have

$$M(\epsilon, \mathcal{F}|_{\mathbf{x}_m}, \|\cdot\|_\infty) \leq M(Q_\alpha(\epsilon), Q_\alpha(\mathcal{F})|_{\mathbf{x}_m}, \|\cdot\|_\infty), \forall \alpha, \epsilon.$$

In particular, if we let $\alpha = \epsilon/2$, then $Q_\alpha(\epsilon) = \epsilon$, so the above inequality reduces to

$$M(\epsilon, \mathcal{F}|_{\mathbf{x}_m}, \|\cdot\|_\infty) \leq M(\epsilon, Q_{\epsilon/2}(\mathcal{F})|_{\mathbf{x}_m}, \|\cdot\|_\infty). \tag{4.2.19}$$

Next, note that if $\alpha < 2\epsilon$, then

$$\text{F-dim}(Q_\alpha(\mathcal{F}), \epsilon) \leq \text{F-dim}(\mathcal{F}, \epsilon - \alpha/2). \tag{4.2.20}$$

To see this, suppose a set S of cardinality n is fat-shattered by the collection $Q_\alpha(\mathcal{F})$ to width ϵ, with witness $\mathbf{c}$. Then there exist 2^n functions $f_\mathbf{e} \in \mathcal{F}$, $\mathbf{e} \in \{0, 1\}^n$ such that at each element $x_i \in S$, we have either

$$Q_\alpha(f_\mathbf{e})(x_i) \geq c_i + \epsilon, \text{ or } Q_\alpha(f_\mathbf{e})(x_i) \leq c_i - \epsilon.$$

So we have either

$$f_\mathbf{e}(x_i) \geq c_i + \epsilon, \text{ or } f_\mathbf{e}(x_i) < c_i - \epsilon + \alpha.$$

Thus the set S is fat-shattered by the family $\mathcal{F}$ to width $\epsilon - \alpha/2$, with the vector whose components are $(c_i + \alpha/2)$ as witness. In particular, if we apply the inequality (4.2.20) with $\alpha = \epsilon$, we get

$$\text{F-dim}(Q_\epsilon(\mathcal{F}), \epsilon) \leq \text{F-dim}(\mathcal{F}, \epsilon/2).$$

Now if we replace ϵ by $\epsilon/2$ we get finally

$$\text{F-dim}(Q_{\epsilon/2}(\mathcal{F}), \epsilon/2) \leq \text{F-dim}(\mathcal{F}, \epsilon/4). \qquad (4.2.21)$$

Now apply Lemma 4.9 to the function family $Q_{\epsilon/2}(\mathcal{F})$. This family maps X into $\{0, \ldots, r\epsilon/2\}$ where as before $r = \lfloor 2R/\epsilon \rfloor$. Moreover, by (4.2.21), we have $\text{F-dim}(Q_{\epsilon/2}(\mathcal{F}), \epsilon/2) \leq d$. Thus Lemma 4.9 implies that whenever $\mathbf{x}_m \in X^m$, we have

$$|Q_{\epsilon/2}(\mathcal{F})|_{\mathbf{x}_m}| \leq 2(mr^2)^{\lceil \lg y \rceil}.$$

So clearly the packing number of this set is also bounded by the same number. Thus

$$M(\epsilon, Q_{\epsilon/2}(\mathcal{F})|_{\mathbf{x}_m}, \|\cdot\|_\infty) \leq |Q_{\epsilon/2}(\mathcal{F})|_{\mathbf{x}_m}| \leq 2(mr^2)^{\lceil \lg y \rceil}.$$

Substituting this estimate in (4.2.19) completes the proof. ∎

4.3 Growth Functions of Iterated Families

Suppose $\mathcal{A}$ is a collection of sets having finite VC-dimension. Define

$$\mathcal{A}\Delta\mathcal{A} := \{A\Delta B : A, B \in \mathcal{A}\}$$

to be another collection of sets formed from $\mathcal{A}$. What, if anything, can be said about the VC-dimension of $\mathcal{A}\Delta\mathcal{A}$? In this section, we prove an extremely general result and several specialized results that can be used to answer such questions.

As has been pointed out previously, there is a natural identification between collections of sets and families of binary-valued functions. For the purposes of this section, it is more convenient to work with families of binary-valued functions.

Suppose $k \geq 2$ is a given integer, and that $u : \{0, 1\}^k \to \{0, 1\}$ is a given function.[4] Suppose $f_1, \ldots, f_k : X \to \{0, 1\}$ are binary-valued functions. Then we define $u(f_1, \ldots, f_k) : X \to \{0, 1\}$ to be the binary-valued function

$$x \mapsto u[f_1(x), \ldots, f_k(x)].$$

Finally, if $\mathcal{A}_1, \ldots, \mathcal{A}_k$ are families of binary-valued functions, we define $\mathcal{U}(\mathcal{A}_1, \ldots, \mathcal{A}_k)$ to be the family of binary-valued functions

$$\mathcal{U}(\mathcal{A}_1, \ldots, \mathcal{A}_k) := \{u(f_1, \ldots, f_k) : f_i \in \mathcal{A}_i \; \forall i\}.$$

For instance, the family $\mathcal{A}\Delta\mathcal{A}$ can be obtained by taking

$$u(a, b) = |a - b|, \; \forall a, b \in \{0, 1\}.$$

Other interesting choices of u are given in the examples below.

---

[4] It is common to refer to such functions as **Boolean functions**.

142   4. Vapnik-Chervonenkis, Pseudo- and Fat-Shattering Dimensions

Theorem 4.5. *Suppose* $\mathcal{A}_1, \ldots, \mathcal{A}_k$ *are families of binary-valued functions, and that* $u : \{0,1\}^k \to \{0,1\}$ *is arbitrary. Finally, suppose VC-dim*$(\mathcal{A}_i)$ *is finite for each* i. *Then* $\mathcal{U}(\mathcal{A}_1, \ldots, \mathcal{A}_k)$ *also has finite VC-dimension. In particular,*

$$\text{VC-dim}[\mathcal{U}(\mathcal{A}_1, \ldots, \mathcal{A}_k)] < \alpha(k) \cdot d, \qquad (4.3.1)$$

where

$$d := \max_{1 \le i \le k} \text{VC-dim}(\mathcal{A}_i),$$

and $\alpha(k) =: \alpha$ *is the smallest integer that satisfies*

$$k < \frac{\alpha}{\lg(e\alpha)}. \qquad (4.3.2)$$

In particular, $\alpha(k) \le 2k \lg(ek)$.

Proof. The proof is based on the following simple observation. Let $\mathcal{A}$ be a family of binary-valued functions, and define the integer $\pi(n; \mathcal{A})$ as in Section 4.2. Under these conditions, if $\pi(n; \mathcal{A}) < 2^n$, then VC-dim$(\mathcal{A}) < n$.

To apply this observation to the problem at hand, we derive an upper bound on the integer $\pi(n; \mathcal{U})$, where $\mathcal{U} := \mathcal{U}(\mathcal{A}_1, \ldots, \mathcal{A}_k)$. For this purpose, let $S = \{x_1, \ldots, x_n\}$ be an arbitrary set of cardinality n, fix an index $i \in \{1, \ldots, k\}$, and examine the set of binary vectors of the form

$$[f(x_1) \ldots f(x_n)]^t \in \{0,1\}^n$$

generated by varying f over $\mathcal{A}_i$. By definition, it follows that the total number of such vectors is no larger than $\pi(S; \mathcal{A}_i)$. Next, given the function $u : \{0,1\}^k \to \{0,1\}$, we can define a corresponding function $\mathbf{u}_n : [\{0,1\}^n]^k \to \{0,1\}^n$ as follows: Given $\mathbf{v}^1, \ldots, \mathbf{v}^k \in \{0,1\}^n$, let

$$\mathbf{u}_n(\mathbf{v}^1, \ldots, \mathbf{v}^k) := [u(v_1^1, \ldots, v_1^k) \ldots u(v_n^1, \ldots, v_n^k)]^t \in \{0,1\}^n.$$

Now we come to the key point: Suppose it is desired to estimate the integer $\pi(n; \mathcal{U})$. By definition, $\pi(n; \mathcal{U})$ is the number of distinct vectors of the form

$$[u(f_1(x_1), \ldots, f_k(x_1)) \ldots u(f_1(x_n), \ldots, f_k(x_n))]^t \in \{0,1\}^n$$

obtained by varying each function f_i over the corresponding family $\mathcal{A}_i$. This is the same as the number of distinct vectors $\mathbf{u}_n(\mathbf{v}^1, \ldots, \mathbf{v}^k)$ obtained by varying each vector $\mathbf{v}^i$ over a set of cardinality $\pi(S; \mathcal{A}_i)$. Hence

$$\pi(S; \mathcal{U}) \le \prod_{i=1}^k \pi(S; \mathcal{A}_i) \le \prod_{i=1}^k \pi(n; \mathcal{A}_i).$$

Now, by Lemma 4.2, whenever $n \ge d$, we have that

$$\pi(n; \mathcal{A}_i) \le \phi(n, d_i) \le \phi(n, d), \; \forall i,$$

where $d_i := \text{VC-dim}(\mathcal{A}_i)$, and as before

$$d := \max_i d_i.$$

Therefore, from Lemma 4.3,

$$\pi(n; \mathcal{A}_i) \leq \left(\frac{en}{d}\right)^d, \quad \forall n \geq d, \forall i.$$

As a result,

$$\pi(S; \mathcal{U}) \leq \left(\frac{en}{d}\right)^{kd}, \quad \forall n \geq d.$$

Finally, since S is arbitrary, it follows that

$$\pi(n; \mathcal{U}) \leq \left(\frac{en}{d}\right)^{kd}, \quad \forall n \geq d.$$

Now, if we can find an integer n such that $\pi(n; \mathcal{U}) < 2^n$, then $\text{VC-dim}(\mathcal{U}) < n$. Thus we look for a solution for n to the inequality

$$\left(\frac{en}{d}\right)^{kd} < 2^n.$$

Let us simplify the problem by looking for solutions of the form $n = \alpha d$, where $\alpha = \alpha(k)$. Then the above inequality becomes

$$(e\alpha)^{kd} < 2^{\alpha d},$$

or, after taking the lg of both sides and dividing by d,

$$k \lg(e\alpha) < \alpha.$$

Hence, if α satisfies (4.3.2), then $\pi(\alpha d; \mathcal{U}) < 2^{\alpha d}$, and $\text{VC-dim}(\mathcal{U}) < \alpha d$. Applying Lemma 4.6 shows that (4.3.2) is satisfied if $\alpha = 2k \lg(ek)$. Hence

$$\text{VC-dim}(\mathcal{U}) < 2k \lg(ek) \max_{1 \leq i \leq k} \text{VC-dim}(\mathcal{A}_i).$$

This completes the proof. ∎

Remarks: It is important to emphasize that the constant α depends only on k and not on d; hence the VC-dimension of $\mathcal{U}(\mathcal{A}_1, \ldots, \mathcal{A}_k)$ is bounded by a *linear* function of d, the maximum VC-dimension of the collections $\mathcal{A}_1, \ldots, \mathcal{A}_k$.

The table below shows, for k between 2 and 10, the *smallest* integer $\alpha(k)$ such that $k \lg(e\alpha) < \alpha$, as well as the number $\lceil 2k \lg(ek) \rceil$, i.e., the number $2k \lg(ek)$ rounded upwards. From this table it can be seen that the estimate $\alpha(k) = \lceil 2k \lg(ek) \rceil$ is not too conservative, and has the advantage of being "in closed form."

k	2	3	4	5	6	7	8	9	10
$\alpha(k)$	10	17	25	33	41	50	59	68	78
$\lceil 2k \lg(ek) \rceil$	10	19	28	38	49	60	72	84	96

Example 4.7. Suppose $\mathcal{A}, \mathcal{B}$ are collections of sets with finite VC-dimension, and define
$$\mathcal{A} \oplus \mathcal{B} := \{A \cup B : A \in \mathcal{A}, B \in \mathcal{B}\},$$
$$\mathcal{A} \odot \mathcal{B} := \{A \cap B : A \in \mathcal{A}, B \in \mathcal{B}\},$$
$$\mathcal{A} \Delta \mathcal{B} := \{A \Delta B : A \in \mathcal{A}, B \in \mathcal{B}\}.$$
Each of these collections has a VC-dimension no larger than 10 times the maximum of VC-dim($\mathcal{A}$), VC-dim($\mathcal{B}$). This can be seen by observing that the sets $\mathcal{A} \oplus \mathcal{B}, \mathcal{A} \odot \mathcal{B}, \mathcal{A} \Delta \mathcal{B}$ correspond to the Boolean functions
$$(a, b) \mapsto \max\{a, b\}, a \cdot b, |a - b|$$
respectively.

Example 4.8. Let $\mathcal{C}_k$ denote the set of closed convex polygons in $\mathbb{R}^l$ with k sides or fewer. Then
$$\text{VC-dim}(\mathcal{C}_k) < \alpha(k)(l + 1).$$
This can be seen by observing that (i) the collection of closed half-spaces in $\mathbb{R}^l$ has VC-dimension $l + 1$ (cf. Example 4.3), (ii) a closed convex polygon of k sides or fewer can be expressed as an intersection of *exactly* k closed half-spaces.[5] Thus, if $\mathcal{H}$ denotes the set of closed half-spaces in $\mathbb{R}^l$, then
$$\mathcal{C}_k = \mathcal{U}(\mathcal{H}, \ldots, \mathcal{H}),$$
where $u : \{0, 1\}^k \to \{0, 1\}$ is defined by
$$u(a_1, \ldots, a_k) = \prod_{i=1}^{k} a_i.$$
Now the bound follows from Theorem 4.5. ∎

Theorem 4.5 is very general in that it applies to *any* Boolean function $u : \{0, 1\}^k \to \{0, 1\}$. The next several results are devoted to deriving tighter bounds for some *specific* Boolean functions.

Lemma 4.10. *Suppose $\mathcal{A} \subseteq \mathcal{S}$ is a collection of sets, and define*
$$\bar{\mathcal{A}} := \{A^c : A \in \mathcal{A}\}.$$
Then
$$\text{VC-dim}(\bar{\mathcal{A}}) = \text{VC-dim}(\mathcal{A}).$$

Proof. Let $S \subseteq X$ be a fixed set, let $B \subseteq S$, and suppose there exists an $A \in \mathcal{A}$ such that $S \cap A = B$. Then clearly $S \cap A^c = S - B$. Hence S is shattered by $\mathcal{A}$ if and only if it is shattered by $\bar{\mathcal{A}}$. ∎

---

[5] Some of these half-spaces may be redundant, but they are introduced so that Theorem 4.5 can be applied.

Example 4.9. The set of *open* half-spaces in $\mathbb{R}^l$ has VC-dimension $l+1$. The set of *open* convex polygons in $\mathbb{R}^l$ with k sides or fewer has VC-dimension less than $\alpha(k)(l+1)$.

Lemma 4.11. *Suppose $\mathcal{A}, \mathcal{B} \subseteq \mathcal{S}$ have finite VC-dimension. Then*

$$VC\text{-}dim(\mathcal{A} \cup \mathcal{B}) \leq VC\text{-}dim(\mathcal{A}) + VC\text{-}dim(\mathcal{B}) + 1.$$

Remarks: Note that $\mathcal{A} \cup \mathcal{B}$ is different from the collection $\mathcal{A} \oplus \mathcal{B}$ defined in Example 4.7. $\mathcal{A} \cup \mathcal{B}$ is the usual union.

Proof. It is easy to see that, for any set $S \subseteq X$,

$$\pi(S; \mathcal{A} \cup \mathcal{B}) \leq \pi(S; \mathcal{A}) + \pi(S; \mathcal{B}).$$

Hence

$$\pi(n; \mathcal{A} \cup \mathcal{B}) \leq \pi(n; \mathcal{A}) + \pi(n; \mathcal{B}).$$

Let $d_a = $ VC-dim$(\mathcal{A})$, $d_b = $ VC-dim$(\mathcal{B})$, and apply Lemma 4.2. This gives

$$\pi(n; \mathcal{A} \cup \mathcal{B}) \leq \phi(n, d_a) + \phi(n, d_b).$$

Now let $n = d_a + d_b + 2$, and observe that

$$\binom{n}{i} = \binom{n}{n-i}.$$

Hence

$$\phi(n, d_a) + \phi(n, d_b) = \sum_{i=0}^{d_a} \binom{n}{i} + \sum_{i=0}^{d_b} \binom{n}{i}$$

$$= \sum_{i=0}^{d_a} \binom{n}{i} + \sum_{i=n-d_b}^{n} \binom{n}{i} = 2^n - \binom{n}{d_a+1} < 2^n,$$

since $n = d_a + d_b + 2$. This shows that

$$\pi(n; \mathcal{A} \cup \mathcal{B}) < 2^n$$

if $n = d_a + d_b + 2$, i.e., that VC-dim$(\mathcal{A} \cup \mathcal{B}) \leq d_a + d_b + 1$. ∎

Example 4.10. Let $X = \mathbb{R}^2$, $\mathcal{A} = $ the set of closed half-planes, and $\mathcal{B} = $ the set of axis-parallel rectangles. Then VC-dim$(\mathcal{A}) = 3$ and VC-dim$(\mathcal{B}) = 4$. Hence VC-dim$(\mathcal{A} \cup \mathcal{B}) \leq 8$. More generally, let $X = \mathbb{R}^n$, $\mathcal{A} = $ the set of closed half-spaces, and $\mathcal{B} = $ the set of axis-parallel hypercubes. Then VC-dim$(\mathcal{A} \cup \mathcal{B}) \leq 3n + 2$.

Lemma 4.12. *Suppose $\mathcal{A} \subseteq \mathcal{S}$ is a collection of sets, and that $C \in \mathcal{A}$. Define*

$$C \Delta \mathcal{A} := \{C \Delta A : A \in \mathcal{A}\}.$$

Then

$$VC\text{-}dim(C \Delta \mathcal{A}) = VC\text{-}dim(\mathcal{A}).$$

146   4. Vapnik-Chervonenkis, Pseudo- and Fat-Shattering Dimensions

Proof. The proof consists of showing that, for any set $S \subseteq X$, the relationship

$$\pi(S; C \Delta \mathcal{A}) = \pi(S; \mathcal{A}) \tag{4.3.3}$$

holds. Once this relationship is established, the lemma follows readily.

In order to establish (4.3.3), we begin with the general relationship

$$F \Delta G = (F \Delta H) \Delta (G \Delta H), \quad \forall F, G, H \subseteq X. \tag{4.3.4}$$

To prove this relationship, it is convenient to use Boolean algebra. Given a set A, let a denote the Boolean variable corresponding to the indicator function of A. Then

$$a + a = a, \ a \cdot a = a. \tag{4.3.5}$$

Moreover, the usual axioms of algebra (e.g., distributivity) hold. Now the sets $F \Delta H$, $G \Delta H$ are represented by the Boolean variables $f + h - fh$ and $g + h - gh$ respectively. Hence the Boolean variable for the right side of (4.3.4) is

$$(f + h - fh) + (g + h - gh) + (f + h - fh) \cdot (g + h - gh).$$

After expansion and repeated use of (4.3.5), this simplifies to $f - g + fg$, thus establishing (4.3.4). A ready consequence of (4.3.4) is that

$$F \Delta H = G \Delta H \Leftrightarrow F = G. \tag{4.3.6}$$

This identity follows upon noting that two sets are equal if and only if their symmetric difference is the empty set. One way of interpreting (4.3.6) is that, for each fixed set H, the map $F \mapsto F \Delta H$ is one-to-one.

Now let us return to the problem at hand, which is to prove (4.3.3). This is done by showing that there exists a one-to-one and onto map between sets of the form $S \cap A$ as A varies over $\mathcal{A}$, and sets of the form $S \cap (C \Delta A)$ as A varies over $\mathcal{A}$. Specifically, define $D := S \cap C$, and consider the map $B \mapsto D \Delta B$. This map is one-to-one, as established by (4.3.6). It maps sets of the form $S \cap A$ into sets of the form $S \cap (C \Delta A)$. If $B = S \cap A$, then

$$D \Delta B = (S \cap C) \Delta (S \cap A) = S \cap (C \Delta A).$$

Finally, the map is onto. Suppose F is a set of the form $S \cap (C \Delta A)$. Then

$$F = S \cap (C \Delta A) = (S \cap C) \Delta (S \cap A)$$

is the image of the set $S \cap A$. This shows that the relationship (4.3.3) holds. ∎

Notes and References The definition of the Vapnik-Chervonenkis dimension is first given in [193], which is a synopsis of [194] without any proofs. Interestingly, in [194], no name is given to this concept; instead, all of the theorems in this paper are stated in terms of (paraphrasing in the present notation) "the smallest integer n such that $\pi(n; \mathcal{A}) < 2^n$," which would be the

VC-dimension *plus one*. The paper by Dudley [61] follows along similar lines, in that he refers to a collection of sets $\mathcal{A}$ as being a "Vapnik-Chervonenkis class" if there exists an integer n such that $\pi(n;\mathcal{A}) < 2^n$, and he defines $V(\mathcal{A})$ to be the smallest such n. Once again, Dudley's $V(\mathcal{A})$ would equal the VC-dimension of $\mathcal{A}$ *plus one*. Neither paper refers to the work of Sauer [171]. The first paper in the empirical process literature that refers to Sauer's paper is that of Steele [179]. But interestingly, while Steele uses the term "shattered" apparently for the first time in the empirical process literature (though the term is of older origin in the combinatorics literature), he does not explicitly give either a name or a symbol to what we now call the VC-dimension (plus one). It was left to subsequent researchers to coin the phrase "Vapnik-Chervonenkis dimension" and to state the definition in the present form (thus getting rid of the "plus one"). Lemma 4.2 is proved independently by both Sauer [171] and Vapnik-Chervonenkis [194]; however, in the computer science literature, this result is usually referred to as "Sauer's Lemma." Interestingly, Sauer himself states in a footnote that the referee of his paper has pointed out that the result is obtained independently by Shelah [173], [174], who was investigating the model theory of real numbers; see the "Notes and References" Section at the end of Chapter 10 for further discussion on this point.

The proof of Lemma 4.2 by double induction as stated here is first given in very compact form by Assouad [11]; the more detailed version given here is reproduced from [32]. Thus the inequality contained in Lemma 4.2 has a long history. In contrast, the very useful "closed-form" inequality contained in Lemma 4.3 is proved rather recently by Blumer *et al.* in [32]. Vapnik ([190], p.154) proves a slightly weaker inequality; he shows that (in the present notation) $\pi(n;\mathcal{A}) \leq 1.5 n^d/d!$. The definition of the P-dimension is given by Pollard [163] and also by Haussler [80]. Theorem 4.3 and its proof are taken from [80]. There are several alternatives to the P-dimension in the case of function classes. For instance, Vapnik [190] gives a definition of the VC-dimension of a family of real-valued functions, which corresponds to taking all components of the vector **c** in the definition of the P-dimension to be equal. Other definitions are due to Dudley [63] and Natarajan [146]. Regarding Theorem 4.5, Dudley [62] shows that if collections $\mathcal{A}_1, \ldots, \mathcal{A}_k$ all have finite VC-dimension, then any collection of sets obtained by performing a finite number of Boolean operations on the sets in each $\mathcal{A}_i$ also has finite VC-dimension. However, no explicit bound is given as is done here. Lemma 4.11 is found in [62], while Lemma 4.12 is found in [32]; but both results might perhaps be of greater antiquity. The fat-shattering dimension and corresponding inequalities are derived in [3]. An expository treatment can also be found in [7]. However, the proofs given here are much simpler and more direct than those in these references.

5. Uniform Convergence of Empirical Means

In this chapter, we study the first of the learning problems introduced in Chapter 3, namely the uniform convergence of empirical means to their true values as the number of samples approaches infinity. Two versions of this problem are studied. In the first version, the probability measure generating the samples is assumed to be known and fixed. In the second version, the probability measure is *not* assumed to be known – only that it belongs to some known *family* of probabilities – and the convergence of the empirical means to their true values is required to be uniform also with respect to the probability (as well as with respect to the function whose mean is being estimated empirically). In each case, several necessary and sufficient conditions for the required uniform convergence are stated and proved.

The main result of this chapter is Theorem 5.3; the remaining results of this chapter are all based on this theorem. Unfortunately, while the statement of this theorem is easy to understand, the proof is rather long and technical. However, if this particular theorem is accepted "on faith", then the proofs of the remaining theorems, based on Theorem 5.3, are relatively simple and straight-forward. Thus a reader who is interested in understanding the main ideas quickly can perhaps accept Theorem 5.3 "on faith" and proceed to the remaining theorems. To follow such a procedure, one should read Sections 5.1 through 5.3, omit Sections 5.4, 5.5 and 5.6, and pick up reading again with Section 5.7.

5.1 Restatement of the Problems Under Study

For the convenience of the reader, the various problems under study are restated here. One is given:

- A set X,
- A σ-algebra $\mathcal{S}$ of subsets of X,
- A family $\mathcal{F}$ of functions mapping X into $[0, 1]$, each of which is measurable with respect to $\mathcal{S}$, or else
- A collection $\mathcal{A}$ of sets belonging to $\mathcal{S}$.
- A family $\mathcal{P}$ of probability measures on $(X, \mathcal{S})$.

150   5. Uniform Convergence of Empirical Means

Consider first the problem of the uniform convergence of empirical means under a *known fixed* probability. In this case, $\mathcal{P}$ is a singleton set $\{P\}$, and P is known. Suppose $x_1, \ldots, x_m \in X$ are i.i.d. samples, drawn in accordance with P, and define $\mathbf{x} := [x_1 \ldots x_m]^t \in X^m$. For each $f \in \mathcal{F}$, define its "true" mean by

$$E(f) := \int_X f(x) \, P(dx),$$

and its "empirical" mean based on the multisample $\mathbf{x}$ by

$$\hat{E}(f; \mathbf{x}) := \frac{1}{m} \sum_{i=1}^{m} f(x_i).$$

Finally, for each $\epsilon > 0$, define[1]

$$q(m, \epsilon) := P^m \{\mathbf{x} \in X^m : \exists f \in \mathcal{F} \text{ s.t. } |\hat{E}(f; \mathbf{x}) - E_P(f)| > \epsilon\}.$$

An equivalent definition of $q(m, \epsilon)$ is:

$$q(m, \epsilon) := P^m \{\mathbf{x} \in X^m : \sup_{f \in \mathcal{F}} |\hat{E}(f; \mathbf{x}) - E_P(f)| > \epsilon\}.$$

The objective is to determine necessary and sufficient conditions under which $q(m, \epsilon) \to 0$ as $m \to \infty$ for each $\epsilon > 0$. If this is true, we say that the family $\mathcal{F}$ has the property of **uniform convergence of empirical means (UCEM)**.

An important special case occurs when every function in $\mathcal{F}$ assumes only the values 0 or 1. If $f : X \to \{0, 1\}$, one can identify f with its **support**

$$\text{supp}(f) := \{x \in X : f(x) = 1\}.$$

Clearly f is just the indicator function of its support, and $E_P(f)$ is the probability measure of this set. Thus, in this special case, one can reformulate the problem in terms of empirically estimating the probabilities of a given family of measurable sets $\mathcal{A} \subseteq \mathcal{S}$. Given a multisample $\mathbf{x} \in X^m$ and a set $A \in \mathcal{A}$, one can think of

$$\hat{P}(A; \mathbf{x}) := \frac{1}{m} \sum_{i=1}^{m} I_A(x_i)$$

as an empirical estimate of the "true" probability $P(A)$. Define

$$q(m, \epsilon) := P^m \{\mathbf{x} \in X^m : \exists A \in \mathcal{A} \text{ s.t. } |\hat{P}(A; \mathbf{x}) - P(A)| > \epsilon\},$$

or equivalently,

---
[1] Actually, it would be more accurate to write $q(m, \epsilon, P)$ instead of $q(m, \epsilon)$. However, so long as P is known and fixed, there is no harm in not explicitly displaying the dependence of q on P.

$$q(m,\epsilon) := P^m\{\mathbf{x} \in X^m : \sup_{A \in \mathcal{A}} |\hat{P}(A;\mathbf{x}) - P(A)| > \epsilon\}.$$

If $q(m,\epsilon) \to 0$ as $m \to \infty$ for each $\epsilon > 0$, we say that the collection of sets $\mathcal{A}$ has the property of **uniform convergence of empirical probabilities (UCEP)**.

As indicated in Chapter 3, it is possible to replace the above notion of "uniform convergence" of empirical means (or probabilities) by the stronger notion of "almost sure convergence" of empirical means (or probabilities). For the benefit of the reader, this more general problem formulation is recalled here.

Let all symbols be as above, and let $(X^\infty, \mathcal{S}^\infty, P^\infty)$ denote the probability space obtained by taking the countably infinite Cartesian product of X and extending $\mathcal{S}, P$ correspondingly. Recall the definition of the stochastic process $\{a_m(\cdot)\}$ from (3.1.3):

$$a_m : \mathbf{x}^* \in X^\infty \mapsto \sup_{f \in \mathcal{F}} |\hat{E}_m(f; \mathbf{x}^*) - E_P(f)|, \qquad (5.1.1)$$

where

$$\hat{E}_m(f; \mathbf{x}^*) := \frac{1}{m} \sum_{i=1}^m f(x_i)$$

is the empirical mean of the function f based on the first m components of the sequence $\mathbf{x}^*$. Now the quantity $q(m,\epsilon)$ defined above is equal to

$$q(m,\epsilon) = P^\infty\{\mathbf{x}^* \in X^\infty : a_m(\mathbf{x}^*) > \epsilon\}.$$

Thus the UCEM property corresponds to the stochastic process $\{a_m(\cdot)\}$ converging to zero *in probability* with respect to P^∞. If, on the other hand, the stochastic process $\{a_m(\cdot)\}$ converges to zero *almost surely*, then we say that the family $\mathcal{F}$ has the property of **almost sure convergence of empirical means (ASCEM)**. Thus the ASCEM property is equivalent to the condition:

$$P^\infty\{\mathbf{x}^* \in X^\infty : \sup_{f \in \mathcal{F}} |\hat{E}_m(f; \mathbf{x}^*) - E_P(f)| \to 0 \text{ as } m \to \infty\} = 1.$$

Clearly the ASCEM property implies the UCEM property. Note that, from Lemma 2.10, if the sequence $\{q(m,\epsilon)\}$ is summable for each fixed $\epsilon > 0$, then the family $\mathcal{F}$ has the ASCEM property. In other words, if empirical means converge uniformly and sufficiently rapidly, then they also converge almost surely. In case all functions in $\mathcal{F}$ are binary-valued so that empirical means of functions are actually empirical probabilities of sets, we can speak of a collection of sets having the property of **almost sure convergence of empirical probabilities (ASCEP)**.

The above problem of the uniform convergence of empirical means pertains to the case where the samples are drawn in accordance with a *known*

152   5. Uniform Convergence of Empirical Means

fixed probability P. The next problem studied in this chapter is a generalization of the above problem, in that P is known only to belong to a a *family* of probabilities $\mathcal{P}$, and the empirical means are required to converge uniformly not only with respect to the functions but also with respect to the probabilities.

Specifically, suppose $P \in \mathcal{P}$ is fixed but unknown, and that $x_1, \ldots, x_m \in X$ are i.i.d. samples drawn in accordance with P. Define $E_P(f)$ and $\hat{E}(f; \mathbf{x})$ as above, and define

$$q(m, \epsilon, P) := P^m\{\mathbf{x} \in X^m : \sup_{f \in \mathcal{F}} |\hat{E}(f; \mathbf{x}) - E_P(f)| > \epsilon\}, \text{ and}$$

$$\bar{q}(m, \epsilon, \mathcal{P}) := \sup_{P \in \mathcal{P}} q(m, \epsilon, P).$$

Note that $q(m, \epsilon, P)$ is the same as the earlier $q(m, \epsilon)$ defined above, except that the dependence of q on P is explicitly identified. A family of functions $\mathcal{F}$ is said to have the property of **uniform convergence of empirical means uniformly in probability (UCEMUP)** with respect to the family $\mathcal{P}$ if

$$\bar{q}(m, \epsilon, \mathcal{P}) \to 0 \text{ as } m \to \infty.$$

In the same spirit, suppose $\mathcal{A}$ is a collection of measurable sets. Given a probability $P \in \mathcal{P}$, define

$$q(m, \epsilon, P) := P^m\{\mathbf{x} \in X^m : \sup_{A \in \mathcal{A}} |\hat{P}(A; \mathbf{x}) - P(A)| > \epsilon\}, \text{ and}$$

$$\bar{q}(m, \epsilon, \mathcal{P}) := \sup_{P \in \mathcal{P}} q(m, \epsilon, P).$$

A collection of sets $\mathcal{A}$ is said to have the property of **uniform convergence of empirical probabilities uniformly in probability (UCEPUP)** with respect to the family $\mathcal{P}$ if

$$\bar{q}(m, \epsilon, \mathcal{P}) \to 0 \text{ as } m \to \infty.$$

In the case where P is known and fixed, we have defined a stronger property than UCEM, that was called ASCEM. However, in the case where P itself varies over a family $\mathcal{P}$, there is no natural way of speaking of "almost sure convergence of empirical means uniformly in probability." We could of course speak about "almost sure convergence of empirical means for *every* probability," which corresponds to the condition

$$P^\infty\{\mathbf{x}^* \in X^\infty : \sup_{f \in \mathcal{F}} |\hat{E}_m(f; \mathbf{x}^*) - E_P(f)| \to 0 \text{ as } m \to \infty\} = 1, \forall P \in \mathcal{P}.$$

However, in contrast with the case where P is fixed and known, this property no longer implies the UCEMUP property, because there is no notion of uniformity *across* probability measures.

5.2 Equivalence of the UCEM and ASCEM Properties

In this section, it is shown that the properties of UCEM and ASCEM are equivalent. In other words, if a family of functions has the property that empirical means converge *uniformly* to their true values, then in fact the empirical means converge *almost surely* to their true values. (The implication in the other direction is obvious.) This rather surprising result is obtained using a powerful notion known as "subadditive" stochastic processes.

Let us now define the notion of a subadditive stochastic process. We shall not give the most general possible definition; for such a treatment, see Kingman [104], [105]. However, the discussion below is quite adequate for the present purposes.

Suppose $\{\zeta_{lm}(\cdot), 0 \leq l < m\}$ is a family of random variables on the probability space $(X^\infty, S^\infty, P^\infty)$, indexed by the nonnegative integers l and m. We say that $\{\zeta_{lm}\}$ is **subadditive** if three conditions are satisfied:

(S1) For all $\mathbf{x}^* \in X^\infty$ and all $l < m < n$, we have

$$\zeta_{ln}(\mathbf{x}^*) \leq \zeta_{lm}(\mathbf{x}^*) + \zeta_{mn}(\mathbf{x}^*).$$

(S2) The collection of random variables $\{\zeta_{lm}\}_{0 \leq l < m}$ has the same joint distribution as the collection of random variables $\{\zeta_{l+1,m+1}\}_{0 \leq l < m}$.

(S3) The expectations $E_{P^\infty}(\zeta_{0m})$ exist for all $m > 0$; moreover, there exists a constant $\mu \geq 0$ such that

$$E_{P^\infty}(\zeta_{0m}) \geq -\mu m, \ \forall m > 0.$$

The central fact about subadditive stochastic processes that is used in the sequel is the following [104]:

Theorem 5.1. *Suppose $\{\zeta_{lm}\}_{0 \leq l < m}$ is a subadditive stochastic process. Then there exists a random variable θ of finite mean such that ζ_{0m}/m converges almost surely to θ.*

As an illustration of the power of this concept, let us study the eventual behaviour of the maximum "discrepancy" of empirical means from their true values. Let $\mathcal{F} \subseteq [0,1]^X$ consist of measurable functions, and define

$$\zeta_{lm}(\mathbf{x}^*) := \sup_{f \in \mathcal{F}} \left| \sum_{i=l+1}^{m} [f(x_i) - E_P(x_i)] \right|.$$

To interpret this stochastic process, observe that

$$\frac{1}{m-l} \sum_{i=l+1}^{m} f(x_i)$$

is just the empirical mean of the function $f(\cdot)$, based on the samples $x_{l+1}, \ldots, x_m$. Thus $\zeta_{lm}(\mathbf{x}^*)$ is the (unnormalized) maximum discrepancy between the empirical mean of f and the true mean of f, as f varies over $\mathcal{F}$.

It is easy to show that the process $\{\zeta_{lm}(\cdot)\}$ is subadditive, by verifying Conditions (S1) through (S3). Condition (S1) follows from the triangle inequality, because for $l < m < n$, we have

$$\left|\sum_{i=l+1}^{n} [f(x_i) - E_P(f)]\right| \leq \left|\sum_{i=l+1}^{m} [f(x_i) - E_P(f)]\right| + \left|\sum_{i=m+1}^{n} [f(x_i) - E_P(f)]\right|,$$

for each function $f \in \mathcal{F}$. Condition (S2) follows because the x_i's are i.i.d., while (S3) is satisfied with $\mu = 0$. This leads to the following useful conclusion.

Lemma 5.1. *The stochastic process*

$$a_m(\mathbf{x}^*) := \sup_{f \in \mathcal{F}} |\hat{E}(f; \mathbf{x}^*) - E_P(f)| \tag{5.2.1}$$

converges almost surely to a constant at $m \to \infty$.

Proof. It has been established above that the stochastic process $\{\zeta_{lm}(\cdot)\}$ is subadditive. Hence by Theorem 5.1, the stochastic process $\zeta_{0m}/m = a_m$ converges almost surely to a random variable θ. It only remains to show that the limit function is in fact *constant* almost everywhere. Since the proof of this part requires some advanced notions from probability theory, it is only sketched here. Clearly the limit is unaffected by replacing $\mathbf{x}^*$ by $T\mathbf{x}^*$, where $T: X^\infty \to X^\infty$ is the shift operation defined by

$$T: (x_1, x_2, \ldots) \mapsto (x_2, x_3, \ldots). \tag{5.2.2}$$

Hence the limit random variable is measurable with respect to the "tail" σ-algebra generated by the x_i's. Since the x_i's are i.i.d., the tail σ-algebra is degenerate by Kolmogorov's 0-1 law (see, e.g., [119], p. 241 or [35]). Hence the limit random variable is constant almost everywhere. ■

Lemma 5.1 states that, whatever be the family $\mathcal{F}$, the worst-case discrepancy between the empirical mean and the true mean converges almost surely to *some* constant. The question is: What is this constant? If the constant is zero, then $\mathcal{F}$ has the property that empirical means converge almost surely to their true values, i.e., the ASCEM property. Using this lemma, we can now prove the following (perhaps surprising) result.

Theorem 5.2. *Suppose $\mathcal{F}$ consists of measurable functions mapping X into $[0,1]$. Then $\mathcal{F}$ has the UCEM property if and only if it has the ASCEM property.*

Proof. "If" Obvious.

"Only if" Suppose $\mathcal{F}$ has the UCEM property. Then, by definition, the stochastic process $\{a_m\}$ defined in (5.2.1) converges *in probability* to zero. By Lemma 5.1, the same stochastic process also converges *almost surely* to some constant. Now this constant can only be zero, because if a stochastic process converges almost surely, then it also converges in probability, and moreover, both limits are the same. Hence we conclude that $\{a_m\}$ converges almost surely to zero, i.e., that $\mathcal{F}$ has the ASCEM property. ∎

In the proofs of Lemma 5.1 and Theorem 5.2, the proof that the limit random variable was in fact a constant depended *only* on the fact that the sample sequence $\{x_i\}$ was ergodic. Thus it is clear that the conclusions of the lemma and the theorem remain valid even if the sample sequence is *not* i.i.d., provided only that it is ergodic.

We shall encounter some other subadditive processes besides $\{\zeta_{lm}(\cdot)\}$ later in this chapter.

5.3 Main Theorems

In this section, we state the main theorems that give necessary and sufficient conditions for a family of bounded measurable functions to have the property that empirical means converge uniformly to their true values. The results cover both the case where $\mathcal{P}$ is a singleton set (i.e., P is known) as well as the case where $\mathcal{P}$ is known but $P \in \mathcal{P}$ is unknown. The proofs of the main theorems are given in the next several sections. The present section contains only the statements of the theorems and a few examples that illustrate the use of the theorems. A little bit of notation is introduced first to facilitate the statement of the theorems.

Recall that the l_∞-norm on $\mathbb{R}^m$ is defined by

$$\| \mathbf{y} \|_\infty = \max_{1 \leq i \leq m} |y_i|, \ \forall \mathbf{y} \in \mathbb{R}^m.$$

Now suppose $S \subseteq \mathbb{R}^m$. By a slight abuse of notation, the symbol $L(\epsilon, S, \| \cdot \|_\infty)$ is used to denote the external ϵ-covering number of the set S with respect to the *metric induced by* the norm $\| \cdot \|_\infty$. The l_∞-norm has a very useful property. Suppose $\mathbf{y} = [\mathbf{y}_1^t \ \mathbf{y}_2^t]^t \in \mathbb{R}^{k+l}$, where $\mathbf{y}_1 \in \mathbb{R}^k$ and $\mathbf{y}_2 \in \mathbb{R}^l$. Then

$$\| \mathbf{y} \|_\infty = \max\{\| \mathbf{y}_1 \|_\infty, \| \mathbf{y}_2 \|_\infty\}.$$

A consequence of this property is the following inequality: Suppose $S = S_1 \times S_2$, where $S_1 \subseteq \mathbb{R}^k$ and $S_2 \subseteq \mathbb{R}^l$. Then

$$L(\epsilon, S, \| \cdot \|_\infty) \leq L(\epsilon, S_1, \| \cdot \|_\infty) \cdot L(\epsilon, S_2, \| \cdot \|_\infty). \quad (5.3.1)$$

This inequality is easy to prove. Suppose $\{\mathbf{y}_1^1, \ldots, \mathbf{y}_m^1\}$ is an external ϵ-cover for S_1, and that $\{\mathbf{y}_1^2, \ldots, \mathbf{y}_M^2\}$ is an external ϵ-cover for S_2; then the set

$$\{(\mathbf{y}_i^1, \mathbf{y}_j^2) : 1 \leq i \leq m, 1 \leq j \leq M\}$$

is an external ϵ-cover for the set $S_1 \times S_2$.

Suppose $\mathbf{x} \in X^m$. For each function $f \in \mathcal{F}$, the vector $\mathbf{f}(\mathbf{x}) \in [0,1]^m$ is defined by

$$\mathbf{f}(\mathbf{x}) := [f(x_1)\ f(x_2) \ldots f(x_m)]^t,$$

while the set $\mathcal{F}|_{\mathbf{x}} \subseteq [0,1]^m$ is defined by

$$\mathcal{F}|_{\mathbf{x}} := \{\mathbf{f}(\mathbf{x}) : f \in \mathcal{F}\}.$$

In other words, $\mathbf{f}(\mathbf{x})$ denotes the vector obtained by evaluating the function f at each component of the vector $\mathbf{x}$, while $\mathcal{F}|_{\mathbf{x}}$ denotes the collection of vectors $\mathbf{f}(\mathbf{x})$ obtained by letting f vary over $\mathcal{F}$. Now let $L(\epsilon, \mathcal{F}|_{\mathbf{x}}, \|\cdot\|_\infty)$ denote the external ϵ-covering number of the set $\mathcal{F}|_{\mathbf{x}}$ with respect to the metric induced by $\|\cdot\|_\infty$. Since $\mathcal{F}|_{\mathbf{x}} \subseteq [0,1]^m$, it is easy to see that

$$L(\epsilon, \mathcal{F}|_{\mathbf{x}}, \|\cdot\|_\infty) \leq \frac{1}{\epsilon^m}, \ \forall \mathbf{x} \in X^m, \forall m. \tag{5.3.2}$$

Now consider the special case where $Y \subseteq [0,1]$ is a fixed *finite* set, and every $f \in \mathcal{F}$ maps X into Y. Then, since Y is finite, there exists a number ϵ_0 such that

$$y_i, y_j \in Y, y_i \neq y_j \Rightarrow |y_i - y_j| \geq 2\epsilon_0.$$

If $Y = \{0,1\}$ so that the problem at hand is one of estimating empirical *probabilities*, then one can take $\epsilon_0 = 0.5$. Now, for every $m \geq 1$, the preceding inequality implies that

$$\mathbf{y}_i, \mathbf{y}_j \in Y^m, \mathbf{y}_i \neq \mathbf{y}_j \Rightarrow \|\mathbf{y}_i - \mathbf{y}_j\|_\infty \geq 2\epsilon_0.$$

Hence, whenever $\epsilon < \epsilon_0$, it follows that $L(\epsilon, \mathcal{F}|_{\mathbf{x}}, \|\cdot\|_\infty)$ is precisely equal to *the number of distinct vectors* $\mathbf{f}(\mathbf{x})$ generated as f varies over $\mathcal{F}$. In other words, $L(\epsilon, \mathcal{F}|_{\mathbf{x}}, \|\cdot\|_\infty)$ equals the *cardinality* of the set $\mathcal{F}|_{\mathbf{x}}$ whenever $\epsilon < \epsilon_0$. Let $\pi(\mathbf{x}; \mathcal{F})$ denote this latter number. Then

$$L(\epsilon, \mathcal{F}|_{\mathbf{x}}, \|\cdot\|_\infty) = \pi(\mathbf{x}; \mathcal{F}) \ \forall \epsilon < \epsilon_0.$$

Now consider the problem of uniform convergence of the empirical *probabilities* of a collection of sets $\mathcal{A} \subseteq \mathcal{S}$. As stated previously, this problem is equivalent to the uniform convergence of the empirical *means* of the associated family of indicator functions $\{I_A, A \in \mathcal{A}\}$. Suppose $\mathbf{x} \in X^m$. By a slight abuse of notation, let $\pi(\mathbf{x}; \mathcal{A})$ denote the number of distinct binary vectors of the form $[I_A(x_1) \ldots I_A(x_m)]^t \in \{0,1\}^m$ generated as A varies over $\mathcal{A}$. It is not difficult to see that $\pi(\mathbf{x}; \mathcal{A})$ is the same as the integer $\pi(S; \mathcal{A})$ defined in Chapter 4 with $S = \{x_1, \ldots, x_m\}$. In general the vector $\mathbf{x}$ need not have all distinct components, and the symbol $\pi(\mathbf{x}; \mathcal{A})$ as defined above makes sense even if $\mathbf{x}$ has some repeated components.

The main theorems of this section are now stated in succession.

5.3 Main Theorems

Theorem 5.3. *Suppose $\mathcal{F} \subseteq [0,1]^X$ consists of measurable functions. Then $\mathcal{F}$ has the property of uniform convergence of empirical means if and only if*

$$\lim_{m \to \infty} \frac{E_{P^m}[\lg L(\epsilon, \mathcal{F}|_{\mathbf{x}}, \|\cdot\|_\infty)]}{m} = 0, \; \forall \epsilon > 0. \qquad (5.3.3)$$

Corollary 5.1. *Let Y be a fixed finite set, and suppose every $f \in \mathcal{F}$ maps X into Y. Then $\mathcal{F}$ has the property of uniform convergence of empirical means if and only if*

$$\lim_{m \to \infty} \frac{E_{P^m}[\lg \pi(\mathbf{x}; \mathcal{F})]}{m} = 0.$$

Corollary 5.2. *A collection of sets $\mathcal{A} \subseteq \mathcal{S}$ has the property of uniform convergence of empirical probabilities if and only if*

$$\lim_{m \to \infty} \frac{E_{P^m}[\lg \pi(\mathbf{x}; \mathcal{A})]}{m} = 0. \qquad (5.3.4)$$

Theorem 5.3 is applicable to any family of functions that assume values in the interval $[0,1]$. However, if we restrict our attention to *binary-valued* functions, i.e., the problem of the uniform convergence of empirical *probabilities* as opposed to empirical means, then it is possible to replace (5.3.4) by a simpler (though of course equivalent) condition. To state this alternative condition, a little notation is required.

Suppose $\mathcal{A} \subseteq \mathcal{S}$ and that $\mathbf{x} \in X^m$. Define $S = \{x_1, \ldots, x_m\} \subseteq X$. Since it is possible for the vector $\mathbf{x}$ to have repeated elements, in general the cardinality of S need not equal m. Look at the collection $\{A \cap S : A \in \mathcal{A}\} \subseteq 2^S$ of subsets of S. Now the integer $d(\mathbf{x})$ is defined as the VC-dimension of the collection $\{A \cap S : A \in \mathcal{A}\}$. Equivalently, $d(\mathbf{x})$ can be defined as the largest integer n such that there exists some subset $\{x_{i_1}, \ldots, x_{i_n}\} \subseteq S$ that is shattered by $\mathcal{A}$. Clearly $d(\mathbf{x}) \leq m$, and in fact, $d(\mathbf{x})$ is no larger than the number of *distinct* elements among $x_1, \ldots, x_m$. With this definition, we are in a position to state an alternative to Theorem 5.3 for the uniform convergence of empirical probabilities.

Theorem 5.4. *The collection of sets $\mathcal{A}$ has the property of uniform convergence of empirical probabilities if and only if*

$$\lim_{m \to \infty} \frac{E_{P^m}[d(\mathbf{x})]}{m} = 0. \qquad (5.3.5)$$

The above theorems have a simple intuitive interpretation. In the case of Theorem 5.3, we have already seen that (5.3.2) holds, whatever be the nature of $\mathcal{F}$. Thus Theorem 5.3 can be interpreted as follows: The family $\mathcal{F}$ has the UCEM property if and only if, "on average" with respect to $\mathbf{x} \in X^m$, the number $L(\epsilon, \mathcal{F}|_{\mathbf{x}}, \|\cdot\|_\infty)$ grows "more slowly" than 2^m as $m \to \infty$. Similarly, Theorem 5.4 requires that, on average, the VC-dimension of the collection $\mathcal{A}$, when restricted to a "random" set of cardinality m or less, should grow more slowly than m.

158   5. Uniform Convergence of Empirical Means

The above theorems and corollaries all pertain to the problem of the uniform convergence of empirical means or probabilities under a *known fixed* probability measure. The next set of results is applicable to the problem of the uniform convergence of empirical means or probabilities uniformly under a *family* of probability measures. Not surprisingly, these results are natural modifications of their counterparts for the case of a known fixed probability.

Theorem 5.5. *The family of functions $\mathcal{F}$ has the UCEMUP property if and only if*
$$\lim_{m \to \infty} \sup_{P \in \mathcal{P}} \frac{E_{P^m}[\lg L(\epsilon, \mathcal{F}|_{\mathbf{x}}, \|\cdot\|_\infty)]}{m} = 0, \ \forall \epsilon > 0. \quad (5.3.6)$$

Corollary 5.3. *Let Y be a fixed finite set, and suppose every $f \in \mathcal{F}$ maps X into Y. Then $\mathcal{F}$ has the UCEMUP property if and only if*
$$\lim_{m \to \infty} \sup_{P \in \mathcal{P}} \frac{E_{P^m}[\lg \pi(\mathbf{x}; \mathcal{F})]}{m} = 0.$$

Corollary 5.4. *A collection of sets $\mathcal{A} \subseteq \mathcal{S}$ has the UCEPUP property if and only if*
$$\lim_{m \to \infty} \sup_{P \in \mathcal{P}} \frac{E_{P^m}[\lg \pi(\mathbf{x}; \mathcal{A})]}{m} = 0. \quad (5.3.7)$$

Theorem 5.6. *The collection of sets $\mathcal{A}$ has the UCEPUP property if and only if*
$$\lim_{m \to \infty} \sup_{P \in \mathcal{P}} \frac{E_{P^m}[d(\mathbf{x})]}{m} = 0. \quad (5.3.8)$$

The application of these results is illustrated through a few examples.

Example 5.1. Let $X \subseteq \mathbb{R}$, $\mathcal{S}$ = the Borel σ-algebra on X, and $\mathcal{A}$ = the collection of all finite subsets of X. We have already seen in Example 3.2 that $\mathcal{A}$ **fails** to have the UCEP property if P is the uniform distribution on X. However, using Theorem 5.4, it is possible to derive a much better result. Let P be *any* probability on X such that the "diagonal" $\{(x, x) : x \in X\} \subseteq X^2$ has zero measure with respect to the product probability P^2. For example, any nonatomic measure satisfies this requirement. Now let m be a fixed integer and let $\mathbf{x} \in X^m$. By virtue of the assumption above that the diagonal in X^2 has zero probability, it follows that all components of $\mathbf{x}$ are distinct, with probability one (with respect to P^m). It is easy to see that *every* finite set is shattered by $\mathcal{A}$. Thus $d(\mathbf{x})$ equals the number of distinct elements in $\mathbf{x}$, which in turn implies that $d(\mathbf{x}) = m$ with probability one (with respect to P^m). Hence
$$E_{P^m}[d(\mathbf{x})] = m \ \forall m \geq 2, \text{ and}$$
$$\lim_{m \to \infty} \frac{E_{P^m}[d(\mathbf{x})]}{m} = 1.$$

Since the condition (5.3.5) is not satisfied, $\mathcal{A}$ does not have the UCEP property.

Example 5.2. Let $X, \mathcal{S}$ be as above, and let $\mathcal{A}$ be as in Example 3.3. Thus $\mathcal{A}$ consists of all unions of the form $A \cup [0, b]$ where $A = \{a_1, \ldots, a_n\}$ with $a_1 > a_2 > \ldots > a_n$, and
$$b = \sum_{i=1}^{n} a_i (0.5)^{i+1}.$$
It is clear that $\mathcal{A}$ shatters every finite set $S = \{x_1, \ldots, x_n\}$ with the property that $x_i \geq 0.5$ for all i, because $b < 0.5$ always. Therefore $d(\mathbf{x})$ is at least equal to the number of distinct components of $\mathbf{x}$ that belong to $[0.5, 1]$. Now let P be any probability on X such that the interval $[0.5, 1]$ has positive measure, and such that the diagonal in X^2 has zero measure with respect to P^2. For convenience, let $\alpha := P\{[0.5, 1]\}$. If m i.i.d. samples are drawn in accordance with P, then all these samples are distinct, with probability one, since the diagonal in X^2 has zero measure. Moreover, the number of samples amongst $x_1, \ldots, x_m$ that belong to the interval $[0.5, 1]$ equals αm on average. Hence
$$E_{P^m}[d(\mathbf{x})] \geq \alpha m, \ \forall m \geq 2, \text{ and}$$
$$\lim_{m \to \infty} \frac{E_{P^m}[d(\mathbf{x})]}{m} \geq \alpha.$$
So once again (5.3.5) is not satisfied, and $\mathcal{A}$ does not have the UCEP property.

Note that the present example is more general than Example 3.3 in that both X and P are more general.

Example 5.3. Let $X = [0, 1]^2$, $\mathcal{S}$ = the Borel $\mathcal{S}$-algebra on X, and $\mathcal{A}$ = the collection of convex sets in X. Let P be the *uniform* probability measure on X. Then $\mathcal{A}$ has the property that empirical probabilities converge uniformly. The proof of this can be found in [162], pp. 22-24, and is reproduced below in the present notation. This example is discussed further in Chapter 6.

Suppose by way of contradiction that the collection $\mathcal{A}$ does not have the UCEP property. Then, by Theorem 5.4, it follows that the condition (5.3.5) is violated. In other words, there exist a number α and a sequence of integers $\{n_i\}$ approaching infinity, such that
$$E_{P^{n_i}}[d(\mathbf{x})/n_i] \geq \alpha, \ \forall i.$$
In turn this implies that, for each i,
$$P^{n_i}\{\mathbf{x} \in X^{n_i} : d(\mathbf{x})/n_i \geq \alpha/2\} \geq \alpha/2.$$
Otherwise, we would have
$$E_{P^{n_i}}[d(\mathbf{x})/n_i] = \int_{d(\mathbf{x})/n_i < \alpha/2} d(\mathbf{x}) P(d\mathbf{x}) + \int_{d(\mathbf{x})/n_i \geq \alpha/2} d(\mathbf{x}) P(d\mathbf{x})$$

160   5. Uniform Convergence of Empirical Means

$$< \frac{\alpha}{2} + \frac{\alpha}{2} = \alpha,$$

which is a contradiction. Let $\epsilon = \alpha/2$, so that the above relationship can be rewritten as

$$P^{n_i}\{\mathbf{x} \in X^{n_i} : d(\mathbf{x}) \geq n_i \epsilon\} \geq \epsilon.$$

Now choose an integer m to be a power of three such that $(8/9)^m < \epsilon/2$. Divide the set $[0,1]^2$ into m^2 squares of size $1/m$ by $1/m$, as shown in Figure 5.1. Let $\mathcal{B}$ denote the collection of all possible unions of such squares. Since $\mathcal{B}$ is a finite collection, it has the UCEP property. Thus there exists an integer m_0 such that

$$P^m\{\mathbf{x} \in X^m : \sup_{A \in \mathcal{B}} |\hat{P}(A;\mathbf{x}) - P(A)| > \epsilon/2\} \leq \epsilon/2, \ \forall m \geq m_0.$$

Now choose an integer $n_i \geq m_0$, and just call it n. Then two inequalities hold:

If $V := \{\mathbf{x} \in X^n : d(\mathbf{x}) \geq n\epsilon\}$, then $P^n(V) \geq \epsilon$, and

If $W := \{\mathbf{x} \in X^n : \sup_{A \in \mathcal{B}} |\hat{P}(A;\mathbf{x}) - P(A)| \leq \epsilon/2\}$, then $P^n(W) \geq 1 - \epsilon/2$.

Thus $P^n(V \cap W) \geq \epsilon/2 > 0$, because $(V \cap W)^c = V^c \cup W^c$, and

$$P^n(V^c \cup W^c) \leq P^n(V^c) + P^n(W^c) \leq 1 - \epsilon + \frac{\epsilon}{2} = 1 - \frac{\epsilon}{2}.$$

In particular, $V \cap W$ is nonempty. Now choose an element $\mathbf{x} \in V \cap W$. Then, by the definition of V, it follows that $d(\mathbf{x}) \geq n\epsilon$. Select a set S of cardinality $n\epsilon$ that is shattered by $\mathcal{A}$, and let H denote the convex hull of S. Now, it is clear that S is shattered by $\mathcal{A}$ (the collection of convex subsets of $[0,1]^2$) if and only if every point in S is an extremal point of H, i.e., no point of S can be expressed as a convex combination of any other subset of points of S (see Figure 5.1 again). Now let A denote those squares in $[0,1]^2$ that

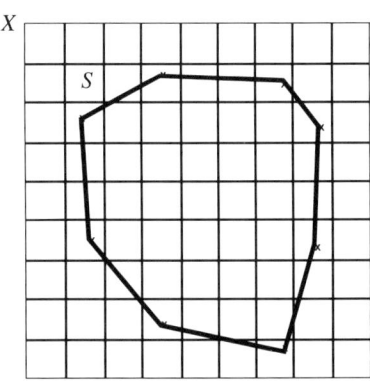

Fig. 5.1. Shattering a Set by Convex Polygons

contain a point of ∂H, where ∂H denotes the boundary of H (see Figure 5.1 again). Then $S \subseteq A$, which implies that $\hat{P}(A; \mathbf{x}) \geq \epsilon$. On the other hand, since $\mathbf{x} \in W$, it follows that $|\hat{P}(A; \mathbf{x}) - P(A)| \leq \epsilon/2$, which in turn implies that $P(A) \geq \epsilon/2$. Now, since m is a power of three, divide $[0, 1]^2$ into nine equal-sized squares as shown in Figure 5.2. It is clear that all nine

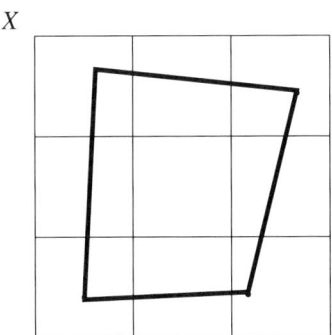

Fig. 5.2. Dividing the Unit Square into Nine Equal Parts

squares cannot contain elements of ∂H, since the middle square is contained in the convex hull of any four points lying one each in each of the four corner squares (see Figure 5.2). Hence at most eight of the nine squares can contain elements of ∂H, i.e., $P(A) \leq 8/9$. Now examine those squares in Figure 5.2 that contain elements of ∂H, and repeat the argument for each of the smaller squares. In any such smaller square, no more than eight out of nine smaller squares can contain part of the boundary ∂H. After repeating this argument $\log_3 m$ times, one arrives at the conclusion that $P(A) \leq (8/9)^m < \epsilon/2$. This contradicts the earlier established fact that $P(A) \geq \epsilon/2$. This contradiction proves that the premise is false, i.e., that (5.3.5) is in fact satisfied, and that the collection of convex sets in $[0, 1]^2$ does indeed have the UCEP property.

A careful examination of the above proof shows that the example can at once be generalized to the following situation: Let $X = [0, 1]^k$, where k is any integer, and let P be any measure on X such that $P = \phi_1 \times \ldots \times \phi_k$, where each ϕ_i is a measure on $[0, 1]$ with a continuous distribution function. To apply the above argument, one would replace the constant $8/9$ by $1 - (1/3)^k$, and divide the set X into squares such that each square has measure $1/3^k$. Figure 5.3 shows how this can be done in the case where $m = 3$ and $k = 2$.

5.4 Preliminary Lemmas

In this subsection we state and prove a few preliminary results that are used in the proof of the main theorem, namely Theorem 5.3.

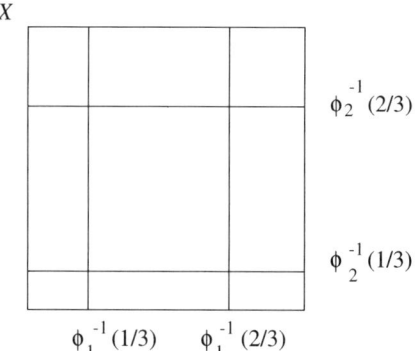

Fig. 5.3. Dividing the Unit Square Under a Nonuniform Distribution

Lemma 5.2. *For each $\epsilon > 0$, the stochastic process*

$$r_m(\epsilon, \mathbf{x}^*) := \frac{\lg L(\epsilon, \mathcal{F}|_{\mathbf{x}}, \|\cdot\|_\infty)}{m} \tag{5.4.1}$$

converges almost surely to a constant, call it $c(\epsilon)$, which is a nondecreasing function of ϵ.

Remarks: Note that $r_m(\epsilon, \mathbf{x}^*)$ in fact depends only on the first m components of $\mathbf{x}^*$.

Proof. The proof is based on the notion of subadditive processes. Let let us define a stochastic process, as follows: Given $0 \leq l < m$, define the set $\mathcal{F}_{lm}|_{\mathbf{x}^*} \subseteq [0,1]^{m-l}$ by

$$\mathcal{F}_{lm}|_{\mathbf{x}^*} := \{[f(x_{l+1})\ldots f(x_m)]^t : f \in \mathcal{F}\}.$$

Thus $\mathcal{F}_{lm}|_{\mathbf{x}^*}$ is the set of values assumed by the various functions f in $\mathcal{F}$ at the components x_{l+1} through x_m, and it is a subset of $[0,1]^{m-l}$. Note that the set $\mathcal{F}|_{\mathbf{x}}$ defined previously is merely $\mathcal{F}_{0m}|_{\mathbf{x}^*}$. Now define

$$r_{lm}(\mathbf{x}^*) := \frac{\lg L(\epsilon, \mathcal{F}_{lm}|_{\mathbf{x}^*}, \|\cdot\|_\infty)}{m-l}, \tag{5.4.2}$$

where $L(\epsilon, \mathcal{F}_{lm}|_{\mathbf{x}^*}, \|\cdot\|_\infty)$ is the external ϵ-covering number of the set $\mathcal{F}_{lm}|_{\mathbf{x}^*}$ with respect to the metric induced by the norm $\|\cdot\|_\infty$. It is shown that $\{(m-l)r_{lm}(\cdot)\}$ is subadditive, by verifying Conditions (S1) through (S3) from Section 5.2. We begin by observing that

$$\mathcal{F}_{ln}|_{\mathbf{x}^*} \subseteq \mathcal{F}_{lm}|_{\mathbf{x}^*} \times \mathcal{F}_{mn}|_{\mathbf{x}^*}.$$

As a consequence, it follows from (5.3.1) that

$$L(\epsilon, \mathcal{F}_{ln}|_{\mathbf{x}^*}, \|\cdot\|_\infty) \leq L(\epsilon, \mathcal{F}_{lm}|_{\mathbf{x}^*}, \|\cdot\|_\infty) \times L(\epsilon, \mathcal{F}_{mn}|_{\mathbf{x}^*}, \|\cdot\|_\infty).$$

Hence

$$\lg L(\epsilon, \mathcal{F}_{ln}|_{\mathbf{x}^*}, \|\cdot\|_\infty) \leq \lg L(\epsilon, \mathcal{F}_{lm}|_{\mathbf{x}^*}, \|\cdot\|_\infty) + \lg L(\epsilon, \mathcal{F}_{mn}|_{\mathbf{x}^*}, \|\cdot\|_\infty).$$

This shows that the Condition (S1) is satisfied. Condition (S2) follows from the fact that the x_i's are i.i.d. random variables, while (S3) is satisfied with $\mu = 0$. Thus the stochastic process $\{(m-l)r_{lm}(\cdot)\}$ is subadditive. Hence, by Theorem 5.1, it follows that the stochastic process $\{r_{0m}\}$ converges almost surely to a random variable. It only remains to show that the limit random variable is in fact a constant almost everywhere. This is achieved exactly as in the proof of Lemma 5.1. Observe that $r_{0m}(\cdot)$ is the same as what we have called $r_m(\epsilon, \cdot)$ above. Hence we have shown that $r_m(\epsilon, \mathbf{x}^*)$ converges almost surely to a constant. Let $c(\epsilon)$ denote this constant. It is obvious that the limit is a nondecreasing function of ϵ. ∎

Hereafter we suppress the dependence of various quantities on ϵ in the interests of brevity.

Since $r_m(\cdot)$ converges almost surely to c, it is clear that $E_{P^m}(r_m)$ also approaches c as $m \to \infty$.

Lemma 5.2 states only that the random variable r_m approaches a constant almost surely. However, for subsequent applications, it is desirable to have an estimate of the *rate* at which this convergence takes place. Such an estimate is provided by the next lemma.

Lemma 5.3. *Let r_m and c be as in Lemma 5.2. Suppose $\eta > 0$. Then*

$$\Pr\{r_m > c + \eta\} \leq \exp(-k\eta^2/8\beta^2), \tag{5.4.3}$$

where $\beta = \lg(1/\epsilon)$, m_0 is an integer selected such that

$$\left| \frac{E_{P^m}(r_m)}{m} - c \right| \leq \frac{\eta}{2} \ \forall m \geq m_0.$$

and k is the integer part of m/m_0.

Proof. Since $E_{P^m}(r_m) \to c$ as $m \to \infty$, it is possible to select an integer m_0 such that the above inequality holds. Let $\mathbf{x} = [\mathbf{x}_1^t \ldots \mathbf{x}_k^t]^t \in X^{km_0}$ for some integer k, and define the k-fold average

$$g_k = \frac{1}{k} \sum_{i=0}^{k-1} r_{im_0,(i+1)m_0},$$

where the random variable $r_{im_0,(i+1)m_0}$ is defined in (5.4.2), that is,

$$r_{im_0,(i+1)m_0} = \frac{\lg L(\epsilon, \mathcal{F}|_{\mathbf{x}_{i+1}}, \|\cdot\|_\infty)}{m_0}.$$

Then the random variables $r_{0,m_0}, \ldots, r_{(k-1)m_0, km_0}$ are independent, and belong to the interval $[0, \beta]$ where $\beta := \lg(1/\epsilon)$. Moreover,

164   5. Uniform Convergence of Empirical Means

$$E(r_{0,m_0}) = E(r_{im_0,(i+1)m_0}) \; \forall i = E(g_k) =: h_0, \text{ say.}$$

Moreover, from the manner in which m_0 was chosen, it follows that $|h_0 - c| \leq \eta/2$. Hence, by Hoeffding's inequality,

$$\Pr\{g_k > c + \eta\} \leq \Pr\{g_k > h_0 + \eta/2\} \leq \exp(-k\eta^2/2\beta^2).$$

Now let us examine the case where m is not an exact multiple of m_0. Suppose $m = km_0 + l$, where $0 \leq l \leq m_0 - 1$. Suppose $\mathbf{z} \in X^m$ and partition it as $\mathbf{z} = \mathbf{xy}$ where $\mathbf{x} \in X^{km_0}$ and $\mathbf{y} \in X^l$. By the subadditivity property of the stochastic process $\{(m-l)r_{lm}\}$, it follows that

$$r_{0m}(\mathbf{z}) \leq \frac{r_{0,km_0}(\mathbf{x}) + r_{km_0,m}(\mathbf{y})}{km_0 + l} \leq g_k(\mathbf{x}) + \frac{1}{k}\frac{\lg(1/\epsilon)}{m_0}.$$

Hence, if m is chosen large enough that k satisfies

$$\frac{1}{k}\frac{\lg(1/\epsilon)}{m_0} \leq \frac{\eta}{2},$$

it follows that

$$\Pr\{r_m > c + \eta\} \leq \Pr\{g_k > c + \eta/2\} \leq \exp(-k\eta^2/8\beta^2),$$

where k is the integer part of m/m_0, and $\beta = \lg(1/\epsilon)$. This is the desired conclusion. ∎

Lemma 5.4. *Suppose $A \subseteq [0,1]^m$, and define*

$$\psi(A) := \frac{1}{2^m} \sum_{\mathbf{y} \in \{-1,1\}^m} \sup_{\mathbf{a} \in A} \frac{|\mathbf{y}^t \mathbf{a}|}{m}. \tag{5.4.4}$$

Suppose that A is convex, and that for some $\epsilon, \gamma > 0$, it is true that

$$M(\epsilon, A, \|\cdot\|_\infty) \geq 2^{2\gamma m}.$$

Choose $\alpha := \alpha(\epsilon, \gamma) > 0$, independent of m, such that[2]

$$2\gamma + \lg \epsilon > \alpha(-\lg \alpha + \lg e) + (1-\alpha)(\gamma + \lg \epsilon) + \alpha \lg(1+\epsilon). \tag{5.4.5}$$

Then

$$\psi(A) \geq \frac{\alpha \epsilon}{2}(2^\gamma - 1) > 0.$$

---

[2] Such an $\alpha > 0$ always exists, because as $\alpha \to 0^+$, the right side of this inequality approaches $\gamma + \lg \epsilon$, which is less than the left side $2\gamma + \lg \epsilon$.

The proof of Lemma 5.4 makes use of two preliminary concepts, namely a quasicube and the ϵ-extension of a set, and is given through a series of additional lemmas.

Let $a > 0$ be a specified number. We define a quasicube in $\mathbb{R}^n$ of side a by induction on n. A one-dimensional quasicube of side a is a closed interval of the form $[c, c+a]$ for some $c \in \mathbb{R}$. A set $S \subseteq \mathbb{R}^n$ is a quasicube of side a if there exists a choice of $n-1$ indices from $\{1, \ldots, n\}$ (which we renumber here as $1, \ldots, n-1$ for convenience) such that the following conditions hold: (i) The projection of S onto its first $n-1$ coordinates, call it S_1, is a quasicube in $\mathbb{R}^{n-1}$ of side a, and (ii) for each $(x_1, \ldots, x_{n-1}) \in S_1$, the set

$$I(x_1, \ldots, x_{n-1}) := \{x_n : (x_1, \ldots, x_{n-1}, x_n) \in S\}$$

is a closed interval of length a.

As an illustration, consider the set S shown in Figure 5.4. The projection

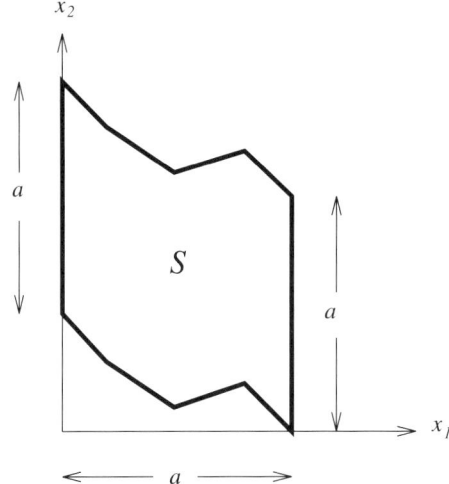

Fig. 5.4. A Quasicube of Side a

of S onto x_1 is an interval of length a, and for each fixed x_1, the set of x_2 such that $(x_1, x_2) \in S$ is an interval of length a.

It is intuitively clear that a quasicube in $\mathbb{R}^n$ has exactly 2^n vertices, which can be placed in one-to-one correspondence with the 2^n bipolar vectors in $\{-1, 1\}^n$, by associating -1 with "minimum" and $+1$ with "maximum." This can be formalized as follows, by induction on n: For $n = 1$, $S = [c, c+a]$ for some c; let $z_{-1} = c$ and $z_1 = c + a$. Now suppose $S \subseteq \mathbb{R}^n$ is a quasicube of side a. Then the projection of S onto some $n-1$ coordinates, which can be assumed to be the *first* $n-1$ coordinates by renumbering if necessary, is a quasicube in $\mathbb{R}^{n-1}$ of side a. Call this projection S_1, and let $\bar{z}_{\mathbf{i}}, \mathbf{i} \in \{-1, 1\}^{n-1}$ denote the 2^{n-1} vertices of S_1. Then, for each i,

166   5. Uniform Convergence of Empirical Means

$$I(\bar{z}_{\mathbf{i}}) := \{x_n : (\bar{z}_{\mathbf{i}}, x_n) \in S\}$$

is a closed interval of length a, say $[c(\bar{z}_{\mathbf{i}}), c(\bar{z}_{\mathbf{i}}) + a]$. Now define

$$\mathbf{z}_{\mathbf{i},-1} := (\bar{z}_{\mathbf{i}}, c(\bar{z}_{\mathbf{i}})), \ \mathbf{z}_{\mathbf{i},1} := (\bar{z}_{\mathbf{i}}, c(\bar{z}_{\mathbf{i}}) + a), \ \mathbf{i} \in \{-1,1\}^{n-1}$$

to be the 2^n vertices of S.

It is easy to show by induction on n that the volume of a quasicube in $\mathbb{R}^n$ of side a equals a^n.

The second concept used here is that of the ϵ-extension of a set. Suppose $A \subseteq \mathbb{R}^m$, and let $\epsilon > 0$. Then the ϵ-**extension** of A is denoted by A_ϵ and is defined as

$$A_\epsilon := \bigcup_{\mathbf{x} \in A} \bar{B}(\epsilon/2, \mathbf{x}, \|\cdot\|_\infty),$$

where $\bar{B}(\epsilon/2, \mathbf{x}, \|\cdot\|_\infty)$ denotes the closed "ball" of radius $\epsilon/2$ in the norm $\|\cdot\|_\infty$ centered at $\mathbf{x}$.[3]

Now we present a series of lemmas that culminate in the proof of Lemma 5.4.

Lemma 5.5. *Suppose $B \subseteq \mathbb{R}^m$, and define, in analogy with (5.4.4),*

$$\psi(B) := \frac{1}{2^m} \sum_{\mathbf{y} \in \{-1,1\}^m} \sup_{\mathbf{a} \in B} \frac{|\mathbf{y}^t \mathbf{a}|}{m}.$$

Suppose that, for some integer $n \le m$, the projection of B onto n coordinates contains a quasicube in $\mathbb{R}^n$ of side b. Then

$$\psi(B) \ge \frac{nb}{2m}.$$

Proof. For convenience, let us renumber the coordinates in $\mathbb{R}^m$ if necessary such that the projection of B onto the *first* n coordinates contains a quasicube in $\mathbb{R}^n$ of side b. Call this quasicube S, and let $\mathbf{z}_{\mathbf{i}}, \mathbf{i} \in \{-1,1\}^n$ denote the 2^n vertices of S. Since S is a projection of B, for each $\mathbf{i} \in \{-1,1\}^n$ there exists a vector $\mathbf{w}_{\mathbf{i}} \in \mathbb{R}^{m-n}$ such that $(\mathbf{z}_{\mathbf{i}}, \mathbf{w}_{\mathbf{i}}) \in B$.[4] Now let

$$V := \{(\mathbf{z}_{\mathbf{i}}, \mathbf{w}_{\mathbf{i}}), \mathbf{i} \in \{-1,1\}^n\}$$

denote a set of preimages in B of the vertices of S. Define $\psi(V)$ in analogy with the definition of $\psi(B)$. Then clearly $\psi(B) \ge \psi(V)$, since $V \subseteq B$. Moreover,

$$m\psi(V) \ge \frac{1}{2^m} \sum_{\mathbf{y} \in \{-1,1\}^m} \sup_{\mathbf{v} \in V} \mathbf{y}^t \mathbf{v},$$

---
[3] Note that a "ball" in the norm $\|\cdot\|_\infty$ is actually a hypercube.
[4] For convenience, let us write $(\mathbf{z}_{\mathbf{i}}, \mathbf{w}_{\mathbf{i}})$ instead of the more cumbersome, though more correct, symbol $[\mathbf{z}_{\mathbf{i}}^t \ \mathbf{w}_{\mathbf{i}}^t]^t$.

where the point to note is that $|\mathbf{y}^t\mathbf{v}|$ has been replaced by $\mathbf{y}^t\mathbf{v}$. To proceed further, let us partition each bipolar vector $\mathbf{y} \in \{-1,1\}^m$ as $\mathbf{ij}$ where $\mathbf{i} \in \{-1,1\}^n$ and $\mathbf{j} \in \{-1,1\}^{m-n}$. Then

$$m\psi(V) \geq \frac{1}{2^m} \sum_{\mathbf{i}\in\{-1,1\}^n} \sum_{\mathbf{j}\in\{-1,1\}^{m-n}} \sup_{\mathbf{v}\in V}[\mathbf{i}\,\mathbf{j}]^t\mathbf{v}$$

$$\geq \frac{1}{2^n} \sum_{\mathbf{i}\in\{-1,1\}^n} \frac{1}{2^{m-n}} \sum_{\mathbf{j}\in\{-1,1\}^{m-n}} \mathbf{i}^t\mathbf{z_i} + \mathbf{j}^t\mathbf{w_i}.$$

The point to note here is that, for each fixed $\mathbf{i} \in \{-1,1\}^n$, the supremum over $\mathbf{v} \in V$ has been replaced by substituting the *particular* value $\mathbf{v} = (\mathbf{z_i}, \mathbf{w_i})$. Let us further rewrite the right side of the above inequality as

$$\frac{1}{2^n} \sum_{\mathbf{i}\in\{-1,1\}^n} \left[\mathbf{i}^t\mathbf{z_i} + \frac{1}{2^{m-n}} \sum_{\mathbf{j}\in\{-1,1\}^{m-n}} \mathbf{j}^t\mathbf{w_i}\right].$$

However,

$$\sum_{\mathbf{j}\in\{-1,1\}^{m-n}} \mathbf{j}^t\mathbf{w} = 0, \; \forall \mathbf{w} \in \mathbb{R}^{m-n}.$$

This can be seen by thinking of $\mathbf{j}$ as a random $(m-n)$-dimensional vector each of whose components assumes the values ± 1 with equal probability. Hence

$$m\psi(B) \geq m\psi(V) \geq \frac{1}{2^n} \sum_{\mathbf{i}\in\{-1,1\}^n} \mathbf{i}^t\mathbf{z_i}.$$

Now the proof is completed by showing that the right side of the above inequality equals $nb/2$. This is established by induction on n. For $n = 1$, the two vertices of S are of the form $c, c + b$ for some c. Thus

$$\frac{1}{2} \sum_{i\in\{-1,1\}} iz_i = \frac{-c + (c+b)}{2} = \frac{b}{2}.$$

Now suppose it has been established that

$$\frac{1}{2^{n-1}} \sum_{\mathbf{i}\in\{-1,1\}^{n-1}} \mathbf{i}\mathbf{z_i} = \frac{(n-1)b}{2}$$

for all quasicubes in $\mathbb{R}^{n-1}$ of side b, and let $S \subseteq \mathbb{R}^n$ be a quasicube of side b. Project S onto $\mathbb{R}^{n-1}$, and let $\bar{z}_\mathbf{k}, \mathbf{k} \in \{-1,1\}^{n-1}$ denote the 2^{n-1} vertices of the projection (which is a quasicube in $\mathbb{R}^{n-1}$ of side b). Then the 2^n vertices of S are of the form $(\bar{z}_\mathbf{k}, c(\bar{z}_\mathbf{k}))$ and $(\bar{z}_\mathbf{k}, c(\bar{z}_\mathbf{k}) + b)$ as $\bar{z}_\mathbf{k}$ varies over the 2^{n-1} vertices of the projection of S. Now

168   5. Uniform Convergence of Empirical Means

$$\frac{1}{2^n} \sum_{\mathbf{i} \in \{-1,1\}^n} \mathbf{i}^t \mathbf{z_i} = \frac{1}{2^{n-1}} \sum_{\mathbf{k} \in \{-1,1\}^{n-1}} \frac{1}{2} \left[\sum_{i_n = \pm 1} \mathbf{k}^t \bar{\mathbf{z}}_\mathbf{k} + i_n z_n \right]$$

$$= \frac{1}{2^{n-1}} \sum_{\mathbf{k} \in \{-1,1\}^{n-1}} \frac{1}{2} [\mathbf{k}^t \bar{\mathbf{z}}_\mathbf{k} - c(\bar{\mathbf{z}}_\mathbf{k}) + \mathbf{k}^t \bar{\mathbf{z}}_\mathbf{k} + c(\bar{\mathbf{z}}_\mathbf{k}) + b]$$

$$= \frac{b}{2} + \frac{1}{2^{n-1}} \sum_{\mathbf{k} \in \{-1,1\}^{n-1}} \mathbf{k}^t \bar{\mathbf{z}}_\mathbf{k}$$

$$= \frac{b}{2} + \frac{(n-1)b}{2} \text{ by the inductive assumption}$$

$$= \frac{nb}{2}.$$

This completes the proof the lemma. ∎

Lemma 5.6. *Suppose $A \subseteq \mathbb{R}^m$, $\epsilon > 0$, and let A_ϵ denote the ϵ-extension of A. Suppose that, for some integer n and some number $b > \epsilon$, the projection of A_ϵ onto n coordinates contains a quasicube of side b. Then*

$$\psi(A) \geq \frac{n(b-\epsilon)}{2m}.$$

Proof. Suppose without loss of generality that the projection of A_ϵ onto the first n coordinates contains a quasicube S of side $b > \epsilon$. Let $\mathbf{z_i}, \mathbf{i} \in \{-1,1\}^n$ denote the 2^n vertices of S. Since S is a projection of A_ϵ, there exist vectors $\mathbf{w_i}, \mathbf{i} \in \{-1,1\}^n$ such that $(\mathbf{z_i}, \mathbf{w_i}) \in A_\epsilon$ for each $\mathbf{i}$. Since A_ϵ is an ϵ-extension of A, every vector in A_ϵ is within a distance $\epsilon/2$ of some vector in A. In particular, for each $\mathbf{i}$, there exists a vector $(\mathbf{a_i}, \mathbf{b_i}) \in A$ such that

$$\left\| \begin{bmatrix} \mathbf{z_i} - \mathbf{a_i} \\ \mathbf{w_i} - \mathbf{b_i} \end{bmatrix} \right\|_\infty \leq \frac{\epsilon}{2}, \quad (5.4.6)$$

and of course $\mathbf{a_i} \in \mathbb{R}^n$, $\mathbf{b_i} \in \mathbb{R}^{m-n}$. Now let V_A denote the set $\{(\mathbf{a_i}, \mathbf{b_i}), \mathbf{i} \in \{-1,1\}^n\}$. Since $V_A \subseteq A$, it follows that

$$m\psi(A) \geq m\psi(V_A) \geq \frac{1}{2^m} \sum_{\mathbf{y} \in \{-1,1\}^m} \sup_{\mathbf{x} \in V_A} \mathbf{y}^t \mathbf{x},$$

where in the last step $|\mathbf{y}^t \mathbf{x}|$ has been replaced by $\mathbf{y}^t \mathbf{x}$. Now we can mimic the reasoning in the proof of Lemma 5.5 to partition $\mathbf{y} \in \{-1,1\}^m$ as $\mathbf{ij}$ and to arrive at

$$m\psi(V_A) \geq \frac{1}{2^n} \sum_{\mathbf{i} \in \{-1,1\}^n} \left[\mathbf{i}^t \mathbf{a_i} + \frac{1}{2^{m-n}} \sum_{\mathbf{j} \in \{-1,1\}^{m-n}} \mathbf{j}^t \mathbf{b_i} \right]$$

$$= \frac{1}{2^n} \sum_{\mathbf{i} \in \{-1,1\}^n} \mathbf{i}^t \mathbf{a_i},$$

because

$$\sum_{\mathbf{j} \in \{-1,1\}^{m-n}} \mathbf{j}^t \mathbf{b} = 0 \ \forall \mathbf{b} \in \mathbb{R}^{m-n}.$$

Now (5.4.6) implies that $\|\mathbf{z_i} - \mathbf{a_i}\|_\infty \leq \epsilon/2$ for each $\mathbf{i}$. Moreover, every vector $\mathbf{i} \in \{-1,1\}^n$ has an l_1-norm of n. Hence

$$|\mathbf{i}^t \mathbf{z_i} - \mathbf{i}^t \mathbf{a_i}| \leq \frac{n\epsilon}{2}, \ \forall \mathbf{i} \in \{-1,1\}^n,$$

and as a result,

$$\frac{1}{2^n} \sum_{\mathbf{i} \in \{-1,1\}^n} \mathbf{i}^t \mathbf{a_i} \geq \frac{1}{2^n} \sum_{\mathbf{i} \in \{-1,1\}^n} \mathbf{i}^t \mathbf{z_i} - \frac{n\epsilon}{2}$$

$$= \frac{nb}{2} - \frac{n\epsilon}{2} = \frac{n(b-\epsilon)}{2},$$

because the $\mathbf{z_i}$ are the vertices of a quasicube of side b. Combining these inequalities gives the desired conclusion. ∎

Lemma 5.7. *Suppose a set $C \subseteq [0,1]^m$ is convex and satisfies*

$$V(C) > \binom{m}{n} a^{m-n} \tag{5.4.7}$$

for some integer n and some $a \in (0,1]$, where $V(C)$ denotes the volume of C. Then there exists a choice of n coordinates such that the projection of C onto these n coordinates contains a quasicube of side a.

Proof. The proof is by induction on n. Suppose first that $n = 1$, and that the projection of C onto any one coordinate axis is contained in an interval of length less than a. Then C itself is contained in a Cartesian product of m intervals, each of which has length less than a. Hence

$$V(C) \leq a^m \leq ma^{m-1},$$

which contradicts (5.4.7) in the case $n = 1$. Hence, if (5.4.7) holds with $n = 1$, then the projection of C onto at least one coordinate axis must contain an interval of length a.

Now suppose m_0, n_0 are given, and suppose by way of induction that the claim is true for all convex subsets of $[0,1]^{m_0-1}$ and all integers $n \leq n_0$, and for all convex subsets of $[0,1]^{m_0}$ and all integers $n \leq n_0 - 1$. The objective is to establish the claim for the case $m = m_0, n = n_0$. Let us drop the subscripts "0" on m_0 and n_0 for convenience. Suppose $C \subseteq [0,1]^m$ is convex and that (5.4.7) holds. Project C onto the first $m-1$ coordinates, and call the projection C_1. Now there are two cases to consider, namely:

(i) $V(C_1) > \binom{m-1}{n} a^{m-n-1}$, or (ii) $V(C_1) \leq \binom{m-1}{n} a^{m-n-1}$.

Consider first Case (i). In this case, by the inductive hypothesis, the projection of C_1 onto some n coordinates contains a quasicube of side a. Since this projection of C_1 is also a projection of C, the claim is established. Now consider Case (ii). For each $\mathbf{x} \in \mathbb{R}^m$, let $\mathbf{x}_1$ denote the projection of $\mathbf{x}$ onto its first $m-1$ coordinates; in other words, $\mathbf{x}_1$ is obtained from $\mathbf{x}$ by dropping the last component. So by definition, $C_1 = \{\mathbf{x}_1 : \mathbf{x} \in C\}$. Now, for each $\mathbf{x}_1 \in C_1$, define

$$\phi_u(\mathbf{x}_1) := \sup\{x_m : (\mathbf{x}_1, x_m) \in C\}, \quad \phi_l(\mathbf{x}_1) := \inf\{x_m : (\mathbf{x}_1, x_m) \in C\},$$

$$\phi(\mathbf{x}_1) := \phi_u(\mathbf{x}_1) - \phi_l(\mathbf{x}_1).$$

Then it is easy to verify that $\phi_u(\cdot)$ is a concave function, $\phi_l(\cdot)$ is convex, and that $\phi(\cdot)$ is concave. Now define

$$C_2 := \{\mathbf{x}_1 \in C_1 : \phi(\mathbf{x}_1) > a\}.$$

Then C_2 is a convex subset of $[0,1]^{m-1}$ because C_1 is a convex set (being a projection of the convex set C), and the function $\phi(\cdot)$ is concave. Now it is claimed that

$$V(C_2) > \binom{m-1}{n-1} a^{m-n}. \tag{5.4.8}$$

Suppose to the contrary that

$$V(C_2) \leq \binom{m-1}{n-1} a^{m-n},$$

and recall that by assumption

$$V(C_1) \leq \binom{m-1}{n} a^{m-n-1}.$$

Now it follows from the definition of $\phi(\cdot)$ that

$$V(C) = \int_{C_1} \phi(\mathbf{x}_1)\, d\mathbf{x}_1 = \int_{C_1 - C_2} \phi(\mathbf{x}_1)\, d\mathbf{x}_1 + \int_{C_1} \phi(\mathbf{x}_1)\, d\mathbf{x}_1.$$

Also by definition, $\phi(\mathbf{x}_1) \leq a$ for all $\mathbf{x}_1 \in C_1 - C_2$, and of course $\phi(\mathbf{x}_1) \leq 1$ for all $\mathbf{x}_1 \in C_2$ because $C \subseteq [0,1]^m$. Hence

$$V(C) \leq aV(C_1 - C_2) + V(C_2) \leq aV(C_1) + V(C_2)$$

$$= a^{m-n}\left[\binom{m-1}{n} + \binom{m-1}{n-1}\right]$$
$$= \binom{m}{n} a^{m-n},$$

where we make use of (4.2.4). However, this last inequality contradicts the assumption (5.4.7). Hence (5.4.8) must be true.

Since C_2 is convex and satisfies (5.4.8), it follows from the inductive hypothesis that the projection of C_2 onto some $n-1$ coordinates contains a quasicube of side a. Renumber the coordinates if necessary so that these are the *last* (not first!) $n-1$ of the $m-1$ coordinates, and call this projection C_3. Observe that C_3 is also contained in the projection of C onto coordinates $m-n$ through $m-1$. Now let S_3 denote the quasicube of side a contained in C_3. For each vector $\mathbf{x}_3 \in \mathbb{R}^{n-1}$ contained in S_3, there is a corresponding vector $\mathbf{x}_1 \in \mathbb{R}^{m-1}$ such that $\mathbf{x}_1 \in C_2$ and such that $\mathbf{x}_3$ is the projection of $\mathbf{x}_1$ onto its last $n-1$ coordinates. Moreover, since $\mathbf{x}_1 \in C_2$, it follows that $\phi(\mathbf{x}_1) > a$. Now define an interval $I(\mathbf{x}_1)$ of length a as follows: $I(\mathbf{x}_1) := [c - a/2, c + a/2]$, where
$$c := [\phi_u(\mathbf{x}_1) + \phi_l(\mathbf{x}_1)]/2.$$

Now the set
$$\{\mathbf{x}_3 \times I(\mathbf{x}_1) : \mathbf{x}_3 \in S_3\}$$
is a quasicube in $\mathbb{R}^n$ (not $\mathbb{R}^{n-1}$) of side a. Moreover, this set is contained in the projection of C onto its last n coordinates. This completes the proof of the inductive step in Case (ii), and completes the proof of the lemma. ∎

Lemma 5.8. *Suppose $B \subseteq [-\epsilon/2, 1+\epsilon/2]^m$ is convex, and satisfies, for some integer n and some $b > 0$, the inequality*
$$V(B) \cdot (1+\epsilon)^{-n} > \binom{m}{n} b^{m-n}. \tag{5.4.9}$$

Then there exists a choice of n coordinates such that the projection of B onto these n coordinates contains a quasicube of side b.

Proof. The property of the projection containing a quasicube of side b is obviously "translation-invariant," so it can be assumed that B is a subset of $[0, 1+\epsilon]^m$ rather than $[-\epsilon/2, 1+\epsilon/2]^m$. Next, let us "contract" B by a factor of $1+\epsilon$, by replacing every $\mathbf{x} \in B$ by $(1+\epsilon)^{-1}\mathbf{x}$; call the resulting set $\bar{B}$. Then $\bar{B} \subseteq [0,1]^m$. Moreover, a projection of B contains a quasicube of side b if and only if the corresponding projection of $\bar{B}$ contains a quasicube of side $b/(1+\epsilon)$. From Lemma 5.7, this will be the case provided
$$V(\bar{B}) > \binom{m}{n} b^{m-n} (1+\epsilon)^{-m+n}.$$

However,

172   5. Uniform Convergence of Empirical Means

$$V(\bar{B}) = V(B) \cdot (1+\epsilon)^{-m}.$$

Hence the preceding inequality is equivalent to

$$V(B) \cdot (1+\epsilon)^{-m} > \binom{m}{n} b^{m-n}(1+\epsilon)^{-m+n},$$

which is the same as the hypothesis of the lemma. ∎

Lemma 5.9. *Suppose $A \subseteq \mathbb{R}^m$. Then*

$$V(A_\epsilon) \geq M(\epsilon, A, \|\cdot\|_\infty) \cdot \epsilon^m.$$

Proof. Let $\mathbf{a}_1, \ldots, \mathbf{a}_M \in A$ be a maximally ϵ-separated set, where $M = M(\epsilon, A, \|\cdot\|_\infty)$. Then the balls $\bar{B}(\epsilon/2, \mathbf{a}_i, \|\cdot\|_\infty)$ are pairwise disjoint, and all of these balls are contained in A_ϵ. Clearly each ball has volume ϵ^m. ∎

At last we come to the proof of the main lemma.

Proof. of **Lemma 5.4**: By assumption, we have

$$M(\epsilon, A, \|\cdot\|_\infty) \geq 2^{2\gamma m}.$$

Hence, by Lemma 5.9, it follows that

$$V(A_\epsilon) \geq 2^{2\gamma m} \cdot \epsilon^m.$$

Also, if A is convex, so is A_ϵ. Now suppose (5.4.5) holds, and apply Lemma 5.8 with $B = A_\epsilon$ and $b = \epsilon 2^\gamma$. For this purpose, observe that

$$\binom{m}{n} = \frac{m(m-1)\cdots(m-n+1)}{n!} \leq m^n e^n n^{-n},$$

where the last step follows from Stirling's approximation. Hence (5.4.9) holds (with A_ϵ replacing B) provided

$$2^{2\gamma m} \epsilon^m (1+\epsilon)^{-n} > m^n e^n n^{-n} b^{m-n},$$

or, after taking binary logarithms,

$$2\gamma m + m \lg \epsilon > n[\lg(m/n) + \lg e] + (m-n)\lg b + n \lg(1+\epsilon).$$

Now divide through by m, and let $n = \alpha m$. Then the above inequality becomes, after noting that $\lg b = \gamma + \lg \epsilon$,

$$2\gamma + \lg \epsilon > \alpha(-\lg \alpha + \lg e) + (1-\alpha)(\gamma + \lg \epsilon) + \alpha \lg(1+\epsilon),$$

which is precisely (5.4.5). Hence, by Lemma 5.8, a projection of A_ϵ onto some n coordinates contains a quasicube of side b. Thus by Lemma 5.6 it follows that

$$\psi(A) \geq \frac{n(b-\epsilon)}{2m} = \frac{\alpha\epsilon}{2}(2^\gamma - 1).$$

This completes the proof of the lemma. ∎

5.5 Theorem 5.1: Proof of Necessity

At last we come to the proof of the main theorem itself. The proof of the necessity part of the theorem is given in this section, while the proof of sufficiency is given in the next section.

Proof. **of the "Only if" Part of Theorem 5.3**: Suppose the family $\mathcal{F}$ has the property that empiricial means converge uniformly. It is desired to show that (5.3.3) holds. The proof is divided into several steps.

Step 1. Let $\mathcal{F}_c$ denote the set of all convex combinations of functions in $\mathcal{F}$. It is not difficult to see that $\mathcal{F}$ has the UCEM property if and only if $\mathcal{F}_c$ has the UCEM property. The "if" part is obvious because $\mathcal{F}$ is a subset of $\mathcal{F}_c$. The "only if" part also follows readily. Suppose $g \in \mathcal{F}_c$ is a convex combination of $f_1, \ldots, f_k \in \mathcal{F}$, and that a multisample $\mathbf{x} \in X^m$ satisfies

$$|E_P(f_i) - \hat{E}(f_i, \mathbf{x})| \leq \epsilon \text{ for } i = 1, \ldots, k.$$

Then clearly

$$|E_P(g) - \hat{E}(g, \mathbf{x})| \leq \epsilon.$$

Hence, for each m,

$$P^m\{\mathbf{x} \in X^m : \sup_{f \in \mathcal{F}} |\hat{E}(f; \mathbf{x}) - E_P(f)| > \epsilon\}$$

$$= P^m\{\mathbf{x} \in X^m : \sup_{f \in \mathcal{F}_c} |\hat{E}(f; \mathbf{x}) - E_P(f)| > \epsilon\}.$$

Since $\mathcal{F}$ has the UCEM property by assumption, it follows that $\mathcal{F}_c$ also has the UCEM property.

Step 2. For each $\mathbf{x} \in X^m$, define

$$\xi_m(\mathbf{x}) := \frac{1}{2^m} \sum_{\mathbf{y} \in \{-1,1\}^m} \sup_{f \in \mathcal{F}_c} \frac{|\mathbf{y}^t \mathbf{f}(\mathbf{x})|}{m}. \quad (5.5.1)$$

It is claimed that $\xi_m(\mathbf{x}) \to 0$ in probability as $m \to \infty$, i.e, that

$$P^m\{\mathbf{x} \in X^m : \xi_m(\mathbf{x}) > \epsilon\} \to 0 \text{ as } m \to \infty, \ \forall \epsilon > 0. \quad (5.5.2)$$

Note that, since $\| \mathbf{y} \|_\infty = 1 \ \forall \mathbf{y} \in \{-1,1\}^m$ and $\mathbf{f}(\mathbf{x}) \in [0,1]^m$ for all $\mathbf{x} \in X^m$ and all $f \in \mathcal{F}_c$, it follows that

$$0 \leq \xi_m(\mathbf{x}) \leq 1, \ \forall \mathbf{x} \in X^m, \ \forall m.$$

Since ξ_m is a family of *uniformly bounded* random variables, if it can be established that

$$E_{P^m}(\xi) \to 0 \text{ as } m \to \infty, \quad (5.5.3)$$

it will then follow that $\xi_m \to 0$ in probability.[5]

It is now shown that the expected value of ξ_m approaches zero, i.e., that, given any $\eta > 0$, there exists an integer m_0 such that

$$E_{P^m}(\xi_m) \leq \eta \;\; \forall m \geq m_0. \tag{5.5.4}$$

To establish this, divide the summation in (5.5.1) into two parts. For a vector $\mathbf{y} \in \{-1,1\}^m$, let $k(\mathbf{y})$ denote the number of $+1$'s in $\mathbf{y}$. The first part of the summation is over all $\mathbf{y} \in \{-1,1\}^m$ such that

$$\left| k(\mathbf{y}) - \frac{m}{2} \right| > m^{2/3},$$

while the second part of the summation is over all $\mathbf{y} \in \{-1,1\}^m$ such that

$$\left| k(\mathbf{y}) - \frac{m}{2} \right| \leq m^{2/3}.$$

Let us refer to these two subsets of $\{-1,1\}^m$ as Y_1 and Y_2 respectively. For the first part, we have

$$\frac{1}{2^m} \sum_{\mathbf{y} \in Y_1} \sup_{f \in \mathcal{F}_c} \frac{|\mathbf{y}^t \mathbf{f}(\mathbf{x})|}{m} \leq \frac{1}{2^m} \sum_{\mathbf{y} \in Y_1} = \frac{|Y_1|}{2^m},$$

where $|Y_1|$ denotes the cardinality of Y_1. Now

$$\frac{|Y_1|}{2^m} = \frac{1}{2^m} \cdot 2 \sum_{k=0}^{(m/2)-m^{2/3}} \binom{m}{k} \to 0 \text{ as } m \to \infty,$$

because the above summation is the size of the two "tails" of the binomial distribution. Hence, given any $\eta > 0$, there exists a number m_1 such that the above summation is less than $\eta/3$ whenever $m \geq m_1$.

For the second part of the summation, let us begin with the observation that, if $\mathcal{F}_c$ has the UCEM property, then

$$P^{k+l}\{\mathbf{xy} \in X^{k+l} : \sup_{f \in \mathcal{F}_c} |\hat{E}(f;\mathbf{x}) - \hat{E}(f;\mathbf{y})| > \epsilon\} \to 0, \text{ as } \min\{k,l\} \to \infty. \tag{5.5.5}$$

The above relationship follows from the triangle inequality. Since $\mathcal{F}_c$ is assumed to have the UCEM property, there exists an integer n_0 such that

$$P^n\{\mathbf{x} \in X^n : \sup_{f \in \mathcal{F}_c} |\hat{E}(f;\mathbf{x}) - E_P(f)| > \epsilon/2\} \leq \delta, \; \forall n \geq n_0.$$

Now suppose that

---

[5] This step is in some sense "obvious." But the proof is shown in great detail to bring out the fact that the probability P is *not* explicitly used anywhere. This feature is of significance in the proof of Theorem 5.5.

$$|\hat{E}(f;\mathbf{x}) - \hat{E}(f;\mathbf{y})| > \epsilon.$$

Then either $|\hat{E}(f;\mathbf{x}) - E_P(x)| > \epsilon/2$, or else $|\hat{E}(f;\mathbf{y}) - E_P(f)| > \epsilon/2$ (or both). Hence, whenever $\min\{k,l\} \geq n_0$, we have that

$$P^{k+l}\{\mathbf{xy} \in X^{k+l} : \sup_{f \in \mathcal{F}_c} |\hat{E}(f;\mathbf{x}) - \hat{E}(f;\mathbf{y})| > \epsilon\}$$

$$\leq P^k\{\mathbf{x} \in X^k : \sup_{f \in \mathcal{F}_c} |\hat{E}(f;\mathbf{x}) - E_P(x)| > \epsilon/2\}$$

$$+ P^l\{\mathbf{y} \in X^l : \sup_{f \in \mathcal{F}_c} |\hat{E}(f;\mathbf{y}) - E_P(f)| > \epsilon/2\}$$

$$\leq 2\delta.$$

This establishes the relation (5.5.5).

Now to estimate the size of the summation in (5.5.1) as $\mathbf{y}$ varies over Y_2, let $\mathbf{y} \in Y_2, \mathbf{x} \in X^m$ be arbitrary, and partition $\mathbf{x}$ into two parts, namely:

$$\mathbf{x}_+ = [x_i : y_i = +1] \in X^{k(\mathbf{y})}, \ \mathbf{x}_- = [x_i : y_i = -1] \in X^{m-k(\mathbf{y})}.$$

In other words, $\mathbf{x}_+$ consists of those components of $\mathbf{x}$ such that the corresponding components of $\mathbf{y}$ satisfy $y_i = +1$, while $\mathbf{x}_-$ consists of those components of $\mathbf{x}$ such that the corresponding components of $\mathbf{y}$ satisfy $y_i = -1$. Then, for arbitrary $f \in \mathcal{F}_c$, we have

$$\mathbf{y}^t \mathbf{f}(\mathbf{x}) = \sum_{y_i=1} f(x_i) - \sum_{y_i=-1} f(x_i) = k(\mathbf{y})\hat{E}(f;\mathbf{x}_+) - [m - k(\mathbf{y})]\hat{E}(f;\mathbf{x}_-).$$

For brevity let us use the shorthand notation

$$\hat{E}_+ = \hat{E}(f;\mathbf{x}_+), \hat{E}_- = \hat{E}(f;\mathbf{x}_-).$$

With this notation, one can write

$$\frac{|\mathbf{y}^t \mathbf{f}(\mathbf{x})|}{m} = \frac{k(\mathbf{y})}{m}\hat{E}_+ - \frac{m - k(\mathbf{y})}{m}\hat{E}_-$$

$$= \frac{\hat{E}_+ - \hat{E}_-}{2} + \left(\frac{k(\mathbf{y})}{m} - \frac{1}{2}\right)\hat{E}_+ - \left(\frac{m - k(\mathbf{y})}{m} - \frac{1}{2}\right)\hat{E}_-$$

$$\leq \frac{\hat{E}_+ - \hat{E}_-}{2} + 2m^{-1/3},$$

because $|k(\mathbf{y}) - m/2| \leq m^{2/3}$, and both $\hat{E}_+$ and $\hat{E}_-$ belong to $[0,1]$. Now, in view of the relationship (5.5.5), it follows that there exists an integer m_2 such that

$$P^m\{\mathbf{x} \in X^m : \sup_{f \in \mathcal{F}_c} \left|\frac{\hat{E}_+ - \hat{E}_-}{2}\right| > \frac{\eta}{6}\} \leq \frac{\eta}{3} \ \forall m \geq m_2,$$

provided of course that $\mathbf{y} \in Y_2$, i.e., that $|k(\mathbf{y}) - m/2| \leq m^{2/3}$. Finally, we can always choose one last integer m_3 such that $2m^{-1/3} \leq \eta/6$ for all $m \geq m_3$. It is now shown that $m_0 = \max\{m_1, m_2, m_3\}$ satisfies the relationship (5.5.4). For every $\mathbf{y} \in Y_2$, we have that

$$\sup_{f \in \mathcal{F}_c} \frac{|\mathbf{y}^t \mathbf{f}(\mathbf{x})|}{m} \leq \sup_{f \in \mathcal{F}_c} \frac{\hat{E}_+ - \hat{E}_-}{2} + 2m^{-1/3} \leq \frac{\eta}{6} + \frac{\eta}{6} = \frac{\eta}{3}$$

with probability of at least $1 - \eta/3$ with respect to $\mathbf{x}$. Hence the expected value of this quantity is no more than $\eta/3 + \eta/3 = 2\eta/3$. Since this is true for *every* $\mathbf{y} \in Y_2$, it follows that

$$E_{P^m} \left[\frac{1}{2^m} \sum_{\mathbf{y} \in Y_2} \sup_{f \in \mathcal{F}_c} \frac{|\mathbf{y}^t \mathbf{f}(\mathbf{x})|}{m} \right] \leq \frac{2\eta}{3}.$$

We have already seen that the corresponding expected value of the summation over $\mathbf{y} \in Y_1$ is no more than $\eta/3$. Thus the expected value of ξ_m is no more than η, provided $m \geq m_0$. This establishes the relationship (5.5.4) and completes the proof of the second step.

Step 3. This is the last step of the proof. Define, as before

$$c_c(\epsilon) := \lim_{m \to \infty} \frac{E_{P^m}[\lg L(\epsilon, \mathcal{F}_c|\mathbf{x}, \|\cdot\|_\infty)]}{m},$$

and note that $\mathcal{F}$ in (5.3.3) has been replaced by $\mathcal{F}_c$. Since $\mathcal{F}|_\mathbf{x}$ is a subset of $\mathcal{F}_c|_\mathbf{x}$ for every $\mathbf{x}$, it follows that

$$L(\epsilon, \mathcal{F}_c|\mathbf{x}, \|\cdot\|_\infty) \geq L(\epsilon, \mathcal{F}|\mathbf{x}, \|\cdot\|_\infty) \; \forall \mathbf{x} \in X^m, \; \forall \epsilon.$$

As a result,

$$c_c(\epsilon) \geq c(\epsilon) \; \forall \epsilon.$$

Now suppose that $c(\epsilon) > 0$ for some $\epsilon > 0$. Then surely $c_c(\epsilon) > 0$. Using this fact, it is shown that the random variable $\xi_m(\mathbf{x})$ *fails* to converge to zero in probability, which contradicts the conclusion of Step 2. This contradiction establishes that $c(\epsilon) = 0$ for every $\epsilon > 0$, thus completing the proof of the "only if" part.

Suppose for the sake of contradiction that $c(\epsilon) > 0$, whence $c_c(\epsilon) > 0$. For convenience, let $c_0 := c(\epsilon)$, and select m_0 large enough that

$$\frac{E_{P^m}[\lg L(\epsilon, \mathcal{F}_c|\mathbf{x}, \|\cdot\|_\infty)]}{m} > \frac{3c_0}{4} \; \forall m \geq m_0. \tag{5.5.6}$$

Such an m_0 exists, because the limit of the left side as $m \to \infty$ is at least equal to c_0. For notational convenience, let $r_m(\mathbf{x})$ denote the random variable

$$r_m(\mathbf{x}) = \frac{\lg L(\epsilon, \mathcal{F}_c|\mathbf{x}, \|\cdot\|_\infty)}{m}.$$

5.5 Theorem 5.1: Proof of Necessity   177

This is the same random variable defined previously in the proof of Lemma 5.3, except that $\mathcal{F}$ is replaced by $\mathcal{F}_c$. Now, since $\mathcal{F}_c|_{\mathbf{x}}$ is the convex hull of $\mathcal{F}|_{\mathbf{x}}$ and $\mathcal{F}|_{\mathbf{x}} \subseteq [0,1]^m$, it follows that $\mathcal{F}_c|_{\mathbf{x}} \subseteq [0,1]^m$ as well. Hence

$$0 \leq r_m(\mathbf{x}) \leq \lg(1/\epsilon) =: \beta, \; \forall m.$$

Now define
$$S_m = \{\mathbf{x} \in X^m : r_m(\mathbf{x}) \geq c_0/2\}.$$
Then (5.5.6) implies that
$$P^m(S_m) \geq \frac{c_0}{4\beta} \; \forall m \geq m_0.$$

Otherwise,
$$E_{P^m}(r_m) = \int_{X^m - S_m} r_m(\mathbf{x}) \, P^m(d\mathbf{x}) + \int_{S_m} r_m(\mathbf{x}) \, P^m(d\mathbf{x})$$
$$\leq \frac{c_0}{2} + \frac{c_0}{4\beta} \cdot \beta = \frac{3c_0}{4},$$

which contradicts (5.5.6). Now, whenever $\mathbf{x} \in S_m$, we have
$$L(\epsilon, \mathcal{F}_c|_{\mathbf{x}}, \|\cdot\|_\infty) \geq 2^{mc_0/2},$$
which in turn implies that
$$M(\epsilon, \mathcal{F}_c|_{\mathbf{x}}, \|\cdot\|_\infty) \geq 2^{mc_0/2}.$$

(See Lemma 2.2.) Also, $\mathcal{F}_c|_{\mathbf{x}}$ is a convex subset of $[0,1]^m$. Now apply Lemma 5.4 with $\gamma = c_0/4$ and ϵ as above, and choose $\alpha := \alpha(\epsilon, \gamma) > 0$ such that (5.4.5) is satisfied. It follows that

$$\psi_m(\mathcal{F}_c|_{\mathbf{x}}) := \frac{1}{2^m} \sum_{\mathbf{y} \in \{-1,1\}^m} \sup_{\mathbf{a} \in \mathcal{F}_c|_{\mathbf{x}}} \frac{|\mathbf{y}^t \mathbf{a}|}{m} \quad (5.5.7)$$

satisfies
$$\psi_m(\mathcal{F}_c|_{\mathbf{x}}) \geq \frac{\alpha \epsilon}{2}(2^\gamma - 1) =: \eta, \; \text{say}.$$
However, comparing (5.5.1) and (5.5.7) shows that
$$\psi_m(\mathcal{F}_c|_{\mathbf{x}}) = \xi_m(\mathbf{x}).$$
Hence
$$P^m\{\mathbf{x} \in X^m : \xi_m(\mathbf{x}) \geq \eta\} \geq \frac{c_0}{4\beta} \; \forall m \geq m_0.$$

In particular, $\xi_m(\mathbf{x})$ does *not* converge to zero as $m \to \infty$, which contradicts Step 2. Hence $c(\epsilon) = 0$ for each $\epsilon > 0$. ∎

5.6 Theorem 5.1: Proof of Sufficiency

This section contains a proof that the condition (5.3.3) is sufficient for a family of functions to have the property that empirical means converge uniformly. Actually much more is established here – explicit estimates are given for the *rate* at which empirical means converge to their true values, and this is done for two different metrics that measure the disparity between the true mean $E_P(f)$ and the empirical mean $\hat{E}(f;\mathbf{x})$.

Throughout the section, we make use of the "averaged" l_1-norm on $\mathbb{R}^m$ defined by

$$\| \mathbf{y} \|_{a1} = \frac{1}{m} \sum_{i=1}^{m} |y_i|.$$

The extra factor $1/m$ distinguishes the "averaged" l_1-norm from the usual l_1-norm. We also use the symbol $L(\epsilon, S, \|\cdot\|_{a1})$ to denote the external ϵ-covering number of a set S with respect to the metric induced by the norm $\|\cdot\|_{a1}$. Note that

$$\| \mathbf{y} \|_{a1} \leq \| \mathbf{y} \|_{\infty}, \forall \mathbf{y} \in \mathbb{R}^m.$$

As a result, it follows that

$$L(\epsilon, S, \|\cdot\|_{a1}) \leq L(\epsilon, S, \|\cdot\|_{\infty}), \forall S \subseteq \mathbb{R}^m, \forall \epsilon > 0. \qquad (5.6.1)$$

Now suppose a family $\mathcal{F}$ has the property that

$$\lim_{m \to \infty} \frac{E_{P^m}[\lg L(\epsilon, \mathcal{F}|_{\mathbf{x}}, \|\cdot\|_{\infty})]}{m} = 0, \forall \epsilon > 0.$$

It is desired to show that $\mathcal{F}$ has the property that empirical means converge uniformly to their true values, i.e., that the error measure

$$q(m, \epsilon) := P^m \{ \mathbf{x} \in X^m : \sup_{f \in \mathcal{F}} |\hat{E}(f;\mathbf{x}) - E_P(f)| > \epsilon \}$$

approaches zero as $m \to \infty$. This is established below. However, rather than merely bound the difference $|\hat{E}(f;\mathbf{x}) - E_P(f)|$ as f varies over $\mathcal{F}$, we give estimates for a more general metric distance between the two numbers $\hat{E}(f;\mathbf{x})$ and $E_P(f)$. It turns out that, by suitably choosing various adjustable "parameters" in this general distance measure, one can prove better bounds on the number of samples m needed to achieve a certain level of accuracy in the estimate of $E_P(f)$.

To make the flow of ideas clear, all the main results are stated in succession, and their significance is discussed, before the proofs are given.

Recall that the objective is to determine how "close" the empirical estimate $\hat{E}(f;\mathbf{x}) =: a$ is to the true mean $E_P(f) =: b$. For this purpose, one can of course just compare a and b directly, and define $\rho(a, b) = |a - b|$. However, there are situations in which some other metrics can give more insight.

5.6 Theorem 5.1: Proof of Sufficiency

For instance, suppose it so happens that $a = 0$, i.e., that $f(x_i) = 0$ at each point in our multisample; what can we say about the confidence that the true mean $E_P(f)$ is less than some number ϵ? Such questions arise in connection with learning problems that employ so-called "consistent" algorithms (see Chapters 6 and 7).

With this motivation, let us define the family of functions $\rho_\alpha : [0,1]^2 \to [0,1]$ as follows:

$$\rho_\alpha(a,b) := \frac{|a-b|}{\alpha + a + b}.$$

It is possible to verify through routine but tedious calculations that ρ_α does indeed satisfy the triangle inequality, and is hence a pseudometric on $[0,1]^2$ for every $\alpha > 0$. Also, it is easy to see that

$$\frac{|a-b|}{\alpha+2} \leq \rho_\alpha(a,b) \leq \frac{|a-b|}{\alpha}. \tag{5.6.2}$$

Given a family of functions $\mathcal{F}$, let us define the modified error measure

$$s_\alpha(m, \beta) := P^m\{\mathbf{x} \in X^m : \exists f \in \mathcal{F} \text{ s.t. } \rho_\alpha[\hat{E}(f;\mathbf{x}), E_P(f)] > \epsilon\},$$

or equivalently,

$$s_\alpha(m, \beta) := P^m\{\mathbf{x} \in X^m : \sup_{f \in \mathcal{F}} \rho_\alpha[\hat{E}(f;\mathbf{x}), E_P(f)] > \epsilon\}.$$

In this section, explicit upper bounds are given for the quantities $q(m, \epsilon)$ and $s_\alpha(m, \beta)$. These two quantities are related as follows: By (5.6.2) above,

$$\rho_2[\hat{E}(f;\mathbf{x}), E_P(f)] \leq \frac{\epsilon}{4} \Rightarrow |\hat{E}(f;\mathbf{x}) - E_P(f)| \leq \epsilon.$$

Hence

$$q(m, \epsilon) \leq s_2(m, \epsilon/4).$$

Thus an upper bound for $s_\alpha(m, \beta)$ can be readily translated into a corresponding upper bound for $q(m, \epsilon)$. But there are other applications as well. Suppose we are interested in bounding the quantity

$$v(m, \epsilon) := P^m\{\mathbf{x} \in X^m : \exists f \in \mathcal{F} \text{ s.t. } \hat{E}(f;\mathbf{x}) = 0 \text{ and } E_P(f) > \epsilon\}. \tag{5.6.3}$$

As mentioned previously, this quantity arises in connection with so-called "consistent" learning algorithms. Of course $v(m,\epsilon) \leq q(m,\epsilon)$, because if a function $f \in \mathcal{F}$ satisfies

$$\hat{E}(f;\mathbf{x}) = 0 \text{ and } E_P(f) > \epsilon, \tag{5.6.4}$$

then it certainly satisfies

$$|\hat{E}(f;\mathbf{x}) - E_P(f)| > \epsilon.$$

180   5. Uniform Convergence of Empirical Means

However, bounding $v(m, \epsilon)$ by $q(m, \epsilon)$ may give an overly conservative estimate. On the other hand, observe that if a function $f \in \mathcal{F}$ satisfies (5.6.4) above, then it also satisfies

$$\rho_\epsilon[\hat{E}(f; \mathbf{x}), E_P(f)] > \frac{1}{2}.$$

Hence

$$v(m, \epsilon) \leq s_\epsilon(m, 1/2).$$

It turns out that the above bound is a considerable improvement over bounding $v(m, \epsilon)$ by $q(m, \epsilon)$. Specifically, the estimate $v(m, \epsilon) \leq q(m, \epsilon)$ leads to a bound of the form

$$v(m, \epsilon) \leq \text{const. } \exp(-\text{const. } m\epsilon^2),$$

whereas the estimate $v(m, \epsilon) \leq s_\epsilon(m, 1/2)$ leads to an bound of the form

$$v(m, \epsilon) \leq \text{const. } \exp(-\text{const. } m\epsilon).$$

The fact that the latter bound contains an $m\epsilon$ in the exponent rather than $m\epsilon^2$ makes it less conservative. Thus there are good reasons for studying the quantity $s_\alpha(m, \beta)$ in addition to $q(m, \epsilon)$.

Now the two main results of the section are stated in succession, and some consequences of these results (including the sufficiency of the condition (5.3.3) for the UCEM property) are stated as corollaries. One symbol is used to facilitate the presentation. Suppose $S \subseteq [0, 1]^{2m}$; then the set $\Delta S \subseteq [-1, 1]^m$ is defined as

$$\Delta S := \{\mathbf{s}_1 - \mathbf{s}_2 : \mathbf{s}_1 \mathbf{s}_2 \in S\}.$$

Here $\mathbf{s}_1, \mathbf{s}_2 \in [0, 1]^m$ denote respectively the "first half" and the "second half" of a vector $\mathbf{s} \in [0, 1]^{2m}$. Thus ΔS is obtained from S by subtracting the second half of each vector in S from its first half.

Theorem 5.7. *Suppose a family* $\mathcal{F} \subseteq [0, 1]^X$ *satisfies the condition (5.3.3). Then*

$$q(m, \epsilon) \leq 2 E_{P^{2m}}[\min\{1, 2\, L(\epsilon/4, \Delta\mathcal{F}|_\mathbf{z}, \|\cdot\|_{a1})\, \exp(-m\epsilon^2/32)\}], \quad (5.6.5)$$

where $\mathbf{z}$ *varies over* X^{2m}.

Theorem 5.8. *Suppose a family* $\mathcal{F} \subseteq [0, 1]^X$ *satisfies the condition (5.3.3). Then*

$$s_\alpha(m, \beta) \leq 2 E_{P^{2m}}[\min\{1, 2\, L(\alpha\beta/8, \mathcal{F}|_\mathbf{z}, \|\cdot\|_{a1})\, \exp(-m\alpha\beta^2/8)\}], \quad (5.6.6)$$

where $\mathbf{z}$ *varies over* X^{2m}.

Corollary 5.5. *Suppose a family* $\mathcal{F} \subseteq [0, 1]^X$ *satisfies the condition (5.3.3). Then* $q(m, \epsilon) \to 0$ *as* $m \to \infty$, *i.e., the family* $\mathcal{F}$ *has the UCEM property.*

Corollary 5.6. *A family $\mathcal{F} \subseteq [0,1]^X$ has the UCEM property if and only if*

$$\lim_{m \to \infty} \frac{E_{P^{2m}}[\lg L(\epsilon, \mathcal{F}|_{\mathbf{z}}, \|\cdot\|_{a1})]}{m} = 0 \; \forall \epsilon > 0, \tag{5.6.7}$$

where $\mathbf{z}$ varies over X^{2m}.

Now the various results are compared and contrasted.

1. Corollary 5.6 *appears* to give a weaker necessary and sufficient condition for the UCEM property compared to Theorem 5.3, because the limit in (5.6.7) is no larger than the limit in (5.3.3). However, in reality both conditions (5.6.7) and (5.3.3) are equivalent, since both are equivalent to the UCEM property.

2. As pointed out previously, $q(m, \epsilon) \leq s_2(m, \epsilon/4)$. Substituting $\alpha = 2, \beta = \epsilon/4$ into (5.6.6) gives the estimate

$$q(m, \epsilon) \leq 2 E_{P^{2m}}[\min\{1, 2\, L(\epsilon/16, \mathcal{F}|_{\mathbf{z}}, \|\cdot\|_{a1})\, \exp(-m\epsilon^2/64)\}].$$

This is substantially worse than the bound given by (5.6.5). It can be easily verified that, for any set $S \subseteq [0,1]^{2m}$,

$$L(\epsilon, \Delta S, \|\cdot\|_{a1}) \leq L(\epsilon/2, S, \|\cdot\|_{a1}).$$

Indeed, if $\{\mathbf{s}^{(1)}, \ldots, \mathbf{s}^{(m)}\}$ is an external $\epsilon/2$-cover for S with respect to $\|\cdot\|_{a1}$, then $\{\mathbf{s}_1^{(1)} - \mathbf{s}_2^{(1)}, \ldots, \mathbf{s}_1^{(m)} - \mathbf{s}_2^{(m)}\}$ is an external ϵ-cover for ΔS. Thus, even if we substitute upper bound

$$L(\epsilon/4, \Delta \mathcal{F}|_{\mathbf{z}}, \|\cdot\|_{a1}) \leq L(\epsilon/8, \mathcal{F}|_{\mathbf{z}}, \|\cdot\|_{a1})$$

into (5.6.5), we would get

$$q(m, \epsilon) \leq 2 E_{P^{2m}}[\min\{1, 2\, L(\epsilon/8, \mathcal{F}|_{\mathbf{z}}, \|\cdot\|_{a1})\, \exp(-m\epsilon^2/32)\}]. \tag{5.6.8}$$

The latter bound obtained from Theorem 5.7 is better than the former bound obtained from Theorem 5.8 for two reasons: (i) the exponent is more negative, because 64 is replaced by 32, and (ii) the covering number is taken with respect to a larger radius, namely $\epsilon/8$ instead of $\epsilon/16$. So in this respect Theorem 5.7 is better than Theorem 5.8.

3. On the other hand, Theorem 5.8 is better than Theorem 5.7 when it comes to estimating the number $v(m, \epsilon)$ defined in (5.6.3). Using the bound $v(m, \epsilon) \leq q(m, \epsilon)$ together with the bound (5.6.8) for $q(m, \epsilon)$ gives

$$v(m, \epsilon) \leq 2 E_{P^{2m}}[\min\{1, 2\, L(\epsilon/8, \mathcal{F}|_{\mathbf{z}}, \|\cdot\|_{a1})\, \exp(-m\epsilon^2/32)\}].$$

In contrast, using the bound $v(m, \epsilon) \leq s_\epsilon(m, 1/2)$ and estimating $s_\epsilon(m, 1/2)$ from (5.6.6) leads to

$$v(m, \epsilon) \leq 2 E_{P^{2m}}[\min\{1, 2\, L(\epsilon/16, \mathcal{F}|_{\mathbf{z}}, \|\cdot\|_{a1})\, \exp(-m\epsilon/32)\}]. \tag{5.6.9}$$

Here the key point to note is that the exponent contains an ϵ instead of an ϵ^2. Of course, the external covering number is taken with respect to a smaller radius, namely $\epsilon/16$ instead of $\epsilon/8$. However, if all functions in $\mathcal{F}$ map X into a *finite* set (for example, in the problem of empirically determining *probabilities* rather than means), then L is independent of ϵ for sufficiently small ϵ, and the advantage is entirely with the second bound.

Now the proofs of the various results are given.

Proof. **of Theorem 5.7**: The proof proceeds in several steps.
Step 1. Suppose $m \geq 2/\epsilon^2$. Define the sets

$$Q := \{\mathbf{x} \in X^m : \exists f \in \mathcal{F} \text{ s.t. } |\hat{E}(f;\mathbf{x}) - E_p(f)| > \epsilon\}, \text{ and}$$

$$R := \{\mathbf{xy} \in X^{2m} : \exists f \in \mathcal{F} \text{ s.t. } |\hat{E}(f;\mathbf{x}) - \hat{E}(f;\mathbf{y})| > \epsilon/2\}.$$

Then it is claimed that

$$P^m(Q) \leq 2P^{2m}(R).$$

Note that $q(m,\epsilon)$ is precisely $P^m(Q)$. Now the set R consists of multisamples of length $2m$ such that the empirical means computed on the basis of the first m samples and the last m samples differ by more than $\epsilon/2$.

To establish this claim, observe that by Chebycheff's inequality, given $f \in \mathcal{F}$, we have

$$P^m\{\mathbf{y} \in X^m : |\hat{E}(f;\mathbf{y}) - E_P(f)| > \epsilon/2\} \leq \frac{1}{4m(\epsilon/2)^2} \leq \frac{1}{2}$$

whenever $m \geq 2/\epsilon^2$. Given $\mathbf{x} \in X^m$, suppose $f \in \mathcal{F}$ satisfies $|\hat{E}(f;\mathbf{x}) - E_p(f)| > \epsilon$. Then by the triangle inequality, with probability of at least $1/2$ with respect to $\mathbf{y}$, we have

$$|\hat{E}(f;\mathbf{x}) - \hat{E}(f;\mathbf{y})| \geq |\hat{E}(f;\mathbf{x}) - E_p(f)| - |\hat{E}(f;\mathbf{y}) - E_p(f)| > \frac{\epsilon}{2}.$$

So

$$P^{2m}\{\mathbf{xy} \in X^{2m} : \exists f \in \mathcal{F} \text{ s.t. } |\hat{E}(f;\mathbf{x}) - \hat{E}(f;\mathbf{y})| > \epsilon/2\}$$
$$\geq P^{2m}\{\mathbf{xy} \in X^{2m} : \exists f \in \mathcal{F} \text{ s.t. } |\hat{E}(f;\mathbf{x}) - E_p(f)| > \epsilon$$
$$\text{and } |\hat{E}(f;\mathbf{y}) - E_P(f)| \leq \epsilon/2\}$$
$$\geq \frac{1}{2}P^m\{\mathbf{x} \in X^m : \exists f \in \mathcal{F} \text{ s.t. } |\hat{E}(f;\mathbf{x}) - E_p(f)| > \epsilon\}.$$

This last inequality is the same as: $P^{2m}(R) \geq P^m(Q)/2$, which establishes the claim.

Step 2. Let Γ_m denote the set of permutations γ on $\{1, \ldots, 2m\}$ such that, for each $i \in \{1, \ldots, m\}$, either $\gamma(i) = i$ and $\gamma(m+i) = m+i$, or else $\gamma(i) = m+i$ and $\gamma(m+i) = i$. Thus Γ_m consists of all permutations that swap some (or all, or no) indices $i \in \{1, \ldots, m\}$ with $m+i$. Clearly there are 2^m permutations in Γ_m. Now it is claimed that

$$P^{2m}(R) = \int_{X^{2m}} \frac{1}{2^m} \sum_{\gamma \in \Gamma_m} I_R(\gamma \mathbf{z}) \, P^{2m}(d\mathbf{z}). \qquad (5.6.10)$$

To establish this claim, observe that

$$\sum_{\gamma \in \Gamma_m} \int_{X^{2m}} I_R(\gamma \mathbf{z}) \, P^{2m}(d\mathbf{z}) = \int_{X^{2m}} \sum_{\gamma \in \Gamma_m} I_R(\gamma \mathbf{z}) \, P^{2m}(d\mathbf{z}),$$

because the summation is finite and can thus be interchanged with the integration. Now, for each fixed $\gamma \in \Gamma_m$, we have

$$\int_{X^{2m}} I_R(\gamma \mathbf{z}) \, P^{2m}(d\mathbf{z}) = \int_{X^{2m}} I_R(\gamma \mathbf{z}) \, P^{2m}(d\gamma \mathbf{z}) = P^{2m}(R),$$

because the permutation γ merely relabels the components of $\mathbf{z}$, and P^{2m} is a product measure. Hence the previous equation becomes

$$2^m P^{2m}(R) = \int_{X^{2m}} \sum_{\gamma \in \Gamma_m} I_R(\gamma \mathbf{z}) \, P^{2m}(d\mathbf{z}),$$

which is the same as (5.6.10).

The integrand in (5.6.10) has a very intuitive interpretation. For a fixed $\mathbf{z} \in X^{2m}$, the integrand is the *fraction* of permutations γ in Γ_m such that $\gamma \mathbf{z} \in R$. This fraction is estimated next.

Step 3. Suppose $\mathbf{a} \in [-1,1]^m$. Then it is claimed that the number of vectors $\mathbf{s} \in \{-1,1\}^m$ such that $|\mathbf{s}^t \mathbf{a}| > m\epsilon/4$ is at most equal to $2^m \cdot 2e^{-m\epsilon^2/32}$. To establish this claim, it is helpful to think of $Y_i = a_i s_i$ as a random variable that assumes the values $\pm a_i$ with equal probability. Obviously Y_i has zero mean, and $-1 \le Y_i \le 1$. Hence, by Hoeffding's inequality,

$$\Pr\{\left|\sum_{i=1}^m Y_i\right| > m\epsilon/4\} \le 2\exp(-m\epsilon^2/32).$$

But this "probability" is precisely the fraction of vectors $\mathbf{s} \in \{-1,1\}^m$ that satisfy $|\mathbf{s}^t \mathbf{a}| > m\epsilon/4$. This establishes the claim.

Step 4. For each fixed $\mathbf{z} \in X^{2m}$, it is claimed that the integrand in (5.6.10) is not more than

$$\min\{1, 2 \, L(\epsilon/4, \Delta\mathcal{F}|_{\mathbf{z}}, \|\cdot\|_{a1}) \exp(-m\epsilon^2/32)\}. \qquad (5.6.11)$$

To establish this claim, select a minimal external $\epsilon/4$-cover for the set $\Delta\mathcal{F}|_{\mathbf{z}}$, and call it $\mathbf{g}^1, \ldots, \mathbf{g}^L$, where $L = L(\epsilon/4, \Delta\mathcal{F}|_{\mathbf{z}}, \|\cdot\|_{a1})$. Suppose the permutation γ is such that $\gamma \mathbf{z} \in R$, i.e., suppose there exists a function $f \in \mathcal{F}$ such that

$$\frac{1}{m}\left|\sum_{i=1}^{m} f(z_{\gamma(i)}) - f(z_{\gamma(m+i)})\right| > \frac{\epsilon}{2}. \quad (5.6.12)$$

This can be expressed in a more manageable form if it is observed that each permutation $\gamma \in \Gamma_m$ either flips or does not flip the index i with $m+i$. Thus there is a one-to-one correspondence between permutations in Γ_m and vectors $\mathbf{s} \in \{-1, 1\}^m$ in the obvious way, namely:

$$s_i = \begin{cases} +1 & \text{if } \gamma(i) = i, \\ -1 & \text{if } \gamma(i) = m+i. \end{cases}$$

Given a function $f \in \mathcal{F}$, define

$$\Delta\mathbf{f}(\mathbf{z}) := [f(z_1) - f(z_{m+1}) \ldots f(z_m) - f(z_{2m})]^t \in [-1, 1]^m.$$

Then (5.6.12) is equivalent to

$$\frac{1}{m}|\mathbf{s}^t \Delta\mathbf{f}(\mathbf{z})| > \frac{\epsilon}{2},$$

where $\mathbf{s} \in \{-1, 1\}^m$ corresponds to γ. Thus $I_R(\gamma\mathbf{z}) = 1$ if and only if there exists a vector $\Delta\mathbf{f}(\mathbf{z}) \in \Delta\mathcal{F}|_\mathbf{z}$ such that the above inequality holds.

Now suppose $\gamma\mathbf{z} \in R$, and select a vector $\Delta\mathbf{f}(\mathbf{z}) \in \Delta\mathcal{F}|_\mathbf{z}$ such that (5.6.12) holds. Using the covering property, select an index i such that

$$\|\Delta\mathbf{f}(\mathbf{z}) - \mathbf{g}^i\|_{a1} \leq \frac{\epsilon}{4}.$$

Then it is routine to verify using the triangle inequality that

$$\frac{1}{m}|\mathbf{s}^t \mathbf{g}^i| \geq \frac{1}{m}|\mathbf{s}^t \Delta\mathbf{f}(\mathbf{z})| - \frac{\epsilon}{4} > \frac{\epsilon}{4}.$$

What has been shown is this: For a fixed $\mathbf{z} \in X^{2m}$, if a permutation $\gamma \in \Gamma_m$ satisfies $\gamma\mathbf{z} \in R$, then there exists an index $i \in \{1, \ldots, L\}$ such that

$$\frac{1}{m}|\mathbf{s}^t \mathbf{g}^i| > \frac{\epsilon}{4}, \quad (5.6.13)$$

where $\mathbf{s}$ corresponds to γ.

Now we are in a position to bound the integrand in (5.6.10). By Step 3, *for each fixed* index i, the number of vectors $\mathbf{s} \in \{-1, 1\}^m$ that satisfy (5.6.13) is no more than $2^m \cdot 2e^{-m\epsilon^2/32}$. Hence the number of $\mathbf{s} \in \{-1, 1\}^m$ for which *there exists* an index $i \in \{1, \ldots, L\}$ such that (5.6.13) holds is no more than

$$2^m \cdot 2 \exp(-m\epsilon^2/32) \, L(\epsilon/4, \Delta\mathcal{F}|_\mathbf{z}, \|\cdot\|_{a1}).$$

So the integrand in (5.6.10) is no more than

$$2 \exp(-m\epsilon^2/32) \, L(\epsilon/4, \Delta\mathcal{F}|_\mathbf{z}, \|\cdot\|_{a1}).$$

Of course, the integrand in (5.6.10) is also never more than one. This establishes (5.6.11).

Now the proof of the theorem can be completed. From (5.6.10) and (5.6.11), it follows that

$$P^{2m}(R) \leq E_{P^{2m}}[\min\{1, 2\,L(\epsilon/4, \Delta\mathcal{F}|_{\mathbf{z}}, \|\cdot\|_{a1})\,\exp(-m\epsilon^2/32)\}].$$

Finally, by Step 1,

$$q(m, \epsilon) = P^m(Q) \leq 2P^{2m}(R)$$
$$\leq 2E_{P^{2m}}[\min\{1, 2\,L(\epsilon/4, \Delta\mathcal{F}|_{\mathbf{z}}, \|\cdot\|_{a1})\,\exp(-m\epsilon^2/32)\}].$$

This is the same as (5.6.5). ∎

Proof. **of Theorem 5.8**: This proof follows along exactly the same lines as that of Theorem 5.7, with a few minor variations caused by the fact that the pseudometric $|\hat{E}(f;\mathbf{x}) - E_P(f)|$ in the definition of $q(m,\epsilon)$ is replaced by the more general pseudometric $\rho_\alpha[\hat{E}(f;\mathbf{x}), E_P(f)]$ in the definition of $s_\alpha(m, \beta)$.

Step 1. Suppose $m \geq 2/\alpha\beta^2$. Define the sets

$$Q := \{\mathbf{x} \in X^m : \exists f \in \mathcal{F} \text{ s.t. } \rho_\alpha[\hat{E}(f;\mathbf{x}), E_P(f)] > \beta\}, \text{ and}$$

$$R := \{\mathbf{xy} \in X^{2m} : \exists f \in \mathcal{F} \text{ s.t. } \rho_\alpha[\hat{E}(f;\mathbf{x}), \hat{E}(f;\mathbf{y})] > \beta/2\}.$$

Then it is claimed that $P^m(Q) \leq 2P^{2m}(R)$.

The proof is exactly the same as in the previous Step 1, and uses the fact that ρ_α (being a pseudometric) satisfies the triangle inequality. The bound $m \geq 2/\alpha\beta^2$ is used to ensure that $\rho_\alpha[\hat{E}(f;\mathbf{y}), E_P(f)] \leq \beta/2$ with a probability of at least $1/2$, and takes the place of the earlier bound $m \geq 2/\epsilon^2$.

Step 2. It is claimed that

$$P^{2m}(R) = \int_{X^{2m}} \frac{1}{2^m} \sum_{\gamma \in \Gamma_m} I_R(\gamma \mathbf{z})\, P^{2m}(d\mathbf{z}). \tag{5.6.14}$$

The proof is exactly as before.

Step 3. Both the claim as well as its proof are somewhat different from the previous Step 3. Suppose $\mathbf{a} \in [0,1]^{2m}$, and let

$$\hat{\mathbf{a}}^{(1)} := \frac{1}{m} \sum_{i=1}^{m} a_i, \quad \hat{\mathbf{a}}^{(2)} := \frac{1}{m} \sum_{i=1}^{m} a_{m+i}$$

denote respectively the "average of the first half" and the "average of the second half" of the vector $\mathbf{a}$. Then the number of permutations $\gamma \in \Gamma_m$ such that

$$\rho_\alpha[\widehat{\gamma\mathbf{a}}^{(1)}, \widehat{\gamma\mathbf{a}}^{(2)}] > \beta \tag{5.6.15}$$

is no more than $2^m \cdot 2e^{-2m\alpha\beta^2}$.

To establish this claim, let Y_i denote a random variable that assumes the values $\pm(a_i - a_{m+i})$ with equal probability. Now

$$\rho_\alpha[\widehat{\gamma \mathbf{a}}^{(1)}, \widehat{\gamma \mathbf{a}}^{(2)}] = \frac{|\sum_{i=1}^{m}(a_{\gamma(i)} - a_{\gamma(m+i)})|}{m\alpha + \sum_{i=1}^{2m} a_i}.$$

The numerator is precisely $|\sum_{i=1}^{m} Y_i|$. Let us therefore use Hoeffding's inequality to bound

$$\Pr\{|\sum_{i=1}^{m} Y_i| > \beta(m\alpha + \sum_{i=1}^{2m} a_i)\}.$$

Now the random variables Y_i are bounded and satisfy

$$-|a_i - a_{m+i}| \leq Y_i \leq |a_i - a_{m+i}|.$$

Hence the above probability is no more than

$$2\exp\left[-\beta^2 \left(m\alpha + \sum_{i=1}^{2m} a_i\right)^2 / 2 \sum_{i=1}^{m}(a_i - a_{m+i})^2\right].$$

Now let $c := \sum_{i=1}^{2m} a_i$, and observe that since each $a_i \in [0,1]$,

$$\sum_{i=1}^{m}(a_i - a_{m+i})^2 \leq \sum_{i=1}^{m} |a_i - a_{m+i}| \leq c.$$

Hence the desired probability is no larger than

$$2\exp[-\beta^2(m\alpha + c)^2 / 2c].$$

Now an elementary calculation shows that the exponent (without the minus sign) is minimized when $c = m\alpha$. Hence the maximum possible value of the above expression is obtained when $c = m\alpha$. We conclude that the desired probability is no more than $2\exp(-2m\alpha\beta^2)$. Hence the number of permutations $\gamma \in \Gamma_m$ that satisfy (5.6.15) is no more than $2^m \cdot 2\exp(-2m\alpha\beta^2)$.

Step 4. It is claimed that the integrand in (5.6.14) is no more than

$$\min\{1, 2L(\alpha\beta/8, \mathcal{F}|_\mathbf{z}, \|\cdot\|_{a1})\exp(-m\alpha\beta^2/8)\}.$$

The proof of this claim is very similar to that of the earlier Step 4. Fix $\mathbf{z} \in X^{2m}, \gamma \in \Gamma_m$, and suppose that $\gamma \mathbf{z} \in R$. Let $\mathbf{h}^1 \ldots, \mathbf{h}^L$ be a minimal external $\alpha\beta/8$-cover for $\mathcal{F}|_\mathbf{z}$, where of course $L = L(\alpha\beta/8, \mathcal{F}|_\mathbf{z}, \|\cdot\|_{a1})$. Then there exists a function $f \in \mathcal{F}$ such that

$$\rho_\alpha\left[\frac{1}{m}\sum_{i=1}^{m} f(z_{\gamma(i)}), \frac{1}{m}\sum_{i=1}^{m} f(z_{\gamma(m+i)})\right] > \frac{\beta}{2}.$$

This can be written in another way. Define

$$\mathbf{f}(\mathbf{z}) := [f(z_1) \ldots f(z_{2m})]^t \in [0,1]^{2m}.$$

Then
$$\rho_\alpha[\widehat{\gamma \mathbf{f}(\mathbf{z})}^{(1)}, \widehat{\gamma \mathbf{f}(\mathbf{z})}^{(2)}] > \frac{\beta}{2}. \qquad (5.6.16)$$

Now select an index j such that $\|\mathbf{f}(\mathbf{z}) - \mathbf{h}^j\|_{a1} \leq \alpha\beta/8$, i.e., such that
$$\frac{1}{2m} \sum_{i=1}^{2m} |f(z_i) - h_i^j| \leq \frac{\alpha\beta}{8}.$$

Then, since γ merely permutes the indices i, we have that
$$\frac{1}{2m} \sum_{i=1}^{2m} |f(z_{\gamma(i)}) - h_{\gamma(i)}^j| \leq \frac{\alpha\beta}{8},$$

or equivalently,
$$\frac{1}{m} \sum_{i=1}^{m} |f(z_{\gamma(i)}) - h_{\gamma(i)}^j| + |f(z_{\gamma(m+i)}) - h_{\gamma(m+i)}^j| \leq \frac{\alpha\beta}{4}. \qquad (5.6.17)$$

The right side changes from $\alpha\beta/8$ to $\alpha\beta/4$ because the summation is being divided by $1/m$ rather than by $1/2m$ as in the preceding inequality. For convenience, define
$$a_1 := \widehat{\gamma\mathbf{f}(\mathbf{z})}^{(1)}, \, a_2 := \widehat{\gamma\mathbf{f}(\mathbf{z})}^{(2)}, \, b_1 := \widehat{\gamma\mathbf{h}^j}^{(1)}, \, b_2 := \widehat{\gamma\mathbf{h}^j}^{(2)}.$$

Then (5.6.16) states that $\rho_\alpha(a_1, a_2) > \beta/2$, while (5.6.17) implies that
$$|a_1 - b_1| + |a_2 - b_2| \leq \frac{\alpha\beta}{4}.$$

Hence
$$\rho_\alpha(a_1, b_1) + \rho_\alpha(a_2, b_2) \leq \frac{1}{\alpha}[|a_1 - b_1| + |a_2 - b_2|] \leq \frac{\beta}{4}.$$

Now by the triangle inequality, it follows that
$$\rho_\alpha(b_1, b_2) \geq \rho_\alpha(a_1, a_2) - [\rho_\alpha(a_1, b_1) + \rho_\alpha(a_2, b_2)] > \frac{\beta}{2} - \frac{\beta}{4} = \frac{\beta}{4}.$$

In other words,
$$\rho_\alpha(\widehat{\gamma\mathbf{h}^j}^{(1)}, \widehat{\gamma\mathbf{h}^j}^{(2)}) > \frac{\beta}{4}. \qquad (5.6.18)$$

What has been shown is this: For a fixed $\mathbf{z} \in X^{2m}$, if $\gamma\mathbf{z} \in R$, then *there exists* an index j such that (5.6.18) holds. Now, by Step 3, *for each index j*, the number of permutations $\gamma \in \Gamma_m$ such that (5.6.18) holds is no more than $2^m \cdot 2 \exp(-m\alpha\beta^2/8)$. Hence the number of $\gamma \in \Gamma_m$ such that *there exists* an index $j \in \{1, \ldots, L\}$ such that (5.6.18) holds is no more than
$$2^m \cdot 2\, L(\alpha\beta/8, \mathcal{F}|_{\mathbf{z}}, \|\cdot\|_{a1})\, \exp(-m\alpha\beta^2/8).$$

188   5. Uniform Convergence of Empirical Means

This shows that the integrand in (5.6.14) is no more than

$$2\, L(\alpha\beta/8, \mathcal{F}|_\mathbf{z}, \|\cdot\|_{a1})\, \exp(-m\alpha\beta^2/8).$$

Of course the integrand in (5.6.14) is also no more than one. Together with Step 1, this completes the proof of the theorem. ∎

Proof. **of Corollary 5.5**: Suppose the family of functions $\mathcal{F}$ satisfies the condition (5.3.3); it is shown that the family $\mathcal{F}$ has the UCEM property. This completes the proof the sufficiency part of Theorem 5.3. The proof is based on the estimates (5.6.5) and (5.4.3). Since

$$L(\epsilon/4, \Delta\mathcal{F}|_\mathbf{z}, \|\cdot\|_{a1}) \leq L(\epsilon/8, \mathcal{F}|_\mathbf{z}, \|\cdot\|_{a1}) \leq L(\epsilon/8, \mathcal{F}|_\mathbf{z}, \|\cdot\|_\infty),$$

it follows from (5.6.8) that

$$q(m,\epsilon) \leq 4 E_{P^{2m}}[\min\{1, L(\epsilon/8, \mathcal{F}|_\mathbf{z}, \|\cdot\|_\infty)\exp(-m\epsilon^2/32)\}]. \quad (5.6.19)$$

Note that the factor 2 multiplying L in (5.6.8) has been moved outside the "min," causing the factor multiplying $E_{P^{2m}}$ to change from 2 to 4. Let $\epsilon > 0$ be specified, and define $\eta = \epsilon^2/64$. Since $\mathcal{F}$ satisfies the condition (5.3.3), there exists an integer m_0 such that

$$\frac{E_{P^{2m}}[\lg L(\epsilon/8, \mathcal{F}|_\mathbf{z}, \|\cdot\|_\infty)]}{2m} \leq \frac{\epsilon^2}{128} = \frac{\eta}{2}, \quad \forall m \geq m_0.$$

Suppose $m \geq m_0$, and divide X^{2m} into two parts:

$$S_1 := \{\mathbf{z} \in X^{2m} : \frac{\lg L(\epsilon/8, \mathcal{F}|_\mathbf{z}, \|\cdot\|_\infty)}{2m} > \eta\}, \text{ and}$$

$$S_2 := \{\mathbf{z} \in X^{2m} : \frac{\lg L(\epsilon/8, \mathcal{F}|_\mathbf{z}, \|\cdot\|_\infty)}{2m} \leq \eta\}.$$

Now the number $P^{2m}(S_1)$ can be estimated using the bound (5.4.3). This gives (upon noting that $c = 0$)

$$P^{2m}(S_1) \leq \exp(-k\eta^2/8\beta^2),$$

where k is the integer part of m/m_0 and $\beta = \lg(1/\epsilon)$. Next, if $\mathbf{z} \in S_2$, then

$$L(\epsilon/8, \mathcal{F}|_\mathbf{z}, \|\cdot\|_\infty) \leq 2^{2m\eta} = 2^{m\epsilon^2/32},$$

so that

$$L(\epsilon/8, \mathcal{F}|_\mathbf{z}, \|\cdot\|_\infty)\exp(-m\epsilon^2/32) \leq \exp[-m(1-\ln 2)\epsilon^2/32] < 1, \quad \forall \mathbf{z} \in S_2.$$

Finally, it follows from (5.6.19) that

$$q(m,\epsilon) \leq \int_{S_1} P^{2m}(d\mathbf{z}) + \int_{S_2} L(\epsilon/8, \mathcal{F}|_\mathbf{z}, \|\cdot\|_\infty)\, e^{-m\epsilon^2/32}\, P^{2m}(d\mathbf{z})$$

$$\leq e^{-k\eta^2/8\beta^2} + e^{-m(1-\ln 2)\epsilon^2/32} \to 0 \text{ as } m \to \infty. \quad (5.6.20)$$

This completes the proof that the family $\mathcal{F}$ has the UCEM property. ∎

5.6 Theorem 5.1: Proof of Sufficiency 189

In Section 5.2, it is shown that the UCEM property implies the ASCEM property, using the notion of subadditive processes. This proof is very simple and elegant, but uses advanced ideas. If one wishes to have an "elementary" (though of course not so elegant) proof of this implication, one could observe that the bound for $q(m, \epsilon)$ given in (5.6.20) above is summable with respect to m for each fixed ϵ. Hence, by Lemma 2.10, it follows that the stochastic process $\{a_m\}$ defined in (5.1.1) converges *almost surely* to zero, i.e., the family $\mathcal{F}$ has the ASCEM property.

Proof. **of Corollary 5.6**: This is based on Theorem 5.3 and Lemma 5.2. As a preliminary step, observe that covering numbers with respect to the norm $\|\cdot\|_{a1}$ have the same "submultiplicativity" property as do covering numbers with respect to $\|\cdot\|_\infty$. In other words, if $S_1 \subseteq \mathbb{R}^k$, $S_2 \subseteq \mathbb{R}^l$, and $S = S_1 \times S_2 \subseteq \mathbb{R}^{k+l}$, then

$$L(\epsilon, S, \|\cdot\|_{a1}) \leq L(\epsilon, S_1, \|\cdot\|_{a1}) \cdot L(\epsilon, S_2, \|\cdot\|_{a1}).$$

This in turn follows from the easily established fact that

$$\mathbf{a} \in \mathbb{R}^k, \mathbf{b} \in \mathbb{R}^l, \|\mathbf{a}\|_{a1} \leq \epsilon, \|\mathbf{b}\|_{a1} \leq \epsilon \Rightarrow \left\| \begin{bmatrix} \mathbf{a} \\ \mathbf{b} \end{bmatrix} \right\|_{a1} \leq \epsilon.$$

By now the reader will have observed that the above submultiplicativity property is the key to the proofs of Lemma 5.2. Thus the lemma remains valid even if $\|\cdot\|_\infty$ is replaced by $\|\cdot\|_{a1}$.

For the sake of clarity, define

$$c_a(\epsilon) := \lim_{m \to \infty} \frac{E_{P^{2m}}[\lg L(\epsilon/2, \Delta\mathcal{F}|_\mathbf{z}, \|\cdot\|_{a1})]}{2m}. \tag{5.6.21}$$

Then, since

$$L(\epsilon/2, \Delta\mathcal{F}|_\mathbf{z}, \|\cdot\|_{a1}) \leq L(\epsilon, \mathcal{F}|_\mathbf{z}, \|\cdot\|_\infty),$$

it follows that

$$c_a(\epsilon) \leq c(\epsilon),$$

where $c(\epsilon)$ denotes the limit in (5.3.3).

"If" Suppose $c_a(\epsilon) = 0$ for all $\epsilon > 0$. Then the random variable

$$\frac{\lg L(\epsilon/2, \Delta\mathcal{F}|_\mathbf{z}, \|\cdot\|_{a1})}{2m}$$

is everywhere dominated by the random variable

$$\frac{\lg L(\epsilon, \mathcal{F}|_\mathbf{z}, \|\cdot\|_{a1})}{2m},$$

which in turn approaches zero in probability, by the analog of Lemma 5.3. In particular, a bound of the form (5.4.3) applies. Hence, from Theorem 5.7 and the bound (5.6.5), it follows (as in the proof of Corollary 5.5) that $\mathcal{F}$ has the UCEM property.

"Only if" This follows from the fact that $c_a(\epsilon) \leq c(\epsilon)$, and the fact that $c(\epsilon) = 0$ for all $\epsilon > 0$ whenever $\mathcal{F}$ has the UCEM property. ∎

5.7 Proofs of the Remaining Theorems

In this section, the proofs of remaining theorems from Section 5.3, other than Theorem 5.3, are given. The reader is reminded of Theorem 5.2, which states that the UCEM property and the ASCEM property are equivalent.

Corollaries 5.1 and 5.2 are obvious consequences of Theorem 5.3.

Before proving Theorem 5.4, a brief digression is made to show that the quantity $d(\mathbf{x})/m$ approaches a constant almost surely as $m \to \infty$, for *every* collection of sets $\mathcal{A}$, i.e., whether or not $\mathcal{A}$ has the UCEP property. The result complements Lemmas 5.1 and 5.2.

Lemma 5.10. *Suppose $\mathcal{A} \subseteq \mathcal{S}$ is a given collection of sets, and define a stochastic process $\{\alpha_m(\cdot)\}$ on X^∞ by*

$$\alpha_m(\mathbf{x}^*) := \frac{d(x_1, \ldots, x_m)}{m} \ \forall m \geq 1,$$

where $d(x_1, \ldots, x_m)$ is the VC-dimension of the collection $\mathcal{A}$ intersected with the set $\{x_1, \ldots, x_m\}$. Then $\{\alpha_m(\cdot)\}$ converges almost surely to a constant as $m \to \infty$.

Proof. The proof is once again based on the notion of subadditive stochastic processes, as in the case of Lemmas 5.1 and 5.2. Specifically, define a doubly indexed stochastic process $\{\beta_{lm}(\cdot)\}$ on X^∞ as follows:

$$\beta_{lm}(\mathbf{x}^*) := d(x_{l+1}, \ldots, x_m),$$

where, as before, $d(x_{l+1}, \ldots, x_m)$ denotes the VC-dimension of the collection $\mathcal{A}$ intersected with the set $\{x_{l+1}, \ldots, x_m\}$. Now it is claimed that the process $\{\beta_{lm}(\cdot)\}$ is subadditive. To establish Condition (S1), let $\mathbf{x}^* \in X^\infty$ be arbitrary. Suppose $l < m < n$, and let

$$S_{ln} := \{x_{l+1}, \ldots, x_m\},$$

and define S_{mn} and S_{ln} analogously. Then clearly $S_{ln} = S_{lm} \cup S_{mn}$. For brevity, let d_{ln} denote the VC-dimension $d(x_{l+1}, \ldots, x_n)$, and define d_{lm} and d_{mn} in a similar fashion. Suppose $\mathcal{A} \subseteq S_{ln}$ is a set of cardinality d_{ln} that is shattered by $\mathcal{A}$, and define $A_1 := S_{lm} \cap A$, $A_2 := S_{mn} \cap A$. Then it is easy to see that both A_1 and A_2 are shattered by $\mathcal{A}$. Hence $d_{lm} \geq |A_1|$, $d_{mn} \geq |A_2|$, which implies that

$$d_{ln} = |A| \leq |A_1| + |A_2| \leq d_{lm} + d_{mn}.$$

Since the above inequality holds for *every* $\mathbf{x}^* \in X^\infty$, Condition (S1) holds. Condition (S2) is immediate since $\{x_i\}$ is an i.i.d. sequence, while (S3) holds with $\mu = 0$. Hence, by Theorem 5.1, it follows that the stochastic process $\{\beta_{0m}/m\} = \{\alpha_m\}$ converges almost surely to a random variable. Finally, it can be shown as in the proof of Lemma 5.1 that the limiting function is constant almost everywhere. ∎

5.7 Proofs of the Remaining Theorems

Proof. **of Theorem 5.4** The proof consists of showing that the conditions (5.3.5) and (5.3.4) are equivalent.

It is first shown that (5.3.5) implies (5.3.4). The proof is based on Theorem 4.1. We begin by observing that $m \geq d(\mathbf{x})$ for all m and all $\mathbf{x} \in X^m$. Hence, from (4.2.1), it follows that for each $\mathbf{x} \in X^m$ we have

$$\pi(\mathbf{x}; \mathcal{A}) \leq \left[\frac{em}{d(\mathbf{x})}\right]^{d(\mathbf{x})},$$

$$\ln \pi(\mathbf{x}; \mathcal{A}) \leq d(\mathbf{x})[1 - \ln(d(\mathbf{x})/m)],$$

$$\frac{\ln \pi(\mathbf{x}; \mathcal{A})}{m} \leq \frac{d(\mathbf{x})}{m}\left[1 - \ln\left(\frac{d(\mathbf{x})}{m}\right)\right].$$

Now note that the function $\phi : x \mapsto x(1 - \ln x)$ satisfies $\phi(0) = 0$, $\phi(1) = 1$, and is concave on $(0, 1)$. Hence by Jensen's inequality (see e.g., [35], p. 80), it follows that for any measurable function $f : X \to [0, 1]$, we have

$$E[\phi(f)] \leq \phi[E(f)].$$

In particular,

$$E_{P^m}\left[\frac{\ln \pi(\mathbf{x}; \mathcal{A})}{m}\right] \leq E_{P^m}\{\phi[d(\mathbf{x})/m]\} \leq \phi\left\{E_{P^m}\left[\frac{d(\mathbf{x})}{m}\right]\right\} \to 0 \text{ as } m \to \infty,$$

by virtue of (5.3.5). Since ln and lg differ only by a constant factor, it follows that (5.3.4) holds.

Now it is shown that if (5.3.5) is violated, then (5.3.4) is also violated. Suppose (5.3.5) does not hold. Then there exists a $\mu > 0$ and a sequence of integers $\{m_i\}$ approaching infinity, such that

$$\frac{E_{P^{m_i}}[d(\mathbf{x})]}{m_i} \geq \mu, \forall i. \tag{5.7.1}$$

To make the notation less cumbersome, let us temporarily drop the subscript i, and suppose

$$\frac{E_{P^m}[d(\mathbf{x})]}{m} \geq \mu \tag{5.7.2}$$

for a fixed integer m. It is now shown that

$$\frac{E_{P^m}[\lg \pi(\mathbf{x}; \mathcal{A})]}{m} \geq \frac{\mu^2}{4}.$$

This inequality, together with (5.7.1), is enough to show that (5.3.4) fails to hold. To prove this inequality, let us begin with the following simple observation: If $f : X^m \to [0, 1]$ and $E_{P^m}[f(\mathbf{x})] \geq \mu$, then

$$P^m\{\mathbf{x} \in X^m : f(\mathbf{x}) \geq \mu/2\} \geq \mu/2.$$

Otherwise, we would have

$$\int_{X^m} f(\mathbf{x})\, P^m(d\mathbf{x}) = \int_{f<\mu/2} f(\mathbf{x})\, P^m(d\mathbf{x}) + \int_{f\geq\mu/2} f(\mathbf{x})\, P^m(d\mathbf{x})$$

$$< \frac{\mu}{2} + \frac{\mu}{2} = \mu,$$

which is a contradiction. Thus (5.7.2) implies that

$$P^m\{\mathbf{x} \in X^m : d(\mathbf{x}) \geq m\mu/2\} \geq \mu/2.$$

However, if $d(\mathbf{x}) \geq m\mu/2$, then $\pi(\mathbf{x}; \mathcal{A}) \geq 2^{m\mu/2}$, and $\lg \pi(\mathbf{x}; \mathcal{A}) \geq m\mu/2$. Hence

$$P^m\{\mathbf{x} \in X^m : \lg \pi(\mathbf{x}; \mathcal{A}) \geq m\mu/2\} \geq \mu/2,$$

and as a result,

$$\frac{E_{P^m}[\lg \pi(\mathbf{x}; \mathcal{A})]}{m} \geq \frac{\mu^2}{4}.$$

Now (5.7.1) implies that the above inequality holds for infinitely many values of m, which in turn implies that (5.3.4) fails to hold. ■

Proof. **of Theorem 5.5** As one would expect, the proof is a fairly straightforward modification of that of Theorem 5.3.

"If" For notational convenience, define

$$\phi(m, P) := E_{P^m}[\lg L(\epsilon, \mathcal{F}|_{\mathbf{x}}, \|\cdot\|_\infty)], \text{ and}$$

$$\bar{\phi}(m, \mathcal{P}) := \sup_{P \in \mathcal{P}} \phi(m, P).$$

By assumption, (5.3.6) holds, i.e.,

$$\lim_{m \to \infty} \sup_{P \in \mathcal{P}} \frac{E_{P^m}[\lg L(\epsilon, \mathcal{F}|_{\mathbf{x}}, \|\cdot\|_\infty)]}{m} = 0, \forall \epsilon > 0.$$

Hence, given any $\eta > 0$, there exists an $m_0 > 0$ such that

$$\frac{\bar{\phi}(m, \mathcal{P})}{m} \leq \frac{\eta}{2}, \forall m \geq m_0.$$

Now we simply mimic the proof of the inequality (5.4.3), being careful to take into account the dependence of the various quantities on P. Let the random variables g_k, $r_{l,m}$ be as in the proof of Lemma 5.3. Then, for each $P \in \mathcal{P}$, we have

$$E_P(g_k) = E_P(r_{im_0, (i+1)m_0}) \, \forall i = \frac{\phi(m_0, P)}{m_0} \leq \frac{\bar{\phi}(m_0, \mathcal{P})}{m_0} \leq \frac{\eta}{2}.$$

Hence, *for each fixed* $P \in \mathcal{P}$, we have

5.7 Proofs of the Remaining Theorems

$$P^{km_0}\{g_k > \eta\} \leq P^{km_0}\{g_k \geq \frac{\phi(m_0, P)}{m_0}\}$$

$$\leq \exp(-k\eta^2/2\beta^2),$$

where $\beta := \lg(1/\epsilon)$. The key point to note is the following: For different probabilities $P \in \mathcal{P}$, the *sets* above will in general be different. Nevertheless, for each P, the *measure* of the corresponding set is uniformly bounded independently of P. Now we can proceed as before to establish the analog of (5.4.3), namely

$$\sup_{P \in \mathcal{P}} \{r_m \geq \eta\} \leq \exp(-k\eta^2/8\beta^2),$$

where k is the integer part of m/m_0. With this inequality established, the remainder of the proof of Corollary 5.5 goes through. In particular, if m_0 is defined as above with $\eta = \epsilon^2/64$, then [compare (5.6.20)]

$$\bar{q}(m, \epsilon, \mathcal{P}) \leq e^{-k\eta^2/8\beta^2} + e^{-m(1-\ln 2)\epsilon^2/32} \to 0 \text{ as } m \to \infty.$$

"Only if" This part also requires nothing more than retracing the necessity part of the proof of Theorem 5.3, ensuring that various convergences are uniform with respect to $P \in \mathcal{P}$ as well.

Step 1. Let $\mathcal{F}_c$ denote the set of convex combinations of functions in $\mathcal{F}$. Since $\mathcal{F}$ is assumed to have the UCEMUP property, so does $\mathcal{F}_c$.

Step 2. Define the random variable ξ_m by [compare (5.5.1)]

$$\xi_m(\mathbf{x}) := \frac{1}{2^m} \sum_{\mathbf{y} \in \{-1,1\}^m} \sup_{f \in \mathcal{F}_c} \frac{|\mathbf{y}^t \mathbf{f}(\mathbf{x})|}{m}.$$

Then it is claimed that $\xi_m \to 0$ *uniformly* in probability, i.e., that

$$\sup_{P \in \mathcal{P}} P^m \{\mathbf{x} \in X^m : \xi_m(\mathbf{x}) > \epsilon\} \to 0 \text{ as } m \to \infty, \forall \epsilon > 0. \quad (5.7.3)$$

Taking the supremum with respect to $P \in \mathcal{P}$ is the added feature here compared to (5.5.2). This relationship can be established if it can be shown that

$$\sup_{P \in \mathcal{P}} E_{P^m}(\xi) \to 0 \text{ as } m \to \infty, \quad (5.7.4)$$

To prove the above, we proceed as follows: Since $\xi_m(\mathbf{x}) \subset [0,1]\ \forall \mathbf{x}, m$, it is clear that

$$P^m\{\mathbf{x} \in X^m : \xi_m(\mathbf{x}) > \epsilon\} \leq \frac{E_{P^m}(\xi_m)}{\epsilon}.$$

Hence (5.7.4) implies (5.7.3). Now (5.7.4) is established exactly as before.

Step 3. This is the only part of the proof that requires a few additional wrinkles. Suppose that, for some $\epsilon > 0$, we have

$$\limsup_{m \to \infty, P \in \mathcal{P}} \frac{E_{P^m}[\lg L(\epsilon_0, \mathcal{F}|_{\mathbf{x}}, \|\cdot\|_\infty)]}{m} =: c_0 > 0.$$

194   5. Uniform Convergence of Empirical Means

Then
$$\limsup_{m \to \infty} \sup_{P \in \mathcal{P}} \frac{E_{P^m}[\lg L(\epsilon, \mathcal{F}_c | \mathbf{x}, \|\cdot\|_\infty)]}{m} \geq c_0.$$

Note that $\mathcal{F}$ has been replaced by $\mathcal{F}_c$. Now choose m_0 large enough that
$$\sup_{P \in \mathcal{P}} \frac{E_{P^m}[\lg L(\epsilon, \mathcal{F}_c | \mathbf{x}, \|\cdot\|_\infty)]}{m} > \frac{3c_0}{4}, \ \forall m \geq m_0.$$

Then, for each $m \geq m_0$, there exists a probability $P_m \in \mathcal{P}$ such that
$$\frac{E_{P^m}[\lg L(\epsilon, \mathcal{F}_c | \mathbf{x}, \|\cdot\|_\infty)]}{m} \geq \frac{3c_0}{4}.$$

Hence, as before, it follows that, *for this particular probability*, we have
$$P_m^m \{\mathbf{x} \in X^m : L(\epsilon, \mathcal{F}_c | \mathbf{x}, \|\cdot\|_\infty) > 2^{mc_0/2}\} \geq \frac{c_0}{4\beta}.$$

Now let $\gamma = c_0/4$, and choose $\alpha := \alpha(\epsilon, \gamma) > 0$ such that (5.4.5) is satisfied. Define
$$\eta := \frac{\alpha \epsilon}{2}(2^\gamma - 1),$$
and apply Lemma 5.4. Then, as before,
$$P_m^m \{\mathbf{x} \in X^m : \xi_m(\mathbf{x}) \geq \eta\} \geq \frac{c_0}{4\beta}.$$

As a consequence, it follows that
$$\sup_{P \in \mathcal{P}} P^m \{\mathbf{x} \in X^m : \xi_m(\mathbf{x}) \geq \eta\} \geq \frac{c_0}{4\beta}, \ \forall m \geq m_0.$$
which contradicts (5.7.3). Hence $c_0 = 0$. ∎

Corollaries 5.3 and 5.4 are immediate consequences of Theorem 5.5. Finally, the proof of Theorem 5.6 consists of showing that the conditions (5.3.7) and (5.3.8) are equivalent. This can be achieved by mimicking the corresponding proof showing the equivalence of (5.3.5) and (5.3.4) and is therefore left to the reader.

5.8 Uniform Convergence Properties of Iterated Families

In Section 4.3 we studied the VC-dimension of Boolean functions of sets. These results are used in the present section to show that if a collection of sets $\mathcal{A}$ has the property that empirical probabilities converge uniformly, then (roughly speaking) every Boolean function of $\mathcal{A}$ also has the UCEP property. Similarly, if a family of functions $\mathcal{F}$ has the property that empirical means converge uniformly, then every uniformly continuous function of $\mathcal{F}$ also has the UCEM property. Finally, if a family of hypothesis functions $\mathcal{H}$ has the UCEM property, so does the associated family of "loss" functions.

5.8.1 Boolean Operations on Collections of Sets

Given a measurable space $(X, \mathcal{S})$, suppose $\mathcal{A} \subseteq \mathcal{S}$ is a given collection of sets. By a slight abuse of notation, one can also think of $\mathcal{A}$ as a family of functions mapping X into $\{0, 1\}$. Suppose k is an integer and that $u : \{0, 1\}^k \to \{0, 1\}$ is a given function. In analogy with Section 4.3, one can define a corresponding collection of sets $\mathcal{U}(\mathcal{A})$ as follows: Suppose $f_1, \ldots, f_k : X \to \{0, 1\}$ are binary-valued functions. Then we define $u(f_1, \ldots, f_k) : X \to \{0, 1\}$ to be the binary-valued function
$$x \mapsto u[f_1(x), \ldots, f_k(x)].$$
Finally, $\mathcal{U}(\mathcal{A})$ is defined as
$$\mathcal{U}(\mathcal{A}) := \{u(f_1, \ldots, f_k) : f_i \in \mathcal{A} \; \forall i\}.$$
This defines $\mathcal{U}(\mathcal{A})$ as a family of binary-valued functions, but there is an obvious interpretation of $\mathcal{U}(\mathcal{A})$ as a collection of measurable sets. A few examples serve to illustrate the definition.

Given $\mathcal{A} \subseteq \mathcal{S}$, define
$$\mathcal{A} \oplus \mathcal{A} := \{A \cup B : A, B \in \mathcal{A}\},$$
$$\mathcal{A} \odot \mathcal{A} := \{A \cap B : A, B \in \mathcal{A}\},$$
$$\mathcal{A} \Delta \mathcal{A} := \{A \Delta B : A, B \in \mathcal{A}\}.$$
These collections of sets can be formed from $\mathcal{A}$ by defining
$$u(a, b) = \max\{a, b\}, \; a \cdot b, \; |a - b|$$
respectively.

Theorem 5.9. *Suppose $\mathcal{A} \subseteq \mathcal{S}$ has the property of uniform convergence of empirical probabilities, and that $u : \{0, 1\}^k \to \{0, 1\}$ is a given function. Then $\mathcal{U}(\mathcal{A})$ also has the UCEP property.*

Proof. The proof consists of showing that the collection $\mathcal{U}(\mathcal{A})$ satisfies the condition (5.3.5) with $\mathcal{A}$ replaced by $\mathcal{U}(\mathcal{A})$, and then appealing to Theorem 5.4. By assumption, $\mathcal{A}$ has the UCEP property. Hence, by Theorem 5.4,
$$\frac{E_{P^m}[d(\mathbf{x}; \mathcal{A})]}{m} = 0,$$
where we use $d(\mathbf{x}; \mathcal{A})$ instead of $d(\mathbf{x})$ to make clear which collection of sets we are talking about. Now, by Theorem 4.5, there exists a constant $\alpha(k)$ that depends only on k (and not on $\mathcal{A}$ or $\mathbf{x}$ or m) such that
$$d(\mathbf{x}; \mathcal{U}(\mathcal{A})) \leq \alpha(k) \, d(\mathbf{x}; \mathcal{A}), \; \forall \mathbf{x} \in X^m, \; \forall m \geq 2.$$
Hence
$$\lim_{m \to \infty} \frac{E_{P^m}[d(\mathbf{x}; \mathcal{U}(\mathcal{A}))]}{m} \leq \alpha(k) \lim_{m \to \infty} \frac{E_{P^m}[d(\mathbf{x}; \mathcal{A})]}{m} = 0.$$
Hence, by Theorem 5.4, it follows that $\mathcal{U}(\mathcal{A})$ also has the UCEP property. ∎

196   5. Uniform Convergence of Empirical Means

Corollary 5.7. *Suppose $\mathcal{A} \subseteq \mathcal{S}$ has the property of uniform convergence of empirical probabilities, and that $\{u_i, i = 1, \ldots, l\}$ is a (finite) collection of functions mapping $\{0,1\}^k$ into $\{0,1\}$. Then the collection of sets $\cup_{i=1}^{l} \mathcal{U}_i(\mathcal{A})$ also has the UCEP property.*

Proof. Observe that the number of distinct functions mapping $\{0,1\}^k$ into $\{0,1\}$ is finite (and equal to 2^{2^k}). Also, by Theorem 5.9, each set $\mathcal{U}_i(\mathcal{A})$ has the UCEP property. It follows that a finite union of such sets also has the UCEP property. ∎

Corollary 5.8. *Suppose $\mathcal{A} \subseteq \mathcal{S}$ has the UCEPUP property with respect to the family of probabilities $\mathcal{P}$, and that $u : \{0,1\}^k \to \{0,1\}$ is a given function. Then $\mathcal{U}(\mathcal{A})$ also has the UCEPUP property with respect to $\mathcal{P}$.*

Example 5.4. It was shown in Example 5.3 that the collection of convex subsets of $[0,1]^l$ (where l is some integer) has the UCEP property if P is the uniform probability measure. Now let k be a given positive integer, and let $\mathcal{A}_k$ consist of all subsets of $[0,1]^l$ that can be expressed as a union of up to k convex sets in $[0,1]^l$. Then it follows from Theorem 5.9 that $\mathcal{A}_k$ also has the UCEP property.

5.8.2 Uniformly Continuous Mappings on Families of Functions

To prove a result analogous to Theorem 5.9 for families of functions, we proceed as follows: Suppose $\mathcal{F}$ is a family of measurable functions mapping X into $[0,1]$, that $k \geq 1$ is a given integer, and that $u : [0,1]^k \to [0,1]$ is a measurable function. Finally, suppose that u is *uniformly continuous*, i.e., that for each $\epsilon > 0$ there exists a $\delta(\epsilon) > 0$ such that, for all vectors $\mathbf{a}, \mathbf{b} \in [0,1]^k$, we have

$$\| \mathbf{a} - \mathbf{b} \|_\infty \leq \delta(\epsilon) \Rightarrow |u(\mathbf{a}) - u(\mathbf{b})| \leq \epsilon. \quad (5.8.1)$$

Now the family $\mathcal{U}(\mathcal{F})$ is defined in a natural way. Given functions $f_1, \ldots, f_k \in \mathcal{F}$, the corresponding function $u(f_1, \ldots, f_k) : X \to [0,1]$ is defined by

$$x \mapsto u[f_1(x), \ldots, f_k(x)].$$

Finally,
$$\mathcal{U}(\mathcal{F}) := \{u(f_1, \ldots, f_k) : f_i \in \mathcal{F} \; \forall i\}.$$

Theorem 5.10. *Suppose the family of functions $\mathcal{F} \subseteq [0,1]^X$ has the property of uniform convergence of empirical means, and let $u : [0,1]^k \to [0,1]$ be uniformly continuous. Then the family $\mathcal{U}(\mathcal{F})$ also has the UCEM property.*

5.8 Uniform Convergence Properties of Iterated Families 197

Proof. Given $\epsilon > 0$, choose a constant $\delta(\epsilon) > 0$ such that (5.8.1) holds. Then it is claimed that

$$L(\epsilon, \mathcal{U}(\mathcal{F})|_{\mathbf{x}}, \|\cdot\|_\infty) \leq [L(\delta(\epsilon), \mathcal{F}|_{\mathbf{x}}, \|\cdot\|_\infty)]^k. \quad (5.8.2)$$

Once this claim is established, it follows that

$$\lg L(\epsilon, \mathcal{U}(\mathcal{F})|_{\mathbf{x}}, \|\cdot\|_\infty) \leq k \lg L(\delta(\epsilon), \mathcal{F}|_{\mathbf{x}}, \|\cdot\|_\infty).$$

Hence, for each $\epsilon > 0$,

$$\lim_{m\to\infty} \frac{E_{P^m}[\lg L(\epsilon, \mathcal{U}(\mathcal{F})|_{\mathbf{x}}, \|\cdot\|_\infty)]}{m} \leq k \lim_{m\to\infty} \frac{E_{P^m}[\lg L(\delta(\epsilon), \mathcal{F}|_{\mathbf{x}}, \|\cdot\|_\infty)]}{m} = 0,$$

since by assumption $\mathcal{F}$ has the UCEM property and thus satisfies (5.3.3). Thus, by Theorem 5.3, it follows that $\mathcal{U}(\mathcal{F})$ also has the UCEM property. So the proof is complete once (5.8.2) is established.

To prove (5.8.2), select a minimal external $\delta(\epsilon)$-cover $\mathbf{g}^1, \ldots, \mathbf{g}^L$ for $\mathcal{F}|_{\mathbf{x}}$, where each $\mathbf{g}^i \in [0,1]^m$, and of course $L = L(\delta(\epsilon), \mathcal{F}|_{\mathbf{x}}, \|\cdot\|_\infty)$. Now the set $\mathcal{U}(\mathcal{F})|_{\mathbf{x}}$ consists of all m-vectors of the form

$$[u(f_1(x_1), \ldots, f_k(x_1)) \ldots u(f_1(x_m), \ldots, f_k(x_m))]^t \in [0,1]^m$$

as $f_1, \ldots, f_k$ vary over $\mathcal{F}$. Now fix some $f_1, \ldots, f_k \in \mathcal{F}$. By assumption, for each index $i \in \{1, \ldots k\}$ there exists a corresponding index $j_i \in \{1, \ldots, L\}$ such that

$$\|\mathbf{f}_i(\mathbf{x}) - \mathbf{g}^{j_i}\|_\infty \leq \delta(\epsilon),$$

where, as before, $\mathbf{f}_i(\mathbf{x})$ denotes the vector

$$\mathbf{f}_i(\mathbf{x}) := [f_i(x_1) \ldots f_i(x_m)]^t \in [0,1]^m.$$

Hence, for each index $l \in \{1, \ldots, m\}$, it follows that[6]

$$\|[f_1(x_l) \ldots f_k(x_l)]^t - [g_l^{j_1} \ldots g_l^{j_k}]^t\|_\infty \leq \delta(\epsilon).$$

Now by the uniform continuity condition (5.8.1), it follows that

$$|u(f_1(x_l), \ldots, f_k(x_l)) - u(g_l^{j_1}, \ldots, g_l^{j_k})| \leq \epsilon, \text{ for } l = 1, \ldots, m.$$

This inequality shows that the collection of m-dimensional vectors

$$[u(g_1^{j_1}, \ldots, g_1^{j_k}) \ldots u(g_m^{j_1}, \ldots, g_m^{j_k})]^t \in [0,1]^m,$$

generated by varying $j_1, \ldots, j_k$ over $\{1, \ldots, L\}$, forms an external ϵ-cover for $\mathcal{U}(\mathcal{F})|_{\mathbf{x}}$. It is clear that the cardinality of this cover is L^k. This establishes (5.8.2). ∎

---

[6] The fact that the norm $\|\cdot\|_\infty$ is used throughout is of some help here.

Corollary 5.9. *Suppose the family of functions $\mathcal{F} \subseteq [0,1]^X$ has the UCEMUP property with respect to the family of probability measures $\mathcal{P}$, and let $u : [0,1]^k \to [0,1]$ be uniformly continuous. Then the family $\mathcal{U}(\mathcal{F})$ also has the UCEMUP property with respect to $\mathcal{P}$.*

In order to prove an analog of Corollary 5.7 for *functions*, it is necessary to take into account the fact that the number of uniformly continuous functions mapping $[0,1]^k$ into $[0,1]$ is *infinite*. One can get around this difficulty by dealing with a *compact* family of continuous functions. One can define a metric ρ on the set of continuous functions mapping $[0,1]^k$ into $[0,1]$ as follows:
$$\rho(u,v) := \max_{\mathbf{a} \in [0,1]^k} |u(\mathbf{a}) - v(\mathbf{a})|.$$

Suppose Φ is a family of uniformly continuous functions mapping $[0,1]^k$ into $[0,1]$. Then one can speak of the covering number $N(\epsilon, \Phi, \rho)$ in the usual manner. In particular, Φ is compact if and only if it is closed and $N(\epsilon, \Phi, \rho)$ is finite for each $\epsilon > 0$. In this connection, it is worthwhile to recall the classical *Arzela-Ascoli theorem* (see e.g., [65], p. 266), which states that a family Φ is compact if and only if it is **equicontinuous**, that is, for each $\epsilon > 0$, there exists a $\delta(\epsilon) > 0$ such that

$$\|\mathbf{a} - \mathbf{b}\|_\infty \leq \delta(\epsilon) \Rightarrow |u(\mathbf{a}) - u(\mathbf{b})| \leq \epsilon, \forall u \in \Phi.$$

Corollary 5.10. *Suppose the family of functions $\mathcal{F}$ has the UCEM property, and let Φ be a compact family of continuous functions mapping $[0,1]^k$ into $[0,1]$. Then the family of functions $\cup_{u \in \Phi} \mathcal{U}(\mathcal{F})$ also has the UCEM property.*

Proof. The idea behind the proof is rather simple. Given $\epsilon > 0$, first find an $\epsilon/2$-cover of *functions* $\{u_1, \ldots, u_s\}$ for Φ, where $s = N(\epsilon/2, \Phi, \rho)$. Then, for each function $u \in \Phi$, there exists an index $t \in \{1, \ldots, s\}$ such that $\rho(u, u_t) \leq \epsilon/2$, that is,
$$|u(\mathbf{a}) - u_t(\mathbf{a})| \leq \epsilon/2, \forall \mathbf{a} \subset [0,1]^k.$$
In particular, if $x \in X$ and $f_1, \ldots, f_k \in \mathcal{F}$, then
$$|u(f_1(x), \ldots, f_k(x)) - u_t(f_1(x), \ldots, f_k(x))| \leq \epsilon/2.$$

Now let $\mathbf{x} \in X^m$. The above inequality shows that an external $\epsilon/2$-cover of vectors for the union $\cup_{t=1}^s \mathcal{U}_t(\mathcal{F})|_\mathbf{x}$ is also an external ϵ-cover for the union $\cup_{u \in \Phi} \mathcal{U}(\mathcal{F})|_\mathbf{x}$. In the course of the proof Theorem 5.10, it was shown that each $\mathcal{U}_t(\mathcal{F})|_\mathbf{x}$ has an external $\epsilon/2$-cover of cardinality $[L(\delta(\epsilon/2), \mathcal{F}|_\mathbf{x}, \|\cdot\|_\infty)]^k$. Therefore

$$\lg L(\epsilon, \cup_{u \in \Phi} \mathcal{U}(\mathcal{F})|_\mathbf{x}, \|\cdot\|_\infty) \leq k \lg N(\epsilon/2, \Phi, \rho) \lg L(\delta(\epsilon/2), \mathcal{F}|_\mathbf{x}, \|\cdot\|_\infty).$$

Now the desired conclusion follows by using (5.3.3) and then appealing to Theorem 5.3. ∎

5.8 Uniform Convergence Properties of Iterated Families

Example 5.5. As a concrete application of the above theorem and corollary, let us investigate the problem of the uniform convergence of empirical *distances*. Let $\mathcal{F} \subseteq [0,1]^X$ be a family of measurable functions mapping X into $[0,1]$, and observe that each $f, g \in \mathcal{F}$, the function $x \mapsto |f(x) - g(x)|$ also maps X into $[0,1]$, and is measurable. Let us define

$$\mathcal{F} \Delta \mathcal{F} := \{|f(\cdot) - g(\cdot)| : f, g \in \mathcal{F}\},$$

where $|f(\cdot) - g(\cdot)|$ is shorthand for the function $x \mapsto |f(x) - g(x)|$. In the case where all functions in $\mathcal{F}$ are binary-valued (i.e., are indicator functions of a collection of sets $\mathcal{A}$), the family $\mathcal{F} \Delta \mathcal{F}$ is just the family of indicator functions of all sets of the form $A \Delta B$ where $A, B \in \mathcal{A}$. Observe that the function $u : (a, b) \mapsto |a - b| : [0, 1]^2 \to [0, 1]$ is uniformly continuous. Hence by Theorem 5.10, it follows that if $\mathcal{F}$ has the property that empirical *means* converge uniformly (almost surely), then the family of functions $\mathcal{F} \Delta \mathcal{F}$ also has the same property.

In practical terms, this means the following: Let P be a given probability measure on $(X, \mathcal{S})$. Then one can define a pseudometric d_P on $[0,1]^X$ in the familiar manner, namely

$$d_P(f, g) := \int_X |f(x) - g(x)| \, P(dx).$$

Since the function $|f(\cdot) - g(\cdot)|$ belongs to $[0,1]^X$, one can empirically estimate $d_P(f, g)$ in the familiar way: Let $x_1, \ldots, x_m \in X$ be i.i.d. samples drawn in accordance with P, and define

$$\hat{d}(f, g; \mathbf{x}) := \frac{1}{m} \sum_{i=1}^{m} |f(x_i) - g(x_i)|$$

as the "empirical distance" between f and g. Now, as earlier, one can ask whether the empirical estimate $\hat{d}(f, g; \mathbf{x})$ converges to the true value $d_P(f, g)$ as the number of samples approaches infinity, and whether the convergence is uniform (almost sure) with respect to the functions involved. To make the question precise, define

$$q_d(m, \epsilon) := P^m \{ \mathbf{x} \in X^m : \sup_{f, g \in \mathcal{F}} |\hat{d}(f, g; \mathbf{x}) - d_P(f, g)| > \epsilon \}.$$

We say that the family $\mathcal{F}$ has the property of **uniform convergence of empirical distances(UCED)** if $q_d(m, \epsilon) \to 0$ as $m \to \infty$, for each fixed ϵ. Clearly this is the same as the family $\mathcal{F} \Delta \mathcal{F}$ having the UCEM property. Now Theorem 5.10 implies that, if $\mathcal{F}$ has the UCEM property, then it also has the UCED property. In other words, if $\mathcal{F}$ has the property that empirical *means* converge uniformly, then it also has the property that empirical *distances* converge uniformly. Finally, in view of Theorem 5.2, the preceding arguments show that empirical distances converge *almost surely*. In other words, $\mathcal{F}$ has the property that

$$P^\infty\{\mathbf{x}^* \in X^\infty : \sup_{f,g \in \mathcal{F}} |\hat{d}_m(f, g : \mathbf{x}^*) - d_P(f,g)| \to 0 \text{ as } m \to \infty\} = 1.$$

5.8.3 Families of Loss Functions

In Section 3.3 we introduced a class of learning problems called "model-free" learning. The essential feature of such problems is that one attempts to fit randomly generated data with a function belonging to a "hypothesis class" $\mathcal{H}$. In Section 3.3 we defined an associated family of functions $\mathcal{L}_\mathcal{H}$ that depends both on the hypothesis class $\mathcal{H}$ and the "loss function" ℓ. In Theorem 3.2 it is shown that if the family of functions $\mathcal{L}_\mathcal{H}$ has the UCEMUP property, then any algorithm that nearly minimizes empirical risk with high probability is PAC. Now that we have available some conditions for a family of functions to have the UCEM and UCEMUP properties, it is of interest to see whether these conditions can be used to shed some light on when the family $\mathcal{L}_\mathcal{H}$ has the UCEMUP property. It turns out that it is possible to prove a very natural result, namely: $\mathcal{L}_\mathcal{H}$ has the UCEMUP property provided (i) the hypothesis class $\mathcal{H}$ has the UCEMUP property, and (ii) the loss function ℓ satisfies an equicontinuity condition. The result is sufficient to cover most practical applications of model-free learning.

Let us recall some notation from Section 3.3. One is given sets X, Y, U and a family $\mathcal{H}$ of measurable functions mapping X into U, known as the **hypothesis class**. One is also given a **loss function** $\ell : Y \times U \to [0, 1]$. The reader is referred to Section 3.3 for the details of the roles played by $\mathcal{H}$ and ℓ in the model-free learning problem. For the present purposes, the relevant entity is an associated family of functions $\mathcal{L}_\mathcal{H}$, defined next. Given a function $h \in \mathcal{H}$, define the corresponding function $\ell_h : X \times Y \to [0, 1]$ by

$$\ell_h(x, y) := \ell(y, h(x)), \ \forall x, y.$$

Finally, define

$$\mathcal{L}_\mathcal{H} := \{\ell_h : h \in \mathcal{H}\}.$$

Thus $\mathcal{L}_\mathcal{H}$ is the collection of functions ℓ_h generated by varying h over $\mathcal{H}$.

In the model-free learning problem, one is also given a family of probability measures $\bar{\mathcal{P}}$ on $X \times Y$. If $\bar{P}$ is a probability measure on $X \times Y$, one can "project" it onto another probability measure $\bar{P}_x$ on X alone, as follows: For each measurable set $A \subseteq X$, define

$$\bar{P}_x(A) := \bar{P}(A \times Y).$$

The measure $\bar{P}_x$ is also called the "marginal" of $\bar{P}$ on X. Let $\mathcal{P}$ denote the collection of projected (or marginal) measures $\{\bar{P}_x : \bar{P} \in \bar{\mathcal{P}}\}$.

Now we come to the main result of this subsection.

5.8 Uniform Convergence Properties of Iterated Families

Theorem 5.11. *Let $\bar{\mathcal{P}}$ be a family of probability measures on $X \times Y$, and let $\mathcal{P}$ denote the corresponding family of marginal probability measures on X. Suppose $Y = U = [0, 1]$, and that the family of functions $\{\ell(y, \cdot) : y \in [0, 1]\}$ is equicontinuous. Suppose also that the hypothesis class $\mathcal{H}$ has the UCEMUP property with respect to $\mathcal{P}$. Then $\mathcal{L}_\mathcal{H}$ has the UCEMUP property with respect to $\bar{\mathcal{P}}$.*

Remarks: The assumption that the family of functions $\{\ell(y, \cdot) : y \in [0, 1]\}$ is equicontinuous means the following: For each $\epsilon > 0$, there exists a $\delta = \delta(\epsilon)$ such that

$$|\ell(y, u_1) - \ell(y, u_2)| \leq \epsilon \; \forall u_1, u_2 \in [0, 1] \text{ with } |u_1 - u_2| \leq \delta, \; \forall y \in [0, 1]. \quad (5.8.3)$$

Commonly used loss functions such as $\ell(y, u) = |y - u|^s$ where $s \in [1, \infty)$ all satisfy this assumption.

Proof. The proof is based on Theorem 5.5. Suppose $m \geq 1$ is an integer, and that $(x_1, y_1), \ldots, (x_m, y_m)$ all belong to $X \times [0, 1]$. For notational convenience, define

$$\mathbf{x} := [x_1 \ldots x_m]^t \in X^m, \; \mathbf{y} := [y_1 \ldots y_m]^t \in [0, 1]^m,$$

$$\mathbf{z} := [(x_1, y_1), \ldots, (x_m, y_m)]^t \in (X \times [0, 1])^m.$$

In analogy with the symbols $\mathbf{f}(\mathbf{x})$ and $\mathcal{F}|_\mathbf{x}$ defined in Section 5.3, for $h \in \mathcal{H}$ define

$$\mathbf{h}(\mathbf{x}) := [h(x_1) \ldots h(x_m)]^t \in [0, 1]^m,$$

$$\mathcal{H}|_\mathbf{x} := \{\mathbf{h}(\mathbf{x}) : h \in \mathcal{H}\} \subseteq [0, 1]^m.$$

Similarly, define

$$\ell_h(\mathbf{z}) := [\ell_h(x_1, y_1) \ldots \ell_h(x_m, y_m)]^t = [\ell(y_1, h(x_1)) \ldots \ell(y_m, h(x_m))]^t \in [0, 1]^m,$$

$$\mathcal{L}_\mathcal{H}|_\mathbf{z} := \{\ell_h(\mathbf{z}) : h \in \mathcal{H}\} \subseteq [0, 1]^m.$$

The key step in the proof is to bound the external covering numbers of the set $\mathcal{L}_\mathcal{H}|_\mathbf{z}$ in terms of those of the set $\mathcal{H}|_\mathbf{x}$. Specifically, given any $\epsilon > 0$, choose a $\delta = \delta(\epsilon)$ such that (5.8.3) holds. Then the claim is that

$$L(\epsilon, \mathcal{L}_\mathcal{H}|_\mathbf{z}, \|\cdot\|_\infty) \leq L(\delta(\epsilon), \mathcal{H}|_\mathbf{x}, \|\cdot\|_\infty). \quad (5.8.4)$$

To prove this claim, suppose $\{\mathbf{v}^1, \ldots, \mathbf{v}^k\}$ is an external $\delta(\epsilon)$-cover for the set $\mathcal{H}|_\mathbf{x}$, where $k := L(\delta(\epsilon), \mathcal{H}|_\mathbf{x}, \|\cdot\|_\infty)$. Then it is shown that the set of k vectors $\{\mathbf{w}^1, \ldots, \mathbf{w}^k\}$ defined by

$$w_i^j := \ell(y_i, v_i^j), \; 1 \leq i \leq m, \; 1 \leq j \leq k,$$

is an external ϵ-cover for $\mathcal{L}_\mathcal{H}|_\mathbf{z}$. To see this, let $h \in \mathcal{H}$ be arbitrary. Then by the covering property there exists an index $j \in \{1, \ldots, k\}$ such that

$$\| \mathbf{h}(\mathbf{x}) - \mathbf{v}^j \|_\infty \le \delta(\epsilon), \text{ or } |h(x_i) - v_i^j| \le \delta(\epsilon) \text{ for } 1 \le i \le m.$$

By the equicontinuity property (5.8.3), it follows that

$$|\ell(y_i, h(x_i)) - \ell(y_i, v_i^j)| \le \epsilon \text{ for } 1 \le i \le m,$$

or equivalently

$$\| \ell_h(\mathbf{z}) - \mathbf{w}^j \|_\infty \le \epsilon.$$

Thus $\{\mathbf{w}^1, \ldots, \mathbf{w}^k\}$ is an external ϵ-cover for $\mathcal{L}_\mathcal{H}|_\mathbf{z}$. This establishes the inequality (5.8.4).

The proof is concluded by appealing to Theorem 5.5. Suppose $\bar P \in \bar{\mathcal P}$ is arbitrary. Then

$$E_{\bar P^m}[\lg L(\epsilon, \mathcal{L}_\mathcal{H}|_\mathbf{z}, \|\cdot\|_\infty)] = \int_{X^m \times Y^m} \lg L(\epsilon, \mathcal{L}_\mathcal{H}|_\mathbf{z}, \|\cdot\|_\infty)\, \bar P^m(d\mathbf{x}, d\mathbf{y})$$

$$\le \int_{X^m \times Y^m} \lg L(\delta(\epsilon), \mathcal{H}|_\mathbf{x}, \|\cdot\|_\infty)\, \bar P^m(d\mathbf{x}, d\mathbf{y})$$

$$= \int_{X^m} \lg L(\delta(\epsilon), \mathcal{H}|_\mathbf{x}, \|\cdot\|_\infty)\, P^m(d\mathbf{x}), \text{ where } P = \bar P_x$$

$$= E_{P^m}[\lg L(\delta(\epsilon), \mathcal{H}|_\mathbf{x}, \|\cdot\|_\infty)].$$

Therefore

$$\lim_{m\to\infty} \sup_{\bar P \in \bar{\mathcal P}} \frac{E_{\bar P^m}[\lg L(\epsilon, \mathcal{L}_\mathcal{H}|_\mathbf{z}, \|\cdot\|_\infty)]}{m}$$

$$\le \lim_{m\to\infty} \sup_{P \in \mathcal P} \frac{E_{P^m}[\lg L(\delta(\epsilon), \mathcal{H}|_\mathbf{x}, \|\cdot\|_\infty)]}{m} = 0,$$

since by assumption the hypothesis class $\mathcal{H}$ has the UCEMUP property with respect to $\mathcal{P}$. Hence it follows from Theorem 5.5 that the family $\mathcal{L}_\mathcal{H}$ has the UCEMUP property with respect to $\bar{\mathcal{P}}$. ∎

Now let us examine the case where $Y = U = \{0, 1\}$, and $\ell(y, u) = |y - u|$. This means that the hypothesis class $\mathcal{H}$ consists of *binary*-valued functions, and also that in each randomly drawn sample (x_i, y_i), the "outcome" y_i is binary. Consequently, the family of loss functions $\mathcal{L}_\mathcal{H}$ is also binary. These features enable us to prove *necessary* as well as sufficient conditions for $\mathcal{L}_\mathcal{H}$ to have the UCEMUP property. This is in contrast to Theorem 5.11, which gives only a sufficient condition.

Theorem 5.12. *Suppose $Y = U = \{0, 1\}$, and that $\ell(y, u) = |y - u|$. Let $\bar{\mathcal{P}}$ be a family of probability measures on $X \times Y$, and let $\mathcal{P}$ denote the corresponding family of marginal probability measures on X. Then $\mathcal{L}_\mathcal{H}$ has the UCEMUP property with respect to $\bar{\mathcal{P}}$ if and only if $\mathcal{H}$ has the UCEMUP property with respect to $\mathcal{P}$.*

5.8 Uniform Convergence Properties of Iterated Families

Proof. The proof is based on Theorem 5.6. Suppose

$$\mathbf{z} := [(x_1, y_1) \ldots (x_m, y_m)]^t \in (X \times Y)^m,$$

and let $\mathbf{x} := [x_1 \ldots x_m]^t \in X^m$ denote its "projection" on X^m. It is shown that

$$d(\mathcal{L}_\mathcal{H}; \mathbf{z}) = d(\mathcal{H}; \mathbf{x}).$$

In other words, the VC-dimension of $\mathcal{L}_\mathcal{H}$ when restricted to $\mathbf{z}$ is the same as the VC-dimension of $\mathcal{H}$ when restricted to $\mathbf{x}$. This is equivalent to the following statement: Suppose

$$S := \{(x_1, y_1), \ldots, (x_n, y_n)\} \subseteq (X \times Y)^m,$$

and let

$$S_x := \{x_1, \ldots, x_n\} \subseteq X$$

denote its projection onto X; then S is shattered by $\mathcal{L}_\mathcal{H}$ if and only if S_x is shattered by $\mathcal{H}$. The proof of the latter statement is based on the following elementary observation: Suppose $\mathbf{y} := [y_1 \ldots y_n]^t \in \{0,1\}^n$ is arbitrary, and for each Boolean vector $\mathbf{b} \in \{0,1\}^n$, define $\mathbf{y} \Delta \mathbf{b} \in \{0,1\}^n$ by

$$(\mathbf{y}\Delta\mathbf{b})_i := |y_i - b_i|, \; i = 1, \ldots, n.$$

In other words, $\mathbf{y}\Delta\mathbf{b}$ is just the exclusive-or function applied componentwise. Then the map $\mathbf{b} \mapsto \mathbf{y}\Delta\mathbf{b}$ is one-to-one and onto. Now S is shattered by $\mathcal{L}_\mathcal{H}$ if and only if, for each $\mathbf{b} \in \{0,1\}^n$, there exists a function $H_\mathbf{b} \in \mathcal{H}$ such that

$$\ell_{H_\mathbf{b}}(x_i, y_i) = b_i, \; \forall i,$$

or equivalently,

$$|y_i - H_\mathbf{b}(x_i)| = b_i, \; \forall i,$$

or equivalently,

$$H_\mathbf{b}(x_i) = |y_i - b_i|, \; \forall i.$$

However, note that $\mathbf{y} \in \{0,1\}^n$ is a *fixed* vector; also, as observed above, the map $\mathbf{b} \mapsto \mathbf{y}\Delta\mathbf{b}$ is one-to-one and onto. Therefore S is shattered by $\mathcal{L}_\mathcal{H}$ if and only if S_x is shattered by $\mathcal{H}$. Returning to the original notation, it follows that $d(\mathcal{L}_\mathcal{H}; \mathbf{z}) = d(\mathcal{H}; \mathbf{x})$. Now note that, given a vector $\mathbf{x} \in X^m$, there are 2^m vectors $\mathbf{z} \in (X \times Y)^m$ that "project" onto $\mathbf{x}$, and $d(\mathcal{L}_\mathcal{H}; \mathbf{z})$ is the same for each of these 2^m vectors.

Next, suppose $\bar{P} \in \bar{\mathcal{P}}$ is arbitrary, and let $P \in \mathcal{P}$ denote its marginal probability on X. It is shown that

$$E_{\bar{P}^m}[d(\mathcal{L}_\mathcal{H}; \mathbf{z})] = E_{P^m}[d(\mathcal{H}; \mathbf{x})].$$

With each $x \in X$, one can associate the conditional probabilities $\bar{p}(1|x)$ and $\bar{p}(0|x)$. Given $\mathbf{z} = (\mathbf{x}, \mathbf{y})$, define

204   5. Uniform Convergence of Empirical Means

$$Q(\mathbf{x},\mathbf{y}) := \prod_{i=1}^{m} \bar{p}(y_i|x_i).$$

Then

$$E_{\bar{P}^m}[d(\mathcal{L}_{\mathcal{H}};\mathbf{z})] = \int_{X^m} \sum_{\mathbf{y}\in\{0,1\}^m} d(\mathcal{L}_{\mathcal{H}};(\mathbf{x},\mathbf{y}))\, Q(\mathbf{x},\mathbf{y})\, P^m(d\mathbf{x})$$

$$= \int_{X^m} d(\mathcal{H};\mathbf{x}) \left[\sum_{\mathbf{y}\in\{0,1\}^m} Q(\mathbf{x},\mathbf{y})\right] P^m(d\mathbf{x})$$

$$= \int_{X^m} d(\mathcal{H};\mathbf{x})\, P^m(d\mathbf{x}) = E_{P^m}[d(\mathcal{H};\mathbf{x})].$$

Note that, in the second equation, we used the fact that $d(\mathcal{L}_{\mathcal{H}};(\mathbf{x},\mathbf{y}))$ is independent of $\mathbf{y}$, while in the third equation we used the fact that the $Q(\mathbf{x},\mathbf{y})$ add up to 1 as $\mathbf{y}$ varies over $\{0,1\}^m$. Finally, it follows from the above that the family $\mathcal{L}_{\mathcal{H}}$ satisfies the condition (5.3.8) with respect to $\bar{P}$ if and only if $\mathcal{H}$ satisfies (5.3.8) with respect to $\mathcal{P}$. The desired conclusion now follows from Theorem 5.6. ∎

Notes and References Most of the material in this chapter is taken from the two classic papers [194] and [196], and the treatment of these two papers in [190], with some simplifications due to Steele [179]. In particular, the statement of the main theorem, namely Theorem 5.3, is from [196] and [190], and the proof of this theorem, including the lemmas in Sections 5.4 and 5.5 and their proofs, are all adapted from the Appendix to Chapter 7 in [190]. The paper [193] contains a statement, without any proofs, of the main theorems in [194]. Theorem 5.4, giving an alternative necessary and sufficient condition in terms of the "average" behaviour of the restricted VC-dimension, is given in [179]. Interestingly, the paper of Vapnik and Chervonenkis [194] does *not* interpret their main theorem in this form, even though all the required preliminary results are already available therein. Theorem 5.2 is more or less contained in [194]; see Theorem 3, p. 271 therein. Thus Vapnik and Chervonenkis were aware that their conditions implied not only the convergence *in probability* of empirical means to their true values, but also *almost sure* convergence. In a roundabout way, this implies that the UCEM property and the ASCEM property are equivalent. However, the contribution of Steele [179] lay in showing that the maximum discrepancy between empirical means and true means converges almost surely to a constant for *every* family of functions; the only question is whether this constant is zero or not. Moreover, Steele's proof of Theorem 5.2 is very direct and simple. The idea of applying the theory of subadditive processes to the problem at hand is due to Steele [179]. Actually, [179] only studies the problem of the almost sure

5.8 Uniform Convergence Properties of Iterated Families

convergence of empirical *probabilities*, but methods therein carry over quite readily to the problem of empirical *means*. Thus Lemma 5.1 is an adaptation from [179]. Theorem 5.5 is not stated explicitly in the English literature, but might perhaps be contained in [195]. A similar condition (but for sets and not functions) is stated without proof as a necessary and sufficient condition for "finite learnability" in [145] as an "abstract of work in progress." Subsequently, the full paper [149] contains only the sufficiency of this condition but not the necessity; Example 6.4 in the next chapter shows that the uniform convergence of empirical means to their true values is in fact *not* a necessary condition for "solid" learnability. Such issues are examined in detail in subsequent chapters. The material in Section 5.6 is essentially from the paper by Haussler [80] but with some improvements. In particular, Theorem 5.7 is an improved version of the corresponding result in [80] in that some of the constants are less conservative. Theorem 5.8 and its proof are taken from [80]. The material in Section 5.8 on iterated families and loss functions is given here for the first time.

In the "Notes and References" section of Chapter 3, it is suggested that there is a case for studying the "one-sided" convergence of empirical means to their true values. Specifically, given the family of functions $\mathcal{F}$, define

$$q_u(m, \epsilon) := P^m \{\mathbf{x} \in X^m : \exists f \in \mathcal{F} \text{ s.t. } \hat{E}(f; \mathbf{x}) - E_P(f) > \epsilon\}.$$

Necessary and sufficient conditions for this one-sided quantity to converge to zero as $m \to \infty$ for each $\epsilon > 0$ are given in [197]. Obviously these conditions are weaker than those in Theorem 5.3.

6. Learning Under a Fixed Probability Measure

6.1 Introduction

In this chapter, we study the problems of concept and function learning in the case where the samples are drawn in accordance with a *known fixed* distribution. Various necessary and/or sufficient conditions are presented for a concept class or a function class to be learnable. The principal results of the chapter can be summarized as follows: Suppose the input sequence to the learning algorithm is i.i.d. Then we have the following:

1. If a function class $\mathcal{F}$ (or a concept class $\mathcal{C}$) has the property of uniform convergence of empirical means, then it is also ASEC learnable. However, the converse is *not* true in general – there exist function classes that are ASEC learnable even though they do not possess the UCEM property.
2. A function class is PUAC learnable if it possesses a property known as the "shrinking width" property. The shrinking width property is also a *necessary* condition in order for every consistent algorithm to be PUAC.
3. Similarly, there is a necessary and sufficient condition for a function family to be consistently PAC learnable.
4. It can be shown that PUAC learnability is equivalent to consistent PUAC learnabilit. In contrast, PAC learnability is not equivalent in general to consistent PAC learnability.
5. A function class (or a concept class) is learnable *if* it satisfies a property known as "finite metric entropy."
6. In order for a *concept* class to be learnable, the finite metric entropy condition is *necessary* as well as sufficient; however, for a *function* class to be learnable, the finite metric entropy condition is sufficient but is not necessary in general.

There are of course several other nuances as well, and these are brought out by and by.

All of the above statements pertain to the case where the input sequence to the learning algorithm is i.i.d. However, with very little effort, many of the preceding results can be extended to the case of α-mixing input sequences.

For the convenience of the reader, the definition of learnability is repeated here. Recall that the basic ingredients of the learning problem under a fixed probability are:

208   6. Learning Under a Fixed Probability Measure

- A set X,
- A σ-algebra $\mathcal{S}$ of subsets of X,
- A fixed known probability measure P on the measurable space $(X, \mathcal{S})$,
- A subset $\mathcal{C} \subseteq \mathcal{S}$, called the *concept class*, or else a family $\mathcal{F}$ of measurable functions mapping X into $[0, 1]$.

Let us reprise the problem of function learning, since the concept learning problem can be thought of as a special case of the function learning problem. There is a fixed but unknown function $f \in \mathcal{F}$, called the "target" function, and i.i.d. samples $x_1, \ldots, x_m \in X$ are drawn in accordance with P. For each sample x_i, an "oracle" returns the value $f(x_i)$ of the unknown target function. An "algorithm" is an indexed family of maps

$$A_m : (X \times [0, 1])^m \to \mathcal{F}.$$

Let $h_m(f; \mathbf{x})$ denote the function in $\mathcal{F}$ produced by the algorithm when the target function is f and the multisample is $\mathbf{x} = [x_1 \ldots x_m]^t \in X^m$. In other words,

$$h_m(f; \mathbf{x}) := A_m[(x_1, f(x_1)), \ldots, (x_m, f(x_m))].$$

Now define

$$r(m, \epsilon) := \sup_{f \in \mathcal{F}} P^m \{ \mathbf{x} \in X^m : d_P[f, h_m(f; \mathbf{x})] > \epsilon \}, \quad (6.1.1)$$

where d_P denotes the pseudometric on $\mathcal{F}$ defined by

$$d_P(f, g) := \int_X |f(x) - g(x)| \, P(dx). \quad (6.1.2)$$

The algorithm $\{A_m\}$ is said to be **probably approximately correct (PAC) to accuracy** ϵ if $r(m, \epsilon) \to 0$ as $m \to \infty$, and **probably approximately correct (PAC)** if $r(m, \epsilon) \to 0$ for every ϵ. The family $\mathcal{F}$ is said to be **PAC learnable** if there exists a PAC algorithm.

One can also define the notion of an algorithm being probably uniformly approximately correct (PUAC) almost surely eventually correct (ASEC), as in Chapter 3. The first step is to define the countable Cartesian product X^∞, the corresponding σ-algebra $\mathcal{S}^\infty$ on X^∞, and the probability measure P^∞ on $(X^\infty, \mathcal{S}^\infty)$. Given the family $\mathcal{F}$, one defines the stochastic process

$$b_m(\mathbf{x}^*) := \sup_{f \in \mathcal{F}} d_P[f, h_m(f; \mathbf{x}^*)],$$

where $h_m(f; \mathbf{x}^*)$ is the output of the algorithm after m steps when f is the target function and $\mathbf{x}^*$ is the sequence of samples, as defined above. The algorithm $\{A_m\}$ is said to be **probably uniformly approximately correct (PUAC)** if the stochastic process $\{b_m(\cdot)\}$ converges to zero in probability; that is, the quantity

$$s(m, \epsilon) := P^\infty \{ \mathbf{x}^* \in X^\infty : \sup_{f \in \mathcal{F}} d_P[f, h_m(f; \mathbf{x}^*)] > \epsilon \}$$

approaches zero as $m \to \infty$, for each $\epsilon > 0$. Since the quantity $d_P[f, h_m(f; \mathbf{x}^*)]$ depends only on the first m components of the sequence $\mathbf{x}^*$, an equivalent definition of the quantity $s(m, \epsilon)$ is

$$s(m, \epsilon) = P^m \{ \mathbf{x} \in X^m : \sup_{f \in \mathcal{F}} d_P[f, h_m(f; \mathbf{x})] > \epsilon \}.$$

The algorithm $\{A_m\}$ is said to be **almost surely eventually correct (ASEC)** if the stochastic process $\{b_m(\cdot)\}$ converges to zero almost surely (with respect to P^∞), or in other words,

$$P^\infty \{ \mathbf{x}^* \in X^\infty : \sup_{f \in \mathcal{F}} d_P[f, h_m(f; \mathbf{x}^*)] \to 0 \text{ as } m \to \infty \} = 1.$$

It is easy to see that the ASEC property implies the PUAC property. However, it is not as yet clear whether the converse is true in general.

6.2 UCEM Property Implies ASEC Learnability

The principal result of this section states that, *if* a family of bounded measurable functions has the property that empirical means converge uniformly, then the family is also PUAC learnable. In other words, the UCEM property studied in the preceding chapter is a *sufficient* condition for a family of functions to be PUAC learnable. However, as we shall see in subsequent sections, the condition is *not* necessary in general. The principal result is proved by relating the UCEM property to another property called the uniform convergence of empirical *distances*.

As always, let $\mathcal{F}$ denote a family of functions mapping X into $[0, 1]$, where each function in $\mathcal{F}$ is measurable with respect to a given σ-algebra $\mathcal{S}$ of subsets of X. Let P be a given probability measure on $(X, \mathcal{S})$. Then one can define a pseudometric d_P on $\mathcal{F}$ as in (6.1.2). Note that the function $x \mapsto |f(x) - g(x)|$ also maps X into $[0, 1]$, and is measurable. Hence, in analogy with Section 5.1, one can "empirically" estimate $d_P(f, g)$ as follows: Let $x_1, \ldots, x_m \in X$ be i.i.d. samples drawn in accordance with P, and define

$$\hat{d}(f, g; \mathbf{x}) = \hat{d}_m(f, g; \mathbf{x}^*) := \frac{1}{m} \sum_{i=1}^m |f(x_i) - g(x_i)| \qquad (6.2.1)$$

as the "empirical distance" between f and g. Note that $\hat{d}(f, g; \mathbf{x})$ is just the empirical mean of the function $x \mapsto |f(x) - g(x)| \in [0, 1]^X$. Now, as before, one can ask whether the empirical estimate $\hat{d}(f, g; \mathbf{x})$ converges to the true value $d_P(f, g)$ as the number of samples approaches infinity, and whether the

210   6. Learning Under a Fixed Probability Measure

convergence is uniform with respect to the functions involved. To make the question precise, define

$$q_d(m, \epsilon) = P^m\{\mathbf{x} \in X^m : \exists f, g \in \mathcal{F} \text{ s.t. } |\hat{d}(f, g; \mathbf{x}) - d_P(f, g)| > \epsilon\}. \quad (6.2.2)$$

We say that the family $\mathcal{F}$ has the property of **uniform convergence of empirical distances(UCED)** if $q_d(m, \epsilon) \to 0$ as $m \to \infty$, for each fixed ϵ.

It is a consequence of Example 5.5 that, if $\mathcal{F}$ has the property that empirical *means* converge uniformly (UCEM property), then $\mathcal{F}$ also has the property that empirical *distances* converge uniformly (UCED property). Conversely, if the zero function belongs to $\mathcal{F}$ and $\mathcal{F}$ has the UCED property, then it is easy to show that $\mathcal{F}$ also has the UCEM property. In the case of a collection of sets $\mathcal{A}$, if the empty set belongs to $\mathcal{A}$ and if $\mathcal{A}$ has the UCED property, then $\mathcal{A}$ also has the UCEP property.

A central idea in what follows is the idea of a "consistent" algorithm, which is defined next. Recall that, in the present context, an "algorithm" is merely an indexed family of maps

$$A_m : (X \times [0, 1])^m \to \mathcal{F}.$$

Let $h_m := h_m(f; \mathbf{x})$ denote the hypothesis generated by the algorithm when the target concept is f and the multisample is $\mathbf{x}$. In other words,

$$h_m := A_m\{[x_1, f(x_1)], \ldots, [x_m, f(x_m)]\}.$$

We say that the hypothesis h_m **agrees with** f **on x** if

$$h_m(x_i) = f(x_i), \ i = 1, \ldots, m.$$

The algorithm is said to be **consistent** if $h_m(f; \mathbf{x})$ agrees with f on $\mathbf{x}$ for every function $f \in \mathcal{F}$ and every multisample $\mathbf{x} \in X^m$, for every $m \geq 1$. To put it into words: An algorithm is consistent if the hypothesis produced by the algorithm always matches the data points.

Note that in the statistics literature the term "consistent" has an entirely different meaning, which is not to be confused with the usage here. In the statistics literature, roughly speaking "consistency" means that if the data is indeed being generated by a "true" model, then the estimated model converges to the true model as the data size approaches infinity. In the context of model-free learning for example, consistency would mean that if the data is generated by some true function $f \in \mathcal{H}$, then the hypothesis h_m would converge to f in an appropriate topology. In contrast, the present usage of "consistent" to mean an algorithm that matches all the available data is quite common in the learning theory literature.

Example 6.1. Let $X = [0, 1]^2$, and let $\mathcal{C}$ consist of all convex sets in X. Let P denote the uniform probability measure on X. Suppose $T \in \mathcal{C}$ is an unknown convex set in X. To learn T, i.i.d. samples $x_1, \ldots, x_m$ are drawn from X, and

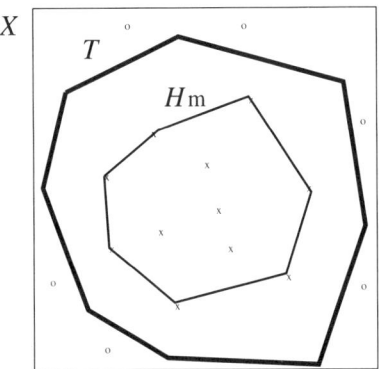

Fig. 6.1. Learning an Unknown Convex Polygon

each sample is labelled as to whether or not it belongs to T. (See Figure 6.1, which is the same as Figure 1.1.) Suppose H_m is defined to be the convex hull of all "positive" examples, that is, all x_i that belong to T. Thus H_m is the *smallest* convex polygon that correctly classifies all the sample points. Then it is easy to see that the algorithm is consistent.

Similary, it can be seen that the algorithm defined in Example 3.5 for learning axis-parallel rectangles is also consistent. ∎

Observe that, if an algorithm is *consistent*, then

$$\hat{d}[f, h_m(f; \mathbf{x})] = 0 \ \forall \mathbf{x} \in X^m, \ \forall m \geq 0, \ \forall f \in \mathcal{F}.$$

This follows from the fact that $h_m(x_i) = f(x_i)$ for all i if the algorithm is consistent. Alternatively, we could also take the above equality as the *definition* of a consistent algorithm.

Now the question arises as to whether there exists a consistent algorithm in *every* learning problem. If one ignores issues of effective computability, computational complexity, and the like, then the answer is always "yes." This can be seen as follows: By the axiom of choice, the set $\mathcal{F}$ can always be well-ordered. Thus, given a labelled sample $\{[x_1, f(x_1)], \ldots, [x_m, f(x_m)]\}$, one can simply scan through all the functions in $\mathcal{F}$ in order until one finds a function that matches the labelled sample. Such a function surely exists, since $f \in \mathcal{F}$. Let h_m denote the first function (in the sense of the ordering on $\mathcal{F}$) that matches the labelled sample. This algorithm is well-defined, and is consistent. It is also "recursive" in the sense that h_{m+1} depends only on h_m and the labelled sample $[x_{m+1}, f(x_{m+1})]$. Of course, this "algorithm" is also purely conceptual and is not claimed to be implementable in any way.[1] The topic of *computational* learning theory is addressed to the development of learning algorithms that are in some sense *effective*, e.g., in terms of being

---

[1] However, if the family $\mathcal{F}$ is *recursively enumerable*, then the above procedure would indeed satisfy most persons as being a true algorithm.

212   6. Learning Under a Fixed Probability Measure

effectively computable, or using only a reasonable amount of resources (storage, time, etc.). In contrast, the present treatment does not concern itself with such issues; rather, the emphasis here is on what is *theoretically* possible, rather than on what is *effectively* possible. Thus the topic of discourse here might perhaps be called "statistical" learning theory, to be contrasted with "computational" learning theory as described above.

In some applications, the requirement that an algorithm be consistent is rather strict. In the next few paragraphs, we introduce several less restrictive versions of "consistency" that are in some sense "good enough" to ensure learnability of various types.

The first notion is that of an algorithm that is "almost surely" consistent. An algorithm is said to be **almost surely consistent** if

$$P^m\{\mathbf{x} \in X^m : \hat{d}[f, h_m(f; \mathbf{x})] = 0\} = 1, \, \forall f \in \mathcal{F}.$$

In other words, for each target function $f \in \mathcal{F}$, the hypothesis $h_m(f; \mathbf{x})$ agrees with f on $\mathbf{x}$ for all $\mathbf{x}$ except those belonging to a set of measure zero.

Thus Example 3.9 serves to show that consistent learnability is a more stringent requirement than solid or potential learnability.

Example 6.2. The purpose of this example is to demonstrate an algorithm that is almost surely consistent but not consistent. In fact we have already encountered this algorithm in Example 3.9, the details of which are recalled here for the convenience of the reader.

Let $X = [0, 1]$, $\mathcal{S}$ = the Borel σ-algebra on X, and let P = the uniform probability measure on X. Let $\mathcal{G}$ denote the collection of all finite subsets of X, and let $\tau : \mathcal{G} \to [0, 0.5)$ be a one-to-one (but not necessarily onto) map. If $a \in [0, 0.5)$ belongs to the range of the map τ, let $\tau^{-1}(a) =: G$ denote the unique finite subset of X such that $\tau(G) = a$; otherwise, let $\tau^{-1}(a) = \emptyset$. Then $\mathcal{C}$ consists of all sets of the form $[0, a] \cup \tau^{-1}(a)$ as a varies over $[0, 0.5]$, together with the set X itself.

For each multisample $\mathbf{x} \in X^m$, let $\mathcal{C}|_\mathbf{x} \subseteq \{0, 1\}^m$ denote the set of labels $[I_T(x_1) \ldots I_T(x_m)]^t$ as T varies over $\mathcal{C}$. This symbol is analogous to $\mathcal{F}|_\mathbf{x}$ defined in Chapter 5. Now the algorithm $\{A_m\}$ consists of an indexed family of maps, where

$$A_m : \bigcup_{\mathbf{x} \in X^m} (\{\mathbf{x}\} \times \mathcal{C}|_\mathbf{x}) \to \mathcal{C}.$$

Note that the above is just an alternative way of viewing the map A_m. The algorithm is as follows: For each $\mathbf{x} \in X^m$, if the label vector $\mathbf{L} \in \mathcal{C}|_\mathbf{x}$ equals $[1 \ldots 1]^t$, then $H_m(\mathbf{x}; \mathbf{L})$ equals X. Otherwise $H_m(\mathbf{x}; \mathbf{L}) = [0, h] \cup \tau^{-1}(h)$, where

$$h := \min\{0.5, \max\{x_i : I_T(x_i) = 1\}\}.$$

It is shown first that the algorithm is *not* consistent. To see this, let $y \in (0.5, 1]$, and let $b = \tau(\{y\})$. Let $m \geq 1$ be arbitrary, and suppose $\mathbf{x} \in X^m$ satisfies two conditions: (i) $x_i = y$ for some i, and (ii) $x_j \in (b, 0.5)$ for some other j. Suppose the target concept is $T = [0, b] \cup \{y\}$. Then $I_T(x_i) = I_T(y) = 1$, and since $x_i = y \geq 0.5$, the output of the algorithm is $H_m = [0, 0.5]$. (Note that, since τ maps $\mathcal{G}$ into $[0, 0.5)$, we have $\tau^{-1}(0.5) = \emptyset$, so that $H_m = [0, 0.5] \cup \tau^{-1}(0.5) = [0, 0.5]$.) However, since $x_j \in (b, 0.5)$, we have $I_T(x_j) = 0$ but $I_{H_m}(x_j) = 1$. Hence the algorithm is not consistent.

On the other hand, given a target concept $T = [0, a] \cup \tau^{-1}(a)$, one can assert with probability one (with respect to P) that a randomly selected $x \in X$ does *not* belong to the "tail" $\tau^{-1}(a)$. Hence, with probability one (with respect to P^m), the hypothesis H_m agrees with T on $\mathbf{x}$, and the algorithm is almost surely consistent. ∎

A closely related concept is that of an "asymptotically" consistent algorithm. Let $h_m(f; \mathbf{x})$ denote the output of the algorithm when the target function is f and the multisample is $\mathbf{x}$. Then the algorithm is said to be **asymptotically consistent** if

$$\sup_{f \in \mathcal{F}} P^m \{\mathbf{x} \in X^m : \hat{d}[f, h_m(f; \mathbf{x}); \mathbf{x}] > \epsilon\} \to 0 \text{ as } m \to \infty, \forall \epsilon > 0.$$

Thus one can think of an asymptotically consistent algorithm as one that produces hypotheses that are "nearly" consistent with the data "with high probability" as more and more samples are drawn. Note that an almost surely consistent algorithm, as in Example 6.2, is also asymptotically consistent. Similarly, one can define an algorithm to be **asymptotically uniformly consistent** if

$$P^m \{\mathbf{x} \in X^m : \sup_{f \in \mathcal{F}} \hat{d}[f, h_m(f; \mathbf{x}); \mathbf{x}] > \epsilon\} \to 0 \text{ as } m \to \infty, \forall \epsilon > 0.$$

In the same vein, the algorithm $\{A_m\}$ is said to be **almost surely eventually consistent** if

$$P^\infty \{\mathbf{x}^* \in X^\infty : \sup_{f \in \mathcal{F}} \hat{d}_m[f, h_m(f; \mathbf{x}^*); \mathbf{x}^*] \to 0 \text{ as } m \to \infty\} = 1.$$

In other words, an algorithm is asymptotically uniformly consistent if the maximal empirical distance between a target function and the hypothesis approaches zero *in probability*. The algorithm is almost surely eventually consistent if the empirical distance between a target function and the hypothesis approaches zero *almost surely*. Note that a consistent algorithm has *all* of the above properties.

Now we come to the main result of this section.

Theorem 6.1. *Suppose a family $\mathcal{F} \subseteq [0, 1]^X$ of measurable functions has the property that empirical means converge uniformly. Then the family is ASEC learnable. In particular,*

- *Every asymptotically consistent algorithm is PAC.*
- *Every asymptotically uniformly consistent algorithm is PUAC.*

214   6. Learning Under a Fixed Probability Measure

– *Every almost surely eventually consistent algorithm is ASEC.*

Remarks: A reader who is not interested in the nuances of the various types of consistency can simplify the above theorem statement to: "Every consistent algorithm is ASEC," which is the strongest form of learnability discussed in this chapter.

Proof. Suppose the family $\mathcal{F}$ has the UCEM property, and let $\{A_m\}$ be any asymptotically consistent algorithm. Define $r(m, \epsilon)$ as in (6.1.1); it is desired to show that $r(m, \epsilon) \to 0$ as $m \to \infty$. For this purpose, let $\epsilon, \delta > 0$ be specified, and define

$$\tilde{q}(m, \epsilon) := \sup_{f \in \mathcal{F}} P^m\{\mathbf{x} \in X^m : \exists g \in \mathcal{F} \text{ s.t. } |\hat{d}(f, g; \mathbf{x}) - d_P(f, g)| > \epsilon\},$$

where $\hat{d}(f, g; \mathbf{x})$ is defined in (6.2.1). Note that $\tilde{q}(m, \epsilon) \leq q_d(m, \epsilon)$, where $q_d(m, \epsilon)$ is defined in (6.2.2). Also, by Example 5.5, the fact that the family $\mathcal{F}$ has the UCEM property implies that $q_d(m, \epsilon) \to 0$ as $m \to \infty$, which in turn implies that $\tilde{q}(m, \epsilon) \to 0$ as $m \to \infty$. Finally, the algorithm $\{A_m\}$ is assumed to be asymptotically consistent. Hence it is possible to choose m_0 large enough that

$$\tilde{q}(m, \epsilon/2) \leq \delta/2 \,\forall m \geq m_0, \text{ and}$$

$$\sup_{f \in \mathcal{F}} P^m\{\mathbf{x} \in X^m : \hat{d}[f, h_m(f; \mathbf{x}); \mathbf{x}] > \epsilon\} \leq \delta/2 \,\forall m \geq m_0.$$

It is now shown that

$$r(m, \epsilon) \leq \delta \,\forall m \geq m_0.$$

To establish this inequality, fix $f \in \mathcal{F}$ and draw a multisample $\mathbf{x} = [x_1 \ldots x_m]^t \in X^m$. Then, with probability at least $1 - \delta/2$ with respect to $\mathbf{x}$, it is true that

$$\hat{d}(f, h_m; \mathbf{x}) \leq \epsilon/2.$$

where h_m is a shorthand for $h_m(f; \mathbf{x})$. Also, with probability $1 - \tilde{q}(m, \epsilon/2) \geq 1 - \delta/2$ with respect to $\mathbf{x}$, it is true that

$$|\hat{d}(f, h_m; \mathbf{x}) - d_P(f, h_m)| \leq \epsilon/2.$$

Hence, with probability at least $1 - \delta$ with respect to $\mathbf{x}$, it is true that

$$d_P(f, h_m) \leq \epsilon.$$

This is the same as saying that $r(m, \epsilon) \leq \delta$, which is precisely the PAC inequality. This shows that every asymptotically consistent algorithm is PAC.

The proof that every asymptotically uniformly consistent algorithm is PUAC is entirely similar and is left to the reader.

The proof that every almost surely eventually consistent algorithm is ASEC is also quite similar. By Example 5.5, the fact that $\mathcal{F}$ has the UCEM

6.2 UCEM Property Implies ASEC Learnability

property implies that empirical distances between pairs of functions in $\mathcal{F}$ converge almost surely to their true values. In other words, $\mathcal{F}$ has the property that

$$P^\infty\{\mathbf{x}^* \in X^\infty : \sup_{f,g \in \mathcal{F}} |\hat{d}_m(f, g : \mathbf{x}^*) - d_P(f, g)| \to 0 \text{ as } m \to \infty\} = 1.$$

Now, if the algorithm is almost surely eventually consistent, it is true that

$$P^\infty\{\mathbf{x}^* \in X^\infty : \sup_{f \in \mathcal{F}} \hat{d}_m[f, h_m(f; \mathbf{x}^*); \mathbf{x}^*] \to 0 \text{ as } m \to \infty\} = 1.$$

By the triangle inequality, it follows that

$$d_P[f, h_m(f; \mathbf{x}^*)] \leq \hat{d}_m[f, h_m(f; \mathbf{x}^*); \mathbf{x}^*]$$
$$+ |\hat{d}_m[f, h_m(f; \mathbf{x}^*); \mathbf{x}^*] - d_P[f, h_m(f; \mathbf{x}^*)]|.$$

However, by assumption, the supremum with respect to $f \in \mathcal{F}$ of both quantities on the right side approaches zero almost surely as $m \to \infty$. This shows that the algorithm is almost surely eventually correct (ASEC). ∎.

Theorem 6.1 has an interesting intuitive appeal. Suppose a family of functions has the property that, by repeatedly drawing i.i.d. samples, one can estimate the *mean value* of each function with high accuracy and high confidence; then in fact it is possible, not merely to make an accurate assessment of the mean value of the function, but of *the function itself*. In the case of estimating probabilities empirically, this result can be interpreted as follows: If a family of measurable sets has the property that the *size* of each set can be estimated with accuracy and confidence by drawing i.i.d. samples; then it is possible to estimate the *set itself*.

Example 6.3. Consider once again the problem of learning the family of convex polygons inside the unit square $[0, 1]^2$. From Example 5.3, this family of sets has the property that probabilities converge almost surely to their true values. Hence, by Theorem 6.1, it follows that this family is also learnable, and that *every* consistent algorithm is ASEC. For instance, one could simply choose H_m to be the *smallest* convex polygon that correctly classifies all the sample points, i.e., the convex hull of all the positive examples (all x_i such that $I_T(x_i) = 1$). This algorithm is ASEC.

More generally, let k, l be fixed positive integers. Suppose $X = [0, 1]^k$, $\mathcal{S} = $ the Borel σ-algebra on X, and let P equal the uniform probability measure on X. Let $\mathcal{C}$ consist of all unions of l *or fewer* convex sets in X. Then one can write

$$\mathcal{C} = \bigcup_{s=1}^{l} \mathcal{C}_s,$$

where the collection $\mathcal{C}_s$ consists of all unions of *exactly* s convex sets in X. By Example 5.4, it follows that each $\mathcal{C}_s$ has the UCEP property. Hence their finite union $\mathcal{C}$ also has the UCEP property. Now it follows from Theorem 6.1 that the collection $\mathcal{C}$ is ASEC learnable, and that in fact every consistent algorithm is ASEC. However, in contrast to the case where $\mathcal{C}$ consists of all convex sets in X, finding a consistent hypothesis is is no longer as straightforward as taking the convex hull of all positive examples.

6.3 Finite Metric Entropy Implies Learnability

In this section, it is shown that a family of functions is learnable if it satisfies a so-called "finite metric entropy" condition. This is established by introducing a "minimum empirical risk" algorithm that is PAC whenever the finite metric entropy condition is satisfied.

Recall that the problem at hand is one of learning a family of functions $\mathcal{F}$ when the samples $x_1, \ldots, x_m$ are drawn in accordance with a known fixed probability measure P. Let d_P denote the pseudometric on $\mathcal{F}$ defined by (6.1.2), and let $N(\epsilon, \mathcal{F}, d_P)$ denote the ϵ-covering number of the family $\mathcal{F}$ with respect to the pseudometric d_P. Note that this is an *entirely different* covering number from the ones encountered in the preceding section. In particular, throughout Chapter 5 and in the preceding section we made heavy use of the number $L(\epsilon, \mathcal{F}|_{\mathbf{z}}, \|\cdot\|_{a1})$, which differs from $N(\epsilon, \mathcal{F}, d_P)$ in at least three significant respects:

- L is an *external* covering number, whereas N is the usual covering number; see Section 2.1.2 for the definitions and a discussion of these two quantities.
- L is a covering number of $\mathcal{F}|_{\mathbf{z}}$, which is a subset of $[0,1]^{2m}$; hence $L(\epsilon, \mathcal{F}|_{\mathbf{z}}, \|\cdot\|_{a1})$ is a function of $\mathbf{z}$. In contrast, N is a covering number of the entire set $\mathcal{F}$, and is just a number (and not a function of $\mathbf{z}$).
- L is a covering number with respect to the norm $\|\cdot\|_{a1}$, whereas N is a covering number with respect to the pseudometric d_P.

The principal result of this section states that a family $\mathcal{F}$ is learnable if the covering number $N(\epsilon, \mathcal{F}, d_P)$ is finite for each ϵ. This condition is sometimes referred to as the **finite metric entropy** condition. It turns out that there exist families that satisfy the finite metric entropy condition but *do not* satisfy the conditions for the uniform convergence of empirical means. Hence the result proved in this section is better than Theorem 6.1. On the other hand, Theorem 6.1 states (among other things) that *every* asymptotically consistent algorithm is PAC, whereas Theorem 6.2 shows only that *a specific* algorithm is PAC. Thus both theorems are valuable in their own right.

We begin by defining the so-called "minimum empirical risk" algorithm. Let $\epsilon > 0$ be specified, and let $g_1, \ldots, g_k$ be an $\epsilon/2$-cover (not necessarily minimal!) of $\mathcal{F}$ with respect to d_P. Observe that each g_i is a *function* belonging to $\mathcal{F}$. Now the algorithm is as follows: Draw i.i.d. samples $x_1, \ldots, x_m \in X$ in accordance with P. After m samples are drawn, define the "cost" function

6.3 Finite Metric Entropy Implies Learnability

$$\hat{J}_i = \frac{1}{m} \sum_{j=1}^{m} |f(x_i) - g(x_i)|$$

for each index $i \in \{1, \ldots, k\}$. Note that $\hat{J}_i$ is merely the *empirical* distance $\hat{d}(f, g_i; \mathbf{x})$ between the unknown target function f and the function g_i, based on the multisample $\mathbf{x}$. Even though the target function f is unknown, one can nevertheless compute $\hat{J}_i$, because the *values* $f(x_1), \ldots, f(x_m)$ are available through the "oracle." Now the output of the algorithm after m samples is a function g_l such that

$$\hat{J}_l = \min_{1 \le i \le k} \hat{J}_i.$$

In case the minimum is achieved for more than one index i, one can use any sensible tie-breaking rule; for example, l can be chosen to be the *smallest* index i that achieves the above minimum.

The minimum empirical risk algorithm is very intuitive. Once the accuracy parameter ϵ is specified, an $\epsilon/2$-cover $g_1, \ldots, g_k$ is constructed. At this point, it is known that (at least) *one* of the functions g_i is within a distance $\epsilon/2$ of the target function f, but it is not known *which one*. Now an empirical risk $\hat{J}_i$ is computed for each index i, which corresponds to the error made in "pretending" that the target function is indeed g_i. The hypothesis h_m after m samples is chosen as a function g_k that minimizes this empirical risk. Of course, there is no guarantee that h_m is the *best* approximation to the unknown target function f among $g_1, \ldots, g_k$. But then, it is not required to be – In order for the above algorithm to be PAC to accuracy ϵ, it is enough if $d_P(f, h_m) \le \epsilon$ with high probability. Note that $\min_{1 \le i \le k} d_P(f, g_i) \le \epsilon/2$. The difference between ϵ and $\epsilon/2$ is the "cushion" used by the algorithm.

In the case where each function in $\mathcal{F}$ assumes values in $\{0, 1\}$ (i.e., the problem is one of *concept* learning), the cost function $\hat{J}_i$ is precisely the fraction of samples misclassified by the function g_i. Hence the output of the minimum empirical risk algorithm in this case is a concept among $g_1, \ldots, g_k$ that *misclassifies the fewest number of samples*.

Note that the minimum empirical risk algorithm is *not* consistent in general.

Theorem 6.2. *The minimum empirical risk algorithm is PAC to accuracy ϵ. In particular, we have that*

$$r(m, \epsilon) \le k \, \exp(-m\epsilon^2/8). \tag{6.3.1}$$

Hence

$$P^m\{\mathbf{x} \in X^m : d_P(f, h_m) > \epsilon\} \le \delta,$$

whenever

$$m \ge \frac{8}{\epsilon^2} \ln \frac{k}{\delta}. \tag{6.3.2}$$

Proof. Since $g_1, \ldots, g_k$ is an $\epsilon/2$-cover for $\mathcal{F}$, there exists an index i such that $d_P(f, g_i) \leq \epsilon/2$. Without loss of generality, suppose that $d_P(f, g_k) \leq \epsilon/2$. Next, if $d_P(f, g_i) \leq \epsilon$ for *all* i, then the PAC inequality is trivially satisfied, because the hypothesis h_m is one of the g_i's. So no further analysis is required unless $d_P(f, g_i) > \epsilon$ for some i. Again, without loss of generality, suppose the g_i's are renumbered such that

$$d_p(f, g_i) > \epsilon \text{ for } i = 1, \ldots, l, \text{ and}$$

$$d_P(f, g_i) \leq \epsilon \text{ for } i = l+1, \ldots, k.$$

Obviously $l \leq k - 1$.

To establish the PAC inequality, observe that $d_P(f, h_m) \leq \epsilon$ if h_m equals one of $g_{l+1}, \ldots, g_k$. This will be the case if (i) $\hat{J}_k \leq 3\epsilon/4$ and (ii) $\hat{J}_i > 3\epsilon/4$ for $i = 1, \ldots, l$. Note that the above is only a *sufficient* condition for $d_P(f, h_m) \leq \epsilon$. Hence, in order for $d_P(f, h_m)$ to exceed ϵ, it is necessary that the sample $\mathbf{x}$ satisfies

$$\hat{J}_k > 3\epsilon/4, \text{ or else } \hat{J}_i \leq 3\epsilon/4 \text{ for some } i \in \{1, \ldots, l\}. \tag{6.3.3}$$

To put it another way, $P^m \{\mathbf{x} \in X^m : d_P(f, h_m) > \epsilon\}$ is no larger than the probability measure of the set of $\mathbf{x} \in X^m$ that satisfy the above condition.

Let us bound the probability of each of the above events separately, using Hoeffding's inequality. Then the sum of these bounds gives an upper bound on the probability that $d_P(f, h_m) > \epsilon$. Observe that, for each index i, the cost $\hat{J}_i$ is the empirical mean of the function $|f(\cdot) - g(\cdot)|$ based on the multisample $\mathbf{x}$, and that the true mean value of $|f(\cdot) - g(\cdot)|$ equals the distance $d_P(f, g_i)$, which exceeds ϵ if $1 \leq i \leq l$, and is at most equal to $\epsilon/2$ if $i = k$. Hence

$$P^m \{\mathbf{x} \in X^m : \hat{J}_i \leq 3\epsilon/4\} \leq \exp(-m\epsilon^2/8), \ i = 1, \ldots, l,$$

and

$$P^m \{\mathbf{x} \in X^m : \hat{J}_k > 3\epsilon/4\} \leq \exp(-m\epsilon^2/8).$$

Hence, the measure of the set of $\mathbf{x} \in X^m$ that satisfy (6.3.3) is no larger than $(l+1) \exp(-m\epsilon^2/8)$, which in turn is no larger than $k \exp(-m\epsilon^2/8)$ because $l \leq k - 1$. Thus we conclude that

$$P^m \{\mathbf{x} \in X^m : d_P(f, h_m) > \epsilon\} \leq k \exp(-m\epsilon^2/8).$$

This proves the inequality (6.3.1). Now setting

$$k \exp(-m\epsilon^2/8) \leq \delta$$

and solving for m leads to the bound (6.3.2). ■

Since concept learning is a special case of function learning, Lemma 6.2 applies also to concept learning. In particular, the estimate given in Lemma

6.3 Finite Metric Entropy Implies Learnability

6.2 for the number of samples m needed to achieve an accuracy of ϵ and a confidence of δ continues to hold even for concept learning. However, by using the multiplicative form of the Chernoff bounds instead of Hoeffding's inequality, it is possible to improve this bound in the case of concept learning. This is done next, with a few obvious notational changes. The covering number is now denoted by $N(\epsilon/2, \mathcal{C}, d_P)$, the target concept by T, and an $\epsilon/2$-cover of $\mathcal{C}$ by $B_1, \ldots, B_k$. For each index $i \in \{1, \ldots, k\}$, the cost function $\hat{J}_i$ now equals the *fraction of samples misclassified* by choosing $H_m = B_i$. Hence the output of the minimum empirical risk algorithm is a concept B_k that misclassifies the fewest samples.

Theorem 6.3. *Suppose $\mathcal{C}$ is a concept class, $\epsilon > 0$, and that $B_1, \ldots, B_k$ is an $\epsilon/2$-cover for $\mathcal{C}$. Then the minimum empirical risk algorithm is PAC to accuracy ϵ. In particular, we have that*

$$r(m, \epsilon) \leq k \, \exp(-m\epsilon/32).$$

As a consequence,

$$P^m\{\mathbf{x} \in X^m : d_P(f, h_m) > \epsilon\} \leq \delta,$$

whenever

$$m \geq \frac{32}{\epsilon} \ln \frac{k}{\delta}. \tag{6.3.4}$$

Remark: Observe that the above bound (6.3.4) is an improvement over (6.3.2) whenever $\epsilon \leq 1/4$.

Proof. In analogy with the proof of Lemma 6.2, renumber the B_i's such that $d_P(T, B_k) \leq \epsilon/2$, and

$$d_P(T, B_i) > \epsilon \text{ for } i = 1, \ldots, l, \text{ and}$$

$$d_P(T, B_i) \leq \epsilon \text{ for } i = l+1, \ldots, k.$$

Clearly $l \leq k - 1$.

Next, observe that for each fixed index i, the number $|I_T(x) - I_{B_i}(x)|$ has only two possible values, namely 0 or 1. Moreover, the expected value of the random variable $|I_T(x) - I_{B_i}(x)|$ is precisely $d_P(T, B_i)$. So by assumption,

$$p_k := E_P(|I_T(x) - I_{B_k}(x)|) \leq \epsilon/2, \text{ and}$$

$$p_i := E_P(|I_T(x) - I_{B_i}(x)|) > \epsilon \text{ for } i = 1, \ldots, l.$$

Now let us compute the probability that a multisample $\mathbf{x} \in X^m$ satisfies

$$\hat{J}_k > 3\epsilon/4 \text{ or } \hat{J}_i \leq 3\epsilon/4 \text{ for some } i \in \{1, \ldots, l\}.$$

If the above event does not occur, then $d_P(T, H_m) \leq \epsilon$. Let us estimate each of the above probabilities separately using the multiplicative form of the

Chernoff bounds. First, since $p_k \leq \epsilon/2$, one can apply (2.3.1) with $\mu = \epsilon/2$ and $\gamma = 1/2$. This gives

$$P^m\{\mathbf{x} \in X^m : \hat{J}_k > 3\epsilon/4\} \leq \exp(-m\epsilon/24).$$

Next, for $i = 1, \ldots, l$, we have

$$P^m\{\mathbf{x} \in X^m : \hat{J}_i \leq 3\epsilon/4\} \leq P^m\{\mathbf{x} \in X^m : \hat{J}_i < 3p_i/4\}$$

$$\leq \exp(-mp_i/32) \leq \exp(-m\epsilon/32).$$

Therefore

$$P^m\{\mathbf{x} \in X^m : \hat{J}_i \leq 3\epsilon/4 \text{ for some } i \in \{1,\ldots,l\} \text{ or } \hat{J}_k > 3\epsilon/4\}$$

$$\leq (l-1)e^{-m\epsilon/32} + e^{-m\epsilon/24} \leq k \exp(-m\epsilon/32),$$

since $l \leq k - 1$. Finally, setting

$$k \exp(-m\epsilon/32) \leq \delta$$

and solving for m establishes (6.3.4) and completes the proof. ∎

At this stage, the reader might wonder why we do not at once choose $k = N(\epsilon/2, \mathcal{F}, d_P)$, i.e., the *minimum* cardinality of an $\epsilon/2$-cover. The reason is that, if $g_1, \ldots, g_{k(\alpha)}$ is a *minimal* α-cover, and if $\beta < \alpha$, then it might not be possible to embed the $g_1, \ldots, g_{k(\alpha)}$ into another *minimal* β-cover. However, if one does not insist on using *minimal* covers, then the difficulty can be alleviated, using the lemma below.

Lemma 6.1. *Suppose (Y, ρ) is a pseudometric space, and that $S \subseteq Y$ is totally bounded. Let $\epsilon > 0$ be specified, and let $\{y_1, \ldots, y_l\}$ be an ϵ-separated subset of S. Then there exists an ϵ-separated ϵ-cover of S containing $\{y_1, \ldots, y_l\}$ as a subset. Moreover, the cardinality of this ϵ-separated ϵ-cover is no larger than $M(\epsilon, S, \rho)$.*

Proof. Given $\{y_1, \ldots, y_l\}$, there are two possibilities: Either it is true that

$$\bigcup_{i=1}^{l} \bar{\mathcal{B}}(\epsilon, y_i, \rho) \supseteq S,$$

or it is not true. In the former case, $\{y_1, \ldots, y_l\}$ is itself the desired ϵ-separated ϵ-cover. In the latter case, there exists an $y_{l+1} \in S$ such that $\rho(y_i, y_{l+1}) > \epsilon$ for $i = 1, \ldots, l$. Now repeat the argument for the ϵ-separated set $\{y_1, \ldots, y_l, y_{l+1}\}$. Since S is totally bounded, the number $M(\epsilon, S, \rho)$ is finite, and as a consequence, the above process of adding new elements must terminate after a finite number of steps, say k steps. Thus $\{y_1, \ldots, y_{l+k}\}$ is the desired ϵ-separated ϵ-cover of S. By the definition of the packing number $M(\epsilon, S, \rho)$, it follows that $l + k \leq M(\epsilon, S, \rho)$. ∎

Now let us get back to the learning problem. Suppose the family of functions $\mathcal{F}$ is totally bounded under the pseudometric d_P. Then we can take advantage of Lemma 6.1 and construct a countable dense subset of $\mathcal{F}$ as follows: First construct a $(1/2)$-separated $(1/2)$-cover of $\mathcal{F}$ (under the pseudometric d_P). Denote this cover by $g_1, \ldots, g_{l_1}$, and observe that $l_1 \leq M(1/2, \mathcal{F}, d_P)$. Now a $(1/2)$-separated set is also $(1/4)$-separated; thus $g_1, \ldots, g_{l_1}$ can be embedded in a $(1/4)$-separated $(1/4)$-cover $g_1, \ldots, g_{l_2}$ of $\mathcal{F}$, where $l_2 \leq M(1/4, \mathcal{F}, d_P)$. By repeating this process, we generate a 2^{-i}-separated 2^{-i}-cover of $\mathcal{F}$ whose cardinality is no more than $M(2^{-i}, \epsilon, d_P)$, which in turn is embedded in a $2^{-(i+1)}$-separated $2^{-(i+1)}$-cover, and so on. This process generates a countable dense sequence in $\mathcal{F}$, which is denoted by $\{g_i\}$. In this sequence, $\{g_1, \ldots, g_{l_i}\}$ is a 2^{-i}-cover of $\mathcal{F}$. Moreover, since this cover is also 2^{-i}-separated, it follows that $l_i \leq M(2^{-i}, \mathcal{F}, d_P)$.

Once a countable dense subset is constructed in this manner, if one is content with an algorithm that is PAC to a specified accuracy ϵ, then one can proceed as follows: Given ϵ, choose an integer i such that $2^{-i} \leq \epsilon/2$, and then apply the minimum empirical risk algorithm to the set of functions $\{g_1, \ldots, g_{l_i}\}$.

Next, let us examine the *sample complexity* of the minimum empirical risk algorithm. At the same time, let us also modify the ϵ-dependent algorithm described above into another that does not explicitly make use of ϵ, and is PAC to arbitrary accuracy. The function learning problem is discussed first, as the modifications to the various formulae in the case of concept learning are obvious. Attention is focused on two cases. In the first case, the family of functions $\mathcal{F}$ satisfies

$$N(\epsilon, \mathcal{F}, d_P) = O(1/\epsilon^\lambda)$$

for some constant λ. In the second case, the family $\mathcal{F}$ satisfies

$$N(\epsilon, \mathcal{F}, d_P) = O[\exp(1/\epsilon^\beta)]$$

for some constant β. Virtually every known instance of a totally bounded set, and in particular every example in [107], falls into one of these two cases. In the first case, it follows from Theorem 6.2 that

$$m_0(\epsilon, \delta) = O\left[\frac{1}{\epsilon^2}\left(\ln\frac{1}{\epsilon} + \ln\frac{1}{\delta}\right)\right]$$

samples are sufficient to learn any function to accuracy ϵ and confidence δ. Note that the exponent λ may affect the constant hidden under the $O(\cdot)$ symbol, but does not affect the rate of growth with respect to ϵ. In the second case,

$$m_0(\epsilon, \delta) = O\left[\frac{1}{\epsilon^2}\left(\frac{1}{\epsilon^\beta} + \ln\frac{1}{\delta}\right)\right]$$

samples are enough to learn any function to accuracy ϵ and confidence δ. In this instance, the constant β appears explicitly in the rate of growth with respect to ϵ. From these expressions it can be said that "confidence is cheaper than accuracy," since m_0 grows as $O(\ln(1/\delta))$ for fixed ϵ. Moreover, it is worth noting that m_0 grows only *polynomially* with respect to $1/\epsilon$ in both cases.

In the concept learning problem,

$$m_0(\epsilon, \delta) = O\left[\frac{1}{\epsilon}\left(\ln\frac{1}{\epsilon} + \ln\frac{1}{\delta}\right)\right]$$

samples are sufficient to learn any concept to accuracy ϵ and confidence δ in the first case, and

$$m_0(\epsilon, \delta) = O\left[\frac{1}{\epsilon}\left(\frac{1}{\epsilon^\beta} + \ln\frac{1}{\delta}\right)\right]$$

samples are enough to learn any concept to accuracy ϵ and confidence δ in the second case.

The minimum empirical risk algorithm described above depends explicitly on the accuracy parameter ϵ. However, by mimicking the proof of Theorem 3.1, it is possible to come up with another algorithm that does not depend explicitly on ϵ, and is PAC to arbitrary accuracy. Moreover, by altering the argument slightly, it is possible to ensure that the sample complexity *remains polynomial* in $1/\epsilon$ and $\ln(1/\delta)$.

To focus the discussion, let us first study the function learning problem. Let $\{g_i\}$ be a countable dense subset of $\mathcal{F}$ constructed in accordance with Lemma 6.1. Thus $\{g_1, \ldots, g_{l_i}\}$ is a 2^{-i}-cover of $\mathcal{F}$. Now define a function $\phi : (0, 1) \to \mathbb{R}$ such that, using $\phi(\alpha)$ samples, it is possible to learn to accuracy α and confidence $e^{-1/\alpha}$.[2] Thus, in the present instance, we can choose

$$\phi(\alpha) := \frac{8}{\alpha^2}\left(\ln l_{i(\alpha)} + \frac{1}{\alpha}\right),$$

where

$$i(\alpha) := \lceil \lg(1/\alpha) \rceil + 1.$$

The definition of the integer $i(\alpha)$ ensures that $i(\alpha) - 1 \geq \lg(1/\alpha)$, i.e., that $2^{-i(\alpha)} \geq \alpha/2$. Now the algorithm can be described. Given m samples, choose the *smallest* number α such that $m \geq \phi(\alpha)$. Then apply the minimum empirical risk algorithm on these m samples to the functions $\{g_1, \ldots, g_{l_{i(\alpha)}}\}$. Now Theorem 6.2 implies that the resulting hypothesis is accurate to within α with a probability of at least $1 - e^{-1/\alpha}$. Since $\alpha \to 0$ as $m \to \infty$, this algorithm is PAC.

To estimate the sample complexity, let us suppose there exist constants M_0 and λ such that

$$N(\epsilon, \mathcal{F}, d_P) \leq \frac{M_0}{\epsilon^\lambda} \text{ for } 0 < \epsilon < 1.$$

---

[2] This is a departure from the proof of Theorem 3.1. In that proof, the accuracy and confidence parameters are equal.

In other words, $\mathcal{F}$ is in Case (i) above. In this case, it follows from Lemma 6.1 that

$$l_i \leq M(2^{-i}, \mathcal{F}, d_P) \leq N(2^{-i+1}, \mathcal{F}, d_P) \leq M_0 2^{\lambda(i-1)}.$$

Since $\lceil \lg(1/\alpha) \rceil \leq \lg(1/\alpha) + 1$, it follows from the definition of $i(\alpha)$ that

$$l_{i(\alpha)} \leq \frac{M_0 2^{\lambda}}{\alpha^{\lambda}},$$

and as a result,

$$\phi(\alpha) = O(1/\alpha^3).$$

Hence the number of samples needed to learn to accuracy ϵ and confidence δ is

$$m_0(\epsilon, \delta) = O\left[\frac{1}{\epsilon^2}\left(\ln\frac{1}{\epsilon} + \ln\frac{1}{\delta}\right)\right].$$

Now suppose $\mathcal{F}$ falls in Case (ii), and suppose there exist constants M_0, β such that

$$N(\epsilon, \mathcal{F}, d_P) \leq M_0 \exp(1/\epsilon^{\beta}).$$

In this case we have

$$l_i \leq N(2^{-i+1}, \mathcal{F}, d_P) \leq M_0 \exp(2^{\beta(i-1)}).$$

Hence

$$l_{i(\alpha)} \leq M_0 \exp(2^{\beta\lceil \lg(1/\alpha) \rceil - 1}) \leq M_0 \exp(2^{\beta \lg(1/\alpha)}) = M_0 \exp(2^{\beta}/\alpha^{\beta}),$$

$$\ln(l_{i(\alpha)}) = O(1/\alpha^{\beta}),$$

and as a consequence,

$$\phi(\alpha) = O(1/\alpha^{\max\{\beta,1\}+2}).$$

Hence the number of samples needed to learn to accuracy ϵ and confidence δ is

$$m_0(\epsilon, \delta) = O\left[\frac{1}{\epsilon^2}\left(\frac{1}{\epsilon^{\beta}} + \ln\frac{1}{\delta}\right)\right].$$

Note that, whether $\mathcal{F}$ falls in Case (i) or Case (ii), the sample complexity is polynomial in $1/\epsilon$ and $\ln(1/\delta)$. However, if $\mathcal{F}$ belongs to Case (ii) so that its covering number blows up exponentially with respect to the desired covering accuracy, then the integer $l_{i(\alpha)}$ increases exponentially with respect to $1/\alpha$. In other words, *the number of functions* to which the minimum empirical risk algorithm needs to be applied blows up exponentially with respect to $1/\epsilon$.

The arguments in the case of *concept* learning are entirely similar, except that one uses Theorem 6.3 instead of Theorem 6.2. If the concept class $\mathcal{C}$ satisfies

$$N(\epsilon, \mathcal{C}, d_P) = O(1/\epsilon^\lambda)$$

for some constant λ, then the sample complexity of the preceding algorithm is

$$m_0(\epsilon, \delta) = O\left[\frac{1}{\epsilon}\left(\ln\frac{1}{\epsilon} + \ln\frac{1}{\delta}\right)\right].$$

If $\mathcal{C}$ satisfies

$$N(\epsilon, \mathcal{C}, d_P) = O[\exp(1/\epsilon^\beta)]$$

for some constant β, then the sample complexity is

$$m_0(\epsilon, \delta) = O\left[\frac{1}{\epsilon}\left(\frac{1}{\epsilon^\beta} + \ln\frac{1}{\delta}\right)\right].$$

In each case, the term $1/\epsilon^2$ in the case of function learning is replaced by $1/\epsilon$. The details are easy and are left to the reader.

6.4 Consistent Learnability

In the preceding two sections, we have derived two distinct kinds of *sufficient* conditions for learnability. In particular, the UCEM property implies that every consistent algorithm is PUAC, while the finite metric entropy property implies that a particular algorithm (known as the minimum empirical risk algorithm) is PAC. In the present section, we study the conditions under which every consistent algorithm is PAC or PUAC.

6.4.1 Consistent PAC Learnability

In the present subsection, we study conditions under which every consistent algorithm is PAC (not PUAC).

Definition 6.1. *A function class $\mathcal{F}$ is said to be* **consistently learnable** *if every consistent algorithm is PAC.*

In the computational learning theory literature, one encounters the notions of "solid learnability" [147] or "potential learnability" [9]. These two notions are equivalent to each other, and the definition of these notions is that every consistent algorithm must be PAC. Thus they are equivalent to the present notion of "consistent learnability."

Let $\mathcal{A}$ denote the set of all consistent algorithms for a given problem. Recall the quantity $r(m, \epsilon, P)$ defined in (3.2.1). Strictly speaking, the quantity $r(m, \epsilon, P)$ also depends on the underlying algorithm. Let us define a function class $\mathcal{F}$ to be "strongly consistently PAC learnable" if

$$\sup_{\{A_m\} \in \mathcal{A}} r(m, \epsilon, P) \to 0 \text{ as } m \to \infty. \qquad (6.4.1)$$

6.4 Consistent Learnability

Thus the extra feature of "strong consistent PAC learnability" is the requirement that, in addition to every consistent algorithm being PAC, the learning rate $r(m, \epsilon, P)$ must converge to zero at a rate that is uniform with respect to the underlying consistent algorithm. The reason for not highlighting this property separately is that, as shown below, this uniformity feature "comes for free."

Theorem 6.4. *Define the quantity*

$$\phi_{m,f}(\mathbf{x}^*) := \sup\{d_P(f, g) : g \in \mathcal{F} \text{ and } \hat{d}_m(f, g; \mathbf{x}^*) = 0\}.$$

Then the following three statements are equivalent:

1. *Every consistent algorithm is PAC.*
2. *Every consistent algorithm is PAC, and moreover, (6.4.1) holds.*
3. *We have*

$$\psi(m, \epsilon) := \sup_{f \in \mathcal{F}} P^\infty\{\mathbf{x}^* \in X^\infty : \phi_{m,f}(\mathbf{x}^*) > \epsilon\} \to 0 \text{ as } m \to \infty. \quad (6.4.2)$$

Remark: The above theorem is essentially proved in [79], though she does not distinguish between items (1) and (2).

Proof. **(2)** $\Rightarrow$ **(1)**. Obvious.

(3) $\Rightarrow$ **(2)**. By the definition of ϕ, for any consistent algorithm, we have

$$d_P[f, h_m(f; \mathbf{x}^*)] \le \phi_{m,f}(\mathbf{x}^*).$$

Now suppose (6.4.2) holds. Then

$$r(m, \epsilon, P) \le \psi(m, \epsilon) \to 0 \text{ as } m \to \infty.$$

Moreover, since the right side is independent of the particular consistent algorithm being used, the learning rate is also uniform with respect to the algorithm.

(1) $\Rightarrow$ **(3)**. We shall prove the contrapositive, namely that if (3) fails to hold, then so does (1). Suppose that (6.4.2) fails to hold. Then there exist sequences $\{m_i\} \to \infty$, $f_i \in \mathcal{F}$ and constants $\epsilon, \delta > 0$ such that

$$P^{m_1}\{\mathbf{x} \in X^{m_i} : \phi_{m_i, f_i}(\mathbf{x}^*) > \epsilon\} > \delta, \ \forall i.$$

In turn this implies that

$$P^{m_1}\{\mathbf{x} \in X^{m_i} : \exists g_i \in \mathcal{F} \text{ s.t. } \hat{d}_{m_i}(f, g; \mathbf{x}^*) = 0 \text{ and } d_P(f_i, g_i) > \epsilon\} > \delta, \ \forall i.$$

Now it is always possible to choose a consistent algorithm such that, for each f_i and for each $\mathbf{x}^*$ in the above set, the algorithm assigns the corresponding g_i as the hypothesis. Such an algorithm is consistent but not PAC. ∎

6.4.2 Consistent PUAC Learnability

In the preceding subsection we studied conditions under which every consistent algorithm is PAC. In the present subsection, we study conditions under which every consistent algorithm is PUAC, which is clearly a stronger requirement. In the process, it is shown that PUAC learnability is equivalent to consistent PUAC learnability (whereas the corresponding statement for PAC learnability is not true).

Definition 6.2. *A function class $\mathcal{F}$ is said to be* **consistently PUAC learnable** *if every consistent algorithm is PUAC.*

Theorem 6.1 leads at once to the following result.

Corollary 6.1. *Suppose a family of functions has the property that empirical means converge uniformly to their true values (UCEM property). Then the family is consistently PUAC learnable.*

Thus the UCEM property is a *sufficient* condition for consistent PUAC learnability. It is shown in this section that the UCEM condition is *not* necessary in general. Rather, a necessary and sufficient condition for consistent PUAC learnability is the so-called "shrinking width" property, which is defined next.

Definition 6.3. *Given a family of functions $\mathcal{F}$, define*

$$w(m, \epsilon) := P^m\{\mathbf{x} \in X^m : \exists f, g \in \mathcal{F} \text{ s.t. } \hat{d}(f, g; x) = 0 \text{ and } d_P(f, g) > \epsilon\}.$$

The family $\mathcal{F}$ is said to have the **shrinking width property** *if $w(m, \epsilon) \to 0$ as $m \to \infty$.*

The shrinking width property can also be interpreted in terms of the convergence of a stochastic process. Given an $\mathbf{x}^* \in X^\infty$, define

$$\phi_m(\mathbf{x}^*) := \sup\{d_P(f, g) : f, g \in \mathcal{F} \text{ and } \hat{d}_m(f, g; \mathbf{x}^*) = 0\},$$

where $\hat{d}_m(f, g; \mathbf{x}^*)$ denotes the empirical distance between the functions f and g based on the first m components of $\mathbf{x}^*$, i.e.,

$$\hat{d}_m(f, g; \mathbf{x}^*) := \frac{1}{m} \sum_{i=1}^{m} |f(x_i) - g(x_i)|.$$

Since $\phi_m(\mathbf{x}^*)$ depends only on the first m components of $\mathbf{x}^*$, one can also write $\phi_m(\mathbf{x})$ instead of $\phi_m(\mathbf{x}^*)$, where $\mathbf{x} \in X^m$ consists of the first m components of $\mathbf{x}^*$. It is easy to see that the shrinking width property is equivalent to the requirement that the stochastic process $\phi_m(\cdot)$ converges *in probability* to the zero function. The ultimate behaviour of this stochastic process is the topic of the next lemma.

6.4 Consistent Learnability

Lemma 6.2. *Given any family of functions $\mathcal{F} \subseteq [0,1]^X$, there exists a constant $c = c(\mathcal{F})$ such that the stochastic process $\{\phi_m(\cdot)\}$ converges almost surely to c as $m \to \infty$.*

Proof. For each fixed $\mathbf{x}^* \in X^\infty$, the sequence $\{\phi_m(\mathbf{x}^*)\}$ is a nonincreasing sequence of real numbers and is bounded below by zero. Hence it converges to a limit, call it $c(\mathbf{x}^*)$. It only remains to show that $c(\mathbf{x}^*)$ is a constant almost everywhere. This is achieved exactly as in the proof of Lemma 5.1. ∎

Now we come to the main result of this section.

Theorem 6.5. *Given a family of functions $\mathcal{F}$, the following statements are equivalent:*

1. *The family $\mathcal{F}$ has the shrinking width property.*
2. *The family $\mathcal{F}$ is consistently PUAC learnable.*
3. *The family $\mathcal{F}$ is PUAC learnable.*

Remark:. Thus Theorem 6.5 shows that in the case of PUAC learnability, there is no distinction between consistent (PUAC) learnability and plain (PUAC) learnability. As shown in Example 6.7 below, the corresponding statement is *not* true for PAC learnability.

Proof. **(1)** $\Rightarrow$ **(2)** Suppose $\mathcal{F}$ has the shrinking width property. Then by assumption the stochastic process $\{\phi_m(\cdot)\}$ defined above converges in probability to zero. Thus, given any $\epsilon, \delta > 0$, there exists an $m_0 = m_0(\epsilon, \delta)$ such that

$$P^m\{\mathbf{x} \in X^m : \phi_m(\mathbf{x}) > \epsilon\} \leq \delta, m \geq m_0.$$

Now suppose $f \in \mathcal{F}$, and as usual let $h_m(f; \mathbf{x})$ denote the hypothesis generated by the algorithm when the target function is f and the multisample is $\mathbf{x}$. If the algorithm is consistent, then f and h_m agree on $\mathbf{x}$; that is, $\hat{d}[f, h_m(f; \mathbf{x}); \mathbf{x}] = 0$. Hence, by the definition of $\phi_m(\cdot)$, it follows that

$$P^m\{\mathbf{x} \in X^m : \sup_{f \in \mathcal{F}} d_P[f, h_m(f; \mathbf{x})] > \epsilon\} \leq \delta, m \geq m_0.$$

This implies that the algorithm is PUAC. Hence $\mathcal{F}$ is consistently PUAC learnable.

(2) $\Rightarrow$ **(1)** Suppose $\mathcal{F}$ fails to have the shrinking width property. Then there exist numbers ϵ, δ and a sequence $\{m_i\}$ approaching infinity such that

$$P^{m_i}\{\mathbf{x} \in X^{m_i} : \phi_{m_i}(\mathbf{x}) > \epsilon\} \geq \delta, \forall i.$$

Let us temporarily drop the subscript "i" on m_i and examine the above inequality. From the definition of $\phi_m(\cdot)$, this inequality is equivalent to:

$$P^m\{\mathbf{x} \in X^m : \exists f, g \in \mathcal{F} \text{ s.t. } \hat{d}(f, g; \mathbf{x}) = 0 \text{ and } d_P(f, g) > \epsilon\} \geq \delta.$$

For convenience, let us define the set $S \subseteq X^m$ by

228   6. Learning Under a Fixed Probability Measure

$$S = \{\mathbf{x} \in X^m : \exists f, g \in \mathcal{F} \text{ s.t. } \hat{d}(f, g; \mathbf{x}) = 0 \text{ and } d_P(f, g) > \epsilon\}.$$

Suppose $\mathbf{x} \in S$, and choose an $f_0 \in \mathcal{F}$ such that

$$\exists g \in \mathcal{F} \text{ s.t. } \hat{d}(f_0, g; \mathbf{x}) = 0 \text{ and } d_P(f_0, g) > \epsilon.$$

It is clear that such an f_0 can be found for each $\mathbf{x} \in S$.[3] As in Chapter 5, let $\mathbf{f}_0(\mathbf{x}) \in [0,1]^m$ denote the vector

$$\mathbf{f}_0(\mathbf{x}) := [f_0(x_1) \ldots f_0(x_m)]^t.$$

If $\{A_m\}$ is a consistent algorithm, then the hypothesis $h_m(f_0; \mathbf{x})$ satisfies $\hat{d}(f_0, h_m; \mathbf{x}) = 0$. In other words, $\mathbf{h}_m(\mathbf{x}) = \mathbf{f}_0(\mathbf{x})$. If we let $\{A_m\}$ vary over *all* consistent algorithms, then at least one of them would return the hypothesis $h_m = g$, where g also satisfies $\mathbf{g}(\mathbf{x}) = \mathbf{f}_0(\mathbf{x})$, and moreover $d_P(f_0, g) > \epsilon$. Hence, for at least one particular consistent algorithm, we have

$$\sup_{f \in \mathcal{F}} d_P[f, h_m(f; \mathbf{x})] \geq d_P(f_0, g) > \epsilon.$$

This argument can be repeated for *every* $\mathbf{x} \in S$. For each $\mathbf{x} \in S$, there exist two functions $f_0, g \in \mathcal{F}$, both dependent on $\mathbf{x}$, such that $f_0(\mathbf{x}) = g(\mathbf{x})$, and $d_P(f_0, g) > \epsilon$. Since an algorithm makes use of only the multisample $\mathbf{x}$ and the corresponding oracle output $\mathbf{f}_0(\mathbf{x})$, there exists a consistent algorithm that such that, given the inputs $\mathbf{x}$ and $\mathbf{f}_0(\mathbf{x})$, it returns the hypothesis g for each $\mathbf{x} \in S$. Hence, for this particular algorithm, we have

$$\sup_{f \in \mathcal{F}} d_P[f, h_m(f; \mathbf{x})] \geq d_P[f_0, h_m(f_0; \mathbf{x})] > \epsilon, \ \forall \mathbf{x} \in S.$$

Now let us restore the subscript "i" on m_i, and label the set S as $S_i \subseteq X^{m_i}$. Since $P^{m_i}(S_i) \geq \delta$ for all i, this particular consistent algorithm is *not* PUAC. Hence $\mathcal{F}$ is not consistently PUAC learnable.

(2) $\Rightarrow$ (3). Obvious.

(3) $\Rightarrow$ (2). Actually, it is shown that (3) implies (1) by proving the contrapositive, namely that if (1) fails then (3) also fails. Since the equivalence of (1) and (2) has already been established, this suffices to establish that (3) implies (2).

Accordingly, suppose (1) fails, and define the set S as in the proof of (2) $\Rightarrow$ (1) above. In other words,

$$S = \{\mathbf{x} \in X^m : \exists f, g \in \mathcal{F} \text{ s.t. } \hat{d}(f, g; \mathbf{x}) = 0 \text{ and } d_P(f, g) > \epsilon\}.$$

Suppose $\mathbf{x} \in S$, and choose $f, g \in \mathcal{F}$ such that $\hat{d}(f, g; \mathbf{x}) = 0$ and $d_P(f, g) > \epsilon$. Let $\{A_m\}$ be *any* algorithm. Then it follows that $h(f; \mathbf{x}) = h(g; \mathbf{x})$. In other

---

[3] Thus, strictly speaking, we should write $f_0(\mathbf{x})$ to indicate the dependence of f_0 on $\mathbf{x}$; however, such a notation would make the subsequent formulae almost unreadable. The reader is therefore requested to keep in mind this dependence, though it is not explicitly displayed.

words, the algorithm returns the same hypothesis with the input sequence $\mathbf{x}$ whether the target function is f or g. This is because $f(x_i) = g(x_i)$ for all i, so that the two functions are indistinguishable on the sample $\mathbf{x}$. Now, since $d_P(f,g) > \epsilon$, it follows from the triangle inequality that either $d_P[f, h(f;\mathbf{x})] > \epsilon/2$, or else $d_P[g, h(g;\mathbf{x})] > \epsilon/2$. In either case, we have

$$\sup_{t\in\mathcal{F}} d_P[t, h(t;\mathbf{x})] \geq \max\{d_P[f, h(f;\mathbf{x})], d_P[g, h(g;\mathbf{x})]\} > \epsilon/2.$$

Thus
$$P^m\{\mathbf{x} \in X^m : \sup_{t\in\mathcal{F}} d_P[t, h(t;\mathbf{x})] > \epsilon/2\} \geq P^m(S) \geq \delta.$$

Now the above argument can be repeated for every m in the sequence $\{m_i\}$ (refer to the proof that (2) implies (1)), which shows that the algorithm is not PUAC. Since this statement holds for *any* algorithm, the family $\mathcal{F}$ is not PUAC learnable. ∎

We shall see in Chapter 8 that the above theorem is quite general and transcends the case of fixed distribution learning. In particular, the above theorem holds for an *arbitrary* family of probability measures $\mathcal{P}$. However, in the case of fixed distribution learning, it is possible to state a stronger result.

Corollary 6.2. *When $\mathcal{P} = \{P\}$, a singleton set, the following statements are equivalent:*

1. *The family $\mathcal{F}$ has the shrinking width property.*
2. *Every consistent algorithm is ASEC.*

Proof. (i) ⇒ (ii) By Lemma 6.2, if the stochastic process $\phi_m(\cdot)$ converges to zero in probability, then it also converges to zero almost surely. Now it is possible to mimic the first part of the proof of Theorem 6.5. Suppose an algorithm is consistent. Then we have

$$\hat{d}_m[f, h_m(f;\mathbf{x}^*);\mathbf{x}^*] = 0, \ \forall m, \ \forall \mathbf{x}^* \in X^\infty, \ \forall f \in \mathcal{F}.$$

Now by the shrinking width property we have

$$P^\infty\{\mathbf{x}^* \in X^\infty : \sup\{d_P(f,g) : \hat{d}_m(f,g;\mathbf{x}^*) = 0\} \to 0 \text{ as } m \to \infty\} = 1.$$

Combining these two identities leads to

$$P^\infty\{\mathbf{x}^* \in X^\infty : \sup_{f\in\mathcal{F}} d_P[f, h_m(f;\mathbf{x})] \to 0 \text{ as } m \to \infty\} = 1.$$

Hence the algorithm is ASEC.

(ii) ⇒ (i) Since an ASEC algorithm is also PUAC, this part follows from Theorem 6.5 above. ∎

6.5 Examples

In this section we gather together several examples that bring out the distinction between the many distinct notions presented thus far.

The first set of examples deal with the shrinking width property and its variations.

Thus far we have seen that the shrinking width property is a necessary and sufficient condition for a family of functions to be consistently PUAC learnable. In Section 6.2 we have seen that *if* a family of functions has the UCEM property, then it is consistently PUAC learnable. The question naturally arises as to whether the UCEM property is equivalent to the shrinking width property (and thus to consistent PUAC learnability). The next several examples show that this is not so (i.e., shrinking width is strictly weaker than the UCEM property), and also relate the shrinking width property to finite VC-dimension.

Example 6.4. Let $X = [0,1]$, $\mathcal{S}$ = the Borel σ-algebra on X, and P = the uniform probability measure on X. Let $\mathcal{C}_1$ = the collection of all finite subsets of X. Then, as we have seen in Example 3.2, $\mathcal{C}_1$ does *not* have the UCEP property. On the other hand, since $d_P(A, B) = 0$ for every pair $A, B \in \mathcal{S}$, it follows that $w(m, \epsilon) = 0$ for every integer m and every $\epsilon > 0$. Hence $\mathcal{C}_1$ *does* have the shrinking width property. As a result, every consistent algorithm is PUAC. This example shows that the shrinking width property is strictly weaker than the UCEP (or UCEM) property. See Example 6.6 below for a less trivial example of a collection of sets that does not have the UCEP property but does have the shrinking width property.

Now define $\mathcal{C}_2 = \mathcal{C}_1 \cup \{X\}$. Thus $\mathcal{C}_2$ consists of all finite subsets of X together with X itself. Now it is claimed that $w(m, \epsilon) = 1$ for every $\epsilon < 1$ and every integer m. To see this, let $m \geq 1$, and let $\mathbf{x} \in X^m$ be arbitrary. Define $S = \{x_1, \ldots, x_m\}$ after deleting repeated elements if necessary. Then $\hat{d}(S, X; \mathbf{x}) = 0$, since each x_i belongs to both S and X. On the other hand, $d_P(S, X) = 1 > \epsilon$ if $\epsilon < 1$. This establishes the claim. So clearly $\mathcal{C}_2$ does *not* have the shrinking width property, and from Theorem 6.5, it follows that not every consistent algorithm is PUAC.

In this simple example, it is easy to construct a consistent algorithm that fails to be PUAC. Given a labelled sample $[((x_1, I_T(x_1)), \ldots, ((x_m, I_T(x_m))]$, define
$$H_m(T; \mathbf{x}) := \bigcup_{I_T(x_i)=1} \{x_i\}.$$

In words, $H_m(T; \mathbf{x})$ consists of those x_i that are classified as belonging to T by the oracle, or equivalently, all "positive" examples of the unknown target concept. The algorithm is clearly consistent. Now suppose the target concept T equals X. Then
$$H_m(T; \mathbf{x}) = \{x_1, \ldots, x_m\},$$

and $d_P[T, H_m(T; \mathbf{x})] = 1$. Since this is true for *every* $\mathbf{x} \in X^m$, it follows that the quantity $r(m, \epsilon)$ defined in (3.2.1) satisfies

$$r(m, \epsilon) = 1 \ \forall m, \text{ if } \epsilon < 1.$$

Thus the algorithm is not even PAC, let alone PUAC.

This example is adapted from [22].

Example 6.5. Let $(X, \mathcal{S})$ be a measurable space, and suppose $\mathcal{C} \subseteq \mathcal{S}$ is a concept class with finite VC-dimension. It is shown that $\mathcal{C}$ has the shrinking width property; moreover, an explicit estimate is given for the "width" function $w(m, \epsilon)$.

The estimate is based on the inequality (5.6.9), with a few refinements. First, since $\| \mathbf{v} \|_{a1} \leq \| \mathbf{v} \|_\infty$ for each vector $\mathbf{v} \in [0, 1]^m$, it is permissible to replace $\| \cdot \|_{a1}$ by $\| \cdot \|_\infty$ in the right side of the inequality. Second, if we are dealing with a *concept* class $\mathcal{A}$ instead of a *function* class $\mathcal{F}$, then all vectors in $\mathcal{A}|_\mathbf{x}$ belong to $\{0, 1\}^m$ and not just $[0, 1]^m$. Since $\epsilon/16 < 1$ whenever $\epsilon \leq 1$, the covering number $L(\epsilon/16, \mathcal{A}|_\mathbf{x}, \| \cdot \|_\infty)$ equals the number $\pi(\mathcal{A}; \mathbf{x})$ of distinct vectors in $\mathcal{A}|_\mathbf{x}$. Finally, the integer $\pi(\mathcal{A}; \mathbf{x})$ can be bounded using Theorem 4.1. If $\mathcal{A}$ has VC-dimension d, then $\pi(\mathcal{A}; \mathbf{x}) \leq (em/d)^d$, $\forall m \geq d$, $\forall \mathbf{x} \in X^m$. Substituting all this into (5.6.9) leads to the estimate

$$v(m, \epsilon; \mathcal{A}) \leq 4 \left(\frac{em}{d} \right)^d e^{-m\epsilon/32}, \ \forall m \geq d,$$

where the symbol $v(m, \epsilon)$ has been replaced by the symbol $v(m, \epsilon; \mathcal{A})$ in the left side of (5.6.9) to make clear which concept class we are speaking about.

Now apply the above inequality to the collection of sets

$$\mathcal{C} \Delta \mathcal{C} := \{A \Delta B : A, B \in \mathcal{C}\}.$$

As shown in Theorem 4.5, if $\mathcal{C}$ has VC-dimension d, then $\mathcal{C} \Delta \mathcal{C}$ has VC-dimension no larger than $10d$. Therefore

$$v(m, \epsilon; \mathcal{C} \Delta \mathcal{C}) \leq 4 \left(\frac{em}{10d} \right)^{10d} e^{-m\epsilon/32}, \ \forall m \geq 10d.$$

Finally, note that the quantity $v(m, \epsilon; \mathcal{C} \Delta \mathcal{C})$ is precisely what is called $w(m, \epsilon, \mathcal{C})$ in the definition of the shrinking width property; that is,

$$w(m, \epsilon, \mathcal{C}) = P^m \{\mathbf{x} \in X^m : \exists A, B \in \mathcal{C} : \hat{d}(A, B; \mathbf{x}) = 0 \text{ and } d_P(A, B) > \epsilon\}.$$

Hence we obtain the useful bound

$$w(m, \epsilon, \mathcal{C}) \leq 4 \left(\frac{em}{10d} \right)^{10d} e^{-m\epsilon/32}, \ \forall m \geq 10d.$$

232   6. Learning Under a Fixed Probability Measure

Example 6.6. Let $X = [0,1]$, $\mathcal{S} =$ the Borel σ-algebra on X, and let $P =$ the uniform probability measure on X. Let $\mathcal{C}$ consist of all unions of the form $[0,a] \cup F$ where $a \leq 0.5$ and F is a *finite* subset of $(0.5, 1]$.

It is shown first that $\mathcal{C}$ *does not* have the UCEP property. The proof of this claim is based on Theorem 5.4. It is clear that every subset of $(0.5, 1]$ is shattered by $\mathcal{C}$. Hence, given a multisample $\mathbf{x} \in X^m$, the restricted VC-dimension $d(\mathbf{x})$ is at least equal to the number of components of $\mathbf{x}$ that lie in $(0.5, 1]$, which equals $m/2$ on average. Hence

$$\lim_{m \to \infty} \frac{E_{P^m}[d(\mathbf{x})]}{m} \geq \frac{1}{2}, \, \forall m \geq 2.$$

Since the condition (5.3.5) of Theorem 5.4 is violated, the collection of sets $\mathcal{C}$ *fails* to have the UCEP property.

Now it is claimed that $\mathcal{C}$ *does* have the shrinking width property. Suppose $A = [0,a] \cup F, B = [0,b] \cup G$ belong to $\mathcal{C}$, where $0 \leq a, b \leq 0.5$, and F, G are finite subsets of $(0.5, 1]$. Suppose $\hat{d}(A, B; \mathbf{x}) = 0$ for some $\mathbf{x} \in X^m$, i.e., suppose A, B agree on a multisample $\mathbf{x}$. Then in particular A and B also agree on all components of $\mathbf{x}$ lying in $[0, 0.5]$. Given $\mathbf{x} \in X^m$, let $\phi(\mathbf{x})$ denote the number of components of $\mathbf{x}$ lying in $[0, 0.5]$. Since these are also uniformly distributed in $[0, 0.5]$, the probability that $\hat{d}(A, B; \mathbf{x}) = 0$ is no larger than $(0.5 - |a-b|)^{\phi(\mathbf{x})} = [0.5 - d_P(A,B)]^{\phi(\mathbf{x})}$, since $0.5 - d_P(A,B)$ is the probability that a randomly selected $x \in [0,1]$ belongs to $[0,0.5]$ but not to $A \Delta B$. Therefore, the probability that $\hat{d}(A, B; \mathbf{x}) = 0$ for a random $\mathbf{x} \in X^m$ is at most

$$\sum_{l=0}^{m} [0.5 - d_P(A,B)]^l \cdot \Pr\{\phi(\mathbf{x}) = l\}.$$

Since the map $\lambda \mapsto (0.5 - \lambda)^l$ is a decreasing function of λ for each l, it follows that the probability that $\hat{d}(A, B; \mathbf{x}) = 0$ given $d_P(A, B) > \epsilon$ is at most

$$\sum_{l=0}^{m} (0.5 - \epsilon)^l \cdot \Pr\{\phi(\mathbf{x}) = l\}.$$

This quantity is an upper bound for $w(m, \epsilon)$. Now note that $(0.5 - \epsilon)^l \leq 0.5^l$ for each l, and also that 0.5^l is a decreasing function of l. Hence, for each m, we have

$$w(m, \epsilon) \leq \Pr\{\phi(\mathbf{x}) < m/3\} + 0.5^{m/3} \Pr\{\phi(\mathbf{x}) \geq m/3\}.$$

Since P is the uniform measure, $\phi(\mathbf{x})$ has the binomial distribution. Hence $\Pr\{\phi(\mathbf{x}) < m/3\} \to 0$ as $m \to \infty$. Also, $0.5^{m/3} \to 0$ as $m \to \infty$. This leads to the conclusion that $w(m, \epsilon) \to 0$ as $m \to \infty$, for each $\epsilon > 0$, i.e., that $\mathcal{C}$ has the shrinking width property.

6.5 Examples

Next, we present several examples of learnable function and concept classes. The source for most of these examples is [107], which is a veritable gold mine of explicit computations of covering and packing numbers. Of the very large number of examples available in [107], only two are discussed here. However, the first example is not from this source, and is intended to show that a function class can be learnable even though it does not have the shrinking width property.

Example 6.7. Consider once again the concept class of Example 6.4, namely: $\mathcal{C}$ consists of all finite subsets of $X := [0, 1]$ together with X itself. In this case, the concept class does *not* have the shrinking width property, so it is not consistently PUAC learnable. Hence by Theorem 6.5, the concept class is also not PUAC learnable. On the other hand, the pair $\{\emptyset, X\}$ is an ϵ-cover of $\mathcal{C}$ for *every* $\epsilon > 0$. Hence $\mathcal{C}$ is learnable using the minimum empirical risk algorithm applied to this pair, to zero accuracy and zero confidence using *just one sample*. To see this, suppose the target concept T is a finite set. Pick an $x \in X$ at random. Then $x \notin T$ with probability one. Hence the minimum empirical risk algorithm returns the hypothesis $H = \emptyset$ with probability one, which is at a distance of zero from the target concept. On the other hand, if the target concept $T = X$, then every x belongs to T, and as a result, the minimum empirical risk algorithm returns the hypothesis $H = X$, which happens to be correct. This example therefore shows that PAC learnability is a strictly weaker notion than PUAC learnability. ∎

Before proceeding to the next set of examples, we make a brief digression to relate the covering numbers of a function class to the covering numbers of a corresponding concept class.

Suppose $(X, \mathcal{S}, P)$ is a probability space, and let $F \subseteq [0,1]^X$ be a family of measurable functions. For each function $f \in \mathcal{F}$, define its *sous-graph* $SG(f)$ as

$$SG(f) := \{(x, y) \in X \times [0, 1] : 0 \leq y \leq f(x)\}.$$

Thus $SG(f)$ is a measurable subset of $X \times [0, 1]$. As f varies over the family of functions $\mathcal{F}$, the corresponding collection of sous-graphs $SG(f)$ can be thought of as a *concept* class $\mathcal{C}(\mathcal{F})$ of subsets of $X \times [0, 1]$. Now one can ask: What is the connection between the covering numbers of $\mathcal{F}$ and of $\mathcal{C}(\mathcal{F})$?

Lemma 6.3. *Let $\bar{P} := P \times U$ denote the product measure on $X \times [0, 1]$ where U denotes the uniform measure on $[0, 1]$. Suppose $f, g \in [0, 1]^X$. Then*

$$\bar{P}[SG(f) \Delta SG(g)] = d_P(f, g).$$

Consequently, for every $\epsilon > 0$, we have

$$N(\epsilon, \mathcal{C}(\mathcal{F}), d_{\bar{P}}) = N(\epsilon, \mathcal{F}, d_P).$$

234   6. Learning Under a Fixed Probability Measure

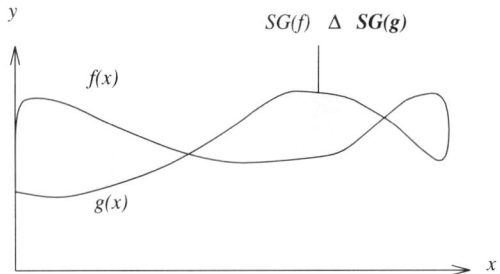

Fig. 6.2. Relating Covering Numbers of Functions Classes and Concept Classes

Proof. Figure 6.2 assists in understanding the statement of the lemma. Suppose $x \in X$ and $y \in [0,1]$. Then (x,y) belongs to the symmetric difference $SG(f) \Delta SG(g)$ if and only if

$$f(x) < y \leq g(x), \text{ or } g(x) < y \leq f(x).$$

Thus the measure of the symmetric difference is given by

$$\bar{P}[SG(f)\Delta SG(g)] = \int_X |f(x) - g(x)|\ P(dx) = d_P(f,g).$$

The statement about the covering numbers follows as a ready consequence. ∎

With the aid of Lemma 6.3, one can effectively turn every available result on the covering numbers of a family of *functions* into a corresponding result on the covering numbers of a *concept* class.

Example 6.8. Suppose $[a,b]$ is a finite interval, L is a finite constant, and let $\mathcal{F}$ consist of all functions $f : [a,b] \to \mathbb{R}$ such that $f(0) = 0$, and $f(\cdot)$ satisfies the Lipschitz condition

$$|f(x) - f(y)| \leq L|x - y|, \ \forall x, y \in [a,b].$$

Define the L_∞-norm $\|\cdot\|_\infty$ on $\mathcal{F}$ by

$$\| f \|_\infty := \sup_{x \in [a,b]} |f(x)|,$$

and define the corresponding metric ρ on $\mathcal{F}$ by

$$\rho(f,g) := \sup_{x \in [a,b]} |f(x) - g(x)|.$$

Then it is shown in [107], Example 2, that

$$\lg N(\epsilon, \mathcal{F}, \rho) \leq \lg M(2\epsilon, \mathcal{F}, \rho) = r(\epsilon),$$

where
$$r(\epsilon) := \frac{L(b-a)}{\epsilon} - 1 \text{ if } L(b-a)/\epsilon \text{ is an integer, and}$$
$$r(\epsilon) := \lfloor \frac{L(b-a)}{\epsilon} \rfloor \text{ if } L(b-a)/\epsilon \text{ is not an integer.}$$

Hence
$$\frac{L(b-a)}{\epsilon} - 1 \leq r(\epsilon) \leq \frac{L(b-a)}{\epsilon}, \forall \epsilon.$$

The above calculation applies to the norm $\|\cdot\|_\infty$ and the associated metric. For the present purposes, let us normalize so that $a = 0$, $b = 1$, and define
$$\| f \|_1 := \int_0^1 |f(x)|\, dx.$$

Then $\| f - g \|_1 = d_P(f, g)$, where P is the uniform probability measure on $[0, 1]$. Since $d_P(f, g) \leq \rho(f, g)$ by virtue of the fact that $\| f \|_1 \leq \| f \|_\infty$, it follows that
$$\lg N(\epsilon, \mathcal{F}, d_P) \leq r(\epsilon).$$

On the other hand, by adapting the argument in this example in [107], especially Figure 3, it follows that
$$\lg M(L\epsilon^2, \mathcal{F}, d_P) \geq \lfloor 1/\epsilon \rfloor,$$
or in other words,
$$\lg M(\epsilon, \mathcal{F}, d_P) \geq \lfloor (L/\epsilon)^{1/2} \rfloor.$$

Thus
$$\exp(O(1/\epsilon^{1/2})) \leq N(\epsilon, \mathcal{F}, d_P) \leq \exp(O(1/\epsilon)).$$

In particular, the metric entropy of $\mathcal{F}$ increases *superpolynomially* with respect to $1/\epsilon$ as $\epsilon \to 0^+$.

Now let $\mathcal{F}$ consist of all functions $f : [0, 1]^k \to \mathbb{R}$ that satisfy $f(\mathbf{0}) = 0$, and
$$|f(\mathbf{x}) - f(\mathbf{y})| \leq L \| \mathbf{x} - \mathbf{y} \|_\infty, \forall \mathbf{x}, \mathbf{y} \in [0, 1]^k.$$

Let P denote the uniform probability measure on $[0, 1]^k$. Then it can be shown that
$$\lg N(\epsilon, \mathcal{F}, d_P) \leq \lg N(\epsilon, \mathcal{F}, \rho) \leq [r(\epsilon)]^k,$$
where $r(\epsilon)$ is as above.

Example 6.9. Suppose a, b, L, C are given finite numbers, and let $\mathcal{F}$ consist of all functions $f : [a, b] \to \mathbb{R}$ such that $\| f \|_\infty \leq C$, and $f(\cdot)$ satisfies a Lipschitz condition with a Lipschitz constant of L. Then, by [107], Example 3, it follows that
$$\frac{L(b-a)}{\epsilon} + \lg C\epsilon - 3 \leq \lg M(2\epsilon, \mathcal{F}, \rho) \leq \frac{L(b-a)}{\epsilon} + \lg C\epsilon + 3,$$

provided
$$\epsilon \leq \min\{C/4, C^2/16L(b-a)\}.$$
As in the preceding example, the upper bound holds with ρ replaced by d_P provided $a = 0$ and $b = 1$.

6.6 Learnable Concept Classes Have Finite Metric Entropy

In this section, we prove the converse of Theorem 6.3, but only for *concept* learning. It turns out that the converse is *false* in general in the case of *function* learning.

The following deceptively simple-looking result forms the basis for the remainder of the section.

Lemma 6.4. *Suppose $B_1, \ldots, B_M \in \mathcal{S}$ have the property that they are pairwise 2ϵ-separated under the pseudometric d_P, i.e., that*
$$d_P(B_i, B_j) > 2\epsilon \text{ if } i \neq j.$$
Suppose $m \geq 1$ is an integer, and that there exists a function $f : X^m \times \{0,1\}^m \to \mathcal{S}$ with the property that
$$P^m\{\mathbf{x} \in X^m : d_P[B_j, f(\mathbf{x}, \mathbf{I}_{B_j}(\mathbf{x}))] > \epsilon\} \leq \delta \text{ for } j = 1, \ldots, M, \quad (6.6.1)$$
where
$$\mathbf{x} = [x_1 \ldots x_m]^t \in X^m, \text{ and }$$
$$\mathbf{I}_{B_j}(\mathbf{x}) = [I_{B_j}(x_1) \ldots I_{B_j}(x_m)]^t \in \{0,1\}^m.$$
Then
$$m \geq \lg M(1-\delta). \quad (6.6.2)$$

Remarks: One can think of the mapping f as an "algorithm" that associates a "hypothesis" in $\mathcal{S}$ with each multisample $\mathbf{x} \in X^m$ and each m-dimensional binary vector. The lemma states that, if the output of the algorithm is within ϵ of B_j with probability at least $1 - \delta$ whenever the input is a random multisample $\mathbf{x} \in X^m$ and the associated binary vector of indicator functions of B_j, and if the algorithm "works" in this sense for *each* of a collection of M pairwise 2ϵ-separated sets, then the number m of samples must at least equal $\lg M(1 - \delta)$. An immediate consequence of this lemma is the following general *necessary* condition for a class of concepts to be learnable:

Theorem 6.6. *Suppose $\mathcal{C}$ is a given concept class, and let $\epsilon > 0$ be specified. Then any algorithm that is PAC to accuracy ϵ requires at least $\lg M(2\epsilon, \mathcal{C}, d_P)$ samples, where $M(2\epsilon, \mathcal{C}, d_P)$ denotes the 2ϵ-packing number of the concept class $\mathcal{C}$ with respect to the pseudometric d_P. Consequently, $\mathcal{C}$ is learnable to accuracy ϵ only if $M(2\epsilon, \mathcal{C}, d_P)$ is finite.*

6.6 Learnable Concept Classes Have Finite Metric Entropy

Proof. Define a function

$$g : \{1, \ldots, M\} \times X^m \times \{0,1\}^m \to \{0,1\}$$

as follows: Suppose $1 \leq j \leq M$, $\mathbf{x} \in X^m$, and $\mathbf{L} \in \{0,1\}^m$. Then

$$g(j, \mathbf{x}, \mathbf{L}) = \begin{cases} 1 & \text{if } d_P[B_j, f(\mathbf{x}, \mathbf{L})] \leq \epsilon, \text{ and} \\ 0 & \text{otherwise.} \end{cases}$$

From the definition of g and the fact that the sets B_j are pairwise 2ϵ-separated, it follows that, for a fixed $\mathbf{x} \in X^m$ and $\mathbf{L} \in \{0,1\}^m$, there is at most one index j such that $g(j, \mathbf{x}, \mathbf{L}) = 1$, because the set $f(\mathbf{x}, \mathbf{L})$ cannot be within a distance ϵ of more than one B_j. Therefore

$$\sum_{j=1}^{M} g(j, \mathbf{x}, \mathbf{L}) \leq 1, \ \forall \mathbf{x} \in X^m, \ \mathbf{L} \in \{0,1\}^m.$$

As a consequence,

$$\int_{X^m} \sum_{\mathbf{L} \in \{0,1\}^m} \sum_{j=1}^{M} g(j, \mathbf{x}, \mathbf{L}) \, P^m(d\mathbf{x}) \leq \sum_{\mathbf{L} \in \{0,1\}^m} \int_{X^m} P^m(d\mathbf{x}) = 2^m. \quad (6.6.3)$$

Note that there is no difficulty about interchanging the order of the integration and the summations, because the summations are finite. Next, observe that

$$\sum_{\mathbf{L} \in \{0,1\}^m} g(j, \mathbf{x}, \mathbf{L}) \geq g[j, \mathbf{x}, \mathbf{I}_{B_j}(\mathbf{x})]$$

because g is always nonnegative. By the "PAC learning" hypothesis (6.6.1), we have that

$$P^m\{\mathbf{x} \in X^m : g[j, \mathbf{x}, \mathbf{I}_{B_j}(\mathbf{x})] = 1\} \geq 1 - \delta.$$

Therefore, for each fixed j, it follows that

$$\int_{X^m} \sum_{\mathbf{L} \in \{0,1\}^m} g(j, \mathbf{x}, \mathbf{L}) \, P^m(d\mathbf{x}) \geq \int_{X^m} g[j, \mathbf{x}, \mathbf{I}_{B_j}(\mathbf{x})] \, P^m(d\mathbf{x}) \geq 1 - \delta.$$

As a result,

$$\sum_{j=1}^{M} \int_{X^m} \sum_{\mathbf{L} \in \{0,1\}^m} g(j, \mathbf{x}, \mathbf{L}) \, P^m(d\mathbf{x}) \geq M(1 - \delta). \quad (6.6.4)$$

Combining the two inequalities (6.6.3) and (6.6.4) shows that

$$2^m \geq M(1 - \delta),$$

which is the same as (6.6.2). ∎

238   6. Learning Under a Fixed Probability Measure

Proof. **of Theorem 6.6**: Given $\mathcal{C}$ and ϵ, choose a maximal 2ϵ-separated set $B_1, \ldots, B_M$, where $M := M(2\epsilon, \mathcal{C}, d_P)$. Now any algorithm that is PAC to accuracy ϵ on $\mathcal{C}$ must also be PAC to accuracy ϵ on the collection $\{B_1, \ldots, B_M\}$, which is a subset of $\mathcal{C}$. Now apply Lemma 6.4 and let $\delta \to 0$. ∎

Now the following general necessary and sufficient condition for *concept* learning can be stated:

Theorem 6.7. *Suppose $\mathcal{C} \subseteq \mathcal{S}$, and that P is a fixed probability measure on $(X, \mathcal{S})$. Then $\mathcal{C}$ is learnable with respect to P if and only if $\mathcal{C}$ is totally bounded with respect to the pseudometric d_P, that is,*

$$N(\epsilon, \mathcal{C}, d_P) < \infty \ \forall \epsilon > 0. \tag{6.6.5}$$

Proof. The "if" part of the theorem is already proven in Theorem 6.3. To prove the "only if" part, suppose $\mathcal{C}$ is learnable. Then it follows from Theorem 6.6 that $M(2\epsilon, \mathcal{C}, d_P)$ is finite for each $\epsilon > 0$. However, by Lemma 2.2,

$$N(2\epsilon, \mathcal{C}, d_P) \leq M(2\epsilon, \mathcal{C}, d_P).$$

Hence $N(2\epsilon, \mathcal{C}, d_P)$ is finite for each $\epsilon > 0$. Throwing away the superfluous factor of 2 in front of the ϵ leads to (6.6.5). ∎

The next example illustrates the application of Theorem 6.6.

Example 6.10. Let $X = [0, 1)$, $\mathcal{S} =$ the Borel σ-algebra on X, and $P =$ the uniform probability on X. As in Example 1.5, define the concept class $\mathcal{C}$ as follows: Each number $x \in X$ has a unique binary representation

$$x = 0.b_1(x)\, b_2(x) \ldots = \sum_{i=1}^{\infty} b_i(x) 2^{-i}.$$

One can in fact define the functions $b_i(\cdot)$ "explicitly" as follows:

$$b_1(x) = \lfloor 2x \rfloor, \text{ and}$$

$$b_i(x) = \lfloor 2^i x - \sum_{j=1}^{i-1} 2^{i-j} b_j(x) \rfloor, \text{ for } i \geq 2,$$

where $\lfloor y \rfloor$ is the largest integer not greater than y. Now define the set $A_i \subseteq X$ as the *support* of the function b_i; in other words,

$$A_i = \{x \in X : b_i(x) = 1\}.$$

Figure 6.3, which is the same as Figure 1.2, illustrates the functions $b_i(\cdot)$. It is not too difficult to verify that

$$\int_X |b_i(x) - b_j(x)|\, dx = 0.5 \text{ if } i \neq j.$$

6.6 Learnable Concept Classes Have Finite Metric Entropy

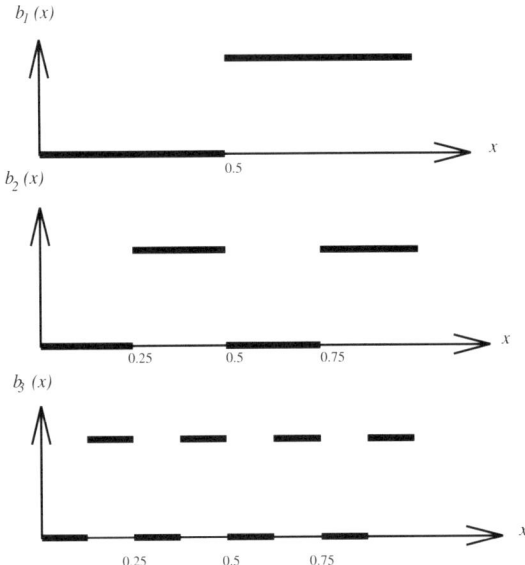

Fig. 6.3. A Nonlearnable Concept Class

Therefore
$$d_P(b_i, b_j) = 0.5, \text{ if } i \neq j.$$
Hence the sets $\{A_i\}$ are pairwise at a distance of 0.5. Since this collection of sets is infinite, the condition of Theorem 6.6 is violated whenever $\epsilon < 0.25$. It follows that this concept class is *not* learnable to any accuracy $\epsilon < 0.25$.

In fact, in this particular example, one can do slightly better, if one insists that the output of the learning algorithm must be one of the concepts A_i.[4] Let T denote the unknown target concept, and let H_m denote the output of a learning algorithm. Then, either $d_P(T, H_m) = 0$ or else $d_P(T, H_m) = 0.5$. In other words, for any $\epsilon < 0.5$, we have that $d_P(T, H_m) < \epsilon$ if and only if $T = H_m$. Now Theorem 6.6 implies that this concept class cannot be learnt to any accuracy less than 0.25. But in view of the above feature, we can strengthen this inference and conclude that this concept class cannot be learnt to any accuracy less than 0.5. This is a formalization of the informal argument advanced in Example 1.5.

In practical terms, this example means the following: The concept class A_i consists of all $x \in [0, 1)$ such that the i-th component of the binary expansion of x equals one. For instance,
$$A_1 = [0.5, 1), \ A_2 = [0.25, 0.5) \cup [0.75, 1),$$

---

[4] Note that Theorem 6.6 is valid *without* this assumption; in other words, the theorem applies even to the case where there is a separate "hypothesis" class $\mathcal{H}$ that can be larger than $\mathcal{C}$.

and so on. In the learning problem, the "target" concept can be thought of as a fixed but unknown integer $i \geq 1$, and the oracle outputs 1 if the i-th term in the binary expansion of the sample x_j equals 1, and zero otherwise. Thus the learning problem consists of making a reasonable guess as to *which component* of the sample x_j the oracle is looking at. Theorem 6.3 tells us that it is impossible to make such a reasonable guess. ∎

Thus it is natural to ask whether the converse of Corollary 6.3 is true, i.e., whether total boundedness with respect to the pseudometric d_P is also a necessary condition for a *function* class to be learnable. The answer is negative, as shown in the next example.

Example 6.11. Let $X = [0,1]$, $\mathcal{S}$ = the Borel σ-algebra on X, and define P as the probability measure on X having the following density function:

$$p(x) = 0.5 + 0.5\delta(x-1).$$

In other words, P equals *half* the uniform probability measure on $[0,1)$, and in addition, has a point mass of 0.5 concentrated at $x = 1$. Now let $\mathcal{F}$ be a countable collection of functions $\{f_i\}_{i \geq 1}$ defined as follows: If $x \in [0,1)$, then $f_i(x) = b_i(x)$, where $b_i(x)$ is the i-th bit in the binary representation of x (see the preceding example). If $x = 1$, then $f_i(1) = 2^{-i}$. Now it is clear that the map $f_i \mapsto f_i(1)$ is one-to-one. In other words, knowledge of $f_i(1)$ is enough to determine f_i uniquely. Now suppose i.i.d. samples $x_1, \ldots, x_m \in X$ are drawn in accordance with P. Then each sample x_i equals 1 with a probability of 0.5. Thus, after m samples are drawn, the probability that *none* of the samples equals 1 is 2^{-m}. Now consider the following naive algorithm: If any one of the samples x_i equals 1, then determine the target function $f(\cdot)$ exactly on the basis of the value $f(1)$ returned by the oracle. If none of the samples equals 1, then declare $h_m = f_1$. Then it follows that $d_P(f, h_m) = 0$ with probability $1 - 2^{-m}$. Hence, given $\epsilon, \delta > 0$, the algorithm is PAC if we draw $m \geq \lg(1/\delta)$ samples. Hence $\mathcal{F}$ is learnable. On the other hand, by adapting the preceding example, one can see that if $i \neq j$, then

$$d_P(f_i, f_j) = \int_{[0,1)} 0.5 \, |b_i(x) - b_j(x)| \, dx + |2^{-i} - 2^{-j}|$$

$$\geq \int_{[0,1)} 0.5 \, |b_i(x) - b_j(x)| \, dx = 0.25.$$

Hence, for every $\epsilon < 0.25$, the ϵ-packing number $M(\epsilon, \mathcal{F}, d_P)$ is infinite. This shows that a family of functions can be learnable even though it does not have finite metric entropy for all ϵ. ∎

Note that the above example is crucially dependent on the oracle output being available with *infinite* precision. It is not known whether there exists an example of a function class with infinite metric entropy that is learnable even if an oracle returns noisy values of $f(x_j)$, where f is the target function and x_j is the sample.

Thus in general the total boundedness of the function class is not necessary for the function class to be learnable. However, in the case where the functions in $\mathcal{F}$ all map X into a fixed *finite* set, the total boundedness condition is once again a necessary condition for learnability. This can be established using the following lemma, which is analogous to Lemma 6.4.

Lemma 6.5. *Suppose $Y \subseteq [0,1]$ is a fixed finite set, and suppose $b_1, \ldots, b_M : X \to Y$ are measurable and 2ϵ-separated with respect to the pseudometric d_P. Suppose $m \geq 1$ is an integer, and that there exists a function $f : (X \times Y)^m \to Y^X$ with the property that*[5]

$$P^m\{\mathbf{x} \in X^m : d_P[b_j, f(\mathbf{x}, \mathbf{b}_j(\mathbf{x}))] > \epsilon\} \leq \delta, \text{ for } j = 1, \ldots, M,$$

where

$$\mathbf{b}_j(\mathbf{x}) = [b_j(x_1) \ldots b_j(x_m)]^t \in Y^m.$$

Then

$$m \geq \log_{|Y|} M(1-\delta).$$

The proof is entirely analogous to that of Lemma 6.4 and is therefore omitted. Using this lemma, one can state the following theorem:

Theorem 6.8. *Suppose $Y \subseteq [0,1]$ is a finite set, and suppose $\mathcal{F}$ is a family of measurable functions mapping X into Y. Then $\mathcal{F}$ is PAC learnable with respect to the probability P if and only if $\mathcal{F}$ is totally bounded with respect to the pseudometric d_P, that is,*

$$N(\epsilon, \mathcal{F}, d_P) < \infty \; \forall \epsilon > 0.$$

The proof is immediate. This theorem means that Theorem 6.3 remains valid if, instead of *concept* classes, we consider measurable functions mapping X into any fixed *finite* set.

We conclude this section with one last result.

Theorem 6.9. *Suppose $(X, \mathcal{S})$ is a measurable space, and that $\mathcal{C} \subseteq \mathcal{S}$. Under these assumptions, if $\mathcal{C}$ has the property that empirical probabilities converge uniformly, then $\mathcal{C}$ is totally bounded with respect to the pseudometric d_P.*

Proof. If $\mathcal{C}$ has the UCEP property, then $\mathcal{C}$ is PAC-learnable, by Theorem 6.1. The total boundedness of $\mathcal{C}$ now follows from Theorem 6.3. ∎

Though the above proof *appears* to be very short and direct, it is in reality very indirect. It is ironic to note that a more direct proof is not known to the author, nor is it known to him whether the above theorem holds with the concept class $\mathcal{C}$ replaced by a function class $\mathcal{F}$.

---

[5] Recall that Y^X denotes the set of all *measurable* functions mapping X into Y.

6.7 Model-Free Learning

In this section, we study the model-free learning problem of Section 3.3 in the case where the underlying probability measure on X is fixed. Results analogous to those in the previous sections are shown to hold in the case of model-free learning as well.

Recall the model-free learning problem as formulated in Section 3.3. One is given

- Sets X, Y, U and a σ-algebra $\bar{S}$ on $X \times Y$.
- A family $\bar{\mathcal{P}}$ of probability measures on $X \times Y$.
- A family $\mathcal{H}$ of measurable functions mapping X into U, called the "hypothesis class."
- A function $\ell : Y \times U \to [0, 1]$, called the "loss function."

With each hypothesis function h and each probability measure $\bar{P}$ one associates the quantity

$$J(h, \bar{P}) := \int_{X \times Y} \ell[y, h(x)] \, \bar{P}(dx, dy).$$

With $\bar{P}$ alone one associates the quantity

$$J^*(\bar{P}) := \inf_{h \in \mathcal{H}} \int_{X \times Y} \ell[y, h(x)] \, \bar{P}(dx, dy).$$

Random samples $(x_1, y_1), \ldots$ are drawn in accordance with an unknown probability measure $\bar{P} \in \bar{\mathcal{P}}$. An "algorithm" is an indexed family of mappings $\{A_m\}_{m \geq 1}$, where $A_m : (X \times Y)^m \to \mathcal{H}$. Define

$$h_m(\mathbf{x}, \mathbf{y}) := A_m[(x_1, y_1), \ldots, (x_m, y_m)]$$

to be the hypothesis produced by the algorithm after m samples. The performance of the algorithm is measured by the quantity

$$r_{\mathrm{mf}}(m, \epsilon) := \sup_{\bar{P} \in \bar{\mathcal{P}}} \bar{P}^m \{(\mathbf{x}, \mathbf{y}) \in X^m \times Y^m : J(h_m, \bar{P}) > J^*(\bar{P}) + \epsilon\}.$$

The algorithm is said to be "probably approximately correct to accuracy ϵ" if $r_{\mathrm{mf}}(m, \epsilon) \to 0$ as $m \to \infty$, and "probably approximately correct" if $r_{\mathrm{mf}}(m, \epsilon) \to 0$ as $m \to \infty$ for each $\epsilon > 0$.

In Section 3.3, an associated family $\mathcal{L}_\mathcal{H}$ is also introduced. For each $h \in \mathcal{H}$, define an associated function $\ell_h : X \times Y \to [0, 1]$ by

$$\ell_h(x, y) := \ell[y, h(x)], \; \forall x, y,$$

and the family of functions $\mathcal{L}_\mathcal{H}$ as

$$\mathcal{L}_\mathcal{H} := \{\ell_h : h \in \mathcal{H}\}.$$

Now it makes sense to ask whether the family $\mathcal{L}_\mathcal{H}$ has the UCEMUP property with respect to $\bar{\mathcal{P}}$. The significance of the UCEMUP property in the context of model-free learning is brought out in Theorem 3.2, the contents of which are repeated here for the convenience of the reader. For each multisample $\mathbf{z} := (\mathbf{x}, \mathbf{y}) \in X^m \times Y^m$ and each $h \in \mathcal{H}$, define

$$\hat{J}(h; \mathbf{z}) := \frac{1}{m} \sum_{i=1}^{m} \ell_h(z_i)$$

to be the empirical estimate of $J(h, \bar{P})$, and define

$$\hat{J}^*(\mathbf{z}) := \inf_{h \in \mathcal{H}} \hat{J}(h; \mathbf{z}),$$

to be the minimum achievable *empirical* risk based on the multisample $\mathbf{z}$. For the algorithm $\{A_m\}$, define

$$\hat{J}[h_m(\mathbf{z}); \mathbf{z}] := \frac{1}{m} \sum_{i=1}^{m} \ell[y_i, h_m(x_i)]$$

to be the actual empirical risk achieved by the algorithm. Now define

$$t(m, \epsilon) := \bar{P}^m \{\mathbf{z} \in Z^m : \hat{J}[h_m(\mathbf{z}); \mathbf{z}] > \hat{J}^*(\mathbf{z}) + \epsilon\}.$$

Thus $t(m, \epsilon)$ is the probability that, after m random samples are drawn, the empirical risk $\hat{J}[h_m(\mathbf{z}); \mathbf{z}]$ is more than ϵ-worse compared to the minimum achievable value $\hat{J}^*(\mathbf{z})$. The algorithm is said to "nearly minimize empirical risk with high probability" if $t(m, \epsilon) \to 0$ as $m \to \infty$. Now Theorem 3.2 states that if the family of functions $\mathcal{L}_\mathcal{H}$ has the UCEMUP property with respect to $\bar{\mathcal{P}}$, then every such algorithm is PAC.

Theorem 5.11 provides a very general theorem relating the UCEMUP properties of the families $\mathcal{H}$ and $\mathcal{L}_\mathcal{H}$. Obviously that theorem continues to apply to the present case as well. However, by restricting the problem slightly, it is possible to prove somewhat better results.

The restriction takes the form of generalizing the "fixed distribution" idea to the present setting. Suppose $\bar{P}$ is a probability measure on $X \times Y$. One can define the corresponding "marginal" probability $\bar{P}_x$ on X alone as follows: Suppose $A \subseteq X$ is measurable. Then

$$\bar{P}_x(A) := \bar{P}(A \times Y).$$

Now, given the family $\bar{\mathcal{P}}$ of probability measures on $X \times Y$, it is assumed that there is a *fixed* probability measure P on X such that

$$\bar{P}_x = P, \ \forall \bar{P} \in \bar{\mathcal{P}}.$$

In other words, it is assumed that, while there might be several probability measures on $X \times Y$, they all have the same marginal probability on X. This

244   6. Learning Under a Fixed Probability Measure

is a natural generalization of the idea of trying to learn a family of functions $\mathcal{F}$ under a fixed probability measure. To illustrate this, it is assumed for notational simplicity that P has a density $p(\cdot)$. In the standard PAC learning formulation with a noise-free oracle as discussed in Example 3.10, one can define a family $\{\bar{P}_f : f \in \mathcal{F}\}$ of probability measures on $X \times Y$ with the density

$$\bar{p}_f(x, y) := p(x)\, \delta(y - f(x)).$$

Suppose the f's are binary-valued and that the oracle makes a mistake with probability α, as in Example 3.11. Then

$$\bar{p}_f(1|x) = \alpha \text{ if } f(x) = 0, \text{ and } 1 - \alpha \text{ if } f(x) = 1,$$

and $\bar{p}_f(0|x)$ is defined analogously. Finally, if $f(\cdot)$ is real-valued and the noise is additive with a density $\phi(\cdot)$ as in Example 3.12, one can choose

$$\bar{p}_f(x, y) := p(x)\, \phi(y - f(x)).$$

In all cases, it is easy to see that the marginal density on x is independent of $f \in \mathcal{F}$.

6.7.1 A Sufficient Condition for Learnability

Thus far in the chapter, two main types of sufficient conditions for learnability have been proven. In Section 6.2, it is shown that if the family $\mathcal{F}$ has the UCEM property, then every consistent algorithm is PAC. An analogous result for model-free learning is already available by combining Theorems 3.2 and 6.1, and need not be stated again. In Section 6.3 it is shown that function classes with finite metric entropy are learnable (even if the class does not possess the UCEM property; see Example 6.7). It is possible to prove an analogous result in the case of model-free learning, provided the loss function satisfies a uniform Lipschitz condition. Specifically, it is assumed that the "decision space" U is a subset of $\mathbb{R}$, and that there exists a finite constant μ such that

$$|\ell(y, u_1) - \ell(y, u_2)| \leq \mu |u_1 - u_2|, \ \forall u_1, u_2 \in \mathbb{R}, \ \forall y \in Y. \tag{6.7.1}$$

The minimum empirical risk algorithm in the case of model-free learning is a natural extension of that introduced previously in Section 6.3. Let $\{g_1, \ldots, g_k\}$ be a finite subset of $\mathcal{H}$. Once samples $(x_1, y_1), \ldots, (x_m, y_m)$ are drawn, define

$$\hat{J}_i := \frac{1}{m} \sum_{j=1}^{m} \ell[y_j, g_i(x_j)], \ 1 \leq i \leq k.$$

Then the hypothesis h_m is chosen as a g_{i_0} such that

$$\hat{J}_{i_0} = \min_{1 \leq i \leq k} \hat{J}_i.$$

Now we can state a result analogous to Theorem 6.2.

6.7 Model-Free Learning

Theorem 6.10. *Suppose*

1. *The family of probabilities $\bar{\mathcal{P}}$ has the property that every $\bar{P} \in \bar{\mathcal{P}}$ has the same marginal measure on X, call it P.*
2. *The hypothesis class $\mathcal{H}$ has the property that*

$$N(\epsilon, \mathcal{H}, d_P) < \infty, \ \forall \epsilon > 0.$$

3. *The loss function ℓ satisfies the uniform Lipschitz condition (6.7.1) above.*

Then the triple $(\mathcal{H}, \bar{\mathcal{P}}, \ell)$ is PAC learnable. In particular, given any $\epsilon > 0$, choose $\{g_1, \ldots, g_k\}$ to be an $\epsilon_0/2\mu$-cover of $\mathcal{H}$ with respect to d_P for some $\epsilon_0 < \epsilon$. Then the minimum empirical risk algorithm applied to $\{g_1, \ldots, g_k\}$ is PAC to accuracy ϵ, and

$$r_{\mathrm{mf}}(m, \epsilon) \leq k \exp(-m\epsilon^2/8).$$

Hence the algorithm is PAC to accuracy ϵ and confidence δ provided at least

$$m \geq \frac{8}{\epsilon^2} \ln \frac{k}{\delta}$$

samples are drawn.

Proof. The first step is to show that, for every $\bar{P} \in \bar{\mathcal{P}}$ and every $f, g \in \mathcal{H}$, we have

$$|J(f, \bar{P}) - J(g, \bar{P})| \leq \mu d_P(f, g). \tag{6.7.2}$$

This is a ready consequence of (6.7.1), since

$$\begin{aligned}
|J(f, \bar{P}) - J(g, \bar{P})| &= \left| \int_{X \times Y} [\ell(y, f(x)) - \ell(y, g(x))] \, \bar{P}(dx, dy) \right| \\
&\leq \int_{X \times Y} |\ell(y, f(x)) - \ell(y, g(x))| \, \bar{P}(dx, dy) \\
&\leq \mu \int_{X \times Y} |f(x) - g(x)| \, \bar{P}(dx, dy) \\
&= \mu \int_X |f(x) - g(x)| \, P(dx) \\
&= \mu d_P(f, g).
\end{aligned}$$

Another useful way of expressing the above inequality is:

$$d_{\bar{P}}(\ell_f, \ell_g) \leq \mu d_P(f, g),$$

and of course

$$|J(f, \bar{P}) - J(g, \bar{P})| \leq d_{\bar{P}}(\ell_f, \ell_g).$$

To prove that the minimum empirical risk algorithm applied to an $\epsilon_0/2\mu$-cover of $\mathcal{H}$ is PAC to accuracy ϵ, let $\bar{P} \in \bar{\mathcal{P}}$ be arbitrary, and select an $h = h(\epsilon, \bar{P})$ such that

246   6. Learning Under a Fixed Probability Measure

$$J(h, \bar{P}) \leq J^*(\bar{P}) + \frac{\epsilon - \epsilon_0}{2}.$$

Such an h exists, by the definition of $J^*(\bar{P})$. Now it is known that h is within a distance $\epsilon_0/2\mu$ (with respect to d_P) of one of the g_i's, though it is not known which one. Assume without loss of generality that the g_i's are renumbered such that $d_P(h, g_k) \leq \epsilon_0/2\mu$, which in turn implies that

$$J(g_k, \bar{P}) \leq J(h, \bar{P}) + \epsilon_0/2 \leq J^*(\bar{P}) + \epsilon/2.$$

Assume that the renumbering is such that

$$J(g_i, \bar{P}) > J^*(\bar{P}) + \epsilon \text{ for } 1 \leq i \leq l, \text{ and}$$

$$J(g_i, \bar{P}) \leq J^*(\bar{P}) + \epsilon \text{ for } l+1 \leq i \leq k.$$

Note that $l \leq k - 1$. Now suppose i.i.d. samples $(x_1, y_1), \ldots, (x_m, y_m)$ are drawn in accordance with $\bar{P}$, and as before let

$$\mathbf{z} := [(x_1, y_1) \ldots (x_m, y_m)]^t.$$

Note that the inequality $J(h_m, \bar{P}) \leq J^*(\bar{P}) + \epsilon$ is satisfied if h_m is one of $g_{l+1}, \ldots, g_k$. This will be the case if

(i) $\hat{J}(g_k; \mathbf{z}) \leq J^*(\bar{P}) + 3\epsilon/4$, and

(ii) $\hat{J}(g_i; \mathbf{z}) > J^*(\bar{P}) + 3\epsilon/4$ for $1 \leq i \leq l$.

Hence, in order for the inequality $J(h_m, \bar{P}) \leq J^*(\bar{P}) + \epsilon$ to be *violated*, it is necessary that

$$\hat{J}(g_k; \mathbf{z}) > J^*(\bar{P}) + 3\epsilon/4, \text{ or}$$

$$\hat{J}(g_i; \mathbf{z}) \leq J^*(\bar{P}) + 3\epsilon/4 \text{ for some } i \leq l.$$

Note that $J(g_i, \bar{P})$ is just the expected value of the function ℓ_{g_i}, while $\hat{J}(g_i; \mathbf{z})$ is its empirical mean based on the multisample $\mathbf{z}$. Hence, by Hoeffding's inequality, each of the above events has a probability no larger than $\exp(-m\epsilon^2/8)$. Hence

$$r_{\mathrm{mf}}(m, \epsilon) = \Pr\{J(h_m, \bar{P}) > J^*(\bar{P}) + \epsilon\} \leq k \exp(-m\epsilon^2/8).$$

Setting

$$k \exp(-m\epsilon^2/8) \leq \delta$$

and solving for m leads to the sample complexity estimate. ∎

As a specific application of the above approach, consider the problem of learning a binary concept class with a noisy oracle, as in Example 3.11. Thus there is a probability space $(X, \mathcal{S}, P)$, and a concept class $\mathcal{C} \subseteq \mathcal{S}$. Given a target concept $T \in \mathcal{C}$ and a random sample $x \in X$, a noisy oracle outputs $I_T(x)$ with a probability $1 - \alpha$ and $1 - I_T(x)$ with a probability of α, where the error probability $\alpha \in [0, 0.5)$ is known. The hypothesis class $\mathcal{H}$ is taken as $\mathcal{C}$ itself, and the collection of probability measures $\bar{\mathcal{P}}$ is taken as $\{\bar{P}_T, T \in \mathcal{C}\}$, where for each $A \subseteq \mathcal{S}$ we have (cf. Example 3.11)

$$\bar{P}_T(A \times \{0\}) := (1 - \alpha)P(A) - (1 - 2\alpha)P(A \cap T), \text{ and}$$

$$\bar{P}_T(A \times \{1\}) := \alpha P(A) + (1 - 2\alpha)P(A \cap T).$$

Also, for each $H \in \mathcal{H}$ and each $P_T \in \bar{\mathcal{P}}$, we have

$$J(H, \bar{P}_T) = \alpha + (1 - 2\alpha)P(H \triangle T) = \alpha + (1 - 2\alpha)\, d_P(H, T), \text{ and}$$

$$J^*(\bar{P}_T) = \inf_{H \in \mathcal{C}} J(H, \bar{P}_T) = \alpha.$$

See Example 3.11 for further details.

Thus far we have merely recapitulated the relevant parts of the problem set-up from Example 3.11. Next, to apply Theorem 6.10, we begin by estimating the Lipschitz constant of the function J. Clearly the loss function $\ell(y, u) = |y - u|$ satisfies a Lipschitz condition with the Lipschitz constant of one, since

$$|\, |y - u_1| - |y - u_2|\, | = |u_1 - u_2|, \; \forall y, u_1, u_2 \in \{0, 1\}.$$

However, by taking advantage of the special nature of the function J, it is possible to obtain a lower Lipschitz constant. Recall that

$$J(H, \bar{P}_T) = \alpha + (1 - 2\alpha)\, d_P(H, T).$$

Now the claim is that

$$|J(H_1, \bar{P}_T) - J(H_2, \bar{P}_T)| \leq (1 - 2\alpha)\, d_P(H_1, H_2).$$

In other words, (6.7.2) is satisfied with $\mu = (1 - 2\alpha)$. Note that

$$|J(H_1, \bar{P}_T) - J(H_2, \bar{P}_T)| = (1 - 2\alpha)\, |d_P(H_1, T) - d_P(H_2, T)|.$$

So the claim is established if it can be shown that

$$|d_P(H_1, T) - d_P(H_2, T)| \leq d_P(H_1, H_2).$$

But this last inequality is immediate, since from the triangle inequality, we have

$$d_P(H_1, T) - d_P(H_2, T) \leq d_P(H_1, H_2), \text{ and}$$

$$d_P(H_2, T) - d_P(H_1, T) \leq d_P(H_1, H_2).$$

Thus, in order for the minimal empirical risk algorithm to be PAC to accuracy ϵ, it is enough to apply the algorithm to an $\epsilon/2(1 - 2\alpha)$-cover of $\mathcal{C}$. Note that, since the infimum $J^*(\bar{P}_T)$ is actually attained for each $T \in \mathcal{C}$ (by choosing $H = T$), it is not necessary to choose an $\epsilon_0 < \epsilon$ as in the proof of Theorem 6.10. Now by the theorem, it follows that

$$r_{\mathrm{mf}}(m, \epsilon) \le k(\epsilon/2(1-2\alpha)) \, \exp(-m\epsilon^2/8),$$

where the notation $k(\epsilon/2(1-2\alpha))$ serves to remind us that k is the cardinality of an $\epsilon/2(1-2\alpha)$-cover of $\mathcal{C}$.

As pointed out in Example 3.11, in the case under study, we have

$$r_{\mathrm{mf}}(m, (1-2\alpha)\epsilon) = r(m, \epsilon).$$

From Theorem 6.2 [cf. (6.3.1)], we have

$$r(m, \epsilon) \le k(\epsilon/2) \, \exp(-m\epsilon^2/8),$$

whereas from the above inequality,

$$r_{\mathrm{mf}}(m, (1-2\alpha)\epsilon) \le k(\epsilon/2) \, \exp[-m(1-2\alpha)^2\epsilon^2/8].$$

The effect of the oracle noise can be gauged from the above two bounds. In the noise-free case, in order to ensure that the hypothesis H_m produced by the algorithm satisfies $d_P(T, H_m) \le \epsilon$ with probability at least $1 - \delta$, it is enough to take

$$m_{\text{noise-free}} = \frac{8}{\epsilon^2} \ln \frac{k(\epsilon/2)}{\delta}$$

samples. In contrast, in the case of noisy oracles, it is enough to take

$$m_{\text{noisy}} = \frac{8}{(1-2\alpha)^2\epsilon^2} \ln \frac{k(\epsilon/2)}{\delta}$$

samples. If $\alpha = 0$ so that the oracle is noise-free, the above bound reduces to its predecessor, whereas if $\alpha \to 0.5^-$, the above bound approaches infinity.

6.7.2 A Necessary Condition

In this subsection, we state and prove a necessary condition for model-free learnability that generalizes Lemma 6.4.

Theorem 6.11. *Suppose $Y = \{0, 1\}$ and consider the model-free learning problem. Given the family $\bar{\mathcal{P}}$, suppose there exists a single fixed probability P on X such that the marginal probability of every $\bar{P} \in \bar{\mathcal{P}}$ equals P. Suppose there exist probabilities $\bar{P}_1, \ldots, \bar{P}_M \in \bar{\mathcal{P}}$ such that no $H \in \mathcal{H}$ can satisfy the inequality*

$$J(H, \bar{P}_j) \le J^*(\bar{P}_j) + \epsilon$$

for more than one value of the index j. Then any algorithm that is PAC to accuracy ϵ and confidence δ requires at least $\lg M(1 - \delta)$ samples.

6.7 Model-Free Learning

Remarks: Suppose, as in Example 3.10, that the family $\bar{\mathcal{P}}$ consists of $\{\bar{P}_T : T \in \mathcal{H}\}$, so that the problem reduces to the standard PAC learning problem. In this case,
$$J(H, \bar{P}_T) = d_P(H, T).$$
Now, if $\{B_1, \ldots, B_M\}$ is a 2ϵ-separated subset of $\mathcal{H}$, then clearly no $H \in \mathcal{H}$ can satisfy $d_P(H, B_j) \leq \epsilon$ for more than one index j. Hence the corresponding set of probabilities $\{\bar{P}_{B_1}, \ldots, \bar{P}_{B_M}\}$ satisfies the hypothesis of the theorem, and the theorem itself reduces to Lemma 6.4.

Proof. Let $A_m : X^m \times \{0,1\}^m \to \mathcal{H}$ denote the algorithm, and let $H_m(\mathbf{x}, \mathbf{L}) \in \mathcal{H}$ denote the hypothesis produced by the algorithm with the input $(\mathbf{x}, \mathbf{L}) \in X^m \times \{0,1\}^m$. Then the "PAC" assumption implies that

$$\bar{P}_j^m\{(\mathbf{x}, \mathbf{L}) \in X^m \times \{0,1\}^m : J[H_m(\mathbf{x}, \mathbf{L}), \bar{P}_j]$$
$$\leq J^*(\bar{P}_j) + \epsilon\} \geq 1 - \delta, \text{ for } j = 1, \ldots, M. \quad (6.7.3)$$

Define a function
$$g : \{1, \ldots, M\} \times X^m \times \{0,1\}^m \to \{0,1\}$$
as follows: Suppose $1 \leq j \leq M$, $\mathbf{x} \in X^m$, and $\mathbf{L} \in \{0,1\}^m$. Then
$$g(j, \mathbf{x}, \mathbf{L}) = \begin{cases} 1 & \text{if } J[H_m(\mathbf{x}, \mathbf{L}), \bar{P}_j] \leq J^*(\bar{P}_j) + \epsilon, \text{ and} \\ 0 & \text{otherwise}. \end{cases}$$

Then the hypothesis of the theorem implies that, for each $\mathbf{x}, \mathbf{L}$, the function $g(j, \mathbf{x}, \mathbf{L})$ can equal 1 for at most one value of j. Hence
$$\sum_{i=1}^{M} g(j, \mathbf{x}, \mathbf{L}) \leq 1, \quad \forall \mathbf{x}, \mathbf{L}. \quad (6.7.4)$$

With each probability $\bar{P}_j$, one can associate the conditional probabilities $q_j(1|x)$ and $q_j(0|x)$. Given $\mathbf{L} \in \{0,1\}^m$ and $\mathbf{x} \in X^m$, define
$$Q_j(\mathbf{x}, \mathbf{L}) := \prod_{i=1}^{m} q_j(b_i|x_i), \text{ where } \mathbf{L} = b_1 \ldots b_m.$$

Thus, given any function $f : X^m \times \{0,1\}^m \to \mathbb{R}$, we have
$$\int_{X^m \times \{0,1\}^m} f(\mathbf{x}, \mathbf{L}) \, \bar{P}_j^m(d\mathbf{x}, d\mathbf{L}) = \int_{X^m} \sum_{\mathbf{L} \in \{0,1\}^m} f(\mathbf{x}, \mathbf{L}) \, Q_j(\mathbf{x}, \mathbf{L}) \, P^m(d\mathbf{x}).$$

In particular,
$$\int_{X^m \times \{0,1\}^m} \sum_{i=1}^{M} g(j, \mathbf{x}, \mathbf{L}) \, \bar{P}_j^m(d\mathbf{x}, d\mathbf{L})$$

$$= \int_{X^m} \sum_{\mathbf{L} \in \{0,1\}^m} \sum_{i=1}^{M} g(j, \mathbf{x}, \mathbf{L}) \, Q_j(\mathbf{x}, \mathbf{L}) \, P^m(d\mathbf{x})$$

$$\leq \int_{X^m} \sum_{\mathbf{L} \in \{0,1\}^m} P^m(d\mathbf{x}) = 2^m$$

from (6.7.4) and the fact that $Q_j(\mathbf{x}, \mathbf{L}) \leq 1$ for all $j, \mathbf{x}, \mathbf{L}$.

On the other hand, the PAC assumption (6.7.3) implies that

$$E_{\bar{P}_j^m}[g(j, \mathbf{x}, \mathbf{L})] \geq 1 - \delta, \text{ for } j = 1, \ldots, M,$$

or equivalently

$$\int_{X^m \times \{0,1\}^m} g(j, \mathbf{x}, \mathbf{L}) \, \bar{P}_j^m(d\mathbf{x}, d\mathbf{L}) \geq 1 - \delta, \text{ for } j = 1, \ldots, M.$$

Consequently

$$\sum_{i=1}^{M} \int_{X^m \times \{0,1\}^m} g(j, \mathbf{x}, \mathbf{L}) \, \bar{P}_j^m(d\mathbf{x}, d\mathbf{L}) \geq M(1 - \delta).$$

Combining the preceding two inequalities shows that

$$2^m \geq M(1 - \delta),$$

which is the desired conclusion. ∎

6.8 Dependent Inputs

Thus far we have studied the learning problem under the assumption that the input samples are independent. However, using the results of Chapter 3, it is straight-forward to extend these results to the case of dependent inputs.

6.8.1 Finite Metric Entropy and Alpha-Mixing Input Sequences

Theorem 6.12. *Suppose a function family $\mathcal{F}$ has finite metric entropy under the pseudometric d_P, and let $\{g_1, \ldots, g_N\}$ be an $\epsilon/2$-cover for $\mathcal{F}$ under d_P. Suppose $\{X_i\}$ is an α-mixing sample sequence. Then the minimal empirical risk algorithm is PAC. Moreover, if $\tilde{P}$ is the joint distribution of the sample sequence, we have*

$$r_\alpha(m, \epsilon, \tilde{P}) \leq N[\exp(-l\epsilon^2/8) + 4\alpha(k)l \exp(4\epsilon l)]$$

whenever $m = lk$. This quantity can be made to approach zero by a judicious choice of k, l.

Proof. Suppose $f \in \mathcal{F}$ is the target concept. With the $\epsilon/2$-cover $\{g_1, \ldots, g_N\}$ specified, renumber them in such a way that $d_P(f, g_N) \leq \epsilon/2$. Now the reasoning proceeds exactly as in the proof of Theorem 6.2. It is only necessary to make an estimate of the probability that the empirical estimate $\hat{d}(f, g_i)$ differs from the true value $d_P(f, g_i)$ by more than $\epsilon/4$. This probability can be bounded from (3.4.10) by replacing ϵ by $\epsilon/4$. Hence, if $m = kl$, then for each i this probability is bounded by

$$\exp(-l\epsilon^2/8) + 4\alpha(k)l\exp(4\epsilon l).$$

Therefore

$$\tilde{P}\{\mathbf{x} \in X^\infty : d_P f(, h_m) > \epsilon\} \leq N[\exp(-l\epsilon^2/8) + 4\alpha(k)l\exp(4\epsilon l)]$$

as required. Now Lemma 3.1 shows that the quantity on the right side can be made to approach zero by a suitable choice of k, l, no matter how slowly $\alpha(k)$ approaches zero. ∎

6.8.2 Consistent Learnability and Beta-Mixing Input Sequences

In this subsection it is shown that the properties of consistent learnability (both PAC and PUAC) are preserved if the i.i.d. sample sequence is replaced by a β-mixing input sequence.

Theorem 6.13. *Suppose a family $\mathcal{F}$ is consistently learnable when the input sequence is i.i.d. Then it continues to be consistently learnable when the input sequence if β-mixing. Suppose $\tilde{P}$ is the joint distribution of the input sequence, and in analogy with (6.4.2) define*

$$\psi_\beta(m, \epsilon) := \sup_{f \in \mathcal{F}} \tilde{P}\{\mathbf{x}^* \in X^\infty : \phi_{m,f}(\mathbf{x}^*) > \epsilon\},$$

$$\psi_{\mathrm{iid}}(m, \epsilon) := \sup_{f \in \mathcal{F}} (\tilde{P}_0)^\infty \{\mathbf{x}^* \in X^\infty : \phi_{m,f}(\mathbf{x}^*) > \epsilon\}.$$

For each m, choose an integer $k_m \leq m$ and let $l_m = \lfloor m/k_m \rfloor$. Then

$$\psi_\beta(m, \epsilon) \leq \max\{\psi_{\mathrm{iid}}(l_m, \epsilon), \psi_{\mathrm{iid}}(l_m + 1, \epsilon)\} + m\beta(k_m).$$

Proof. With k_m, l_m defined as above, let $r_m = m - k_m l_m$. Let us fix $f \in \mathcal{F}$ and examine the stochastic process

$$\phi_{m,f}(\mathbf{x}^*) := \sup\{d_P(f, g) : \hat{d}_m(f, g; \mathbf{x}^*) = 0\}.$$

In analogy with the notion of quasi-subadditivity introduced in Chapter 3, let us say that this stochastic process is quasi-subadditive if the following condition is satisfied:

252   6. Learning Under a Fixed Probability Measure

$$\phi_{m,f}(\mathbf{x}^*) \leq \frac{1}{k_m} \sum_{i=1}^{k_m} b_i,$$

where

$$b_i := \phi_{l_m+1,f}(x_i, x_{i+k_m}, \ldots, x_{i+l_m k_m}), \ i = 1, \ldots, r_m,$$
$$b_i := \phi_{l_m,f}(x_i, x_{i+k_m}, \ldots, x_{i+(l_m-1)k_m}), \ i = 1, \ldots, r_m.$$

If the stochastic process $\{\phi_{m,f}(\cdot)\}$ is indeed quasi-subadditive, then it follows as in the proof of Theorem 3.7 that

$$\tilde{P}\{\phi_{m,f}(\mathbf{x}^*) > \epsilon\} \leq \max\{(\tilde{P}_0)^\infty\{\phi_{l_m,f}(\mathbf{x}^*) > \epsilon\},$$
$$(\tilde{P}_0)^\infty\{\phi_{l_m+1,f}(\mathbf{x}^*) > \epsilon\}\} + m\beta(k_m). \quad (6.8.1)$$

From the above inequality the desired conclusion follows readily. Thus the proof is complete once the quasi-subadditivity of $\{\phi_{m,f}\}$ is established.

Now, if $\hat{d}_m(f, g; \mathbf{x}^*) = 0$, then $f(x_i) = g(x_i)$ for all i. Therefore the empirical distance $\hat{d}(f, g; \mathbf{y})$ equals zero whenever $\mathbf{y}$ is a substring of $(x_1, \ldots, x_m) =: \mathbf{x}_m$. Therefore, whenever $\mathbf{y}$ is a substring of $\mathbf{x}_m$, we have

$$\sup\{d_{\tilde{P}_0}(f, g) : \hat{d}(f, g; \mathbf{y}) = 0\} \geq \sup\{d_{\tilde{P}_0}(f, g) : \hat{d}(f, g; \mathbf{x}_m) = 0\}.$$

In particular then, we have

$$\phi_{m,f} \leq b_i \ \forall i.$$

So certainly $\phi_{m,f}$ is bounded by the *average* of the b_i's. Hence the stochastic process is quasi-subadditive. ∎

Theorem 6.14. *Suppose a family $\mathcal{F}$ is consistently PUAC learnable when the input sequence is i.i.d. Then it continues to be consistently PUAC learnable when the input sequence is β-mixing. Let $\tilde{P}$ denote the joint distribution of the input sequence and define*

$$w_\beta(m, \epsilon) := \tilde{P}\{\mathbf{x}^* \in X^\infty : \exists f, g \in \mathcal{F} \ \text{s.t.} \ \hat{d}_m(f, g; \mathbf{x}^*) = 0 \ \text{and} \ d_{\tilde{P}_0}(f, g) > \epsilon\},$$
$$w_{\text{iid}}(m, \epsilon) := (\tilde{P}_0)^\infty\{\mathbf{x}^* \in X^\infty : \exists f, g \in \mathcal{F} \ \text{s.t.} \ \hat{d}_m(f, g; \mathbf{x}^*) = 0, d_{\tilde{P}_0}(f, g) > \epsilon\}.$$

Then

$$w_\beta(m, \epsilon) \leq \max\{w_{\text{iid}}(l_m, \epsilon), w_{\text{iid}}(l_m+1, \epsilon)\} + m\beta(k_m).$$

The proof is entirely analogous to that of Theorem 6.13 and is therefore omitted.

Notes and References Theorem 6.1 is a ready consequence of the fact that if a family of functions $\mathcal{F}$ has the UCEM property, then so does the family $\mathcal{F}\Delta\mathcal{F}$. This fact is stated explicitly by Vapnik-Chervonenkis in [196], [190], but is not interpreted in terms of learnability. The further interpretation of this property in terms of *almost surely eventually correct* learnability is new. The shrinking width property and its equivalence to consistent PUAC learnability was established for the first time in the first edition of this book. The equivalence of consistent PUAC learnability and plain PUAC learnability is established in [79]. The key to the results in Section 6.3 are Theorem 6.2 and Theorem 6.3 due to Benedek-Itai. The second of these theorems is given in [22], with an earlier version given in [23] (but the earlier version appears in print later!). Actually, Benedek and Itai only state the result for concept learning (Theorem 6.3), but the case of function learning (Theorem 6.2) is easily handled using their methods. Lemma 6.4, leading to the necessity of finite metric entropy for learnability, is proved by Benedek and Itai in [22]. The extension of the results of [22] to model-free learning is new.

7. Distribution-Free Learning

In the preceding chapter we studied the learning problem in the case where the random samples were generated by a *known fixed* probability measure. The problem studied in the present chapter can in some sense be thought of as being at the other end of the spectrum. The focus here is on so-called *distribution-free* learning; that is, the probability measure generating the samples can be *any* probability measure on the underlying measurable space. In other words, there is a *complete absence* of any prior knowledge about the underlying probability measure. In many ways this assumption is somewhat extreme – it is perhaps reasonable to assume at least a little prior knowledge about the probability measure that is generating the learning samples. Nevertheless, as a learning problem the distribution-free case is very "clean" in that it is possible to derive simple necessary and/or sufficient conditions for learnability involving the VC-dimension or the P-dimension. Moreover, we will see in Chapter 8 that, under suitable conditions, a little prior knowledge about the probability does not really help, in the following sense: Learning when there is so-called nonparametric uncertainty about the underlying probability measure is as difficult as distribution-free learning. One could perhaps argue that the positive results in distribution-free learnability are more meaningful than the negative results. If a concept (or function) class is distribution-free learnable, this means that even if one does not have any prior knowledge about the underlying probability measure, one can nevertheless learn unknown target concepts on the basis of randomly selected samples. If one actually has some prior knowledge about the underlying probability measure, then the concept class continues to be learnable. On the other hand, if a concept class is *not* distribution-free learnable, perhaps all it means is that the problem formulation is too unrealistic and restrictive.

Throughout this chapter, the symbol $\mathcal{P}^*$ is used to denote the set of *all* probability measures on a given measurable space $(X, \mathcal{S})$.

7.1 Uniform Convergence of Empirical Means

In this section, we study the problem of the uniform convergence of empirical means (and empirical probabilities) to their true values, in three important cases, namely: function classes, concept classes, and classes of loss functions.

7.1.1 Function Classes

Let us recall the problem under study. Suppose $(X, \mathcal{S})$ is a given measurable space, and that $\mathcal{F}$ is a family of measurable functions mapping X into $[0, 1]$. Finally, let $\mathcal{P}^*$ denote the set of all probability measures on $(X, \mathcal{S})$. For a given probability measure $P \in \mathcal{P}^*$, a function $f \in \mathcal{F}$, and a multisample $\mathbf{x} = [x_1 \ldots x_m]^t \in X^m$, define

$$E_P(f) := \int_X f(x) \, P(dx),$$

$$\hat{E}(f; \mathbf{x}) := \frac{1}{m} \sum_{i=1}^{m} f(x_i), \text{ and}$$

$$q(m, \epsilon, P) := P^m \{ \mathbf{x} \in X^m : \sup_{f \in \mathcal{F}} |\hat{E}(f; \mathbf{x}) - E_P(f)| > \epsilon \}.$$

Finally, define

$$q^*(m, \epsilon) := \sup_{P \in \mathcal{P}^*} q(m, \epsilon, P).$$

We say that the family $\mathcal{F}$ has the property of **distribution-free uniform convergence of empirical means** if $q^*(m, \epsilon) \to 0$ as $m \to \infty$ for each $\epsilon > 0$.

Now we state and prove three distinct *sufficient* conditions for a function family to have the distribution-free UCEM property. The first two theorems are based on the pseudo-dimension, but give two distinct bounds. The last theorem is based on the fat-shattering dimension.

Theorem 7.1. *Suppose the family $\mathcal{F}$ has finite P-dimension, say d. Suppose $0 < \epsilon < e/(4 \lg e) \approx 0.47$. Then*

$$q^*(m, \epsilon) \leq 8 \left(\frac{16e}{\epsilon} \ln \frac{16e}{\epsilon} \right)^d \exp(-m\epsilon^2/32), \, \forall m. \tag{7.1.1}$$

Thus $\mathcal{F}$ has the property of distribution-free uniform convergence of empirical means.

Proof. After all the hard work done in the preceding chapters, the proof of this important theorem is immediate. Suppose $0 < \epsilon < e/(4 \lg e) \approx 0.47$. From Corollary 4.2, we have that, for *any* probability measure $P \in \mathcal{P}^*$,

$$L(\epsilon, \mathcal{F}, d_P) \leq 2 \left(\frac{2e}{\epsilon} \ln \frac{2e}{\epsilon} \right)^d.$$

In particular, let m be any integer, let $\mathbf{z} \in X^{2m}$ be arbitrary, and let $P_{\mathbf{z}}$ be the purely atomic measure concentrated uniformly on the components of $\mathbf{z}$. Then it is easy to see that

$$L(\epsilon, \mathcal{F}, d_{P_{\mathbf{z}}}) = L(\epsilon, \mathcal{F}|_{\mathbf{z}}, \|\cdot\|_{a1}),$$

where the symbol $\mathcal{F}|_{\mathbf{z}}$ is defined in Section 5.3, and $\|\cdot\|_{a1}$ is the "averaged" l_1-norm. This leads to the useful inequality

$$L(\epsilon, \mathcal{F}|_{\mathbf{z}}, \|\cdot\|_{a1}) \le 2\left(\frac{2e}{\epsilon} \ln \frac{2e}{\epsilon}\right)^d, \forall \mathbf{z} \in X^{2m}, \forall m. \tag{7.1.2}$$

Now substitute this into the inequality (5.6.8) and let $P \in \mathcal{P}^*$ be arbitrary. Disregarding the "min" in the right side of (5.6.8) leads to

$$q(m, \epsilon, P) \le 4 E_{P^{2m}}[L(\epsilon/8, \mathcal{F}|_{\mathbf{z}}, \|\cdot\|_{a1})] \exp(-m\epsilon^2/32)$$

$$\le 8\left(\frac{16e}{\epsilon} \ln \frac{16e}{\epsilon}\right)^d \exp(-m\epsilon^2/32),$$

where we also use (7.1.2). Since the right side of the above inequality is independent of P, it follows that $q^*(m, \epsilon)$ is also bounded by the same quantity. This establishes the inequality (7.1.1). Since the right side of (7.1.1) approaches zero as $m \to \infty$ for each $\epsilon > 0$, it follows that $\mathcal{F}$ has the property of distribution-free uniform convergence of empirical means. ∎

Theorem 7.2. *Suppose the family $\mathcal{F}$ assumes values in $[0, R]$ and has finite P-dimension, say d. Then*

$$q^*(m, \epsilon) \le 4\left(\frac{8emR}{\epsilon d}\right)^d \exp(-m\epsilon^2/32), \forall m, \epsilon. \tag{7.1.3}$$

Hence $\mathcal{F}$ has the property of distribution-free uniform convergence of empirical means.

Proof. The proof is based on the bound (5.6.8) and the estimate for covering numbers contained in Theorem 4.2. Replacing ϵ by $\epsilon/8$ in (4.2.10) shows that

$$N(\epsilon/8, \mathcal{F}|_{\mathbf{x}}, \|\cdot\|_\infty) \le \left(\frac{8emR}{\epsilon d}\right)^d.$$

Obviously
$$L(\epsilon/8, \mathcal{F}|_{\mathbf{x}}, \|_{a1}) \le N(\epsilon/8, \mathcal{F}|_{\mathbf{x}}, \|\cdot\|_\infty).$$

Substituting this bound into (5.6.8) and disregarding the "min" leads to the desired conclusion. ∎

Comparing the estimates in (7.1.1) and (7.1.3) shows that the estimate in (7.1.1) consists of a constant term multiplied by a term decaying geometrically in m, whereas in (7.1.3) the estimate consists of a *polynomial in m* multiplied by a term decaying geometrically in m. In this sense (7.1.1) is a better bound for extremely large values of m. On the other hand, simple

258   7. Distribution-Free Learning

algebra shows that the "constant term" in (7.1.1) is larger than the "polynomial term" in (7.1.1) unless $m \geq 2d\ln(16e/\epsilon)$. Thus it is worthwhile having both bounds at our disposal. Moreover, as we shall see subsequently, the bound (7.1.1) is very similar to the bound for concept classes having finite VC-dimension.

Theorem 7.3. *Suppose the function family $\mathcal{F}$ assumes values in $[0, R]$ and finite fat-shattering dimension for each nonzero width. Then $\mathcal{F}$ has the property of distribution-free uniform convergence of empirical means. Moreover,*

$$q^*(m,\epsilon) \leq 8(mr^2)^{\lceil d\lg(emr/d)\rceil} \exp(-m\epsilon^2/32), \qquad (7.1.4)$$

where $d = $ F-dim$(\mathcal{F}, \epsilon/4)$ and $r = \lfloor 16R/\epsilon \rfloor$.

Proof. The proof is based on Theorem 4.4 and (5.6.8). By (4.2.18), we have that

$$L(\epsilon/8, \mathcal{F}|_{\mathbf{x}}, \|\cdot\|_{a1}) \leq N(\epsilon/8, \mathcal{F}|_{\mathbf{x}}, \|\cdot\|_{\infty}) \leq 2(mr^2)^{\lceil \lg y \rceil},$$

where $r = \lfloor 16R/\epsilon \rfloor$ and

$$y = \sum_{i=1}^{d} \binom{m}{i} r^i \leq \left(\frac{emr}{d}\right)^d,$$

where the second step comes about by replacing r^i by r^d and then applying Lemma 4.3 and (4.2.8). Therefore

$$\lg y \leq d\lg(emr/d).$$

The desired conclusion now follows from (5.6.8). ∎

7.1.2 Concept Classes

In this subsection we study the problem of the uniform convergence of empirical *probabilities* of a concept class, as opposed to the uniform convergence of empirical *means* of a function class studied in the preceding subsection. Since a concept class is also a function class, the results of Theorem 7.1 continue to apply. However, it is possible to improve upon the estimate (7.1.1) of the rate at which empirical means converge to their true values. Moreover, it can be shown that the finiteness of the VC-dimension is *necessary* as well as sufficient for the uniform convergence property to hold. This is an improvement over Theorem 7.1.

Let us recall the problem under study. Let $\mathcal{A} \subseteq \mathcal{S}$ be a given collection of measurable sets. As before, for a given set $A \in \mathcal{A}$ and multisample $\mathbf{x} = [x_1 \ldots x_m]^t \in X^m$, define

$$\hat{P}(A; \mathbf{x}) = \frac{1}{m}\sum_{j=1}^{m} I_A(x_j)$$

7.1 Uniform Convergence of Empirical Means

to be the empirical probability of A, and define

$$q(m, \epsilon, P) := P^m\{\mathbf{x} \in X^m : \sup_{A \in \mathcal{A}} |\hat{P}(A; \mathbf{x}) - P(A)| > \epsilon\}.$$

Finally, define

$$q^*(m, \epsilon) := \sup_{P \in \mathcal{P}^*} q(m, \epsilon, P).$$

We say that the collection of sets $\mathcal{A}$ has the property of **distribution-free uniform convergence of empirical probabilities** if $q^*(m, \epsilon) \to 0$ as $m \to \infty$ for each $\epsilon > 0$.

Theorem 7.4. 1. *Suppose $\mathcal{A}$ has finite VC-dimension, say d. Then*

$$q^*(m, \epsilon) \leq 4 \left(\frac{2em}{d}\right)^d \exp(-m\epsilon^2/8), \ \forall m, \epsilon. \tag{7.1.5}$$

Thus $\mathcal{A}$ has the properties of distribution-free uniform convergence of empirical probabilities.

2. *Conversely, suppose $\mathcal{A}$ has the property of distribution-free uniform convergence of empirical probabilities; then the VC-dimension of $\mathcal{A}$ is finite.*

Remarks: Since one can identify the collection $\mathcal{A}$ with the corresponding family of $\{0,1\}$-valued functions $\{I_A(\cdot), A \in \mathcal{A}\}$, and the P-dimension of this family is the same as the VC-dimension of $\mathcal{A}$, it follows that Theorems 7.1 and 7.2 apply also to the case of concept classes. Thus the bounds (7.1.1) and (7.1.3) hold in the present instance as well. Note that the bound in (7.1.3) looks very similar to (7.1.5). The bound (7.1.5) is somewhat better than (7.1.1), because the exponent in the former equation is $-m\epsilon^2/8$ compared to the exponent of $-m\epsilon^2/32$ in the latter equation. On the other hand, (7.1.5) contains a *polynomial in m* premultiplying the decaying exponential, whereas (7.1.1) contains only a constant term.

Proof. **Statement 1.** The proof simply follows that of Theorem 5.7 with only the last step being different. For the convenience of the reader, the various steps are reproduced below. Note that the notation in the various steps has been recast to reflect the fact that we are dealing with concept classes, whereas Theorem 5.7 deals with function classes.
Step 1. Suppose $m \geq 2/\epsilon^2$. Define the sets

$$Q := \{\mathbf{x} \in X^m : \exists A \in \mathcal{A} \text{ s.t. } |\hat{P}(A; \mathbf{x}) - P(A)| > \epsilon\}, \text{ and}$$

$$R := \{\mathbf{xy} \in X^{2m} : \exists A \in \mathcal{A} \text{ s.t. } |\hat{P}(A; \mathbf{x}) - \hat{P}(A; \mathbf{y})| > \epsilon/2\}.$$

Then it is claimed that

$$P^m(Q) \leq 2P^{2m}(R).$$

This step is the same as in Theorem 5.7, and the same proof applies here as well.

Step 2. Let Γ_m denote the set of permutations γ on $\{1, \ldots, 2m\}$ such that, for each $i \in \{1, \ldots, m\}$, either $\gamma(i) = i$ and $\gamma(m+i) = m+i$, or else $\gamma(i) = m+i$ and $\gamma(m+i) = i$. Thus Γ_m consists of all permutations that swap some (or all, or no) indices $i \in \{1, \ldots, m\}$ with $m+i$. Clearly there are 2^m permutations in Γ_m. Now it is claimed that

$$P^{2m}(R) = \int_{X^{2m}} \frac{1}{2^m} \sum_{\gamma \in \Gamma_m} I_R(\gamma \mathbf{z}) \, P^{2m}(d\mathbf{z}). \tag{7.1.6}$$

This step is also the same as in Theorem 5.7, and the same proof applies here as well.

Step 3. Suppose $\mathbf{a} \in [-1,1]^m$. Then it is claimed that the number of vectors $\mathbf{s} \in \{-1,1\}^m$ such that $|\mathbf{s}^t \mathbf{a}| > m\epsilon/2$ is at most equal to $2^m \cdot 2e^{-m\epsilon^2/8}$. This step is also the same as in Theorem 5.7, except that $\epsilon/4$ has been replaced by $\epsilon/2$, and the bound has been amended correspondingly.

Step 4. For each $\mathbf{z} \in X^{2m}$, it is claimed that the integrand in (7.1.6) is not more than

$$2 \left(\frac{2em}{d}\right)^d \exp(-m\epsilon^2/8).$$

This is the step that is a little different. Fix $\mathbf{z} \in X^{2m}$. Then the number of distinct vectors of the form $\mathbf{a}(\mathbf{z}) := [I_A(z_1) \ldots I_A(z_{2m})]^t \in \{0,1\}^{2m}$ generated by varying A over $\mathcal{A}$ can be bounded by $\pi(2m; \mathcal{A})$, which in turn can be bounded by $(2em/d)^d$ using Theorem 4.1. For each such distinct vector $\mathbf{a}(\mathbf{z})$, the number of vectors $\mathbf{s} \in \{-1,1\}^m$ such that $|\mathbf{s}^t \mathbf{a}(\mathbf{z})| > m\epsilon/2$ is at most equal to $2^m \cdot 2e^{-m\epsilon^2/8}$, by Step 3. Hence the number of vectors $\mathbf{s} \in \{-1,1\}^m$ such that *there exists some vector* $\mathbf{a}(\mathbf{z})$ such that $|\mathbf{s}^t \mathbf{a}(\mathbf{z})| > m\epsilon/2$ is at most equal to

$$2^m \cdot 2 \left(\frac{2em}{d}\right)^d \exp(-m\epsilon^2/8).$$

Dividing through by 2^m proves the claim.

Now the proof of the inequality (7.1.5) consists of combining the above steps (see the proof of Theorem Theorem 5.7 for complete details). From the inequality (7.1.5), it follows that $q^*(m, \epsilon) \to 0$ as $m \to \infty$ for each $\epsilon > 0$. Hence the collection of sets $\mathcal{A}$ has the property of distribution-free uniform convergence of empirical probabilities.

Statement 2. We appeal to Theorem 5.6. Suppose $\mathcal{A}$ has infinite VC-dimension. It is shown that the condition (5.3.8) *fails* to hold with $\mathcal{P} = \mathcal{P}^*$, so that $\mathcal{A}$ does not have the property of distribution-free uniform convergence of empirical probabilities. Let m be arbitrary. By assumption, there exists a set $S = \{y_1, \ldots, y_m\}$ of cardinality m that is shattered by $\mathcal{A}$. Now choose P_m to be the purely atomic measure concentrated uniformly on the set S. Thus P_m assigns a weight of $1/m$ to each y_i and a weight of zero to all other points in X. Now choose a vector $\mathbf{x} \in X^m$ at random in accordance with P_m. Then every component of $\mathbf{x}$ belongs to S with probability one. Further, the integer $d(\mathbf{x})$ (defined as the VC-dimension of the collection $\mathcal{A}$ intersected with the set $\{x_1, \ldots, x_m\}$) equals the number of *distinct* components of $\mathbf{x}$. Accordingly, let us obtain a lower bound for $d(\mathbf{x})$.

Let $k < m$, and let T equal a *fixed* subset of S containing k elements. Then the probability that a randomly chosen x_i belongs to T is k/m, whence the probability that *all* x_i belong to T is $(k/m)^m$. Now let $\mathcal{T}$ denote the collection of all subsets of S containing exactly k elements. There are $\binom{m}{k}$ such sets, and they are not disjoint in general. Each such subset has a probability measure of $(k/m)^m$. Hence the union of all subsets in $\mathcal{T}$ has measure no larger than $\binom{m}{k}(k/m)^m$. By Stirling's approximation, $k! \geq k^k e^{-k}$, so that

$$\binom{m}{k} = \frac{m(m-1)\ldots(m-k+1)}{k!} \leq \frac{m^k e^k}{k^k}.$$

By the previous discussion, the probability that $\mathbf{x}$ contains no more than k distinct elements is no larger than $\binom{m}{k}(k/m)^m$. Hence the probability that $\mathbf{x}$ contains at least k distinct elements is at least equal to

$$1 - \frac{m^k e^k}{k^k}\left(\frac{k}{m}\right)^m = 1 - \left(\frac{k}{m}\right)^{m-k} e^k.$$

Now suppose $m = 3k$. Then $d(\mathbf{x}) \geq k$ with a probability at least

$$1 - \left(\frac{1}{3}\right)^{2k} e^k = 1 - \left(\frac{e}{9}\right)^k.$$

Therefore

$$\frac{E_{P^m}[d(\mathbf{x})]}{m} \geq \frac{1 - (e/9)^{m/3}}{3},$$

and as a result,

$$\sup_{P \in \mathcal{P}^*} \frac{E_{P^m}[d(\mathbf{x})]}{m} \geq \frac{1 - (e/9)^{m/3}}{3},$$

$$\lim_{m \to \infty} \sup_{P \in \mathcal{P}^*} \frac{E_{P^m}[d(\mathbf{x})]}{m} \geq \frac{1}{3}.$$

Thus the condition (5.3.8) is violated. ∎

7.1.3 Loss Functions

In this section, we use the results of Section 5.8 to obtain *explicit* estimates of the rates at which empirical means (or probabilities) of loss functions approach their true values, in the case where the underlying hypothesis space

262   7. Distribution-Free Learning

has finite VC-dimension or finite P-dimension. In turn, these results enable us to obtain sample complexity estimates for model-free learning in the distribution-free case.

We begin with the case of real-valued hypothesis functions. All notation is as in Section 5.8.

Theorem 7.5. *Suppose the family $\mathcal{H}$ has finite P-dimension, say d, and that the loss function ℓ satisfies the uniform Lipschitz condition*

$$|\ell(y, u_1) - \ell(y, u_2)| \leq \mu |u_1 - u_2|, \ \forall y, u_1, u_2 \in [0, 1]$$

for some constant μ. Then the family $\mathcal{L}_\mathcal{H}$ has the property of distribution-free uniform convergence of empirical means. Moreover,

$$q^*(m, \epsilon, \mathcal{L}_\mathcal{H}) \leq 8 \left(\frac{16e\mu}{\epsilon} \ln \frac{16e\mu}{\epsilon} \right)^d \exp(-m\epsilon^2/32), \ \forall m, \epsilon.$$

Remarks: The notation $q^*(m, \epsilon, \mathcal{L}_\mathcal{H})$ is used to make explicit which family of functions is under consideration. Theorem 7.1 implies that

$$q^*(m, \epsilon, \mathcal{H}) \leq 8 \left(\frac{16e}{\epsilon} \ln \frac{16e}{\epsilon} \right)^d \exp(-m\epsilon^2/32), \ \forall m, \epsilon.$$

Hence the only difference between the two bounds is the presence of the constant μ in the $(\cdot)^d$ term (but *not* in the exponentially decaying term!). For the commonly used loss function $\ell(y, u) = |y - u|$, one can take $\mu = 1$, in which case both bounds are identical.

Proof. As in the proof of Theorem 7.1 above, it follows that, for each $\mathbf{x} \in X^{2m}$, we have

$$L(\epsilon, \mathcal{H}|_\mathbf{x}, \|\cdot\|_{a1}) \leq 2 \left(\frac{2e}{\epsilon} \ln \frac{2e}{\epsilon} \right)^d.$$

Now it is shown that for each $\mathbf{z} = (\mathbf{x}, \mathbf{y}) \in (X \times Y)^{2m}$, we have

$$L(\epsilon, \mathcal{L}_\mathcal{H}|_\mathbf{z}, \|\cdot\|_{a1}) \leq 2 \left(\frac{2e\mu}{\epsilon} \ln \frac{2e\mu}{\epsilon} \right)^d. \quad (7.1.7)$$

This is achieved by showing that, if $\{\mathbf{v}^1, \ldots, \mathbf{v}^k\}$ is an external ϵ/μ-cover for the set $\mathcal{H}|_\mathbf{x}$ with respect to $\|\cdot\|_{a1}$, then the set of k vectors $\{\mathbf{w}^1, \ldots, \mathbf{w}^k\}$ defined by

$$w_i^j := \ell(y_i, v_i^j), \ 1 \leq i \leq 2m, \ 1 \leq j \leq k,$$

is an external ϵ-cover for $\mathcal{L}_\mathcal{H}|_\mathbf{z}$ with respect to $\|\cdot\|_{a1}$. This is a direct consequence of the Lipschitz condition satisfied by the loss function ℓ. Suppose $h \in \mathcal{H}$ is arbitrary. Then by assumption there exists an index j such that

$$\|\mathbf{h}(\mathbf{x}) - \mathbf{v}^j\|_{a1} \leq \epsilon/\mu,$$

or equivalently
$$\frac{1}{2m}\sum_{i=1}^{2m}|h(x_i)-v_i^j|\le \epsilon/\mu.$$
Now the Lipschitz condition implies that
$$\frac{1}{2m}\sum_{i=1}^{2m}|\ell(y_i,h(x_i))-\ell(y_i,v_i^j)|\le \frac{1}{2m}\sum_{i=1}^{2m}\mu\,|h(x_i)-v_i^j|\le \epsilon.$$
This establishes the bound (7.1.7). Now substitute this inequality into (5.6.8) and let $\bar{P}$ be an arbitrary probability measure on $X\times Y$. Disregarding the "min" in the right side of (5.6.8) leads to
$$q(m,\epsilon,\bar{P},\mathcal{L}_\mathcal{H})\le 4E_{\bar{P}^{2m}}[L(\epsilon/8,\mathcal{L}_\mathcal{H}|_\mathbf{z},\|\cdot\|_{a1})]\,\exp(-m\epsilon^2/32)$$
$$\le 8\left(\frac{16e\mu}{\epsilon}\ln\frac{16e\mu}{\epsilon}\right)^d\exp(-m\epsilon^2/32).$$
Since the right side of the inequality is independent of $\bar{P}$, the desired conclusion follows. ∎

In the case where $Y=U=\{0,1\}$, so that the hypothesis class consists of *binary-valued* functions, and the loss function is $\ell(y,u)=|y-u|$, Theorem 7.5 can be somewhat improved.

Theorem 7.6. *Suppose $Y=U=\{0,1\}$, $\ell(y,u)=|y-u|$, and that $\mathcal{H}$ has finite VC-dimension, say d. Then the family of loss functions $\mathcal{L}_\mathcal{H}$ has the property of distribution-free uniform convergence of empirical means. Moreover,*
$$q^*(m,\epsilon,\mathcal{L}_\mathcal{H})\le 4\left(\frac{2em}{d}\right)^d\exp(-m\epsilon^2/8),\ \forall m,\epsilon.$$

Proof. As shown in the proof of Theorem 5.12, we have
$$\text{VC-dim}(\mathcal{L}_\mathcal{H})=\text{VC-dim}(\mathcal{H}).$$
The desired conclusion now follows from Theorem 7.4, and in particular, (7.1.5). ∎

7.2 Function Learning

In this section we study the function learning problem in a distribution-free setting. The problem of learning concept classes, as opposed to function classes, is taken up in the next section. Only PUAC and PAC learning problems are studied here. Note that the model-free learning problem need not be studied separately, since Theorems 7.5 and 3.1 together show that if (i) the hypothesis class $\mathcal{H}$ has finite P-dimension and (ii) the loss function is uniformly Lipschitz-continuous, then every algorithm that nearly minimizes empirical risk with high probability is PAC.

7.2.1 Finite P-Dimension Implies PAC and PUAC Learnability

In Theorem 6.1 it is shown that if a family $\mathcal{F}$ has the property that empirical means converge uniformly, then $\mathcal{F}$ is also PUAC learnable. In particular, every consistent algorithm is PUAC (and, of course, PAC). A perusal of the proof of Theorem 6.1 reveals that the theorem continues to be valid if the hypothesis is strengthened to the assumption that empirical means converge uniformly also with respect to the probability measure. This leads to the following result.

Theorem 7.7. *Suppose the family $\mathcal{F} \subseteq [0,1]^X$ has finite P-dimension; then $\mathcal{F}$ is PUAC learnable with respect to $\mathcal{P}^*$. Let d denote the P-dimension of $\mathcal{F}$, and let $\{A_m\}$ be any consistent algorithm. Then the algorithm is PUAC. Moreover, the quantity $r(m, \epsilon)$ defined in (3.2.1) is bounded by*

$$r(m,\epsilon) \leq 8 \left(\frac{32e}{\epsilon} \ln \frac{32e}{\epsilon} \right)^d \exp(-m\epsilon/32). \quad (7.2.1)$$

Any consistent algorithm is PAC to accuracy ϵ and confidence δ provided at least

$$m \geq \frac{32}{\epsilon} \left[d \left(\ln \frac{32e}{\epsilon} \ln \ln \frac{32e}{\epsilon} \right) + \ln \frac{8}{\delta} \right] \quad (7.2.2)$$

samples are used. Similarly, the quantity $s(m, \epsilon)$ defined in (3.2.13) is bounded by

$$s(m,\epsilon) \leq 8 \left(\frac{64e}{\epsilon} \ln \frac{64e}{\epsilon} \right)^{2d} \exp(-m\epsilon/32). \quad (7.2.3)$$

Any consistent algorithm is PUAC to accuracy ϵ and confidence δ provided at least

$$m \geq \frac{32}{\epsilon} \left[2d \left(\ln \frac{64e}{\epsilon} \ln \ln \frac{64e}{\epsilon} \right) + \ln \frac{8}{\delta} \right] \quad (7.2.4)$$

samples are used.

Proof. It follows from Theorem 7.1 that $\mathcal{F}$ has the UCEMUP property with respect to $\mathcal{P}^*$. By mimicking the proof of Theorem 6.1 and permitting P to vary over $\mathcal{P}^*$, one can show that $\mathcal{F}$ is also PUAC learnable, and that every consistent algorithm is PUAC. Thus all that remains is to derive the bounds for $r(m, \epsilon)$ and $s(m, \epsilon)$, and the corresponding sample complexity estimates.

Let us first derive the bound (7.2.1) for $r(m, \epsilon)$. Suppose $f \in \mathcal{F}$ is a fixed but arbitrary target function. For each $g \in \mathcal{F}$, define the function $f \Delta g : X \to [0,1]$ by[1]

$$(f \Delta g)(x) := |f(x) - g(x)|, \; \forall x \in X.$$

---

[1] The notation $f \Delta g$ can be justified by the fact that if f and g are indicator functions of the sets A and B respectively, then $f \Delta g$ is the indicator function of the set $A \Delta B$.

Then $f \Delta g$ is also measurable. Now define

$$f \Delta \mathcal{F} := \{f \Delta g : g \in \mathcal{F}\}.$$

Let $\{A_m\}$ be any consistent algorithm, and let $h_m(f; \mathbf{x})$ denote the hypothesis produced by the algorithm after m steps when the multisample is $\mathbf{x}$; that is,

$$h_m(f; \mathbf{x}) = A_m[(x_1, f(x_1)), \ldots, (x_m, f(x_m))].$$

Now the function $f \Delta h_m$ belongs to $f \Delta \mathcal{F}$ (where h_m is a shorthand for $h_m(f; \mathbf{x})$). Moreover, since the algorithm is consistent, we have that

$$h_m(x_i) = f(x_i), \; i = 1, \ldots, m.$$

With this information, the probability that $d_P(f, h_m) > \epsilon$ can be estimated using the bound (5.6.9).

Recall the quantity $v(m, \epsilon)$ defined in (5.6.3), namely

$$v(m, \epsilon, f \Delta \mathcal{F}) := P^m \{\mathbf{x} \in X^m : \exists g \in f \Delta \mathcal{F} \text{ s.t. } \hat{E}(g; \mathbf{x}) = 0 \text{ and } E_P(g) > \epsilon\}.$$

Note that the notation $v(m, \epsilon, f \Delta \mathcal{F})$ is intended to make clear which family of functions is under consideration. For *any* consistent algorithm, the function $g := f \Delta h_m(f; \mathbf{x})$ belongs to $f \Delta \mathcal{F}$ and satisfies $\hat{E}(g; \mathbf{x}) = 0$. Hence

$$r(m, \epsilon) \leq v(m, \epsilon, f \Delta \mathcal{F})$$

for every consistent algorithm. Now it follows from (5.6.9), after disregarding the "min" on the right side of the inequality, that

$$v(m, \epsilon, f \Delta \mathcal{F}) \leq 4 E_{P^{2m}} [L(\epsilon/16, (f \Delta \mathcal{F})|_\mathbf{z}, \| \cdot \|_{a1})] \cdot \exp(-m\epsilon/32). \quad (7.2.5)$$

In order to apply this inequality, let us estimate the covering number $L[\epsilon/16, (f \Delta \mathcal{F})|_\mathbf{z}, \| \cdot \|_{a1}]$. It is claimed that, for each $\mathbf{z} \in X^{2m}$,

$$L[\epsilon, (f \Delta \mathcal{F})|_\mathbf{z}, \| \cdot \|_{a1}] \leq L(\epsilon, \mathcal{F}|_\mathbf{z}, \| \cdot \|_{a1}). \quad (7.2.6)$$

To show this, suppose $\mathbf{v}^1, \ldots, \mathbf{v}^L$ is a minimal external ϵ-cover for $\mathcal{F}|_\mathbf{z}$. Then it is claimed that the set of L vectors

$$\mathbf{f}(\mathbf{z}) \Delta \mathbf{v}^1, \ldots, \mathbf{f}(\mathbf{z}) \Delta \mathbf{v}^L$$

is an external ϵ-cover for $(f \Delta \mathcal{F})|_\mathbf{z}$. Here we use the obvious notation for the "symmetric difference" of two vectors: If $\mathbf{u}, \mathbf{v} \in [0, 1]^{2m}$, then $\mathbf{u} \Delta \mathbf{v} \in [0, 1]^{2m}$ is defined by

$$(\mathbf{u} \Delta \mathbf{v})_i = |u_i - v_i|.$$

To establish this last claim, which in turn proves (7.2.6), suppose $g \in \mathcal{F}$ is arbitrary, and select an index j such that $\| \mathbf{g}(\mathbf{z}) - \mathbf{v}^j \|_{a1} \leq \epsilon$. Then

$$\| (f\Delta g)(\mathbf{z}) - \mathbf{f}(\mathbf{z})\Delta \mathbf{v}^j \|_{a1} = \frac{1}{2m} \sum_{i=1}^{2m} \left| |f(z_i) - g(z_i)| - |f(z_i) - v_i^j| \right|$$

$$\leq \frac{1}{2m} \sum_{i=1}^{2m} |g(z_i) - v_i^j| = \| \mathbf{g}(\mathbf{z}) - \mathbf{v}^j \|_{a1} \leq \epsilon.$$

This establishes (7.2.6). Next, from (7.2.6) and (7.1.2), it follows that

$$L[\epsilon, (f\Delta\mathcal{F})|_{\mathbf{z}}, \|\cdot\|_{a1}] \leq 2\left(\frac{2e}{\epsilon} \ln \frac{2e}{\epsilon}\right)^d, \quad \forall \mathbf{z} \in X^{2m}, \forall m.$$

Substituting the above into (7.2.5) and noting that $r(m,\epsilon) \leq v(m,\epsilon, f\Delta\mathcal{F})$ establishes the bound (7.2.1). To derive the sample complexity bound (7.2.2), suppose ϵ and δ are specified. It is enough to choose m such that

$$8\left(\frac{32e}{\epsilon} \ln \frac{32e}{\epsilon}\right)^d \exp(-m\epsilon/32) \leq \delta,$$

or equivalently,

$$\exp(m\epsilon/32) \geq \frac{8}{\delta}\left(\frac{32e}{\epsilon} \ln \frac{32e}{\epsilon}\right)^d.$$

The bound (7.2.2) follows readily from the above inequality.

The bound for $s(m,\epsilon)$ is proved in an entirely analogous manner; as such, only a sketch of the proof is given here. Given $\mathcal{F}$, define

$$\mathcal{F}\Delta\mathcal{F} := \{f\Delta g : f, g \in \mathcal{F}\}.$$

Then, since $\hat{d}[f, h_m(f; \mathbf{x})] = 0$ for all $f \in \mathcal{F}$, all $\mathbf{x} \in X^m$, and every consistent algorithm $\{A_m\}$, it follows that

$$s(m,\epsilon) \leq v(m,\epsilon, \mathcal{F}\Delta\mathcal{F}).$$

Next, it is claimed that for each $\mathbf{z} \in X^{2m}$, we have

$$L[\epsilon, (\mathcal{F}\Delta\mathcal{F})|_{\mathbf{z}}, \|\cdot\|_{a1}] \leq [L(\epsilon/2, \mathcal{F}|_{\mathbf{z}}, \|\cdot\|_{a1})]^2.$$

To prove this, suppose $\mathbf{v}^1, \ldots, \mathbf{v}^L$ is a minimal external $\epsilon/2$-cover for $\mathcal{F}|_{\mathbf{z}}$. Then it is claimed that the set of L^2 vectors

$$\mathbf{v}^j \Delta \mathbf{v}^k, \quad 1 \leq j, k \leq L$$

forms an external ϵ-cover for $(\mathcal{F}\Delta\mathcal{F})|_{\mathbf{z}}$. To see this, suppose $f, g \in \mathcal{F}$ are arbitrary, and select indices j, k such that

$$\| \mathbf{f}(\mathbf{z}) - \mathbf{v}^j \|_{a1} \leq \epsilon/2, \text{ and } \| \mathbf{g}(\mathbf{z}) - \mathbf{v}^k \|_{a1} \leq \epsilon/2.$$

Then

7.2 Function Learning   267

$$\| (\mathbf{f}\Delta\mathbf{g})(\mathbf{z}) - \mathbf{v}^j \Delta \mathbf{v}^k \|_{a1} = \frac{1}{2m}\sum_{i=1}^{2m}\Big||f(z_i) - g(z_i)| - |v_i^j - v_i^k|\Big|$$

$$\leq \frac{1}{2m}\sum_{i=1}^{2m}|f(z_i) - v_i^j| + \frac{1}{2m}\sum_{i=1}^{2m}|g(z_i) - v_i^k|$$

$$\leq \| \mathbf{f}(\mathbf{z}) - \mathbf{v}^j \|_{a1} + \| \mathbf{g}(\mathbf{z}) - \mathbf{v}^k \|_{a1} \leq \epsilon/2 + \epsilon/2 = \epsilon.$$

Thus it follows from (5.6.9) that

$$v(m,\epsilon,\mathcal{F}\Delta\mathcal{F}) \leq 4E_{P^{2m}}\left\{[L(\epsilon/32,\mathcal{F}|_{\mathbf{z}},\|\cdot\|_{a1})]^2\right\}\cdot \exp(-m\epsilon/32).$$

Substituting for $L(\epsilon/32,\mathcal{F}|_{\mathbf{z}},\|\cdot\|_{a1})$ from (7.1.1) leads to

$$v(m,\epsilon,\mathcal{F}\Delta\mathcal{F}) \leq 8\left(\frac{64e}{\epsilon}\ln\frac{64e}{\epsilon}\right)^{2d}\exp(-m\epsilon/32).$$

This is enough to prove (7.2.3). The sample complexity estimate follows as before. ∎

7.2.2 Finite P-Dimension is not Necessary for PAC Learnability

In the preceding subsection, we have seen (cf. Theorem 7.7) that if a function class has finite P-dimension, then it is distribution-free PUAC learnable. Thus it is natural to ask whether the converse is true; that is: Does the distribution-free PUAC (or PAC) learnability of a function class imply that the function class has finite P-dimension? It is shown here through example that the answer is "No." There exists a function class that is distribution-free PUAC learnable, and yet has infinite P-dimension.

Example 7.1. (M. Krichman) Let $X = [1,\infty)$, $\mathcal{S}$ = the Borel σ-algebra on X, and let $\mathcal{P}^*$ denote the set of all probability measures on the pair $(X,\mathcal{S})$. We construct a function family that maps X into $[-2,2]$ that has infinite P-dimension, and yet is PUAC learnable. In fact, it is *exactly* learnable by drawing a *single* random sample. By defining $g(x) = (f(x)+2)/4$, the family of functions can be rescaled into another family that maps X into $[0,1]$. However, to keep the formulae "clean," this rescaling is not done here.

Define a bijection between $[0,1)$ and the countably infinite Cartesian product $\{0,1\}^{\mathbf{N}}$ minus all sequences that end with an infinite string of 1's in the usual manner, namely:

$$q(\xi) := \sum_{i=1}^{\infty} e_i^{\xi} 2^{-i}, \ \forall \xi \in [0,1).$$

For each $\xi \in [0,1)$, write $q(\xi) = \{e_i^{\xi}\}_{i=1}^{\infty}$, and let $q(\xi,i)$ denote e_i^{ξ}. For each $x \in [1,\infty)$, let $\lfloor x \rfloor$ denote the integer part of x, and let $\{x\}$ denote $x - \lfloor x \rfloor$, the noninteger part of x. Now define the function $a : [1,\infty) \to [-2,2]$ by

$$a(x) = \begin{cases} 1 + \{x\} & \text{if } q(\{x\}, \lfloor x \rfloor) = 1, \text{ and} \\ -1 - \{x\} & \text{if } q(\{x\}, \lfloor x \rfloor) = 0. \end{cases}$$

A graph of the function $a(\cdot)$ is shown in Figure 7.1. Note that, in the interval

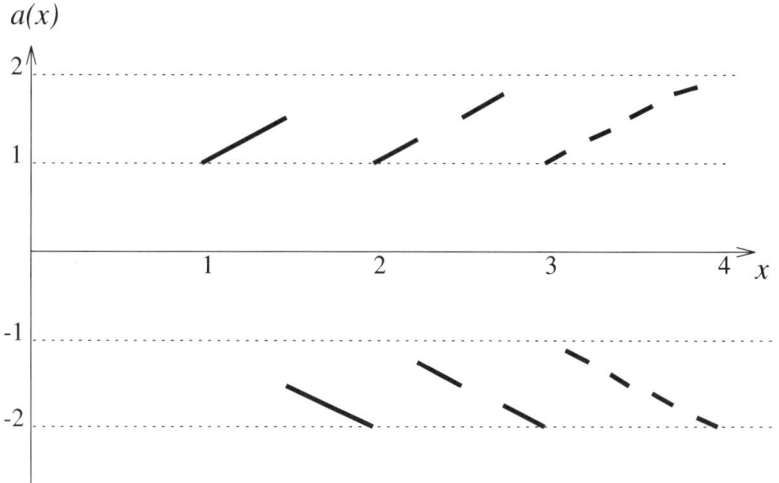

Fig. 7.1. A Learnable Function Class with Infinite P-Dimension

$[n, n+1]$, the function $a(x)$ changes sign 2^n times, and each "segment" has width 2^{-n}. Also note that if $a(x) \geq 1$, then $|a(x) - 1| = \{x\}$, whereas if $a(x) \leq -1$, then $|a(x) + 1| = \{x\}$. Thus, in either case, given the value of $a(x)$, it is possible to determine $\{x\}$ uniquely.

Now define the family of functions $\mathcal{F}$ as $\{a(x + \xi), \xi \in [0, 1)\}$. Thus $\mathcal{F}$ consists of the functions obtained by translating $a(\cdot)$ by every real number $\xi \in [0, 1)$. For notational convenience, denote the function $x \mapsto a(x + \xi)$ by a_ξ. It is claimed that (i) $\mathcal{F}$ has infinite P-dimension, and (ii) $\mathcal{F}$ is PUAC learnable. To establish the first claim, let n be an arbitrary integer, and define $S = \{1, \ldots, n\}$. It is shown that S is P-shattered by $\mathcal{F}$. Let $\mathbf{0}_n$ denote the n-dimensional zero vector. Given any binary vector $\mathbf{e} \in \{0, 1\}^n$, define

$$\xi := \sum_{i=1}^{n} e_i 2^{-i}.$$

Then the function a_ξ satisfies $a_\xi(i) = 1$ if $e_i = 1$, and $a_\xi(i) = -1$ if $e_i = 0$. Thus S is P-shattered by $\mathcal{F}$. Since n is arbitrary, $\mathcal{F}$ has infinite P-dimension. In case it is desired to scale the functions so that they map X into $[0, 1]$, the same scaling must be applied to the zero vector as well. Thus the family of functions $\{(a_\xi + 2)/4\}$ P-shatters S if the vector $\mathbf{c}$ is chosen as the vector whose components are all equal to 0.25. To establish the second claim, pick a single sample $x_0 \in X$; then the oracle returns the number $r := a_\xi(x_0) = a(\xi + x_0)$, where $\xi \in [0, 1)$ is the unknown number to be "learnt." (Note that learning the unknown target function is equivalent to learning the number ξ.) However, as shown above, given the value $a(\xi + x_0)$ one can uniquely determine the value of $\{x_0 + \xi\}$; and since x_0 is known, this is enough to determine ξ uniquely. Hence the map from ξ to r is one-to-one. Therefore a single measurement at a random value of x suffices to determine ξ, and therefore the target function, *exactly*. This shows that $\mathcal{F}$ is PUAC learnable.

7.3 Concept Learning

In this section, we derive several useful results on the problem of concept learning in a distribution-free setting. First, the sample complexity bound of Theorem 7.7 is improved, and a better bound is obtained. Second, a universal *lower* bound on the number of samples needed for learning is obtained. In contrast to the case of concept learning under a fixed probability measure (studied in Chapter 6), in the present case the upper and lower bounds grow at roughly the same rate. Next, it is shown that the finiteness of the VC-dimension is a *necessary* condition for a concept class to be learnable in a distribution-free setting. This is in contrast with the situation in Section 7.2, where it is shown that a function class need not have finite P-dimension in order to be PAC (or even PUAC) learnable. The model-free learning problem is not discussed here, since Theorems 7.6 and 3.1 together show that if (i) a hypothesis class $\mathcal{H}$ consisting of binary-valued functions has finite VC-dimension, and (ii) the loss function equals $\ell(y, u) = |y - u|$, then every algorithm that nearly minimizes empirical risk with high probability is PAC.

7.3.1 Improved Upper Bound for the Sample Complexity

Suppose $\mathcal{C}$ is a concept class and that $\mathcal{C}$ has finite VC-dimension. Then, by Theorem 7.7, every consistent algorithm PAC learns $\mathcal{C}$, and moreover, the error bound (7.2.1) applies. In this subsection, improved bounds are obtained for $r(m, \epsilon)$ and the sample complexity of a learning algorithm, taking advantage of the fact that the learning problem involves concepts and not functions. As for PUAC (as opposed to PAC) learning, the bounds (7.2.3) for $s(m, \epsilon)$ and (7.2.4) for the sample complexity bound (7.2.4) continue to apply. It would be interesting to see whether these bounds could be improved when the problem at hand is one of concept learning.

Theorem 7.8. *Suppose the concept class $\mathcal{C}$ has finite VC-dimension; then $\mathcal{C}$ is learnable with respect to $\mathcal{P}^*$. Let d denote the VC-dimension of $\mathcal{C}$, and let $\{A_m\}$ be any consistent algorithm. Then the algorithm is PAC; moreover, the quantity $r(m, \epsilon)$ defined in (3.2.1) is bounded by*

270   7. Distribution-Free Learning

$$r(m, \epsilon) \leq 2 \left(\frac{2em}{d} \right)^d 2^{-m\epsilon/2}. \tag{7.3.1}$$

Any consistent algorithm learns to accuracy ϵ and confidence δ provided at least

$$m \geq \max \left\{ \frac{8d}{\epsilon} \lg \frac{8e}{\epsilon}, \frac{4}{\epsilon} \lg \frac{2}{\delta} \right\} \tag{7.3.2}$$

samples are used.

Remark Note that the bound (7.2.3) continues to apply to the quantity $s(m, \epsilon)$.

Proof. The only thing that needs proof is the revised bound for $r(m, \epsilon)$. Let $T \in \mathcal{C}$ be a fixed target concept, and let $P \in \mathcal{P}^*$ be a fixed probability measure. Define two sets as follows:

$$Q = Q(m, \epsilon) := \{ \mathbf{x} \in X^m : \exists H \in \mathcal{C} \text{ consistent with } T \text{ on } \mathbf{x},$$
$$\text{and } d_P(T, H) > \epsilon \}$$

and

$$R = R(m, \epsilon) := \{ \mathbf{xy} \in X^{2m} : \exists H \in \mathcal{C} \text{ consistent with } T \text{ on } \mathbf{x}, d_P(T, H) > \epsilon$$
$$\text{and } I_H(y_i) \neq I_T(y_i) \text{ for at least } m\epsilon/2 \text{ indices } i \}.$$

Note that $Q \subseteq X^m$ while $R \subseteq X^{2m}$. We can think of Q as the set of "malicious" samples: Even if a hypothesis H correctly classifies all components of a multisample $\mathbf{x} \in X^m$, it may still be more than ϵ-far away from the target concept T. Thus, if it could be established that $P^m(Q)$ is bounded by the right side of (7.3.1) for every $T \in \mathcal{C}$ and every $P \in \mathcal{P}^*$, it would then follow that for any consistent algorithm, the bound (7.3.1) holds.

Step 1. It is claimed that

$$P^m(Q) \leq 2P^{2m}(R) \text{ if } m > 8/\epsilon.$$

To establish this claim, observe that if $\mathbf{xy} \in R$, then $\mathbf{x} \in Q$, and define

$$\psi_{\mathbf{x}}(\mathbf{y}) = \begin{cases} 1 & \text{if } \mathbf{xy} \in R, \\ 0 & \text{if } \mathbf{xy} \notin R. \end{cases}$$

Then

$$I_R(\mathbf{xy}) = I_Q(\mathbf{x}) \cdot \psi_{\mathbf{x}}(\mathbf{y}).$$

Thus

$$P^{2m}(R) = \int_{X^{2m}} I_R(\mathbf{xy}) \, P^{2m}(d\mathbf{x}, d\mathbf{y})$$
$$= \int_{X^m} I_Q(\mathbf{x}) \left[\int_{X^m} \psi_{\mathbf{x}}(\mathbf{y}) \, P^m(d\mathbf{y}) \right] P^m(d\mathbf{x}).$$

Now it is shown that

$$\int_{X^m} \psi_{\mathbf{x}}(\mathbf{y})\, P^m(d\mathbf{y}) \geq 0.5 \; \forall \mathbf{x} \in Q \text{ if } m > 8/\epsilon. \tag{7.3.3}$$

Together with the preceding equation, this is enough to establish the claim.

Suppose $\mathbf{x} \in Q$, and suppose $H \in \mathcal{C}$ satisfies $d_P(H,T) > \epsilon$. Then the probability that H misclassifies a randomly selected $y \in X$ is at least ϵ. The probability that H misclassifies at most $m\epsilon/2$ components of $\mathbf{y} \in X^m$ can be estimated using the multiplicative form of the Chernoff bound, and is no larger than $e^{-m\epsilon/8}$. Hence, if $m > 8/\epsilon$, any H with $d_P(H,T) > \epsilon$ misclassifies at least $m\epsilon/2$ components of $\mathbf{y}$ with a probability of at least $1 - e^{-1} > 0.5$. Hence, whatever be $\mathbf{x} \in Q$, we have that $\psi_{\mathbf{x}}(\mathbf{y}) = 1$ with probability at least 0.5 with respect to $\mathbf{y}$. This proves (7.3.3) and establishes the claim.

Step 2. It is claimed that

$$P^{2m}(R) \leq \phi(2m, d)\, 2^{-m\epsilon/2},$$

where the function $\phi(\cdot, \cdot)$ is defined in (4.2.2).

To establish this claim, let Γ_m denote the family of permutations defined in the proof of Theorem 5.7 in Section 5.6. Thus, Γ_m consists of all permutations γ of $\{1, \ldots, 2m\}$ such that $\gamma(i) = i$ or $\gamma(i) = m+i$ for all $i \in \{1, \ldots, m\}$. In other words, Γ_m consists of those permutations that interchange some (or all, or no) $i \in \{1, \ldots, m\}$ with $m+i$. clearly $|\Gamma_m| = 2^m$. Now as in Step 2 of the proof of Theorem 5.7, it follows that

$$P^{2m}(R) = \int_{X^{2m}} \frac{1}{2^m} \sum_{\gamma \in \Gamma_m} I_R(\gamma \mathbf{z})\, P^{2m}(d\mathbf{z}), \tag{7.3.4}$$

because P^{2m} is a product measure and γ merely relabels the components of $\mathbf{z}$.

The integrand in (7.3.4) can be interpreted as the fraction of permutations $\gamma \in \Gamma_m$ such that $\gamma \mathbf{z} \in R$. Note that $\mathbf{z} \in R$ if and only if there exists a $H \in \mathcal{C}$ such that (i) H correctly classifies the first half of $\mathbf{z}$, and (ii) H misclassifies at least $m\epsilon/2$ components of the second half of $\mathbf{z}$. For a moment, fix $H \in \mathcal{C}$, and suppose H is consistent with T on the first half of $\mathbf{z}$, and misclassifies at least $m\epsilon/2$ cmponents of the second half of $\mathbf{z}$. If γ is a permutation that interchanges an incorrectly classified component of $\mathbf{z}$ with a correctly classified component, then $\gamma \mathbf{z} \notin R$, because H would not be consistent with T on the first half of $\gamma \mathbf{z}$. So the number of permutations (for this particular H) such that H is consistent with T on the first half of $\gamma \mathbf{z}$ is at most $2^{m - m\epsilon/2} = 2^{m(1-\epsilon/2)}$. Next, for a fixed $\mathbf{z} \in X^{2m}$, two concepts $G, H \in \mathcal{C}$ are indistinguishable if $I_G(z_i) = I_H(z_i)$ for all i, $1 \leq i \leq 2m$. Let us denote this equivalence relationship by $\sim$, and replace $\mathcal{C}$ by the collection of equivalence classes $\mathcal{C}/\sim$. Then $\mathcal{C}/\sim$ consists of at most $\pi(2m; \mathcal{C})$ elements, where $\pi(n; \mathcal{A})$ is defined in Chapter 4. Hence, for a fixed $\mathbf{z} \in X^{2m}$, the number of permutations $\gamma \in \Gamma_m$ such that $\gamma \mathbf{z} \in R$ is no more than

272   7. Distribution-Free Learning

$$\pi(2m;\mathcal{C})\, 2^{m(1-\epsilon/2)} \leq \phi(2m,d)\, 2^{m(1-\epsilon/2)},$$

where the last inequality follows from Theorem 4.1 and the fact that the VC-dimension of $\mathcal{C}$ equals d. Hence the integrand in (7.3.4) is no larger than 2^{-m} times the above, which proves the claim.

To complete the proof of the estimate for $r(m,\epsilon)$, observe from (4.2.1) that

$$\phi(2m,d) \leq \left(\frac{2em}{e}\right)^d, \quad \forall m \geq d/2.$$

Hence

$$P^{2m}(R) \leq \left(\frac{2em}{e}\right)^d 2^{-m\epsilon/2},$$

and finally

$$P^m(Q) \leq 2P^{2m}(R) \leq 2\left(\frac{2em}{e}\right)^d 2^{-m\epsilon/2}.$$

Since the above bound holds for *any* target concept $T \in \mathcal{C}$ and *any* probability measure $P \in \mathcal{P}^*$, the bound (7.3.1) is proved. This completes the proof of the first part of the theorem.

To prove the upper bound on the sample complexity, suppose ϵ and δ are given. We would like to choose m such that

$$2\left(\frac{2em}{e}\right)^d 2^{-m\epsilon/2} \leq \delta,$$

or equivalently,

$$2^{m\epsilon/2} \geq \frac{2}{\delta}\left(\frac{2em}{d}\right)^d, \quad \text{or}$$

$$m \geq \frac{2}{\epsilon}\left(d\lg\frac{2em}{d} + \lg\frac{2}{\delta}\right).$$

This inequality is satisfied if

$$\frac{m}{2} \geq \frac{2}{\epsilon}\lg\frac{2}{\delta} \quad \text{and} \quad \frac{m}{2} \geq \frac{2d}{\epsilon}\lg\frac{2em}{d}.$$

The first of these inequalities is satisfied if

$$m \geq \frac{4}{\epsilon}\lg\frac{2}{\delta}. \tag{7.3.5}$$

Rewrite the second inequality as

$$m \geq \frac{4d}{\epsilon}\lg\frac{2em}{d}.$$

By Lemma 4.6, the above inequality is satisfied if

$$m \geq \frac{8d}{\epsilon}\lg\frac{8e}{\epsilon}. \tag{7.3.6}$$

Combining the bounds (7.3.5) and (7.3.6) establishes (7.3.2). ∎

In the above proof, we have used multisamples of length $2m$ and swapped elements from the first half with corresponding elements from the second half. In [10], a more general technique is proposed, namely: A sample of length $m + k$ is drawn, and some of the last k samples are swapped with some of the first m samples. By choosing the constant k in a careful fashion, the sample complexity can be reduced to

$$m_0(\epsilon, \delta) = \frac{1}{\epsilon(1 - \sqrt{\epsilon})} \left[\lg\left(\frac{d/(d-1)}{\delta}\right) + 2d \lg \frac{6}{\epsilon} \right].$$

See [10], Theorem 2.1 for details.

7.3.2 A Universal Lower Bound for the Sample Complexity

In the preceding subsection, we have seen that a concept class with finite VC-dimension (call it d) is learnable by any consistent algorithm, and that it is enough to take

$$m = O\left(\frac{d}{\epsilon} \lg \frac{1}{\epsilon} + \frac{1}{\epsilon} \lg \frac{1}{\delta}\right) \tag{7.3.7}$$

samples. Moreover, if one ignores issues of computational feasibility, a consistent algorithm always exists. Hence the above may be thought of as a "universal" upper bound for the sample complexity. In the present subsection, a corresponding "universal" lower bound of the form

$$m = \Omega\left(\frac{d}{\epsilon} + \frac{1}{\epsilon} \lg \frac{1}{\delta}\right) \tag{7.3.8}$$

is derived. Comparing these two bounds, one observes that the only difference is that the upper bound contains $d/\epsilon \lg(1/\epsilon)$, whereas the lower bound contains d/ϵ. Hence, for all practical purposes, both bounds have essentially the same rate of growth as $\epsilon \to 0^+$.

Let us contrast the situation in the case of fixed-distribution learning (see Chapter 6) with the present. In the case of fixed distribution learning, the upper bound on the sample complexity is of the form $O(1/\epsilon \ln N(\epsilon/2))$, where $N(\epsilon/2)$ denotes the $\epsilon/2$-covering number (cf. (6.3.4); this is of course for a fixed value of δ. Thus, if $N(\epsilon)$ is bounded by a *polynomial* in $1/\epsilon$, this upper bound is of the same order of growth as (7.3.7). On the other hand, the lower bound on the sample complexity given in Theorem 6.6 is $\lg N(\epsilon)$. If $N(\epsilon)$ is bounded by a polynomial in ϵ, this term is $O(\lg(1/\epsilon))$. Thus the upper and lower bounds are of widely disparate orders. The reason for the contrasting behaviour is not far to seek. In the case of fixed-distribution learning, the probability measure P is fixed, whereas in distribution-free learning, we are at full liberty to choose P so as to make learning as difficult as possible. Indeed, a perusal of the proof of Theorem 7.9 below shows that it depends crucially on an "adverse" choice of P.

Let us now state and prove the lower bound. Before that, an assumption is introduced to rule out pathological cases. A concept class $\mathcal{C}$ is said to be **trivial** if either (i) $\mathcal{C}$ contains only a single element (subset of X), or (ii) $\mathcal{C}$ consists of two sets C_1, C_2 such that $C_1 \cap C_2 = \emptyset$ and $C_1 \cup C_2 = X$. In Case (i), no "learning" is necessary, as $\mathcal{C}$ contains only one element. In Case (ii), it is necessary to select only *one* sample, call it x. If $x \in C_1$ and $I_T(x) = 1$, then the target concept is C_1; otherwise the target concept is C_2; the situation is reversed if $x \in C_2$. Thus the sample complexity equals 1 for every ϵ, δ.

Theorem 7.9. *Suppose $\mathcal{C}$ is a concept class of finite VC-dimension $d \geq 2$. Then for $\delta \leq 0.01$, any algorithm requires at least*

$$m = \max\left\{\frac{d-1}{32\epsilon}, \frac{1-\epsilon}{\epsilon}\ln\frac{1}{\delta}\right\} \tag{7.3.9}$$

samples in order to be PAC to accuracy ϵ and confidence δ.

Proof. Note that the assumption that the VC-dimension d is at least two implies that $\mathcal{C}$ is a nontrivial concept class.

First, let us construct a probability measure P such that any algorithm requires at least

$$m_a := \frac{1-\epsilon}{\epsilon}\ln\frac{1}{\delta}$$

samples in order to be PAC to accuracy ϵ and confidence δ.

Since $\mathcal{C}$ is nontrivial, $\mathcal{C}$ contains at least two sets C_1, C_2 such that either (i) $C_1 \cap C_2 \neq \emptyset$, or else (ii) $C_1 \cup C_2 \neq X$. Let us suppose that we are in Case (i), so that $C_1 \cap C_2 \neq \emptyset$. Let $a \in C_1 \cap C_2$, $b \in C_2 \Delta C_1$. Suppose without loss of generality that $b \in C_2$, $b \notin C_1$. If on the other hand $b \in C_1$ and $b \notin C_2$, interchange the labels 1 and 2 below. Choose P to be a purely atomic measure concentrated on $\{a, b\}$, with $p(a) = 1-\epsilon, p(b) = \epsilon$. Since P is concentrated on $\{a, b\}$, one can assume without loss of generality that $X = \{a, b\}$, $C_1 = \{a\}$, and $C_2 = \{a, b\}$. Let us suppose also that $\mathcal{C} = \{C_1, C_2\}$, because learning $\mathcal{C}$ is at least as difficult as learning $\{C_1, C_2\}$. Now suppose that m samples are drawn from X in accordance with P. Then $x_i = a$ for all i with a probability of $(1-\epsilon)^m$. This set of samples is consistent with *both* C_1 and C_2. Thus, whatever be the output of the algorithm to this multisample, the worst-case error of the algorithm is $P(C_1 \Delta C_2) = p(b) = \epsilon$. Hence, with this particular choice of P, we have[2]

$$r(m, \epsilon) \geq (1-\epsilon)^m.$$

Hence, in order for any algorithm to be PAC to accuracy ϵ and confidence δ, it is necessary that $(1-\epsilon)^m \leq \delta$, or equivalently,

---

[2] In order for the next inequality to hold, it is necessary to modify the definition (3.2.1) of $r(m, \epsilon)$ by changing $d_P[T, H_m(T; \mathbf{x})] > \epsilon$ to $d_P[T, H_m(T; \mathbf{x})] \geq \epsilon$. If one wishes to use the definition as given in Chapter 3, the above argument must be adjusted slightly. But this is a minor detail, and the necessary adjustments are easy and are left to the reader.

7.3 Concept Learning

$$m \geq \frac{\ln(1/\delta)}{\ln[1/(1-\epsilon)]}.$$

Now it is readily verified that

$$\frac{1}{\ln[1/(1-\epsilon)]} > \frac{1-\epsilon}{\epsilon}, \ \forall \epsilon \in (0,1).$$

Hence

$$m \geq \frac{1-\epsilon}{\epsilon} \ln \frac{1}{\delta}.$$

This proves the first half of the bound (7.3.9) in Case (i). In Case (ii), $C_1 \cup C_2 \neq X$. Let $a \in X - (C_1 \cup C_2)$, $b \in C_1 \Delta C_2$, and let P be as above. Then, with probability $(1-\epsilon)^m$, the multisample $\mathbf{x}$ consists of all a's, which does not help the algorithm at all in discriminating between C_1 and C_2. The rest of the analysis is exactly as above.

Next, let us construct a probability measure P such that any algorithm requires at least

$$m_b := \frac{d-1}{32\epsilon}$$

samples in order to be PAC to accuracy ϵ and confidence δ.

Since $\mathcal{C}$ has VC-dimension $d \geq 2$, there exists a set $S = \{y_0, y_1, \ldots, y_k\}$ that is shattered by $\mathcal{C}$, where $k = d - 1$. Given ϵ, choose P to be a purely atomic measure concentrated on S, and let

$$p(y_0) = 1 - 8\epsilon, \ p(y_i) = 8\epsilon/k \text{ for } i = 1, \ldots, k.$$

Since P is concentrated on S, we can as well take $X = S$ and also $\mathcal{C} = 2^S$, since S is shattered by $\mathcal{C}$. Now define

$$\mathcal{C}_0 = \{C \in \mathcal{C} : y_0 \in C\}.$$

Learning $\mathcal{C}$ is at least as difficult as learning $\mathcal{C}_0$. So let us suppose $\mathcal{C}_0$ is the concept class to be learnt. Let m denote the number of samples, and let $H(T; \mathbf{x}) \in \mathcal{C}_0$ denote the hypothesis returned by the algorithm when the target concept is $T \in \mathcal{C}_0$ and the multisample is $\mathbf{x} \in X^m$. Finally, let U denote the set of all $\mathbf{x} \in X^m$ such that $\mathbf{x}$ contains at most $k/2$ distinct elements from $\{y_1, \ldots, y_k\}$.

Step 1. Let $p(\mathbf{x}) = \prod_{i=1}^m p(x_i)$ denote the weight of the multisample $\mathbf{x} \in X^m$. Then it is claimed that there exists a concept $T_0 \in \mathcal{C}_0$ such that

$$\sum_{\mathbf{x} \in U} d_P[T_0, H(T_0; \mathbf{x})] \, p(\mathbf{x}) \geq 2\epsilon \, P^m(U).$$

In other words, the claim is that there exists a concept $T_0 \in \mathcal{C}_0$ such that the expected value of the error $d_P[T_0, H(T_0; \mathbf{x})]$ (when $\mathbf{x}$ is restricted to U) is at least 2ϵ.

To prove the claim, let us first *fix* $\mathbf{x} \in U$, and try to estimate the error $d_P[T, H(T; \mathbf{x})]$ as T varies over $\mathcal{C}_0$. For this purpose, let us suppose that each $T \in \mathcal{C}_0$ is equally likely; that is, T is uniformly distributed over $\mathcal{C}_0$, and each $T \in \mathcal{C}_0$ has a probability of 2^{-k}. (Recall that $|\mathcal{C}_0| = 2^k$.) For a fixed $\mathbf{x} \in U$, the hypothesis $H(T; \mathbf{x})$ depends only on the values $I_T(x_i)$ and nothing else.[3] Let us define $T_1, T_2 \in \mathcal{C}_0$ to be equivalent, denoted by $T_1 \sim_\mathbf{x} T_2$, if $I_{T_1}(x_i) = I_{T_2}(x_i)$ for all i. Then $H(T; \mathbf{x})$ depends only on the equivalence class $[T]$. Further, if $l \leq k/2$ denotes the number of distinct elements in $\{y_1, \ldots, y_k\}$ contained in $\mathbf{x}$, then each equivalence class consists of 2^{k-l} elements (sets) of $\mathcal{C}_0$, and there are 2^l such equivalence classes. Now, within each equivalence class $[T]$, the hypothesis $H([T]; \mathbf{x})$ will, on average, disagree with T on exactly half the points in $\{y_1, \ldots, y_k\}$ not contained in $\mathbf{x}$, and possibly also on some points in $\{y_1, \ldots, y_k\}$ contained in $\mathbf{x}$ (unless the algorithm is consistent). Since each y_i has weight $8\epsilon/k$ for $i \geq 1$, we conclude that

$$E_T(d_P[T, H(T; \mathbf{x})]) \geq \frac{1}{2} \frac{8\epsilon}{k}(k - l) = \frac{4\epsilon(k - l)}{k} \geq 2\epsilon,$$

where T denotes the average value with respect to T. Since the above inequality holds for *every* $\mathbf{x} \in U$, it follows that

$$\sum_{\mathbf{x} \in U} E_T(d_P[T, H(T; \mathbf{x})]) \, p(\mathbf{x}) \geq 2\epsilon \sum_{\mathbf{x} \in U} p(\mathbf{x}) = 2\epsilon \, P^m(U).$$

Interchanging the expectation with respect to T and the (finite) summation over $\mathbf{x}$ gives

$$E_T \left(\sum_{\mathbf{x} \in U} d_P[T, H(T; \mathbf{x})] \, p(\mathbf{x}) \right) \geq 2\epsilon \, P^m(U).$$

Hence *there exists* at least one specific concept $T_0 \in \mathcal{C}_0$ such that the integrand of the $E_T(\cdot)$ exceeds the right side; i.e., there exists a $T_0 \in \mathcal{C}_0$ such that

$$\sum_{\mathbf{x} \in U} d_P[T_0, H(T_0; \mathbf{x})] \, p(\mathbf{x}) \geq 2\epsilon \, P^m(U). \quad (7.3.10)$$

This is precisely the claim.

From now onwards, fix $T_0 \in \mathcal{C}_0$ such that (7.3.10) holds. Now (7.3.10) can be rewritten differently. Define a probability P_U on U by assigning the weight $p(\mathbf{x})/P^m(U)$ to each $\mathbf{x} \in U$. (The division by $P^m(U)$ acts as a normalizing factor.) Then (7.3.10) becomes

$$E_{P_U}(d_P[T_0, H(T_0; \mathbf{x})]) \geq 2\epsilon. \quad (7.3.11)$$

Step 2. It is claimed that

---

[3] The argument still holds good in the case of a so-called "randomized" algorithm, in which H depends on $T, \mathbf{x}$, and also a random vector that is independent of $\mathbf{x}$. For details, see [66].

7.3 Concept Learning

$$P_U\{\mathbf{x} \in U : d_P[T_0, H(T_0; \mathbf{x})] \geq \epsilon\} \geq 1/7. \quad (7.3.12)$$

To establish this claim, observe that both T_0 and $H(T_0; \mathbf{x})$ belong to $\mathcal{C}_0$. Hence

$$d_P[T_0, H(T_0; \mathbf{x})] \leq 8\epsilon \; \forall \mathbf{x} \in U \; \text{(and indeed, } \forall \mathbf{x} \in X^m). \quad (7.3.13)$$

The issue becomes clearer if we define the random variable

$$f(\mathbf{x}) = d_P[T_0, H(T_0; \mathbf{x})]$$

on U. Now (7.3.11) states that $E_{P_U}(f) \geq 2\epsilon$, while (7.3.13) states that f is everywhere bounded by 8ϵ. The claim is that

$$\alpha := P_U\{\mathbf{x} \in U : f(\mathbf{x}) \geq \epsilon\} \geq 1/7.$$

Suppose to the contrary that $\alpha < 1/7$. Then

$$E_{P_U}(f) = \int_{f<\epsilon} f(\mathbf{x}) \, P_U(d\mathbf{x}) + \int_{f\in[\epsilon,8\epsilon]} f(\mathbf{x}) \, P_U(d\mathbf{x})$$

$$\leq \epsilon(1-\alpha) + 8\epsilon\alpha < 2\epsilon \text{ if } \alpha < 1/7.$$

Hence (7.3.12) holds. Since $P_U(\cdot) = P^m(\cdot)/P^m(U)$, (7.3.12) can be rewritten as

$$P^m\{\mathbf{x} \in U : d_P[T_0, H(T_0; \mathbf{x})] \geq \epsilon\} \geq P^m(U)/7.$$

Since U is a subset of X^m, this in turn implies that

$$P^m\{\mathbf{x} \in X^m : d_P[T_0, H(T_0; \mathbf{x})] \geq \epsilon\} \geq P^m(U)/7. \quad (7.3.14)$$

Step 3. It is claimed that

$$P^m(U) \geq 7\delta \text{ if } \delta \leq 0.01 \text{ and } m \leq k/32\epsilon. \quad (7.3.15)$$

To establish this claim, we use the multiplicative form of the Chernoff bound. If $\mathbf{x} \notin U$, then $x_i \in \{y_1, \ldots, y_k\}$ at least $k/2$ times. The probability that a random $x_i \in \{y_1, \ldots, y_k\}$ is 8ϵ, so the probability that this happens at least $k/2$ times out of m samples is no larger than $\exp(-\gamma^2 mp/3)$, where $p = 8\epsilon$ and $\gamma = (k/2mp) - 1$. If $m = k/32\epsilon$, then $mp = k/4$ and $\gamma = 1$, so that

$$P^m(U^c) \leq e^{-k/12} \leq e^{-1/12},$$

or, in other words,

$$P^m(U) \geq 1 - e^{-1/12} \geq 0.07 \geq 7\delta \text{ if } \delta \leq 0.01.$$

The same relationship holds if $m \leq k/32\epsilon$, since the average number of distinct elements in $\mathbf{x} \in X^m$ is a nondecreasing function of m. This establishes the claim.

The conclusion of the proof consists merely of combining (7.3.14) and (7.3.15). This gives

$$P^m\{\mathbf{x} \in X^m : d_P[T_0, H(T_0; \mathbf{x})] \geq \epsilon\} \geq \delta$$

if $\delta \leq 0.01$ and $m \leq k/32\epsilon$. This proves the other half of the bound (7.3.9). ■

In the present theorem and proof, it is assumed that (i) the hypothesis H always belongs to the concept class, and (ii) the algorithm is deterministic. However, with a little care, the proof can be amended so as to dispense with both assumptions, and the same bounds continue to apply. For details, see [66].

7.3.3 Learnability Implies Finite VC-Dimension

Thus far it has been shown that, *if* a concept class has finite VC-dimension, then it is distribution-free learnable. Further, "universal" upper and lower bounds for the sample complexity have been derived. In the present subsection, it is shown that the finiteness of the VC-dimension is also a *necessary* condition for a concept class to be learnable. This is in contrast with the situation in function learning, where it is possible for a function class to have infinite P-dimension and yet be PUAC learnable (see Example 7.1).

We begin with a "universal" necessary condition for a concept class to be learnable, based on Theorem 6.7. This necessary condition is applicable to *all* families of probability measures $\mathcal{P}$, and does not assume that $\mathcal{P} = \mathcal{P}^*$ as in the remainder of the chapter.

Lemma 7.1. *Suppose $(X, \mathcal{S})$ is a measurable space, $\mathcal{C} \subseteq \mathcal{S}$ is a concept class, and $\mathcal{P}$ is a family of probability measures on $(X, \mathcal{S})$. Then $\mathcal{C}$ is learnable to accuracy ϵ with respect to the family $\mathcal{P}$ only if*

$$\sup_{P \in \mathcal{P}} M(2\epsilon, \mathcal{C}, d_P) < \infty. \tag{7.3.16}$$

Proof. Fix $P \in \mathcal{P}$. By Theorem 6.7, even if P were known explicitly, any algorithm (including one that could exploit the knowledge of P) would require at least $\lg M(2\epsilon, \mathcal{C}, d_P)$ samples in order to PAC learn $\mathcal{C}$ to an accuracy of ϵ. Hence an algorithm that uses only $\mathcal{P}$ requires at least

$$\sup_{P \in \mathcal{P}} \lg M(2\epsilon, \mathcal{C}, d_P)$$

samples in order to PAC learn $\mathcal{C}$ to accuracy ϵ with respect to $\mathcal{P}$. Hence $\mathcal{C}$ is learnable to accuracy ϵ with respect to $\mathcal{P}$ only if

$$\sup_{P \in \mathcal{P}} \lg M(2\epsilon, \mathcal{C}, d_P) < \infty.$$

However, it is clear that the supremum of $\lg f(\epsilon)$ is finite if and only if the supremum of $f(\epsilon)$ is finite. Hence we can drop the "lg" from the above expression, which gives (7.3.16). ∎

Lemma 7.2. *Suppose $\mathcal{C}$ shatters a set S of cardinality d. Then there exists a probability measure P on S such that, for each $\epsilon \leq 0.25$,*

$$M(2\epsilon, \mathcal{C}, d_P) \geq \exp[2(0.5 - 2\epsilon)^2 d].$$

Proof. Suppose the set $S = \{x_1, \ldots, x_d\}$ is shattered by $\mathcal{C}$. Choose P to be the purely atomic measure concentrated uniformly on the points of S. Thus, under this probability, it can be assumed that $X = S$ and $\mathcal{C} = 2^S$. Represent each concept in $\mathcal{C}$ by a string of d bits in the obvious fashion. Then, for $A, B \in \mathcal{C}$, we have

$$d_P(A, B) = \frac{\rho_h(A, B)}{d},$$

where $\rho_h(A, B)$ is the Hamming distance between the bit strings A and B, that is, the number of bits in which A and B differ.

Given $A \in \mathcal{C}$, the number of $B \in \mathcal{C}$ such that $d_P(A, B) = k/d$ is precisely $\binom{d}{k}$, for each integer $k \in \{0, \ldots, d\}$. So the number of concepts B such that $d_P(A, B) \leq 2\epsilon$ equals

$$\sum_{k \leq 2\epsilon d} \binom{d}{k}.$$

So the number of pairwise 2ϵ-separated concepts is at least

$$\frac{2^d}{\sum_{k \leq 2\epsilon d} \binom{d}{k}} = \frac{1}{\sum_{k \leq 2\epsilon d} \binom{d}{k} 2^{-d}}.$$

Now apply the Chernoff-Okamoto bound with $m = d, p = 0.5$ and $r = 2\epsilon$. This gives (upon noting that $r \leq p$ if $\epsilon \leq 0.25$)

$$\sum_{k \leq 2\epsilon d} \binom{d}{k} 2^{-d} \leq \exp[-2(0.5 - 2\epsilon)^2 d].$$

Substituting this bound into the previous fraction proves the desired result. ∎

Now we come to the main result of this subsection.

Theorem 7.10. *If $\mathcal{C}$ has infinite VC-dimension, then it is not distribution-free learnable to an accuracy $\epsilon < 0.25$.*

Proof. If $\mathcal{C}$ has infinite VC-dimension, then there exist sets of arbitrarily large cardinality that are shattered by $\mathcal{C}$. Hence it follows from Lemma 7.2 that

$$\sup_{P \in \mathcal{P}^*} M(2\epsilon, \mathcal{C}, d_P) = \infty \; \forall \epsilon < 0.25.$$

Hence, by Lemma 7.1, $\mathcal{C}$ cannot be learnt to an accuracy $\epsilon < 0.25$. ∎

Corollary 7.1. *Suppose a concept class $\mathcal{C}$ satisfies*

$$\sup_{P \in \mathcal{P}^*} N(\epsilon, \mathcal{C}, d_P) < \infty \; \forall \epsilon > 0.$$

Then $\mathcal{C}$ has finite VC-dimension.

Proof. The hypothesis, together with Lemma 2.2 implies that

$$\sup_{P \in \mathcal{P}^*} M(2\epsilon, \mathcal{C}, d_P) \leq \sup_{P \in \mathcal{P}^*} N(\epsilon, \mathcal{C}, d_P) < \infty \; \forall \epsilon > 0.$$

By Theorem 7.10, this implies that $\mathcal{C}$ has finite VC-dimension. ∎

7.4 Learnability of Functions with a Finite Range

Thus far we have seen that a concept class is learnable *if and only if* it has finite VC-dimension, whereas a function class is learnable *if* it has finite P-dimension, but the converse is not true in general. In the present section, we examine the case of a function class $\mathcal{F}$ where each function in $\mathcal{F}$ maps X into a fixed finite subset Y of $[0, 1]$. In this case, it turns out that finite P-dimension is necessary as well as sufficient for $\mathcal{F}$ to be PUAC or PAC learnable. The result is stated precisely next.

Theorem 7.11. *Suppose Y is a fixed finite subset of $[0, 1]$, and that $\mathcal{F}$ is a class of measurable functions mapping X into Y. Then the following three statements are equivalent.*

1. *$\mathcal{F}$ has finite P-dimension.*
2. *$\mathcal{F}$ is PUAC learnable.*
3. *$\mathcal{F}$ is PAC learnable.*

Proof. **(i)** $\Rightarrow$ **(ii)** This is a consequence of Theorem 7.7.

(ii) $\Rightarrow$ **(iii)** This is a consequence of the definitions of PUAC and PAC learnability.

(iii) $\Rightarrow$ **(i)** Suppose $\mathcal{P}$ is some subset of $\mathcal{P}^*$, the set of all probability measures on $(X, \mathcal{S})$. Now, for *each fixed* $P \in \mathcal{P}$, Theorem 6.8 states that $\mathcal{F}$ is learnable to accuracy ϵ only if

$$M(2\epsilon, \mathcal{C}, d_P) < \infty.$$

7.4 Learnability of Functions with a Finite Range

Thus, by mimicking the proof of Lemma 7.1, it is possible to establish the following result: The function class $\mathcal{F}$ is learnable to accuracy ϵ with respect to $\mathcal{P}$ only if
$$\sup_{P \in \mathcal{P}} M(2\epsilon, \mathcal{C}, d_P) < \infty \ \forall \epsilon > 0. \tag{7.4.1}$$

The proof of the implication (iii) $\Rightarrow$ (i) consists of showing that if $\mathcal{F}$ has infinite P-dimension, then the above condition is violated for sufficiently small ϵ, which in turn implies that the function class $\mathcal{F}$ is *not* learnable.

For this purpose, define
$$\alpha := \min\{|y_i - y_j| : y_i, y_j \in Y, y_i \neq y_j\}.$$

Then $\alpha > 0$ since Y is a finite set. Define the "discrete" metric β on Y by
$$\beta(y_i, y_j) := \begin{cases} 1, & \text{if } y_i \neq y_j, \text{ and} \\ 0, & \text{if } y_i = y_j. \end{cases}$$

Then it is easy to see that
$$|y_i - y_j| \leq \beta(y_i, y_j) \leq \frac{1}{\alpha}|y_i - y_j|, \ \forall y_i, y_j \in Y.$$

Next, define a metric β_P on $\mathcal{F}$ by
$$\beta_P(f, g) := \int_X \beta(f(x), g(x)) \ P(dx).$$

Then it follows readily that
$$d_P(f, g) \leq \beta_P(f, g) \leq \frac{1}{\alpha} d_P(f, g), \ \forall f, g \in \mathcal{F}. \tag{7.4.2}$$

Now suppose $\mathcal{F}$ P-shatters a set $S = \{x_1, \ldots, x_d\}$ of cardinality d. Thus there exists a vector $\mathbf{c} \in [0, 1]^d$ such that, for each binary vector $\mathbf{e} \in \{0, 1\}^d$, there exists a corresponding function $f_\mathbf{e} \in \mathcal{F}$ such that
$$f_\mathbf{e}(x_i) \geq c_i \text{ if } e_i = 1, \text{ and } f_\mathbf{e}(x_i) < c_i \text{ if } e_i = 0.$$

Now let P denote the purely atomic measure concentrated uniformly on the set S. Suppose $\mathbf{e}_1, \mathbf{e}_2 \in \{0, 1\}^d$, and consider the corresponding functions $f_{\mathbf{e}_1}, f_{\mathbf{e}_2} \in \mathcal{F}$. Let ρ_h denote the Hamming metric on $\{0, 1\}^d$; thus $\rho_h(\mathbf{e}_1, \mathbf{e}_2)$ equals the number of components in which $\mathbf{e}_1$ and $\mathbf{e}_2$ differ. Then
$$\beta_P(f_{\mathbf{e}_1}, f_{\mathbf{e}_2}) = \frac{1}{m} \sum_{i=1}^m \beta[f_{\mathbf{e}_1}(x_i), f_{\mathbf{e}_2}(x_i)] \geq \frac{1}{m} \sum_{i=1}^m |f_{\mathbf{e}_1}(x_i) - f_{\mathbf{e}_2}(x_i)|$$
$$\geq \frac{1}{\alpha m} \rho_h(\mathbf{e}_1, \mathbf{e}_2),$$

since $|f_{\mathbf{e}_1}(x_i) - f_{\mathbf{e}_2}(x_i)| \geq \alpha$ whenever $e_{1i} \neq e_{2i}$. Now it is already known from Lemma 7.2 that, with respect to the metric ρ_h/m on $\{0,1\}^d$, the 2ϵ-packing number of the set $\{0,1\}^d$ is at least equal to $\exp[2(0.5 - 2\epsilon)^2 d]$, whenever $\epsilon \leq 0.25$. Therefore the set of 2^d functions $\{f_\mathbf{e}, \mathbf{e} \in \{0,1\}^d\}$, call it $\mathcal{F}_S$, satisfies

$$M(2\alpha\epsilon, \mathcal{F}_S, \beta_P) \geq \exp[2(0.5 - 2\epsilon)^2 d], \ \forall \epsilon \leq 0.25.$$

This fact, together with (7.4.2), implies that

$$M(2\alpha^2\epsilon, \mathcal{F}_S, d_P) \geq \exp[2(0.5 - 2\epsilon)^2 d], \ \forall \epsilon \leq 0.25.$$

Hence, if $\mathcal{F}$ has infinite P-dimension, it follows that the condition (7.4.1) is *violated* for every $\epsilon < 0.25\alpha^2$. Consequently, the function class $\mathcal{F}$ cannot be learnt to any accuracy smaller than $0.25\alpha^2$. ∎

Notes and References Theorem 7.1 is a ready consequence of the results of Haussler [80], though the theorem is not explicitly stated therein. The same remark applies to Theorem 7.7. The upper bounds for the quantity $q^*(m, \epsilon)$ have been refined over the years by several researchers, with the objective of reducing the coefficient dividing $-m\epsilon^2$ in the exponent. In (7.1.5), which is the result originally proved in [194], the "constant" equals 8. The best value for this constant to date is 1, as shown in [158]. See [191], p. 85 for a colourful description of the "race for the constant." The fact that a collection of sets has the distribution-free UCEPUP property if and only if it has finite VC-dimension is apparently proved in [195]; however, the present author is not competent to consult this particular reference. Theorem 7.5 is new. Previously, Haussler [80] showed that if a hypothesis class $\mathcal{H}$ has finite P-dimension and if the loss function has a property that he calls "monotonicity," then the family of functions $\mathcal{L}_\mathcal{H}$ also has finite P-dimension – in fact, P-dim($\mathcal{L}_\mathcal{H}$) and P-dim($\mathcal{H}$) are equal. However, Haussler's theorem does not apply to the widely-used loss function $\ell(y,u) = |y-u|$ in the case where both Y and U equal the interval $[0,1]$. In the present case, nothing is said about the finiteness or otherwise of the P-dimension of the family of loss functions $\mathcal{L}_\mathcal{H}$. Instead, the theorem only provides estimates of the rates of convergence of empirical means to their true values, which serves the same purpose. Theorem 7.7 showing that finite P-dimension is sufficient for a function class to be PAC learnable is due to Haussler [80]; the rather minor extension to PUAC learnability is stated here for the first time. Example 7.1 showing that finite P-dimension is *not* necessary for a function class to be learnable is due to M. Krichman (personal communication). In this connection, see the paper by Bartlett *et al.* [16], which shows that if the oracle outputs are corrupted by noise, then the finiteness of the P-dimension is *necessary* as well as sufficient for a function class to be learnable. Thus, unlike in the case of concept learning, the presence or absence of noise can make a substantial difference to the learnability of function classes. This is not altogether surprising, since "perfect" measurements permit the oracle to encode an infinite amount of information about the target function into its output.

Theorem 7.8 established, apparently for the first time, a connection between the VC-dimension and PAC learnability. This theorem is due to Blumer *et al.*, and is stated in [32]. Somewhat better estimates for the sample complexity are obtained by Anthony *et al.* in [10]. An entirely different approach to estimating loss functions, referred to as "simultaneous estimation," is proposed by Buescher and Kumar in [37], [38]. They too are able to improve upon the sample complexity bounds of Blumer *et al.* Theorem 7.9 on a universal lower bound for the sample complexity is stated by Ehrenfucht *et al.* in [66], and improves upon an earlier bound given by Blumer *et al.* in [32]. Lemma 7.2 and its consequence Theorem 7.10 are stated by Kulkarni in [109], and improve upon earlier results of Benedek-Itai in [22]. The learning problem for functions mapping X into a fixed finite set Y is a natural generalization of the concept learning problem, and has been studied by several researchers. At a very basic level, this problem can be studied by encoding the elements of Y using a binary vector of at most $\lceil \lg |Y| \rceil$ bits. However, such an approach makes the problem more difficult than it really is. A comprehensive comparison of various available results can be found in the paper by Ben-David *et al.* [20].

Finally, note that, by using Theorems 3.3 and 3.4, all of the results in the present chapter can be extended to the case of β-mixing input sequences.

8. Learning Under an Intermediate Family of Probabilities

The two preceding chapters have addressed what may be thought of as the two "extreme" situations in learning, namely: the case where the learning samples are generated by a known fixed probability (Chapter 6), and the case where the learning samples are generated by a probability measure that is itself *completely* unknown (Chapter 7). In the present chapter, we study the intermediate situation, namely: the learning samples are generated by a probability measure P belonging to a family $\mathcal{P}$ that is neither a singleton set, nor the set of *all* probability measures. If no assumptions at all are made regarding $\mathcal{P}$ (what might be termed the "general" case), very few results are available. Some such results are summarized in Section 8.1. However, the situation is markedly different if some metric structure is imposed on $\mathcal{P}$. For this purpose, we define a metric ρ on the set $\mathcal{P}^*$ of all probability measures on a measurable space $(X, \mathcal{S})$, as follows: If $P, Q \in \mathcal{P}^*$, then

$$\rho(P, Q) := \sup_{A \in \mathcal{S}} |P(A) - Q(A)|.$$

Note that ρ is called the **total variation** metric. For instance, if X is a subset of $\mathbb{R}$, and both P and Q have densities $p(\cdot)$ and $q(\cdot)$ with respect to the Lebesgue measure, then

$$\rho(P, Q) = \int_X |p(x) - q(x)|\, dx.$$

If $A \in \mathcal{S}$, $P, Q \in \mathcal{P}^*$, then a ready consequence of the definition of ρ is that

$$|P(A) - Q(A)| \leq \rho(A, B).$$

More generally, suppose $f : X \to [0, 1]$ is measurable. Then

$$|E_P(f) - E_Q(f)| \leq \rho(P, Q).$$

Section 8.2 is devoted to the case where the family of probabilities $\mathcal{P}$ is totally bounded with respect to the metric ρ. This situation can be thought of as an extension of the case where $\mathcal{P}$ is a singleton set. Thus the results in this section are based on the results in Chapter 6. Section 8.3 is devoted to the case where the family of probabilities $\mathcal{P}$ has a nonempty interior,

286   8. Learning Under an Intermediate Family of Probabilities

i.e., contains a ball of probabilities. This situation can be thought of as an extension of the case where $\mathcal{P} = \mathcal{P}^*$. In this case, it turns out that a collection of sets needs to have finite VC-dimension either for empirical probabilities to converge uniformly to the true values, or to be learnable. Of course, even with the results in this chapter, the learning problem is far from being resolved in that there still exists a large "gap," corresponding to the case where $\mathcal{P}$ is noncompact but has an empty interior, for example, $\mathcal{P}$ is the set of all nonatomic measures on $[0, 1]$.

For the convenience of the reader, the relevant notation from the preceding chapters is recalled. One is given

- A set X,
- A σ-algebra $\mathcal{S}$ of subsets of X,
- A family $\mathcal{P}$ of probability measures on $(X, \mathcal{S})$, and
- A family $\mathcal{F}$ of measurable functions mapping X into $[0, 1]$ (the function class), or else a collection $\mathcal{C}$ of sets belonging to $\mathcal{S}$ (the concept class).

Two types of problems are studied, namely: uniform convergence of empirical means, and learnability.

Let us discuss first the problem of the uniform convergence of empirical means. Given a probability measure P, a function $f \in \mathcal{F}$, and a multisample $[x_1 \ldots x_m]^t \in X^m$, define the "true" mean of f with respect to P by

$$E_P(f) := \int_X f(x) \, P(dx),$$

and its "empirical" mean based on the multisample $\mathbf{x}$ by

$$\hat{E}(f; \mathbf{x}) := \frac{1}{m} \sum_{i=1}^m f(x_i).$$

Next, define

$$q(m, \epsilon, P) := P^m \{ \mathbf{x} \in X^m : \sup_{f \in \mathcal{F}} |\hat{E}(f; \mathbf{x}) - E_P(f)| > \epsilon \}, \text{ and}$$

$$\bar{q}(m, \epsilon, \mathcal{P}) := \sup_{P \in \mathcal{P}} q(m, \epsilon, P).$$

The family of functions $\mathcal{F}$ is said to have the property of **uniform convergence of empirical means uniformly in probability (UCEMUP)** with respect to $\mathcal{P}$ if $\bar{q}(m, \epsilon, \mathcal{P}) \to 0$ as $m \to \infty$, for each $\epsilon > 0$.

Now consider the learning problem. An "algorithm" is an indexed family of maps

$$A_m : (X \times [0, 1])^m \to \mathcal{F}.$$

Given $\mathbf{x} \in X^m$ and $f \in \mathcal{F}$, define

$$h_m(f; \mathbf{x}) := A_m[(x_1, f(x_1)), \ldots, (x_m, f(x_m))].$$

In other words, $h_m(f;\mathbf{x})$ is the hypothesis returned by the algorithm after m steps when the target function is f and the multisample is $\mathbf{x}$. Next, define

$$r(m,\epsilon) := \sup_{P \in \mathcal{P}} \sup_{f \in \mathcal{F}} P^m\{\mathbf{x} \in X^m : d_P[f, h_m(f;\mathbf{x})] > \epsilon\}, \text{ and}$$

$$s(m,\epsilon) := \sup_{P \in \mathcal{P}} P^m\{\mathbf{x} \in X^m : \sup_{f \in \mathcal{F}} d_P[f, h_m(f;\mathbf{x})] > \epsilon\}.$$

The algorithm $\{A_m\}$ is said to be **probably approximately correct (PAC)** if $r(m,\epsilon) \to 0$ as $m \to \infty$ for each $\epsilon > 0$, and is said to be **probably uniformly approximately correct (PUAC)** if $s(m,\epsilon) \to 0$ as $m \to \infty$ for each $\epsilon > 0$.

8.1 General Families of Probabilities

In this section, we describe the available results in the case where no assumptions are made on the nature of the family of probability measures $\mathcal{P}$. The results proved in Chapter 5 concerning the UCEMUP property are recapitulated, and it is recalled that the UCEMUP property implies consistent learnability, as shown in Chapter 6. Next, the result proved in Chapter 6 is extended to show that the so-called shrinking width property is a necessary and sufficient condition for a family of functions to be consistently learnable. Finally, a universal *necessary* condition for a *concept* class to be PAC learnable is proved, known as the uniform boundedness of metric entropy (UBME) condition. It is shown that the UBME condition is also sufficient for a concept class to be PAC learnable in the two "extreme" conditions, namely when $\mathcal{P}$ is a singleton set, and when $\mathcal{P} = \mathcal{P}^*$. Then it is shown by example that the UBME condition is *not* sufficient in general for intermediate families of probabilities.

8.1.1 Uniform Convergence of Empirical Means

In Chapter 5, a general necessary and sufficient condition is proved for a family $\mathcal{F}$ to have the UCEMUP property. This result is recapitulated here for convenience. Given $\mathbf{x} \in X^m$ and $f \in \mathcal{F}$, let

$$\mathbf{f}(\mathbf{x}) := [f(x_1) \ldots f(x_m)]^t \in [0,1]^m, \text{ and}$$

$$\mathcal{F}|_\mathbf{x} := \{\mathbf{f}(\mathbf{x}) : f \in \mathcal{F}\} \subseteq [0,1]^m.$$

Given an $\epsilon > 0$, let $L(\epsilon, \mathcal{F}|_\mathbf{x}, \|\cdot\|_\infty)$ denote the external ϵ-covering number of the set $\mathcal{F}|_\mathbf{x}$ with respect to the metric induced by the norm $\|\cdot\|_\infty$. Then (cf. Theorem 5.5) the family of functions $\mathcal{F}$ has the UCEMUP property with respect to $\mathcal{P}$ if and only if

288   8. Learning Under an Intermediate Family of Probabilities

$$\lim_{m\to\infty} \sup_{P\in\mathcal{P}} \frac{E_{P^m}[\lg L(\epsilon, \mathcal{F}|_{\mathbf{x}}, \|\cdot\|_\infty)]}{m} = 0, \ \forall \epsilon > 0.$$

Suppose now that $\mathcal{F}$ consists only of functions that map X into $\{0,1\}$, as opposed to $[0,1]$. Then every function in $\mathcal{F}$ can be thought of as the indicator function of a set $A \in \mathcal{S}$. Thus the empirical mean of the indicator function $I_A(\cdot)$ is the same as the empirical probability of the set A. So in this case the problem at hand becomes one of the uniform convergence of empirical probabilities to their true values. Suppose $\mathcal{A} \subseteq \mathcal{S}$, and given an m-tuple $\mathbf{x} \in X^m$, let $d(\mathbf{x})$ denote the VC-dimension of the collection $\mathcal{A}$ intersected with $\{x_1, \ldots, x_m\}$. Then (cf. Theorem 5.6) the collection of sets $\mathcal{A}$ has the UCEPUP property with respect to $\mathcal{P}$ if and only if

$$\lim_{m\to\infty} \sup_{P\in\mathcal{P}} \frac{E_{P^m}[d(\mathbf{x})]}{m} = 0.$$

Using the above criterion for the UCEMUP property, one can readily establish the following result:

Lemma 8.1. *Suppose $\mathcal{P}_1, \ldots, \mathcal{P}_k$ are families of probabilities on $(X, \mathcal{S})$, and let $\mathcal{P} = \cup_{i=1}^k \mathcal{P}_i$. Given a family of functions $\mathcal{F}$, the following statements are equivalent:*

1. *$\mathcal{F}$ has the UCEMUP property with respect to $\mathcal{P}$.*
2. *$\mathcal{F}$ has the UCEMUP property with respect to $\mathcal{P}_i$ for each i.*

Proof. (i) $\Rightarrow$ (ii) Obvious, since each $\mathcal{P}_i$ is a subset of $\mathcal{P}$.
    (ii) $\Rightarrow$ (i) For each integer m, we have that

$$\sup_{P\in\mathcal{P}} E_{P^m}[\lg L(\epsilon, \mathcal{F}|_{\mathbf{x}}, \|\cdot\|_\infty)] = \max_{1\leq i\leq k} \sup_{P\in\mathcal{P}_i} E_{P^m}[\lg L(\epsilon, \mathcal{F}|_{\mathbf{x}}, \|\cdot\|_\infty)].$$

Now the desired conclusion follows from Theorem 5.5. ∎

Note that the above proof depends crucially on the fact that $\mathcal{P}$ is expressed as a *finite* union, since we have implicitly used the reasoning step

$$\lim_{m\to\infty} \max_{1\leq i\leq k} \cdots = \max_{1\leq i\leq k} \lim_{m\to\infty} \cdots.$$

This step is no longer valid in general if $\mathcal{P}$ were to be expressed as an *infinite* union.

8.1.2 Function Learning

Recall from Chapter 6 that a family of functions $\mathcal{F}$ is said to be **consistently learnable** if every consistent algorithm is PUAC. Now (cf. Theorem 6.1), we have the following result:

8.1 General Families of Probabilities

Theorem 8.1. *Suppose $\mathcal{F}$ has the UCEMUP property with respect to $\mathcal{P}$. Then $\mathcal{F}$ is consistently learnable.*

The proof is merely a restatement of that of Theorem 6.1, taking care to ensure that various quantities are now bounded with respect to P as well.

Recall that the "shrinking width" property is defined in Section 6.4 for the case of a *fixed* probability measure. With a little care, the definition can be extended to the case where P is itself variable. Given a family of functions $\mathcal{F}$, define

$$w(m, \epsilon, P) := P^m \{ \mathbf{x} \in X^m : \exists f, g \in \mathcal{F} \text{ s.t. } \hat{d}(f, g; x) = 0 \text{ and } d_P(f, g) > \epsilon \}.$$

This is the same as the quantity $w(m, \epsilon)$ defined in Section 6.4, except that the dependence on P is now explicitly identified. Now let

$$\bar{w}(m, \epsilon, \mathcal{P}) := \sup_{P \in \mathcal{P}} w(m, \epsilon, P).$$

The family $\mathcal{F}$ is said to have the **shrinking width property** with respect to $\mathcal{P}$ if $\bar{w}(m, \epsilon, \mathcal{P}) \to 0$ as $m \to \infty$, for each $\epsilon > 0$.

In the case where $\mathcal{P}$ is a singleton set, the shrinking width property can be interpreted in terms of the convergence of a stochastic process to zero; see Section 6.4. However, if $\mathcal{P}$ is not a singleton set, such an interpretation is not possible. Nevertheless, we can still prove the following useful result (cf. Theorem 6.5):

Theorem 8.2. *Given a family of functions $\mathcal{F}$ and a family of probabilities $\mathcal{P}$, the following statements are equivalent:*

1. *The family $\mathcal{F}$ has the shrinking width property.*
2. *The family $\mathcal{F}$ is consistently learnable.*

Proof. **(i) $\Rightarrow$ (ii)** This part of the proof exactly follows the corresponding part of the proof of Theorem 6.5, except that the various quantities are bounded with respect to P as well. The details are straight-forward and are left to the reader.

(ii) $\Rightarrow$ (i) This part also closely follows the corresponding part of the proof of Theorem 6.5. Suppose $\mathcal{F}$ fails to have the shrinking width property. Then there exist numbers $\epsilon, \delta > 0$, a sequence of integers $\{m_i\}$ approaching infinity, *and a sequence of probabilities* $\{P_i\}$ in $\mathcal{P}$ such that

$$P_i^{m_i} \{ \mathbf{x} \in X^{m_i} : \exists f, g \in F \text{ s.t. } \hat{d}(f, g; \mathbf{x}) = 0 \text{ and } d_{P_i}(f, g) > \epsilon \} > \delta.$$

The fact that P itself may depend on i is the extra twist, compared to the (ii) $\Rightarrow$ (i) part of the proof of Theorem 6.5. Nevertheless, the remainder of the proof goes through. The details are left to the reader. ∎

The results presented thus far are rather obvious modifications of previously proven results from Chapter 6. But the next two results require genuinely different methods of proof. The first theorem is somewhat similar to Theorem 8.1, but the proof is much more intricate.

Theorem 8.3. *Suppose $\mathcal{F}$ is a given family of functions, and that $\mathcal{P}_1, \ldots, \mathcal{P}_k$ are families of probabilities. Define $\mathcal{P} = \cup_{i=1}^k \mathcal{P}_i$. Then the following statements are equivalent:*

1. *$\mathcal{F}$ is PAC learnable with respect to $\mathcal{P}$.*
2. *$\mathcal{F}$ is PAC learnable with respect to $\mathcal{P}_i$ for each i.*

Proof. (i) $\Rightarrow$ (ii) Obvious, since each $\mathcal{P}_i$ is a subset of $\mathcal{P}$.

(ii) $\Rightarrow$ (i) Suppose $\mathcal{F}$ is PAC learnable with respect to $\mathcal{P}_i$ for each i, and suppose further that $\{A_m^i\}$, $i = 1, \ldots, k$ are algorithms such that $\{A_m^i\}$ is PAC when the family of probabilities is $\mathcal{P}_i$. Further, let $m_i(\epsilon, \delta)$ denote the sample complexity of the algorithm $\{A_m^i\}$ when the family of probabilities is $\mathcal{P}_i$. This means the following: Using obvious notation, let $h_m^i(f; \mathbf{x})$ denote the output of the algorithm $\{A_m^i\}$ when the target function is f and the multisample is $\mathbf{x} \in X^m$. Then

$$\sup_{P \in \mathcal{P}_i} \sup_{f \in \mathcal{F}} P^m\{\mathbf{x} \in X^m : d_P[f, h_m^i(f; \mathbf{x})] > \epsilon\} \leq \delta, \ \forall m \geq m_i(\epsilon, \delta).$$

Now we describe an algorithm that, given $\epsilon, \delta > 0$, returns a hypothesis that is within ϵ of the target function at least $1 - \delta$ of the time, provided the number of samples m exceeds

$$m_0(\epsilon, \delta) \geq \max_{1 \leq i \leq k} m_i(\epsilon/2, \delta/2) + \frac{8}{\epsilon^2} \ln \frac{2k}{\delta}.$$

As such, the algorithm is "ϵ, δ-dependent." However, by Theorem 3.1, it is possible to produce another PAC algorithm that does not explicitly make use of ϵ and δ.

The algorithm is as follows: Suppose $f \in \mathcal{F}, P \in \mathcal{P}$ are otherwise unknown. Draw $m \geq m_0(\epsilon, \delta)$ i.i.d. samples according to P. Using the first

$$\bar{m}(\epsilon, \delta) := \max_{1 \leq i \leq k} m_i(\epsilon/2, \delta/2)$$

labelled samples, run *each* of the k algorithms $\{A_m^1\}$ through $\{A_m^k\}$, and generate corresponding hypotheses h^1 through h^k. Since $P \in \mathcal{P}$, it follows that $P \in \mathcal{P}_i$ for some i. Hence we know that, with confidence $1 - \delta/2$, *one* of the h^i is within $\epsilon/2$ of f, but we do not know *which one*. Now, using the next

$$\frac{8}{\epsilon^2} \ln \frac{2k}{\delta} = \frac{8}{\epsilon^2} \ln \frac{k}{\delta/2}$$

or more samples, run the minimum empirical risk algorithm on $h^1, \ldots, h^k$, and call the resulting output h_m. Then, by Theorem 6.2, it follows that

$d_P(f, h_m) \leq \epsilon$ with confidence at least $1 - \delta/2$. Of course, this conclusion is based on the assumption that $d_P(f, h^i) \leq \epsilon/2$ for some i, which itself holds with confidence at least $1 - \delta/2$. Combining these two statements shows that $d_P(f, h_m) \leq \epsilon$ at least $1 - \delta$ of the time. ∎

In the case of concept learning, one can use Theorem 6.3 instead of Theorem 6.2, and thereby improve the estimate for $m_o(\epsilon, \delta)$ to

$$m_0(\epsilon, \delta) = \max_{1 \leq i \leq k} m_i(\epsilon/2, \delta/2) + \frac{32}{\epsilon} \ln \frac{2k}{\delta}.$$

The subsection is concluded with another useful result. Given a family $\mathcal{P}$ of probabilities, define a function $\bar{d}_\mathcal{P} : \mathcal{S} \times \mathcal{S} \to [0, 1]$ by

$$\bar{d}_\mathcal{P}(A, B) := \sup_{P \in \mathcal{P}} d_P(A, B).$$

Then $\bar{d}_\mathcal{P}$ is also a pseudometric on $\mathcal{S}$, even though it might not correspond to any probability on $(X, \mathcal{S})$. In the same way, one can define a corresponding pseudometric on the set of measurable functions mapping X into $[0, 1]$ by

$$\bar{d}_\mathcal{P}(f, g) := \sup_{P \in \mathcal{P}} d_P(f, g).$$

Now we can state the following "universal" sufficient condition for PAC learnability.

Theorem 8.4. *Suppose $\mathcal{F}$ is a given function class, $\mathcal{P}$ is a given family of probabilities, and that $\mathcal{F}$ is totally bounded with respect to the metric $\bar{d}_\mathcal{P}$. Then $\mathcal{F}$ is PAC learnable with respect to $\mathcal{P}$.*

Proof. Suppose $\{g_1, \ldots, g_k\}$ is an $\epsilon/2$-cover of $\mathcal{F}$ with respect to $\bar{d}_\mathcal{P}$. This means that $\{g_1, \ldots, g_k\}$ is an $\epsilon/2$-cover of $\mathcal{F}$ with respect to d_P for *each* $P \in \mathcal{P}$. Now draw at least

$$m \geq \frac{8}{\epsilon^2} \ln \frac{k}{\delta}$$

samples, and apply the minimum empirical risk algorithm to the set of functions $\{y_1, \ldots, y_k\}$. Then, by Theorem 6.2, it follows that such an algorithm is PAC to accuracy ϵ for *every* $P \in \bar{P}$.. ∎

The key to the above proof lies in the fact that the *same* set of functions $\{g_1, \ldots, g_k\}$ is an $\epsilon/2$-cover of $\mathcal{F}$ for *every* $P \in \mathcal{P}$. Such a cover is referred to as a "common" cover.

Now let us restrict ourselves to *concept* learning, and examine how close the above universal sufficient condition is to being necessary. In the case where the set $\mathcal{P}$ is a singleton, it is obvious that the condition of Theorem 8.4 is indeed necessary, since it reduces to the bounded metric entropy condition of Theorem 6.6. On the other extreme, let $\mathcal{P} = \mathcal{P}^*$, the set of *all* probability measures. Then it is claimed that $\bar{d}_{\mathcal{P}^*}$ is the so-called "discrete" metric, that is:

$$\bar{d}_{\mathcal{P}^*}(A, B) = \begin{cases} 1 & \text{if } A = B, \\ 0 & \text{if } A \neq B. \end{cases}$$

To see this, suppose $A \neq B$, and choose an element $x \in A \Delta B$. Let P equal the atomic measure concentrated at x. Then $d_P(A, B) = 1$. This establishes the claim. Therefore, in the extreme case where $\mathcal{P} = \mathcal{P}^*$, concept class $\mathcal{C}$ is totally bounded with respect to $\bar{d}_{\mathcal{P}^*}$ if and only if it is a finite set. This is rather far from the known necessary and sufficient condition, which is that the VC-dimension of $\mathcal{C}$ be finite. Hence one might conclude that the above sufficient condition is rather far from being necessary in general. However, we shall see in Section 8.2 that the condition is in fact necessary when the family of probabilities $\mathcal{P}$ is totally bounded.

8.1.3 Concept Learning

Suppose $\mathcal{C}$ is a given concept class, and $\mathcal{P}$ is a given family of probability measures. Then it is possible to state the following universal *necessary* condition for $\mathcal{C}$ to be PAC learnable with respect to $\mathcal{P}$.

Theorem 8.5. *Given a concept class $\mathcal{C}$ and a family of probabilities $\mathcal{P}$, $\mathcal{C}$ is PAC learnable with respect to $\mathcal{P}$ only if*

$$\sup_{P \in \mathcal{P}} N(\epsilon, \mathcal{C}, d_P) < \infty, \ \forall \epsilon > 0.$$

Remarks: Recall that $N(\epsilon, \mathcal{C}, d_P)$ is the ϵ-covering number of the concept class under the pseudometric d_P. Thus the above condition is referred to as the **uniformly bounded metric entropy (UBME)** condition.

Proof. From Theorem 6.6, even if $P \in \mathcal{P}$ were to be known, learning $\mathcal{C}$ to accuracy ϵ requires at least $\lg M(2\epsilon, \mathcal{C}, d_P) \geq \lg N(2\epsilon, \mathcal{C}, d_P)$ samples. Hence, if $\mathcal{C}$ is learnable for each $P \in \mathcal{P}$, and if in addition, the number of samples is required to be bounded independently of P, it is necessary that

$$\sup_{P \in \mathcal{P}} N(2\epsilon, \mathcal{C}, d_P) < \infty, \ \forall \epsilon > 0.$$

This is the same as the UBME condition, except for the extraneous factor of 2. ∎

Theorem 8.5 states that the UBME condition is a "universal" *necessary* condition for PAC learnability. Thus Theorem 8.5 is to be contrasted with Theorem 8.4 which presents a universal sufficient condition for PAC learnability. The difference is that in Theorem 8.4, *the same* elements of $\mathcal{F}$ (or $\mathcal{C}$) serve as cover elements with respect to d_P for *each* $P \in \mathcal{P}$. However, in the UBME condition of Theorem 8.5, the *number of elements* in an ϵ-cover is uniformly bounded, but the *cover elements themselves* could vary with P.

8.1 General Families of Probabilities 293

It is natural to enquire how close the UBME condition is to being sufficient. For this purpose, let us study the two "extreme" cases, namely, where $\mathcal{P}$ is a singleton set, and where $\mathcal{P}$ is the set of *all* probability measures. In the former case, where $\mathcal{P} = \{P\}$, the condition

$$N(\epsilon, \mathcal{C}, d_P) < \infty \ \forall \epsilon > 0$$

is sufficient as well as necessary for $\mathcal{C}$ to be PAC learnable (cf. Theorem 6.7). In the latter case, where $\mathcal{P} = \mathcal{P}^*$, the set of *all* probabilities on $(X, \mathcal{S})$, it follows from Corollary 7.1 that the UBME condition implies that $\mathcal{C}$ has finite VC-dimension, which in turn implies that $\mathcal{C}$ is PAC learnable. Hence, if $\mathcal{P} = \mathcal{P}^*$, the UBME condition is once again sufficient as well as necessary for $\mathcal{C}$ to be PAC learnable.

In view of the above, it is tempting to conjecture that the UBME condition is sufficient as well as necessary for $\mathcal{C}$ to be PAC learnable for an *arbitrary* family of probabilities $\mathcal{P}$. Unfortunately, while this conjecture is appealing, it is *false*, as shown in the next, rather elaborate, example.

Example 8.1. ([64]) Let $X = \{0, 1\}^\infty$, the set of all binary sequences indexed over the natural numbers (beginning with 1). Let $\mathcal{S}$ equal the Borel σ-algebra over X. Define the sequence

$$p_i = \frac{1}{\lg(i+1)},$$

where lg denotes the logarithm to the base 2. A product measure P_I can be induced on X by identifying $p_i = P(x_i = 1)$. Let $\sigma : \mathbf{N} \to \mathbf{N}$ denote a permutation (possibly infinite) of the integers; thus σ is a one-to-one and onto map on $\mathbf{N}$. Let Σ denote the set of all such permutations. Let P_σ denote the probability measure on X defined by $P_\sigma(x_{\sigma(i)} = 1) = p_i$. Now let $\mathcal{P} = \{P_\sigma, \sigma \in \Sigma\}$. This specifies the family of probability measures. Next, let $C_i = \{x \in X : x_i = 1\}$, and define $\mathcal{C} = \{C_i, i \in \mathbf{N}\} \cup \{\emptyset\}$. Since any C_i with $p_{\sigma^{-1}(i)} < \epsilon$ satisfies $d_{P_\sigma}(C_i, \emptyset) < \epsilon$, it is easy to see that the sets $\{C_{\sigma(1)}, \ldots, C_{\sigma(n)}, \emptyset\}$ form an ϵ-cover for $\mathcal{C}$ with respect to the pseudometric d_{P_σ} provided $n \geq 2^{1/\epsilon}$. It follows therefore that the class $\mathcal{C}$ satisfies the UBME condition with respect to the family $\mathcal{P}$, and that

$$N(\epsilon, \mathcal{C}, \mathcal{P}) \leq 2^{1/\epsilon}.$$

Nevertheless, $\mathcal{C}$ is *not* PAC learnable with respect to $\mathcal{P}$.

Before proving that the concept class is not learnable, let us try to understand the nature of the learning problem at hand. For the sake of simplicity, suppose the unknown target concept is not the empty set. Thus the unknown target concept can be identified with an unknown *integer*, call it n. A random element $x \in X$ is actually an *infinite sequence* in $\{0, 1\}^\infty$. The oracle then returns the n-th component of x as its output. The problem is to guess *which component* of x the oracle is looking at. The nonlearnability of the concept class means that this is not possible.

To show that the concept class at hand is not learnable, suppose to the contrary that there exists an algorithm $\{A_m\}$ such that

$$P^m\{\mathbf{x} \in X^m : d_P[T, H_m(T; \mathbf{x})] > \epsilon\} < \delta, \ \forall T \in \mathcal{C}, \ \forall P \in \mathcal{P}. \quad (8.1.1)$$

In particular, fix an integer k, and let

$$\mathcal{C}_k := \{C_1, \ldots, C_k\} \subseteq \mathcal{C},$$

$$\Sigma_k := \{\sigma \in \Sigma : \sigma(i) = i \ \forall i > k\}, \text{ and}$$

$$\mathcal{P}_k := \{P_\sigma : \sigma \in \Sigma_k\}.$$

Thus $\mathcal{C}_k$ consists of the first k concepts in $\mathcal{C}$; Σ_k consists of those permutations that leave the $(k+1)$-st coordinate onwards unaffected; and $\mathcal{P}_k$ consists of all permutations of $\{p_1, p_2, \ldots\}$ that involve only the first k coordinates. For each $\sigma \in \Sigma_k$ and each $T \in \mathcal{C}_k$, define the set $S(\sigma, T) \subseteq X^m$ by

$$S(\sigma, T) := \{\mathbf{x} \in X^m : d_{P_\sigma}[T, H_m(T; \mathbf{x})] > \epsilon\}.$$

By the PAC assumption (8.1.1), it follows that

$$P_\sigma^m[S(\sigma, T)] < \delta, \ \forall T \in \mathcal{C}_k, \ \forall \sigma \in \Sigma_k.$$

In particular, if Q is any probability measure on the finite set $\Sigma_k \times \mathcal{C}_k$, then

$$E_Q\{P_\sigma^m[S(\sigma, T)]\} < \delta.$$

Now choose Q to be the following probability measure:

$$Q(\sigma, T) = \frac{1}{k!} \text{ if } T = C_{\sigma(1)}, \text{ otherwise}.$$

Note that

$$\min_{j>1} d_{P_I}(C_1, C_j) = 1 - \frac{1}{\lg 3} =: \epsilon_0, \text{ say}.$$

Hence, if $\epsilon < \epsilon_0$, then $d_{P_\sigma}(C_{\sigma(1)}, C_j) < \epsilon$ if and only if $C_j = C_{\sigma(1)}$, for all $\sigma \in \Sigma_k$. In particular, when (σ, T) is distributed according to Q as defined above, it follows that $T = C_{\sigma(1)}$ Q-almost surely (abbreviated hereafter as Q-a.s.). Hence

$$d_{P_\sigma}(C_i, C_j) < \epsilon \Rightarrow C_j = C_i = C_{\sigma(1)} \ Q\text{-a.s.}$$

Next, for each $\mathbf{x} \in X^m$ let $\sigma \mathbf{x} \in X^m$ denote the vector $[x_{\sigma^{-1}(1)} \ldots x_{\sigma^{-1}(m)}]^t \in X^m$, and let $L_\sigma(T; \mathbf{x})$ denote the corresponding labelled sample. Thus

$$L_I(T; \mathbf{x}) := [(x_1, I_T(x_1)), \ldots, (x_m, I_T(x_m))] \in [X \times \{0, 1\}]^m,$$

8.1 General Families of Probabilities

$$L_\sigma(T; \mathbf{x}) := [((\sigma \mathbf{x})_1, I_T((\sigma \mathbf{x})_1)), \ldots, ((\sigma \mathbf{x})_m, I_T((\sigma \mathbf{x})_m))].$$

In the present set-up, $T = C_{\sigma(1)}$ Q-a.s. Therefore

$$L_\sigma(T; \mathbf{x}) = [((\sigma \mathbf{x})_1, I_{C_{\sigma(1)}}((\sigma \mathbf{x})_1)), \ldots, ((\sigma \mathbf{x})_m, I_{C_{\sigma(1)}}((\sigma \mathbf{x})_m))]$$
$$= [((\sigma \mathbf{x})_1, I_{C_1}(x_1)), \ldots, ((\sigma \mathbf{x})_m, I_{C_1}(x_m))]$$
$$=: L_\sigma(\mathbf{x}), \text{ say}. \qquad (8.1.2)$$

Therefore, recalling that $A_m : [X \times \{0,1\}]^m \to \mathcal{C}$ is the m-th algorithmic map, we have

$$E_Q\{P_\sigma^m[S(\sigma, T)]\} = E_Q\{P_\sigma^m[T \neq H_m(T; \mathbf{x})]\}$$
$$= E_Q\{P_\sigma^m[C_{\sigma(1)} \neq A_m(L_I(C_{\sigma(1)}, \mathbf{x}))]\}$$
$$= E_Q\{P_I^m[C_{\sigma(1)} \neq A_m(L_\sigma(\mathbf{x}))]\}$$
$$= E_{P_I^m} E_Q[I_{C_{\sigma(1)} \neq A_m(L_\sigma(\mathbf{x}))}]. \qquad (8.1.3)$$

In the last step, we are able to interchange $E_{P_I^m}$ and E_Q without difficulty since E_Q is just a finite summation. Also, the symbol $I_{C_{\sigma(1)} \neq A_m(L_\sigma(\mathbf{x}))}$ denotes the indicator function of the set $\{\mathbf{x} \in X^m : C_{\sigma(1)} \neq A_m(L_\sigma(\mathbf{x}))\}$.

Next, given vectors $\mathbf{x}, \mathbf{y} \in X^m$, denote by $S(\mathbf{x}, \mathbf{y})$ the set of permutations $\sigma \in \Sigma_k$ such that $\sigma \mathbf{x} = \mathbf{y}$. Note that for many pairs $(\mathbf{x}, \mathbf{y})$ the set $S(\mathbf{x}, \mathbf{y})$ could be empty. By definition, if $\sigma \in S(\mathbf{x}, \mathbf{y})$, then

$$L_\sigma(C_{\sigma(1)}, \mathbf{y}) = [(y_1, I_{C(1)}(x_1)), \ldots, (y_m, I_{C(1)}(x_m))].$$

By the construction of Q, the distribution of σ conditioned on $S(\mathbf{x}, \mathbf{y})$ is uniform. Now let

$$J^\mathbf{x} := \{j \leq k : x_{ij} = 1 \, \forall i = 1, \ldots, m\}, \text{ and}$$
$$J^\mathbf{y} := \{j \leq k : y_{ij} = 1 \, \forall i = 1, \ldots, m\}.$$

Then $S(\mathbf{x}, \mathbf{y})$ is nonempty only if $|J^\mathbf{x}| = |J^\mathbf{y}|$. When X has distribution P_I, we have that $x_{1j} = 1$ almost surely, so $1 \in J^\mathbf{x}$ almost surely; hence $|J^\mathbf{x}| \geq 1$. Let $\sigma_c \in \Sigma_k$ be a fixed permutation such that $\sigma_c(i) \in J^\mathbf{y}$ if $i \in J^\mathbf{x}$. Decompose each permutation $\sigma \in S(\mathbf{x}, \mathbf{y})$ as $\sigma = \sigma_c \cdot \sigma_b \cdot \sigma_a$, where $\sigma_a : J^\mathbf{x} \to J^\mathbf{x}$ and σ_a equals the identity on $\{1, \ldots, k\} \setminus J^\mathbf{x}$, while σ_b maps $\{1, \ldots, k\} \setminus J^\mathbf{x}$ into itself and equals the identity on $J^\mathbf{x}$. This is always possible since every permutation in $S(\mathbf{x}, \mathbf{y})$ satisfies $\sigma \mathbf{x} = \mathbf{y}$. Note that if $S(\mathbf{x}, \mathbf{y})$ is nonempty then $|\Sigma_A| = |J^\mathbf{x}|!$, where

$$\Sigma_A := \{\sigma_a : \sigma \in S(\mathbf{x}, \mathbf{y})\}, \text{ and } \Sigma_B := \{\sigma_b : \sigma \in S(\mathbf{x}, \mathbf{y})\}.$$

Now using (8.1.3) gives

$$E_Q\{P_\sigma^m[S(\sigma,T)]\} = E_{P_I^m}\left[\sum_{\mathbf{y}} E_Q[I_{C_{\sigma(1)} \neq A_m(L_\sigma(\mathbf{x}))}|\ \sigma \in S(\mathbf{x},\mathbf{y})]\ Q(S(\mathbf{x},\mathbf{y}))\right]$$

$$= E_{P_I^m}\left[\sum_{\mathbf{y}} Q(S(\mathbf{x},\mathbf{y})) \frac{\sum_{\sigma_b \in \Sigma_B}\sum_{\sigma_a \in \Sigma_A} I_{C_{\sigma(1)} \neq A_m(L_\sigma(\mathbf{x}))}}{|\Sigma_B| \cdot |\Sigma_A|}\right], \quad (8.1.4)$$

where in the last inequality we have used the uniformity of the conditional distribution over $S(\mathbf{x},\mathbf{y})$, and the sum over $\mathbf{y}$ is taken over all *different* vectors in X^m. By (8.1.2), $L_\sigma(\mathbf{x})$ is constant for all $\sigma \in S(\mathbf{x},\mathbf{y})$, so

$$A_m(L_\sigma(\mathbf{x})) = C_T \in \mathcal{C},$$

where $C_T = C_T(\mathbf{x},\mathbf{y}) \in \mathcal{C}$ does not depend on $\sigma \in S(\mathbf{x},\mathbf{y})$. Thus

$$\sum_{\sigma_a \in \Sigma_A} I_{C_{\sigma(1)} \neq A_m(L_\sigma(\mathbf{x}))} \geq (|J^\times| - 1) \cdot (|J^\times| - 1)!,$$

whereas

$$|\Sigma_A| = |J^\times|!.$$

It follows that, for any integer $\eta > 1$,

$$E_Q\{P_\sigma^m[S(\sigma,T)]\} \geq E_{P_I^m}\left[\frac{(|J^\times| - 1) \cdot (|J^\times| - 1)!}{|J^\times|!}\right]$$

$$= 1 - E_{P_I^m}\left[\frac{1}{|J^\times|!}\right] \geq 1 - \frac{1}{\eta} - P_I^m\{\mathbf{x} \in X^m : |J^\times| \leq \eta\}.$$

Here we make use of the obvious fact that $|J^\times|! \geq 1$ for all $\mathbf{x} \in X^m$, and $\eta! \geq \eta$, which in turn implies that

$$E_{P_I^m}\left[\frac{1}{|J^\times|!}\right] \leq \frac{1}{\eta!} + P_I^m\{\mathbf{x} \in X^m : |J^\times| \leq \eta\} \leq \frac{1}{\eta} + P_I^m\{\mathbf{x} \in X^m : |J^\times| \leq \eta\}.$$

To complete the proof, it is shown that by choosing k sufficiently large, the quantity $|J^\times|$ can be made arbitrarily large with high probability. This follows readily, since, for a fixed index $j \in \mathbf{N}$, we have

$$P_I^m\{\mathbf{x} \in X^m : x_{ij} = 1\ \forall i = 1,\ldots,m\} = p_j^m, \text{ and}$$

$$\sum_{j=1}^{\infty} P_I^m\{\mathbf{x} \in X^m : x_{ij} = 1\ \forall i = 1,\ldots,m\} = \sum_{j=1}^{\infty} p_j^m = \infty.$$

Thus, by the Borel-Cantelli lemma [74], p. 263,[1] it follows that

$$P_I^m\{\mathbf{x} \in X^m : x_{ij} = 1\ \forall i \text{ for infinitely many } j\} = 1.$$

Thus, for any η, it is possible to find a k large enough that $P_I^m\{\mathbf{x} \in X^m : |J^\times| \leq \eta\}$ is arbitrarily small.

---

[1] Note that this is the "other half" of the lemma, and is more difficult than the "easy half" proved in Section 2.4.

8.2 Totally Bounded Families of Probabilities

In this section, we study the PAC learnability of both function and concept classes in the case where the family of probabilities $\mathcal{P}$ is totally bounded with respect to the metric ρ, that is,

$$N(\epsilon, \mathcal{P}, \rho) < \infty \ \forall \epsilon > 0.$$

Of course, every finite set (and in particular, a singleton set) is totally bounded. Hence it is not surprising that the results in this section draw heavily on Chapter 6.

Given a probability measure P, define $\bar{\mathcal{B}}(\epsilon, P, \rho)$ to be the closed ball of radius ϵ (in the metric ρ) centered at P. In other words,

$$\bar{\mathcal{B}}(\epsilon, P, \rho) := \{Q \in \mathcal{P}^* : \rho(P, Q) \leq \epsilon\}.$$

The simple-looking lemma below is the basis of all the results in this section.

Lemma 8.2. *Suppose $\mathcal{F}$ is a given function class, P_0 is a given probability measure, and that the covering number $N(\epsilon/4, \mathcal{F}, d_{P_0})$ is finite. Let $\{g_1, \ldots, g_k\}$ be an $\epsilon/4$-cover of $\mathcal{F}$ with respect to d_{P_0}. Then the minimum empirical risk algorithm applied to $\{g_1, \ldots, g_k\}$ is PAC to accuracy ϵ with respect to the family of probabilities $\bar{\mathcal{B}}(\epsilon/4, P_0, \rho)$.*

Proof. Observe that whenever $P \in \bar{\mathcal{B}}(\epsilon/4, P_0, \rho)$, we have that $\rho(P, P_0) \leq \epsilon/4$, and as a consequence $E_P(h) \leq E_{P_0}(h) + \epsilon/4$ for all $h \in \mathcal{F}$. In particular, whenever $f, g \in \mathcal{F}$, we have that

$$d_P(f, g) \leq d_{P_0}(f, g) + \epsilon/4.$$

This inequality shows that, for *every* probability $P \in \bar{\mathcal{B}}(\epsilon/4, P_0, \rho)$, the finite collection $\{g_1, \ldots, g_k\}$ is an $\epsilon/2$-cover (not $\epsilon/4$) of $\mathcal{F}$. Hence, by Theorem 6.2, the minimum empirical risk algorithm applied to $\{g_1, \ldots, g_k\}$ is PAC to accuracy ϵ. ∎

The key to the above proof is the fact that the *same* collection $\{g_1, \ldots, g_k\}$ is an $\epsilon/2$-cover of $\mathcal{F}$ for *every* probability $P \in \bar{\mathcal{B}}(\epsilon/4, P_0, \rho)$. In other words, $\{g_1, \ldots, g_k\}$ is a "common" $\epsilon/2$-cover for each $P \in \bar{\mathcal{B}}(\epsilon/4, P_0, \rho)$.

Theorem 8.6. *Suppose $\mathcal{P}$ is a totally bounded family of probabilities, and that the family of functions $\mathcal{F}$ satisfies the UMBE condition*

$$\sup_{P \in \mathcal{P}} N(\epsilon, \mathcal{F}, d_P) < \infty \ \forall \epsilon > 0.$$

Then $\mathcal{F}$ is PAC learnable with respect to the family $\mathcal{P}$.

We give two different proofs of this theorem. The first proof gives a better estimate of the sample complexity, while the second proof is more easily extended to more general situations.

298   8. Learning Under an Intermediate Family of Probabilities

Proof. **Proof No. 1** For convenience, define

$$\bar{N}(\epsilon, \mathcal{F}, \mathcal{P}) := \sup_{P \in \mathcal{P}} N(\epsilon, \mathcal{F}, d_P).$$

We give an algorithm that takes a given ϵ and δ, and returns a hypothesis that is accurate to within ϵ at least $1 - \delta$ of the time. Using Theorem 3.1, one can then construct another algorithm that is also PAC, but does not explicitly depend on ϵ and δ.

Given ϵ and δ, first let $\mu := N(\epsilon/8, \mathcal{P}, \rho)$, and choose an $\epsilon/8$-cover $\{P_1, \ldots, P_\mu\}$ of $\mathcal{P}$ with respect to the metric ρ. Then, for each i between 1 and μ, choose an $\epsilon/8$-cover $\{g_1^i, \ldots, g_{N_i}^i\}$ for $\mathcal{F}$ with respect to the pseudometric d_{P_i}, such that

$$N_i \leq \bar{N}(\epsilon/8, \mathcal{F}, \mathcal{P}) =: \bar{N}(\epsilon/8).$$

Now choose at least

$$m_0(\epsilon, \delta) := \frac{32}{\epsilon^2} \ln \frac{2\bar{N}(\epsilon/8)}{\delta} + \frac{8}{\epsilon^2} \ln \frac{2\mu}{\delta}$$

i.i.d. samples. Using the first $(32/\epsilon^2) \ln(2\bar{N}(\epsilon/8)/\delta)$ samples, run the minimum empirical risk algorithm on the set of functions $\{g_1^i, \ldots, g_{N_i}^i\}$ for each i between 1 and μ, and denote the resulting output by h^i. Then, using the last $(8/\epsilon^2) \ln(2\mu/\delta)$ or more samples, run the minimum empirical risk algorithm on the set of functions $\{h^1, \ldots, h^\mu\}$, and denote the resulting output by h_m. It is claimed that $d_P(f, h_m) \leq \epsilon$ with probability at least $1 - \delta$, where f is the target function.

To prove the claim, we proceed as follows: As in the proof of Lemma 8.2, it follows that for each index i between 1 and μ, the set of functions $\{g_1^i, \ldots, g_{N_i}^i\}$ forms an $\epsilon/4$-cover (not $\epsilon/8$) of $\mathcal{F}$ with respect to d_P whenever $\rho(P, P_i) \leq \epsilon/8$. If P denotes the probability generating the i.i.d. samples, then the inequality $\rho(P, P_i) \leq \epsilon/8$ holds for *some* index i, though we do not know which i. Since

$$\frac{32}{\epsilon^2} \ln \frac{2\bar{N}(\epsilon/8)}{\delta} = \frac{8}{(\epsilon/2)^2} \ln \frac{\bar{N}(\epsilon/8)}{\delta/2},$$

it follows from Theorem 6.2 that *one* of the intermediate outputs h^i satisfies $d_P(f, h^i) \leq \epsilon/2$ with confidence at least $1 - \delta/2$, though it is not known which one. Running the minimum empirical risk algorithm once again on $\{h^1, \ldots, h^\mu\}$ using the last $(8/\epsilon^2) \ln(2\mu/\delta)$ or more samples ensures that $d_P(f, h_m) \leq \epsilon$ with probability at least $1 - \delta$. ∎

Proof No. 2 The proof consists of showing that, if $\mathcal{F}$ satisfies the UBME condition with respect to $\mathcal{P}$, and in addition $\mathcal{P}$ is totally bounded, then in fact $\mathcal{F}$ is totally bounded with respect to the pseudometric $\bar{d}_\mathcal{P}$. The desired conclusion then follows from Theorem 8.4.

8.2 Totally Bounded Families of Probabilities

For this purpose, let $\mu := N(\epsilon/4, \mathcal{P}, \rho)$, and select an $\epsilon/4$-cover $\{P_1, \ldots, P_\mu\}$ of $\mathcal{P}$ with respect to the metric ρ. (Note that this is a different μ from the one in Proof No. 1.) Define, as in Proof No. 1, $\bar{N} = \bar{N}(\epsilon/8, \mathcal{F}, \mathcal{P})$. For each index i between 1 and μ, select an $\epsilon/8$-cover $\{f_1^i, \ldots, f_{\bar{N}}^i\}$ of $\mathcal{F}$ with respect to d_{P_i}. Note that, in order to make the cardinality of the cover *exactly* equal to $\bar{N}$, one may have to repeat a few elements; this does not affect the argument below. Next, for each $i \in \{1, \ldots, \mu\}$ and each $j \in \{1, \ldots, \bar{N}\}$, define

$$\mathcal{F}_{ij} := \{f \in \mathcal{F} : d_{P_i}(f, f_j^i) \leq \epsilon/8\}.$$

For each vector $\mathbf{k} = [k_1 \ldots k_\mu]^t \in \{1, \ldots, \bar{N}\}^\mu$, define

$$\mathcal{G}_{\mathbf{k}} := \bigcap_{i=1}^{\mu} \mathcal{F}_{i,k_i}.$$

Note that some of these sets could be empty. Finally, choose (if possible) an arbitrary element $g_{\mathbf{k}} \in \mathcal{G}_{\mathbf{k}}$. The total number of such elements is at most $\bar{N}^\mu$. Now define a pseudometric $\bar{d}_\epsilon$ on $\mathcal{F}$ by

$$\bar{d}_\epsilon(f, g) := \bar{d}_{\{P_1, \ldots, P_\mu\}}(f, g) = \max_{1 \leq i \leq \mu} d_{P_i}(f, g).$$

It is claimed that the collection $\{g_{\mathbf{k}}\}$ is an $\epsilon/4$-cover of $\mathcal{F}$ with respect to the pseudometric $\bar{d}_\epsilon$. Suppose $f \in \mathcal{F}$ is arbitrary. Then for each index i there exists a corresponding index k_i such that $f \in \mathcal{F}_{i,k_i}$, since $\{f_1^i, \ldots, f_{\bar{N}}^i\}$ is an $\epsilon/8$-cover for each i. This means that, for each $f \in \mathcal{F}$, there exists a vector $\mathbf{k}$ such that $f \in \mathcal{G}_{\mathbf{k}}$. In other words, the sets $\{\mathcal{G}_{\mathbf{k}}\}$ cover $\mathcal{F}$. Further, each set $\mathcal{G}_{\mathbf{k}}$ has a "diameter" no greater than $\epsilon/4$ in *each* pseudometric d_{P_i}, and hence in the pseudometric $\bar{d}_\epsilon$. As a result, the collection of elements $\{g_{\mathbf{k}}\}$ is an $\epsilon/4$-cover of $\mathcal{F}$ with respect to the pseudometric $\bar{d}_\epsilon$.

Finally, it is shown that the same collection $\{g_{\mathbf{k}}\}$ is also an $\epsilon/2$-cover (not $\epsilon/4$) of $\mathcal{F}$ with respect to the pseudometric $\bar{d}_{\mathcal{P}}$. To see this, let $P \in \mathcal{P}$ be arbitrary, and select an index $i \in \{1, \ldots, \mu\}$ such that $\rho(P, P_i) \leq \epsilon/4$. Then for each $f, g \in \mathcal{F}$ we have

$$d_P(f, g) \leq d_{P_i}(f, g) + \frac{\epsilon}{4}.$$

Therefore

$$\bar{d}_{\mathcal{P}}(f, g) \leq \bar{d}_\epsilon(f, g) + \frac{\epsilon}{4}.$$

As a consequence, an $\epsilon/4$-cover of $\mathcal{F}$ with respect to $\bar{d}_\epsilon$ is also an $\epsilon/2$-cover of $\mathcal{F}$ with respect to $\bar{d}_{\mathcal{P}}$. This shows that $\{g_{\mathbf{k}}\}$ is an $\epsilon/2$-cover of $\mathcal{F}$ with respect to $\bar{d}_{\mathcal{P}}$. ■

Each of the two proofs leads to a different estimate for the sample complexity. Let $\mu(\epsilon)$ denote the covering number $N(\epsilon, \mathcal{P}, \rho)$ of the family of probabilities $\mathcal{P}$ with respect to ρ. Then Proof No. 1 leads to the estimate

$$m_0(\epsilon,\delta) = \frac{32}{\epsilon^2} \ln \frac{2\bar{N}(\epsilon/8)}{\delta} + \frac{8}{\epsilon^2} \ln \frac{2\mu(\epsilon/8)}{\delta}.$$

Proof No. 2 shows that the covering number of $\mathcal{F}$ with respect to $\bar{d}_{\mathcal{P}}$ is no greater than $[\bar{N}(\epsilon/8)]^{\mu(\epsilon/4)}$. Applying Theorem 6.2 leads to the sample complexity estimate

$$m_0(\epsilon,\delta) = \frac{8}{\epsilon^2} \ln \frac{[\bar{N}(\epsilon/8)]^{\mu(\epsilon/4)}}{\delta} = \frac{8}{\epsilon^2}[\mu(\epsilon/4) \ln \bar{N}(\epsilon/8) + \ln(1/\delta)].$$

The first estimate is in general less conservative, since the factor $\mu(\cdot)$ appears as the argument of a logarithmic function, rather than as a multiplicative factor. However, the second estimate is very convenient, and is easily extended to more general situations, as we shall see shortly.

Note that, in the case of concept classes, the above two estimates for the sample complexity can be improved using Theorem 6.3. The revised estimates become

$$m_0(\epsilon,\delta) = \frac{64}{\epsilon} \ln \frac{2\bar{N}(\epsilon/8)}{\delta} + \frac{32}{\epsilon} \ln \frac{2\mu(\epsilon/8)}{\delta},$$

and

$$m_0(\epsilon,\delta) = \frac{32}{\epsilon}[\mu(\epsilon/4) \ln \bar{N}(\epsilon/8) + \ln(1/\delta)],$$

respectively. One can also state the following result.

Corollary 8.1. *Suppose $\mathcal{C}$ is a concept class, and that $\mathcal{P}$ is a totally bounded family of probabilities. Then $\mathcal{C}$ is PAC learnable with respect to $\mathcal{P}$ if and only if $\mathcal{C}$ satisfies the UBME condition with respect to $\mathcal{P}$.*

Proof. The "if" part follows from Theorem 8.6, while the "only if" part follows from Theorem 8.5. ∎

Next we address the following questions: Suppose a family of functions $\mathcal{F}$ is PAC learnable with respect to a totally bounded family of probabilities $\mathcal{P}$. Is $\mathcal{F}$ also PAC learnable with respect to the closure $\bar{\mathcal{P}}$, the convex hull $\mathbf{C}(\mathcal{P})$, and the closed convex hull $\bar{\mathbf{C}}(\mathcal{P})$? In general, the answers are not known. However, if $\mathcal{F}$ is PAC learnable *by virtue of satisfying a UBME condition,* then the answer is "yes" in all cases. Specifically, for *concept* classes, since PAC learnability is equivalent to the UBME condition, the answer is always "yes."

Let us now define the **convex hull** of $\mathcal{P}$, denoted by $\mathbf{C}(\mathcal{P})$. This consists of all probability measures of the form

$$Q = \sum_{i=1}^{l} \lambda_i P_i, \ P_i \in \mathcal{P} \ \forall i, \ \lambda_i \geq 0 \ \forall i, \text{ and } \sum_{i=1}^{l} \lambda_i = 1.$$

The **closed convex hull** of $\mathcal{P}$, denoted by $\overline{\mathbf{C}(\mathcal{P})}$, is the closure of the set $\mathbf{C}(\mathcal{P})$ under the metric ρ. Finally there is the set $\mathbf{C}(\bar{\mathcal{P}})$, the convex hull of the closure of $\mathcal{P}$.

8.2 Totally Bounded Families of Probabilities

Lemma 8.3. *Given $\mathcal{P}$, we have*

$$\bar{\mathcal{P}} \subseteq \mathbf{C}(\bar{\mathcal{P}}) \subseteq \overline{\mathbf{C}(\mathcal{P})}.$$

Proof. The left containment is obvious. To prove the right containment, suppose $Q \in \mathbf{C}(\bar{\mathcal{P}})$. Then there exist constants $\lambda_1, \ldots, \lambda_l$ and probability measures $Q_1, \ldots, Q_l \in \bar{\mathcal{P}}$ such that

$$\lambda_i \geq 0 \ \forall i, \ \sum_{i=1}^{l} \lambda_i = 1, \text{ and } Q = \sum_{i=1}^{l} \lambda_i Q_i.$$

Since each $Q_i \in \bar{\mathcal{P}}$, there exists a sequence $\{P_{ij}\}_{j \geq 1}$ in $\mathcal{P}$ converging to Q_i, for each i. Hence

$$Q = \sum_{i=1}^{l} \lambda_i \lim_{j \to \infty} P_{ij} = \lim_{j \to \infty} \sum_{i=1}^{l} \lambda_i P_{ij} \in \overline{\mathbf{C}(\mathcal{P})}.$$

This completes the proof. ∎

Lemma 8.4. *Suppose $\mathcal{P}$ is totally bounded. Then $\bar{\mathcal{P}}$, $\mathbf{C}(\mathcal{P})$, $\mathbf{C}(\bar{\mathcal{P}})$, and $\overline{\mathbf{C}(\mathcal{P})}$ are all totally bounded.*

Proof. It is obvious that $\bar{\mathcal{P}}$ is totally bounded (in fact, it is compact). Similarly, once it is shown that $\mathbf{C}(\mathcal{P})$ is totally bounded, it follows readily that $\overline{\mathbf{C}(\mathcal{P})}$ is also totally bounded. Finally, since $\mathbf{C}(\bar{\mathcal{P}}) \subseteq \overline{\mathbf{C}(\mathcal{P})}$, it too is totally bounded.

It is a standard fact that the convex hull of a totally bounded set is itself totally bounded; see [92], Theorem 3, p. 70. However, a complete proof is given here for convenience.

Suppose $\epsilon > 0$ is specified; then an ϵ-cover for $\mathbf{C}(\mathcal{P})$ can be constructed as follows: Choose an $\epsilon/2$-cover $\{P_1, \ldots, P_\mu\}$ for $\mathcal{P}$ with respect to the metric ρ. Then every $Q \in \mathbf{C}(\mathcal{P})$ is within a distance of $\epsilon/2$ from the convex hull of the finite set $\{P_1, \ldots, P_\mu\}$. To see this, suppose

$$Q = \sum_{i=1}^{n} \lambda_i Q_i, \ Q_i \in \mathcal{P} \ \forall i, \ \lambda_i \geq 0 \ \forall i, \ \sum_{i=1}^{n} \lambda_i = 1.$$

Now, for every i, there is an index j_i such that $\rho(Q_i, P_{j_i}) \leq \epsilon/2$, because $\{P_1, \ldots, P_\mu\}$ is an $\epsilon/2$-cover for $\mathcal{P}$. Define

$$P = \sum_{i=1}^{n} \lambda_i P_{j_i} \in \mathbf{C}(\{P_1, \ldots, P_\mu\}).$$

Then

$$\rho(P, Q) \leq \sum_{i=1}^{n} \lambda_i \, \rho(Q_i, P_{j_i}) \leq \epsilon/2.$$

Next, observe that the set $\Lambda^\mu \subseteq \mathbb{R}_+^\mu$ defined by

$$\Lambda^\mu = \{(\lambda_1, \ldots, \lambda_\mu) : \lambda_i \geq 0 \ \forall i, \ \sum_{i=1}^\mu \lambda_i = 1\}$$

is compact. Choose an $\epsilon/2$-cover $\{\mathbf{a}^1, \ldots, \mathbf{a}^k\}$ for Λ^μ with respect to the l_1-norm, where each $\mathbf{a}^j$ is a μ-dimensional vector. Thus, given any $\mathbf{b} \in \Lambda^\mu$, there exists an index j such that

$$\sum_{i=1}^\mu |b_i - a_i^j| \leq \epsilon/2.$$

Now define

$$P^j = \sum_{i=1}^\mu a_i^j P_i, \ j = 1, \ldots, k,$$

and observe that each P^j belongs to the convex hull $\mathbf{C}(\{P_1, \ldots, P_\mu\})$. It is claimed that the set $\{P^1, \ldots, P^k\}$ is an $\epsilon/2$-cover for $\mathbf{C}(\{P_1, \ldots, P_\mu\})$. Once this is established, it follows from the preceding paragraph that $\{P^1, \ldots, P_k\}$ is also an ϵ-cover for $\mathbf{C}(\mathcal{P})$. To establish this claim, suppose P belongs to the convex hull of $P_1, \ldots, P_\mu$. To be specific, suppose that

$$P = \sum_{i=1}^\mu \lambda_i P_i, \ \lambda_i \geq 0 \ \forall i, \ \sum_{i=1}^\mu \lambda_i = 1.$$

Choose a vector $\mathbf{a}^j \subset \Lambda^\mu$ such that

$$\sum_{i=1}^\mu |\lambda_i - a_i^j| \leq \epsilon/2,$$

and define the corresponding probability P^j as above. Then, for each $A \in \mathcal{S}$, we have that

$$|P(A) - P^j(A)| = |\sum_{i=1}^\mu (\lambda_i - a_i^j) P_i(A)| \leq \sum_{i=1}^\mu |\lambda_i - a_i^j| \leq \epsilon/2.$$

This completes the proof. ∎

Theorem 8.7. *Suppose $\mathcal{F}$ is a given family of functions, and $\mathcal{P}$ is a totally bounded family of probability measures. Suppose $\mathcal{F}$ satisfies the UBME condition with respect to $\mathcal{P}$. Then $\mathcal{F}$ also satisfies the UBME condition with respect to each of $\bar{\mathcal{P}}$, $\mathbf{C}(\mathcal{P})$, $\mathbf{C}(\bar{\mathcal{P}})$, and $\overline{\mathbf{C}(\mathcal{P})}$. In particular,*

$$N(\epsilon, \mathcal{F}, \bar{d}_\mathcal{P}) = N(\epsilon, \mathcal{F}, \bar{d}_{\mathbf{C}(\mathcal{P})}), \ \forall \epsilon > 0, \ and \quad (8.2.1)$$

$$N(\epsilon, \mathcal{F}, \bar{d}_{\bar{\mathcal{P}}}) = N(\epsilon, \mathcal{F}, \bar{d}_{\mathbf{C}(\bar{\mathcal{P}})}) = N(\epsilon, \mathcal{F}, \bar{d}_{\overline{\mathbf{C}(\mathcal{P})}}) \leq \lim_{\alpha \to \epsilon^-} N(\alpha, \mathcal{F}, \bar{d}_\mathcal{P}). \quad (8.2.2)$$

Finally, $\mathcal{F}$ is PAC learnable with respect to each of $\bar{\mathcal{P}}$, $\mathbf{C}(\mathcal{P})$, $\mathbf{C}(\bar{\mathcal{P}})$, and $\overline{\mathbf{C}(\mathcal{P})}$.

8.2 Totally Bounded Families of Probabilities

Proof. The first relation is established by showing that $\bar{d}_{\mathcal{P}} = \bar{d}_{\mathbf{C}(\mathcal{P})}$. Clearly
$$\bar{d}_{\mathcal{P}}(f,g) \leq \bar{d}_{\mathbf{C}(\mathcal{P})}(f,g), \ \forall f,g \in \mathcal{F},$$
since $\mathcal{P} \subseteq \mathbf{C}(\mathcal{P})$. To prove the reverse inequality, suppose $Q \in \mathbf{C}(\mathcal{P})$ is arbitrary, and suppose
$$Q = \sum_{i=1}^{l} \lambda_i P_i, \ P_i \in \mathcal{P} \ \forall i, \ \lambda_i \geq 0 \ \forall i, \text{ and } \sum_{i=1}^{l} \lambda_i = 1.$$
Then for $f,g \in \mathcal{F}$, we have
$$d_Q(f,g) = \sum_{i=1}^{l} \lambda_i d_{P_i}(f,g) \leq \left(\sum_{i=1}^{l} \lambda_i\right) \bar{d}_{\mathcal{P}}(f,g) = \bar{d}_{\mathcal{P}}(f,g).$$
Since $Q \in \mathbf{C}(\mathcal{P})$ is arbitrary, this establishes the reverse inequality and proves (8.2.1). In the same way it follows that $\bar{d}_{\bar{\mathcal{P}}} = \bar{d}_{\overline{\mathbf{C}(\mathcal{P})}}$. Finally, since $\bar{\mathcal{P}} \subseteq \mathbf{C}(\bar{\mathcal{P}}) \subseteq \overline{\mathbf{C}(\mathcal{P})}$, it follows that $\bar{d}_{\bar{\mathcal{P}}} = \bar{d}_{\mathbf{C}(\bar{\mathcal{P}})} = \bar{d}_{\overline{\mathbf{C}(\mathcal{P})}}$. It only remains to establish the rightmost relationship in (8.2.2).

Suppose $\mathcal{P}, \mathcal{Q} \subseteq \mathcal{P}^*$, and suppose in addition that
$$\mathcal{Q} \subseteq \bar{B}(\lambda, \mathcal{P}, \rho) := \bigcup_{P \in \mathcal{P}} \bar{B}(\lambda, P, \rho).$$
In other words, for every $Q \in \mathcal{Q}$, there exists a $P \in \mathcal{P}$ such that
$$\rho(P, Q) \leq \lambda.$$
Then, for every $f, g \in \mathcal{F}$, we have
$$d_Q(f,g) \leq d_P(f,g) + \lambda.$$
It follows that
$$\bar{d}_\mathcal{Q}(f,g) \leq \bar{d}_\mathcal{P}(f,g) + \lambda, \ \forall f, g \in \mathcal{F}.$$
Consequently, given an $\epsilon > 0$, every $(\epsilon - \lambda)$-cover of $\mathcal{F}$ with respect to $\bar{d}_\mathcal{P}$ is also an ϵ-cover of $\mathcal{F}$ with respect to $\bar{d}_\mathcal{Q}$. Therefore
$$N(\epsilon, \mathcal{F}, \bar{d}_\mathcal{Q}) \leq N(\epsilon - \lambda, \mathcal{F}, \bar{d}_\mathcal{P}).$$
Now let $\mathcal{Q} = \bar{\mathcal{P}}$. Then
$$\bar{\mathcal{P}} \subseteq \bar{B}(\lambda, \mathcal{P}, \rho) \ \forall \lambda > 0.$$
As a result,
$$N(\epsilon, \mathcal{F}, \bar{d}_{\bar{\mathcal{P}}}) \leq N(\epsilon - \lambda, \mathcal{F}, \bar{d}_\mathcal{P}) \ \forall \lambda > 0.$$
Now let $\alpha = \epsilon - \lambda$, and observe that the right side of the above inequality is a nonincreasing function of α. Let $\lambda \to 0^+$, or equivalently, let $\alpha \to \epsilon^-$. This leads to
$$N(\epsilon, \mathcal{F}, \bar{d}_{\bar{\mathcal{P}}}) \leq \lim_{\alpha \to \epsilon^-} N(\alpha, \mathcal{F}, \bar{d}_\mathcal{P}),$$
which is exactly (8.2.2). ∎

Corollary 8.2. *Suppose a concept class $\mathcal{C}$ is PAC learnable with respect to a totally bounded family of probability measures $\mathcal{P}$. Then $\mathcal{C}$ is PAC learnable with respect to each of $\bar{\mathcal{P}}$, $\mathbf{C}(\mathcal{P})$, $\mathbf{C}(\bar{\mathcal{P}})$, and $\overline{\mathbf{C}(\mathcal{P})}$.*

The proof is a ready consequence of Theorem 8.7 and Corollary 8.1.

All of the preceding theory depends in a crucial manner on the assumption that the family of probabilities $\mathcal{P}$ is totally bounded with respect to the metric ρ. If this assumption does not hold, then the various steps in the proofs are no longer valid in general. This is illustrated in the next few examples.

Example 8.2. The purpose of this example is to demonstrate a concept class that is PAC learnable under a family of probability measures $\mathcal{P}$ but is *not* learnable when $\mathcal{P}$ is replaced by its convex hull. Obviously, in view of Corollary 8.2, the family $\mathcal{P}$ is not compact.

Suppose $(X, \mathcal{S})$ is a measurable space where X is infinite, and let $\mathcal{C}$ consist of all measurable maps from X into $\{0,1\}$. Let P_z denote the point measure concentrated at the point $z \in X$, and let $\mathcal{P}$ consist of all such point measures. Thus $\mathcal{P} = \{P_z, z \in X\}$.

First it is shown that the pair $(\mathcal{C}, \mathcal{P})$ has the UCEMUP property, which in turn implies that the pair $(\mathcal{C}, \mathcal{P})$ is PAC learnable. Let $C \in \mathcal{C}, z \in X$ be arbitrary, and let $P = P_z$. Draw i.i.d. samples $x_1, \ldots, x_m$ in accordance with P_z. Then almost surely we have that $x_i = z$ for all i, whence it follows that

$$\hat{P}(C; \mathbf{x}) = I_C(z) = P_z(C), \text{a.s.}$$

Now suppose $\mathcal{P}$ is replaced by its convex hull $\mathbf{C}(\mathcal{P})$. It is claimed that the pair $(\mathcal{C}, \mathbf{C}(\mathcal{P}))$ is *not* PAC learnable. Clearly the concept class $\mathcal{C}$ has infinite VC-dimension, since it shatters the infinite set X. Now let us appeal to Theorem 7.10, which states that $\mathcal{C}$ is therefore not distribution-free PAC learnable. In fact, a perusal of the proof of Theorem 7.10 shows that $\mathcal{C}$ is not PAC learnable with respect to the set of all finitely supported purely atomic measures on X (since these are the only measures used in the proof of Theorem 7.10). Now note that every finitely supported purely atomic measure on X is a convex combination of point measures. Thus $(\mathcal{C}, \mathbf{C}(\mathcal{P}))$ is *not* PAC learnable.

Example 8.3. Once again let $X, \mathcal{P}, \mathcal{C}$ be as in Example 8.1. Proof No. 1 of Theorem 8.6 depends on being able to cover $\mathcal{P}$ with a finite number of balls of radius $\epsilon/8$, for each $\epsilon > 0$. Obviously this property is *equivalent* to $\mathcal{P}$ being totally bounded with respect to the metric ρ. Proof No. 2 is based on the fact that if $\mathcal{C}$ satisfies the UBME condition with respect to $\mathcal{P}$, then $\mathcal{C}$ is totally bounded with respect to $\bar{d}_\mathcal{P}$. The present example serves to show that this argument is not valid in general if $\mathcal{P}$ is not totally bounded. In the present case, $\mathcal{C}$ *does* satisfy the UBME condition, and yet is *not* PAC learnable, which implies that $\mathcal{C}$ is *not* totally bounded with respect to $\bar{d}_\mathcal{P}$ (take the contrapositive of Theorem 8.4).

8.2 Totally Bounded Families of Probabilities   305

More interestingly, Theorem 8.7 shows that if $\mathcal{P}$ is totally bounded, and if $\mathcal{C}$ satisfies the UBME condition with respect to $\bar{d}_\mathcal{P}$, then $\mathcal{C}$ also satisfies the UBME condition with respect to $\bar{d}_{\mathbf{C}(\mathcal{P})}$. One can ask whether or not this statement holds if $\mathcal{P}$ is not assumed to be totally bounded. To put it another way: Suppose $\mathcal{C}$ is PAC learnable with respect to a (not-totally bounded) family $\mathcal{P}$ by virtue of satisfying the sufficient condition of Theorem 8.4. Is it possible to conclude that $\mathcal{C}$ is also PAC learnable with respect to the family $\mathbf{C}(\mathcal{P})$? In general, the answer is "No." The present $\mathcal{C}, \mathcal{P}$ provide an illustration of this.

Lemma 8.5. *Define*

$$\alpha = 1 - \frac{1}{\lg 3} \approx 0.36907, \quad d = 2^\alpha \approx 1.2915.$$

For each sufficiently small $\epsilon < \alpha$ and each integer n, there exists a probability measure $P \in \mathbf{C}(\mathcal{P})$ such that $\mathcal{C}$ contains a set of cardinality $nd^{1/\epsilon}$ that is ϵ-separated with respect to d_P. Therefore, for each sufficiently small $\epsilon < \alpha/2$,

$$\sup_{P \in \mathbf{C}(\mathcal{P})} N(\epsilon, \mathcal{C}, \mathbf{C}(\mathcal{P})) = \infty.$$

The proof of the lemma makes use of the following preliminary result.

Lemma 8.6. *For each sufficiently small $\delta > 0$ and each sufficiently large integer n, there exists another integer $M = 2^{c(\delta,n)/\delta}$, where $c(\delta, n) \to 1$ as $n \to \infty, \delta \to 0$, such that*

$$\frac{1}{n} \sum_{i=1}^{n} \frac{1}{\lg iM} \geq \delta.$$

Proof. of Lemma 8.6 Let $x = \lg M$. Then the above summation can be written as

$$\frac{1}{n} \sum_{i=1}^{n} \frac{1}{\lg iM} = \frac{1}{n} \sum_{i=1}^{n} \frac{1}{x + \lg i} = \frac{N(x)}{D(x)},$$

where $N(x)$ and $D(x)$ are polynomials in x. Specifically,

$$D(x) = n \prod_{i=1}^{n}(x + \lg i) = nx^n + n\left(\sum_{i=1}^{n} \lg i\right) x^{n-1} + \ldots + \left(\prod_{i=2}^{n} \lg i\right) x$$

after observing that $\lg 1 = 0$. Note that there is no constant term (x^0) in $D(x)$. Similarly,

$$N(x) = \sum_{i=1}^{n} \prod_{j \neq i}(x + \lg j) = nx^{n-1} + \left(\sum_{i=1}^{n} \sum_{j \neq i} \lg j\right) x^{n-2} + \ldots + \prod_{i=2}^{n} \lg i.$$

Now note that $\sum_{i=1}^{n} \lg i = \lg n!$. If we define

$$\beta_n = \sum_{i=1}^{n} \sum_{j \neq i} \lg j,$$

then $\beta_n < n \lg n!$, because $\sum_{j \neq i} \lg j < \lg n!$ for all $i > 1$. Now rewrite the desired inequality as

$$N(x) \geq \delta D(x).$$

Observe that $D(0) = 0$, while $N(0) > 0$. Hence the polynomial

$$\phi(x) := \delta D(x) - N(x)$$

satisfies $\phi(0) < 0$, and $\phi(x) \to \infty$ as $x \to \infty$ (because the degree of $D(x)$ is higher than that of $N(x)$). Let $r(\delta, n)$ denote the smallest positive root of the equation $\phi(x) = 0$. It is claimed that

$$r(\delta, n) \approx 1/\delta$$

for sufficiently large n and sufficiently small δ. To show this, we proceed by establishing that (i) there is a root of the form

$$x_0 = \frac{c(\delta, n)}{\delta},$$

where $c(\delta, n) \to 1$ as $\delta \to 0, n \to \infty$, and (ii) $\phi(x) < 0 \ \forall x < x_0$. To prove (i), substitute $x = c/\delta$ into $\phi(x)$. This gives

$$\phi(c/\delta) = n \frac{c^n}{\delta^{n-1}} + n \lg n! \frac{c^{n-1}}{\delta^{n-2}} + \ldots - n \frac{c^{n-1}}{\delta^{n-1}} - \beta_n \frac{c^{n-2}}{\delta^{n-2}} \ldots$$

For $\delta \to 0$, these are the dominant terms. Observe first that

$$\phi(1/\delta) = \frac{n \lg n! - \beta_n}{\delta^{n-2}} + \ldots > 0$$

because $\beta_n < n \lg n!$. So the root x_0 is less than $1/\delta$ as $\delta \to 0$. However, as $\delta \to 0$,

$$\phi(c/\delta) = \frac{1}{\delta^{n-1}} n(c^n - c^{n-1}) + \ldots$$

equals zero when $c \approx 1$. The same argument shows that if $c < 1$, then $c^n < c^{n-1}$, so that $\phi(c/\delta) < 0$ when $\delta \to 0, n \to \infty$. So the *smallest* positive root of the equation $\phi(x) = 0$ roughly equals $1/\delta$. ■

Proof. of Lemma 8.5 Observe that if $C_i, C_j \in \mathcal{C}$, then

$$C_i \Delta C_j = \{x \in X : x_i = 1 \text{ and } x_j = 0, \text{ or } x_i = 0 \text{ and } x_j = 1\}.$$

Therefore

$$P(C_i \Delta C_j) = \phi_i(1 - \phi_j) + (1 - \phi_i)\phi_j,$$

where
$$\phi_i = P(x_i = 1).$$
Let us use d_σ as an abbreviation for d_{P_σ}. If $P = P_I$, then
$$d_I(C_i, C_j) = p_i(1 - p_j) + (1 - p_i)p_j.$$
If $P = P_\sigma$, then
$$d_\sigma(C_i, C_j) = p_{\sigma(i)}(1 - p_{\sigma(j)}) + (1 - p_{\sigma(i)})p_{\sigma(j)}.$$
Given ϵ, n, choose $M \approx 2^{\alpha/\epsilon} = d^{1/\epsilon}$ such that
$$\frac{1}{n}\sum_{i=1}^{n}\frac{1}{\lg i(M+1)} > \frac{\epsilon}{\alpha}.$$
This is possible by Lemma 8.6. Now define a permutation σ on the natural numbers as follows:
$$\sigma(i) = i + M \text{ for } 1 \leq i \leq (n-1)M,$$
$$\sigma(i) = i - (n-1)M \text{ for } (n-1)M < i \leq nM,$$
$$\sigma(i) = i \text{ for } i > nM.$$
In other words, σ is a "block"-cyclic permutation, and $\sigma^n = I$. Now define
$$P = \frac{1}{n}\sum_{l=0}^{n-1} P_{\sigma^l} \in \mathbf{C}(\mathcal{P}),$$
where σ^0 is taken as the identity permutation. It is now shown that the set $\{C_1, \ldots, C_{nM}\}$ is ϵ-separated with respect to the metric d_P. For this purpose, let us compute $d_P(C_i, C_j)$. There are two cases to consider, namely: (i) i and j belong to the same "block" of length M, that is, $(k-1)M + 1 \leq i, j \leq kM$ for some integer k, and (ii) i and j belong to different "blocks."

Case (i) Suppose i, j belong to the same block. In this case, it is easy to see that $\sigma^l(i), \sigma^l(j)$ also belong to the same block for each l. Also, as l varies from 0 to $n-1$, $\sigma^l(i)$ and $\sigma^l(j)$ will visit each of the n blocks $\{1, \ldots, M\}, \ldots, \{(n-1)M+1, \ldots, nM\}$ exactly once. So it may be assumed without loss of generality that $i, j \in \{1, \ldots, M\}$. In this case, we have
$$d_I(C_i, C_j) = p_i(1 - p_j) + (1 - p_i)p_j.$$
If $i = 1, j > 1$, then
$$d_I(C_i, C_j) = 1 - p_j = 1 - \frac{1}{\lg(j+1)} \geq 1 - \frac{1}{\lg(M+1)} \geq \frac{2\alpha}{\lg(M+1)}$$
whenever $M \geq 2^{2\alpha+1} = 2d^2 \approx 3.3360$. If $i, j > 1$, then

$$1 - p_i, 1 - p_j \geq 1 - p_2 = \alpha,$$

$$p_i, p_j \geq \frac{1}{\lg(M+1)},$$

$$d_I(C_i, C_j) \geq \frac{2\alpha}{\lg(M+1)}.$$

Next, $\sigma(i), \sigma(j) \in \{M+1, \ldots, 2M\}$. So

$$d_\sigma(C_i, C_j) = p_{\sigma(i)}(1 - p_{\sigma(j)}) + (1 - p_{\sigma(i)})p_{\sigma(j)} \geq \frac{2\alpha}{\lg(2M+1)}.$$

Similarly,

$$d_{\sigma^l}(C_i, C_j) \geq \frac{2\alpha}{\lg[(l+1)M+1]}, \quad l = 0, 1, \ldots, n-1.$$

Hence

$$d_P(C_i, C_j) = \frac{1}{n}\sum_{l=0}^{n-1} d_{\sigma^l}(C_i, C_j) \geq \frac{1}{n}\sum_{i=1}^{n} \frac{2\alpha}{\lg(iM+1)} > 2\epsilon.$$

Case (ii) Suppose i, j belong to different blocks. By the same logic as in Case (i), it can be assumed without loss of generality that $i \in \{1, \ldots, M\}$ and $j \in \{M+1, \ldots, nM\}$. In this case

$$d_I(C_i, C_j) \geq p_i(1 - p_j) \geq p_M(1 - p_2) = \frac{\alpha}{\lg(M+1)},$$

because $p_i \geq p_M$ and $1 - p_j \geq 1 - p_2$. Similarly

$$d_{\sigma^l}(C_i, C_j) \geq \frac{\alpha}{\lg[(l+1)M+1]},$$

and as a consequence,

$$d_P(C_i, C_j) \geq \frac{1}{n}\sum_{i=1}^{n} \frac{\alpha}{\lg(iM+1)} > \epsilon.$$

This shows that the set $\{C_1, \ldots, C_{nM}\}$ is ϵ-separated with respect to the pseudometric d_P. ∎

8.3 Families of Probabilities with a Nonempty Interior

In this section we study the problem of *concept* learning in the case where the underlying family of probabilities $\mathcal{P}$ has a nonempty interior with respect to ρ. It is shown that, in this case, the concept class is PAC learnable if and only if

8.3 Families of Probabilities with a Nonempty Interior

it has finite VC-dimension. This result can be interpreted to mean that, even if there is an *arbitrarily small amount of nonparametric uncertainty* about the probability measure generating the learning samples, concept learning essentially reduces to distribution-free learning.

Suppose $P_0 \in \mathcal{P}^*$ and $\lambda \in [0, 1]$. Then we define

$$\bar{\mathcal{B}}_c(\lambda, P_0) := \{(1 - \mu)P_0 + \mu Q : Q \in \mathcal{P}^* \text{ and } \mu \in [0, \lambda]\}.$$

Clearly $\bar{\mathcal{B}}_c(\lambda, P_0)$ consists of all probability measures that are convex combinations of P_0 and an *arbitrary* element $Q \in \mathcal{P}^*$, where the coefficient of Q does not exceed λ. One can think of $\bar{\mathcal{B}}_c(\lambda, P_0)$ as consisting of all probabilities that are essentially equal to P_0, but contain a "nonparametric" uncertainty up to λ in extent. It is obvious that

$$\bar{\mathcal{B}}_c(\lambda, P_0) \subseteq \bar{\mathcal{B}}(\lambda, P_0, \rho),$$

$$\bar{\mathcal{B}}_c(0, P_0) = \bar{\mathcal{B}}(0, P_0, \rho) = \{P_0\}, \text{ and } \bar{\mathcal{B}}_c(1, P_0) = \bar{\mathcal{B}}(1, P_0, \rho) = \mathcal{P}^*.$$

Now we come to the main result of this section.

Theorem 8.8. *Suppose $\mathcal{C} \subseteq \mathcal{S}$ is a given concept class, and that $P_0 \in \mathcal{P}$. Then the following statements are equivalent:*

1. *$\mathcal{C}$ is PAC learnable with respect to $\bar{\mathcal{B}}(\lambda, P_0, \rho)$ for some $\lambda > 0$.*
2. *$\mathcal{C}$ has finite VC-dimension.*
3. *$\mathcal{C}$ is PAC learnable with respect to $\mathcal{P}^*$.*

Proof. (2) $\Rightarrow$ (3) is shown in Chapter 7, and (3) $\Rightarrow$ (1) is obvious since $\bar{\mathcal{B}}(\lambda, P_0, \rho)$ is a subset of $\mathcal{P}^*$. Thus it only remains to show that (1) $\Rightarrow$ (2).

Suppose $\mathcal{C}$ is PAC learnable with respect to $\bar{\mathcal{B}}(\lambda, P_0, \rho)$ for some $\lambda > 0$. Then, by Theorem 8.5, it follows that $\mathcal{C}$ satisfies the UBME condition with respect to $\bar{\mathcal{B}}(\lambda, P_0, \rho)$; that is,

$$\sup_{Q \in \bar{\mathcal{B}}(\lambda, P_0, \rho)} N(\epsilon, \mathcal{C}, d_Q) < \infty, \forall \epsilon > 0.$$

Now let $P \in \mathcal{P}^*$ be arbitrary and let $Q = \lambda P + (1 - \lambda)P_0$. Then, for every $A, B \subset \mathcal{S}$, we have

$$d_Q(A, B) = \lambda P(A \Delta B) + (1 - \lambda)P_0(A \Delta B) \geq \lambda d_P(A, B).$$

Therefore, for every $\epsilon > 0$, an ϵ-cover of $\mathcal{C}$ with respect to d_Q is also an ϵ/λ-cover of $\mathcal{C}$ with respect to d_P. This shows that

$$\sup_{P \in \mathcal{P}^*} N(\epsilon/\lambda, \mathcal{C}, d_P) < \infty, \forall \epsilon > 0.$$

Now apply Corollary 7.1, after throwing away the extraneous factor $1/\lambda$. This shows that $\mathcal{C}$ has finite VC-dimension. ∎

Corollary 8.3. *Suppose $\mathcal{C}$ is a given concept class, and that $\mathcal{P}$ is a family of probability measures with a nonempty interior. Then $\mathcal{C}$ is PAC learnable with respect to $\mathcal{P}$ if and only if $\mathcal{C}$ is distribution-free PAC learnable.*

Proof. The "if" part is obvious, since $\mathcal{P}$ is a subset of $\mathcal{P}^*$. To prove the "only if" part, observe that if $\mathcal{P}$ has a nonempty interior, then it contains a ball $\bar{\mathcal{B}}(\lambda, P_0, \rho)$ for some $\lambda > 0$. Moreover, if $\mathcal{C}$ is PAC learnable with respect to $\mathcal{P}$, then it is also PAC learnable with respect to $\bar{\mathcal{B}}(\lambda, P_0, \rho)$, which is a subset of $\mathcal{P}$. By Theorem 8.8, this implies that $\mathcal{C}$ has finite VC-dimension, and is thus distribution-free PAC learnable. ∎

Corollary 8.4. *Suppose $\mathcal{A} \subseteq \mathcal{S}$, $\mathcal{P} \subseteq \mathcal{P}^*$, and that $\mathcal{P}$ has a nonempty interior. Then $\mathcal{A}$ has the UCEPUP property with respect to $\mathcal{P}$ if and only if VC-dim$(\mathcal{A}) < \infty$.*

Proof. The "if" part follows from Theorem 7.1. To prove the "only if" part, suppose $\mathcal{A}$ has the UCEPUP property with respect to $\mathcal{P}$. Then, by Theorem 8.1, it follows that $\mathcal{A}$ is PAC learnable with respect to $\mathcal{P}$. Now apply Corollary 8.3. ∎

Notes and References The characterization of consistent learnability in terms of the shrinking width property is given here for the first time. Theorem 8.3 is due to Kulkarni [109] and is reproduced in [112]. The fact that the UBME condition is a universal necessary condition for a concept class to be PAC learnable was observed by Benedek-Itai [22], who also conjectured that the UBME condition is sufficient as well as necessary for a concept class to be PAC learnable. The counterexample (Example 8.1) showing that this is not so is due to Dudley *et al.* [64]. Lemma 8.2, 8.6 and 8.7 are all taken from [112], as is Example 8.3. Example 8.2 is from [204]. Finally, Theorem 8.8 is due to Kulkarni [109] and is also reproduced in [112].

Other authors have also studied learning under an intermediate family of probability measures; see for example [18].

9. Alternate Models of Learning

Up to now, we have examined a more or less standard model of learning, which has three characteristic features:

1. The "algorithm" used to map the data into the hypothesis is viewed merely as some function mapping an appropriate "data space" into the hypothesis class. In particular, no restrictions are placed on the *nature* of this function, for example, requiring that the function be efficiently computable.
2. The data that forms the input to the algorithm is assumed to be generated *at random* according to some (possibly unknown) probability measure. In particular, the learner is "passive," and does not have the option of *choosing* the next input to the oracle, with a view towards speeding up the learning process.
3. The efficacy of learning, as measured by the quantity $r(m, \epsilon)$ defined in (3.2.1), is essentially a *worst-case* estimate, since a supremum is taken both with respect to the target concept T as well as the probability measure P. This definition of the speed of learning does not cater to the situation where there exists a prior probability distribution on the target concepts themselves, and a learning algorithm works reasonably well for "most" target concepts.

In this chapter, we study learning problems in the case where one or the other of these features is absent. We begin by introducing the notion of "efficient" learnability, which corresponds roughly to the requirement that the concept class should be learnable using an algorithm that is efficiently computable, and that the sample complexity should grow at a polynomial rate as both the accuracy and the confidence parameters approach zero. This additional requirement that the learning algorithm be *efficiently* computable is precisely the feature that distinguishes *computational* learning theory as studied in the computer science literature (see, e.g., [147], [9], [99]) from *statistical* learning theory as discussed in the preceding chapters of the present book. Then we discuss the notion of "active" learning, in contrast to the "passive" brand of learning discussed in the preceding chapters. As the name implies, in active learning the learner is able to exercise some control over the generation of the data that forms the input to the learning algorithm. At the simplest level of active learning, the learner may be permitted to select an element of the input space, and query the oracle as to whether or not the chosen element belongs to the unknown target concept. A more powerful, and in a certain sense the *most* powerful, form of active learning consists of permitting the learner to make *arbitrary* binary queries. It turns out that the results in active learning are quite satisfactory, in the following sense: If a concept class is actively learnable, then it continues to be learnable even if the learning is passive; the only difference is that the number of samples required to achieve a specified level of accuracy might become significantly larger. In other words, the mere ability to direct the learning process does not result in a previously unlearnable concept class suddenly becoming learnable. At best, the ability to direct the learning can only speed up the learning process. Now it has already been shown in Chapters 6 and 7 that, if a concept class is (passively) learnable with a perfect oracle, then it continues to be learnable even if the oracle occasionally gives out an incorrect output. In other words, the presence of measurement noise is not enough to destroy learnability, though it might slow down the learning process. Taken together, these facts mean that learnability is in some sense *intrinsic*; that is, a concept class is learnable in *any one* of the three models of learning (active, passive with perfect oracles, and passive with noisy oracles) if and only if it is learnable in *all* of them – the only difference is in the sample complexity. Finally, we also examine the case where there is some *prior information* about the target concept. It turns out that the conditions for learnability in this case are qualitatively different from those in other types of learning.

Throughout the chapter, attention is restricted to *concept* learning. It may perhaps be possible to extend most of the results presented here to problems of *function* learning. However, this would be achieved at the cost of increasing the complexity of the arguments considerably.

9.1 Efficient Learning

In this section, we introduce the notion of efficient learning, and present some examples of concept classes that *are* efficiently learnable, as well as other examples of concept classes that are believed *not* to be efficiently learnable. The reader will observe that the results presented in this section are in sharp contrast to the previously stated results on statistical learning theory. In the latter case, it is possible to obtain "universal" necessary and/or sufficient conditions for learnability, as in Chapters 6 through 8. However, adding the requirement that learning must take place through an *efficient* algorithm makes it difficult to obtain universal results, and forces us to proceed on a case by case basis.

9.1.1 Definition of Efficient Learnability

The aim of this subsection is to formulate a definition of an efficiently learnable concept class. We begin with the notion of an efficient *algorithm*, since this notion is central to the definition.

The branch of computer science known as "complexity theory" deals with the *rate* at which the number of operations required by an algorithm grows with the "size" of the input. The reader is directed to any of the standard texts in complexity theory, e.g. [73], for an introduction to the subject; what follows is the most desultory of discussions. Ideally, one would like to measure the "time" taken by the algorithm in a manner that does not depend on the computing speed of the machine on which the algorithm is implemented. One way of achieving such a normalization is to count the number of "unit operations" performed by the algorithm in order to generate its output. This convention still leaves open the question of just what constitutes a unit operation. For instance, one could define a unit operation to be the addition of two bits, or the multiplication of two bits. Thus, if x, y are two numbers represented to b bits of accuracy, then computing $x + y$ requires $3b$ operations or fewer, while computing $x \cdot y$ requires $b^2 + b$ or fewer operations. It is customary *not* to go into such excruciating detail, and to assume that all data in the problem is specified to a prespecified finite number of bits of accuracy (the number b above). With this assumption, it is possible to treat both addition and multiplication of two elements in the given data structure as unit operations, and to measure the time taken by the algorithm in terms of the number of *arithmetical* operations performed by the algorithm. For instance, let X denote the set of integers between 0 and $2^b - 1$, i.e., the integers that can be represented using b or fewer bits, and suppose $\mathbf{x}, \mathbf{y} \in X^n$. Then finding the vector sum $\mathbf{x} + \mathbf{y}$ requires n arithmetic operations on X. Similarly, if $\mathbf{A}, \mathbf{B} \in X^{n \times n}$, then finding the matrix product $\mathbf{A} \cdot \mathbf{B}$ requires $2n^3$ arithmetic operations on X, if the product is computed according to the familiar rule

$$(AB)_{ij} = \sum_{k=1}^{n} a_{ik} b_{kj}.$$

An algorithm is said to be **polynomial-time** if there exist constants M and α such that, for *all* data of size n, the algorithm requires Mn^α or fewer operations to produce its output. Since the above bound is required to hold for *all* data, the requirement is that, *in the worst case*, the algorithm should not require more than Mn^α operations. It is customary to ignore the constant M, and to speak of the algorithm as requiring $O(n^\alpha)$ operations; this convention is consistent with the philosophy that complexity theory concerns itself only with the *rate* at which the number of operations grows as a function of the input size. Thus an algorithm is polynomial-time if its running time grows no faster than a polynomial in the size of the input. For instance, in the example of vector addition described above, if one takes the number of components of the vectors $\mathbf{x}, \mathbf{y}$ as the size of the input, then $\alpha = 1$, whereas in the case of matrix multiplication, $\alpha = 3$. Note that a polynomial-time algorithm is also referred to as an **efficient** algorithm.

Now let us discuss the concept learning problem. For the convenience of the reader, the relevant notation is recalled from Chapter 3. One is given

- A set X,
- A σ-algebra $\mathcal{S}$ of subsets of X,
- A family $\mathcal{P}$ of probability measures on the pair $(X, \mathcal{S})$, and
- A collection of sets $\mathcal{C} \subseteq \mathcal{S}$.

Recall that $\mathcal{C}$ is called the **concept class** that is to be learnt. Suppose $P \in \mathcal{P}$ is a fixed and possibly unknown probability measure on $(X, \mathcal{S})$, and $T \in \mathcal{C}$ is a fixed but unknown target concept. Learning takes place as follows: Independent and identically distributed samples $x_1, \ldots, x_m$ are generated from X in accordance with P, and for each sample x_i, an "oracle" returns the value of the indicator function $I_T(x_i)$. The "labelled sample"

$$[(x_1, I_T(x_1)), \ldots, (x_m, I_T(x_m))] \in [X \times \{0,1\}]^m$$

is fed into an "algorithm" A_m, which is a map from $[X \times \{0,1\}]^m$ into $\mathcal{C}$. Let $\mathbf{x} := [x_1 \ldots x_m]^t \in X^m$ denote the multisample generated at random, and let $H(T; \mathbf{x}) \in \mathcal{C}$ denote the hypothesis generated by the algorithm when the target concept is T and the multisample is $\mathbf{x}$. Then the quantity

$$d_P[T, H(T; \mathbf{x})] := P[T \Delta H(T; \mathbf{x})]$$

gives a quantitative measure of the disparity between the target concept T and the hypothesis $H(T; \mathbf{x})$. Finally, the quantity

$$r(m, \epsilon) := \sup_{P \in \mathcal{P}} \sup_{T \in \mathcal{C}} P^m \{\mathbf{x} \in X^m : d_P[T, H_m(T; \mathbf{x})] > \epsilon\}$$

measures the efficacy of the learning algorithm. Note that $r(m, \epsilon)$ is the probability of generating a multisample of length m that leads to a hypothesis which is at a distance of more than ϵ from the unknown target concept.

What is described above is the very general formulation of the learning problem studied in the preceding chapters. In order to introduce the notion of "efficiency" into the problem formulation, it is now assumed that the various entities above are "graded," as described next. Suppose

$$X = \{X_n\}_{n=1}^{\infty}, \ \mathcal{S} = \{\mathcal{S}_n\}_{n=1}^{\infty}, \ \mathcal{C} = \{\mathcal{C}_n\}_{n=1}^{\infty}, \ \mathcal{P} = \{\mathcal{P}_n\}_{n=1}^{\infty}, \quad (9.1.1)$$

where $\mathcal{S}_n$ is a σ-algebra of subsets of X_n for each n, $\mathcal{C}_n \subseteq \mathcal{S}_n$ for each n, and $\mathcal{P}_n$ is a family of probability measures on $(X_n, \mathcal{S}_n)$ for each n. The integer n provides a measure of the "size" of the inputs to the learning problem; the various examples given below bring out the role of n. It is assumed that the value of n is known to the learner, and that all samples $x_1, \ldots, x_m$ belong to the *same* X_n. Thus, instead of a single learning problem, one is actually given an *indexed sequence* of learning problems, one for each value of n. This being the case, it makes sense to think of the algorithm not as a *singly* indexed family of maps $\{A_m\}$, but as a *doubly* indexed family of maps $\{A_{m,n}\}$, where $A_{m,n}$ maps $[X_n \times \{0,1\}]^m$ into $\mathcal{C}_n$. Similarly, the quantity $r(m,\epsilon)$ is now replaced by

$$r_n(m,\epsilon) := \sup_{P \in \mathcal{P}_n} \sup_{T \in \mathcal{C}_n} P^m\{\mathbf{x} \in X_n^m : d_P[T, H_{m,n}(T;\mathbf{x})] > \epsilon\}.$$

With this revised notation we are in a position to define efficient learnability.

Definition The algorithm $\{A_{m,n}\}$ is said to **efficiently learn** the graded collection $(X, \mathcal{S}, \mathcal{C}, \mathcal{P})$ if the following two properties hold:

(i) The algorithm is polynomial-time with respect to m and n; that is, for each fixed m, n, the number of operations used by the map

$$A_{m,n} : [X_n \times \{0,1\}]^m \to \mathcal{C}_n$$

to produce its output is bounded by a polynomial in m, n.

(ii) For each fixed $n \geq 1$, $\epsilon > 0$, and $\delta > 0$, there exists a number $m_n(\epsilon, \delta)$ that is bounded by a polynomial in n, $1/\epsilon$, and $\ln(1/\delta)$ such that

$$r_n(m, \epsilon) \leq \delta \ \forall m \geq m_n(\epsilon, \delta).$$

The graded collection $(X, \mathcal{S}, \mathcal{C}, \mathcal{P})$ is said to be **efficiently learnable** if there exists an efficient learning algorithm.

Comparing the definition of efficient learnability given above with the definition of PAC learnability given in Chapter 3, one can spot two additional requirements that a PAC algorithm must satisfy in order to qualify as an efficient learning algorithm. First, given an n (problem size), ϵ (accuracy), and δ (confidence), the algorithm must be capable of producing a hypothesis that is PAC to accuracy ϵ and confidence δ using a number of learning samples that is polynomially bounded not only with respect to $1/\epsilon$ and $\ln(1/\delta)$ but also with respect to n. Second, for each fixed n (problem size) and m (sample size), the algorithm must "run" (i.e., produce its output) within polynomial-time with respect to m and n.

Before proceeding to several examples that illustrate the definition of efficient learnability, and in particular the role of the "problem size" parameter n, a slight detour is made to clear up a small technical point. In the computer science literature, it is common to require that the sample complexity $m_n(\epsilon, \delta)$ be bounded by a polynomial in n, $1/\epsilon$ and $1/\delta$, as opposed to $\ln(1/\delta)$ as is the case here. Since $\ln(1/\delta)$ grows much more slowly than $1/\delta$ as $\delta \to 0^+$, the present requirement appears at first glance to be more stringent than the other one prevalent in the computer science community. The point of the next lemma is to show that this is not so – in fact, both notions are equivalent. Actually, the lemma establishes a much stronger result, namely that efficient learnability to a *fixed* nonzero confidence implies efficient learnability to *arbitrarily small* confidence.

Lemma 9.1. [66] *Given the graded collection* $(X, \mathcal{S}, \mathcal{C}, \mathcal{P})$, *suppose there exists an algorithm* $\{B_{m,n}\}$ *where*

$$B_{m,n} : [X_n \times \{0,1\}]^m \to \mathcal{C}_n$$

such that the following two properties hold:

(i) *The algorithm* $\{B_{m,n}\}$ *is polynomial-time with respect to* m, n.
(ii) *There exist a number* $\delta_0 \in (0,1)$ *and a polynomial* $\phi : \mathbb{R} \to \mathbb{R}$ *such that, for each* $\epsilon > 0$, *we have*

$$r_n(m, \epsilon) \leq \delta_0 \; \forall m \geq \phi(n/\epsilon).$$

Then the collection $(X, \mathcal{S}, \mathcal{C}, \mathcal{P})$ *is efficiently learnable.*

Proof. The idea is to define another algorithm $\{A_{m,n}\}$ that repeatedly calls $\{B_{m,n}\}$ as a "subroutine," and efficiently learns $(X, \mathcal{S}, \mathcal{C}, \mathcal{P})$. Let $\psi : \mathbb{R} \to \mathbb{R}$ be a polynomial such that the running time of $B_{m,n}$ is bounded by $\psi(mn)$. Suppose ϵ, δ are specified, and define the algorithm as follows: First, define $k := \lceil \log_{(1/\delta_0)}(2/\delta) \rceil$, and $m_0 := \lceil \phi(2n/\epsilon) \rceil$. Run off k sets of m_0 i.i.d. samples $\{x_i^j, 1 \leq i \leq m_0, 1 \leq j \leq k\}$. Using the m_0 samples $x_1^j, \ldots, x_{m_0}^j$, generate a hypothesis $H_j \in \mathcal{C}_n$ using the map $B_{m_0,n}$. Then run off another $m_e := (32/\epsilon) \ln(2k/\delta)$ i.i.d. samples, apply the minimal empirical risk algorithm on these samples to the set $\{H_1, \ldots, H_k\}$, and declare the "winner" H to be the hypothesis produced by the algorithm $\{A_{m,n}\}$. Now it is necessary to establish three properties:

(i) The number of samples m drawn by the algorithm is polynomial in $n, 1/\epsilon$, and $\ln(1/\delta)$.
(ii) The number of operations taken by $A_{m,n}$ is polynomial in m and n.
(iii) The hypothesis H produced by the algorithm is PAC to accuracy ϵ and confidence δ.

Each of these properties is established in turn. First, the number of samples drawn by the algorithm is

$$m := km_0 + m_e = \lceil \log_{(1/\delta_0)}(2/\delta) \rceil \cdot \lceil \phi(2n/\epsilon) \rceil + \frac{32}{\epsilon} \ln \frac{2k}{\delta},$$

which is clearly polynomial in $n, 1/\epsilon$ and $\ln(1/\delta)$. Second, the algorithm $A_{m,n}$ makes k calls to $B_{m_0,n}$ and then performs one empirical risk minimization. Since it is assumed that $B_{m_0,n}$ runs in polynomial time in m_0, n, and since $k, m_0 \leq m$, it follows that $A_{m,n}$ runs in polynomial time in m, n. Finally, suppose $P \in \mathcal{P}_n, T \in \mathcal{C}_n$ are fixed, and define

$$S := \{\mathbf{x} \in X_n^{m_0} : d_P[T, H_B(T; \mathbf{x})] > \epsilon/2\}.$$

In other words, S is the set of multisamples of length m_0 such that, when the algorithm $B_{m_0,n}$ is applied to the corresponding labelled multisample with the target concept T, the resulting hypothesis $H_B(T;\mathbf{x})$ is more than $\epsilon/2$-far from T. Since $m_0 := \lceil \phi(2n/\epsilon) \rceil$, it follows that $P^{m_0}(S) \leq \delta_0$. Consequently,

$$P^{km_0}(S^k) = [P^{m_0}(S)]^k \leq \delta_0^k \leq \delta/2,$$

since $k := \lceil \log_{(1/\delta_0)}(2/\delta) \rceil$. In other words, it can be asserted with confidence $\geq 1 - \delta/2$ that for at least one value of the integer j, the multisample $(x_1^j, \ldots, x_{m_0}^j)$ does *not* belong to S. Consequently, it can be asserted with confidence $\geq 1 - \delta/2$ that *at least one* of the intermediate hypotheses $H_1, \ldots, H_k$ is within a distance of $\epsilon/2$ from T. If this is so, then it follows from Theorem 6.3 that the output H of the minimum empirical risk algorithm satisfies $d_P(T, H) \leq \epsilon$ with confidence $\geq 1 - \delta/2$. Putting these two facts together shows that the inequality $d_P(T, H) \leq \epsilon$ holds with confidence $\geq 1 - \delta$. ∎

9.1.2 The Complexity of Finding a Consistent Hypothesis

In this subsection, we present several examples of graded concept classes, of which some are demonstrably efficiently learnable while others are believed not to be efficiently learnable. The contents of this subsection highlight the fact that one of the major barriers to efficient learnability is the computational complexity of finding a hypothesis that is consistent with the given data. The examples also serve to highlight the role played by the *representation* of the concept class in making the problem easier or more difficult.

Let us begin by discussing the role played by the VC-dimension in efficient learnability. As before, let $(X, \mathcal{S}, \mathcal{C}, \mathcal{P})$ be a graded collection, so that each of the above objects is an indexed collection, as in (9.1.1). Further, suppose that for each n, the family $\mathcal{P}_n$ equals $\mathcal{P}_n^*$, the set of *all* probability measures on $(X_n, \mathcal{S}_n)$. Thus the problem at hand is one of *distribution-free* efficient learnability of a graded concept class. Suppose for the sake of discussion that, for each value of n, the concept class $\mathcal{C}_n$ has finite VC-dimension, call it d_n. It is often the case that $\mathcal{C}_n$ has finite cardinality for each n, so the above assumption is not very restrictive. Now one can conceive of two possible scenarios: (i) d_n increases faster than any polynomial in n, and (ii) d_n is bounded by a polynomial in n. In the first scenario, the problem at hand is definitely *not* efficiently learnable, because by Theorem 7.9, learning $\mathcal{C}_n$ to accuracy ϵ and confidence δ requires at least

$$\bar{m}_n(\epsilon, \delta) := \max \left\{ \frac{d_n - 1}{32\epsilon}, \frac{1-\epsilon}{\epsilon} \ln \frac{1}{\delta} \right\}$$

samples, irrespective of the algorithm used. Since the number d_n/ϵ increases faster than any polynomial in n, efficient learning is not possible. Now let us examine the second scenario, wherein d_n is bounded by a polynomial in n. In this case, Theorem 7.8 implies that if we generate at least

318   9. Alternate Models of Learning

$$m_n(\epsilon, \delta) := \max\left\{\frac{8d_n}{\epsilon} \lg \frac{8e}{\epsilon}, \frac{4}{\epsilon} \lg \frac{2}{\delta}\right\}$$

samples, *and choose a hypothesis consistent with the data*, then such a hypothesis will be PAC to accuracy ϵ and confidence δ. In the present case, the sample complexity $m_n(\epsilon, \delta)$ is bounded by a polynomial in $n, 1/\epsilon$ and $\ln(1/\delta)$. Thus, if it were possible to generate a hypothesis consistent with the data *using a polynomial-time (in m, n) algorithm*, then the graded concept class would be efficiently learnable. Note that this is only a *sufficient* condition for efficient learnability.

Unfortunately, in many situations, the problem of finding a consistent hypothesis turns out to be NP-complete or NP-hard. It is widely believed that no polynomial-time algorithms exist for such problems. The reader is referred to [73] for a thorough discussion of NP-completeness, but the basic idea is this: A decision problem is said to be "NP" (Nondeterministic Polynomial) if, roughly speaking, it is possible to determine in polynomial time whether or not a proposed "candidate" solution to the problem in fact *is* a solution. A decision problem is said to be "NP-complete" if (i) the problem is NP, and (ii) every other NP problem can be "reduced" to it in such a way that the size of the reduced problem is bounded by a polynomial in the size of the original problem. Thus, if there were to exist a polynomial-time algorithm for an NP-complete problem, then in fact there would exist a polynomial-time algorithm for *every* NP problem. Roughly speaking, an NP-complete problem can be thought as being a "most difficult" one amongst NP problems. A problem is said to be "NP-hard" if every NP-problem can be reduced to this problem; in other words, an NP-hard problem need not itself be NP. One of the most famous open questions in complexity theory can be paraphrased as: "P = NP?" In other words, can every NP problem be solved in polynomial-time? From the discussion above, it can be seen that the above question is equivalent to asking: Does there exist a polynomial-time algorithm for an NP-complete problem? It is widely believed that the answer is "No." Hence, for all practical purposes, an NP-complete problem is considered to be intractable. There is another intermeditate family of decision problems known as RP (Randomized Polynomial), consisting of problems that can be solved in polynomial-time using a randomized algorithm. It is easy to show that P is a subset of RP, which in turn is a subset of NP. However, it is still an open question as to whether the two containments are strict. Once again, it is widely believed that both containments are indeed strict, but these conjectures are open at present.

Various examples are now presented. All except one of these examples have to do with learning Boolean formulas, so we begin with a discussion of Boolean formulas in general. For each integer n, let $U_n := \{u_1, \ldots, u_n\}$ denote the universe of discourse. The elements u_1 through u_n are called **Boolean variables**, in that they can assume the values 1 (for "true") or 0 (for "false"). A **literal** is either a Boolean variable u_i or its negation, denoted by $\neg u_i$. Thus there are $2n$ literals in all. A **valuation** on U_n is a map $v : U_n \to \{0,1\}$; one can think of a valuation as an assignment of "true" or "false" values to each Boolean variable. Let X_n denote the set of valuations on U_n. It is obvious that $|X_n| = 2^n$, and it is natural to identify X_n with $\{0,1\}^n$. A **Boolean function** on U_n is a map $f : X_n \to \{0,1\}$. Thus a Boolean function assigns a true or false value to each assignment of truth values to each of the Boolean variables. It is clear that there are exactly 2^{2^n} Boolean functions on U_n. Since we are also interested in the issue of *representation*, it is necessary to make a distinction between a Boolean *function* and a Boolean *formula*. Leaving aside minor technicalities, a **Boolean formula** is a legal string containing the $2n$ literals $u_1, \ldots, u_n, \neg u_1, \ldots, \neg u_n$, the connectives $\wedge$ (and), $\vee$ (or), and the parenthesis symbols (,). It is clear that the same *function* can be represented by several different *formulas*. For example, the familiar distributive laws state that the formulas $u_1 \wedge (u_2 \vee u_3)$ and $(u_1 \wedge u_2) \vee (u_1 \wedge u_3)$ are equivalent, and similarly the formulas $u_1 \vee (u_2 \wedge u_3)$ and $(u_1 \vee u_2) \wedge (u_1 \vee u_3)$ are equivalent.

Among all possible ways of representing a function as a formula, two types or representations are widely used. A formula is said to be in **conjunctive normal form (CNF)** if it is a conjunction of several "clauses," each of which is a disjunction of some literals. For example,

$$(u_1 \vee \neg u_2 \vee u_7) \wedge (\neg u_1 \vee u_4 \vee \neg u_5) \wedge (\neg u_3 \vee u_6 \vee u_7)$$

is a formula in CNF over the universe $U_7 = \{u_1, \ldots, u_7\}$. Similarly, a formula is said to be in **disjunctive normal form (DNF)** if it is a disjunction of several clauses, each of which is a conjunction of some literals. For example,

$$(u_2 \wedge u_5 \wedge \neg u_7) \vee (u_1 \wedge u_3 \wedge \neg u_4 \wedge u_6) \vee (\neg u_2 \wedge u_4 \wedge \neg u_6)$$

is a formula in DNF over U_7. The clauses in a CNF are referred to as "disjunctive clauses," while the clauses in a DNF are referred to as "conjunctive clauses." An advantage of these normal forms is that they are essentially unique. In other words, given a Boolean function $f : X_n \to \{0,1\}$, there exists a more or less unique *formula* in CNF that is equivalent to f. We say "more or less unique," since it is always possible to permute the order of the clauses, and/or permute the order of the literals within each clause, and still have an equivalent formula. It turns out that this is the only possible source of nonuniqueness. In other words, if ϕ_1 and ϕ_2 are two equivalent formulas, each of which is in CNF (or DNF), it is possible to transform one of the formulas into the other by repeatedly applying one or the other of the above permutations.

In all the examples below involving Boolean formulas, the integer n representing the size of the problem is equal to the number of Boolean variables. The set X_n equals the set of valuations on U_n, and is identified with $\{0,1\}^n$ in a natural way. The σ-algebra $\mathcal{S}_n$ equals the collection of all subsets of X_n, and has cardinality 2^{2^n}. The family $\mathcal{P}_n$ equals the set of *all* probability measures on X_n. Finally, the concept class $\mathcal{C}_n$ varies from one example to another. As is common practice, instead of dealing with a *collection of subsets* $\mathcal{C}_n$ of X_n, we deal with a *family of binary-valued functions* $\mathcal{F}_n$ on X_n. Thus, for each $f \in \mathcal{F}_n$, the associated concept is the set of all valuations on U_n that cause f to evaluate to 1 ("true"). Throughout, the symbol d_n is used to denote the VC-dimension of $\mathcal{F}_n$. Since the set X_n is finite for each n, so is $\mathcal{F}_n$ for each n. In all the examples below, it is enough to use the crude estimate

$$d_n \leq \lfloor \lg |\mathcal{F}_n| \rfloor.$$

Example 9.1. We begin with a rather trivial example. Suppose $\mathcal{F}_n$ equals the set of *all* Boolean functions on X_n. Then it is clear that the set X_n itself is shattered by $\mathcal{F}_n$. Hence $d_n := \text{VC-dim}(\mathcal{F}_n) = |X_n| = 2^n$. Since d_n increases faster than any polynomial in n, it follows from the discussion above that the class $\mathcal{F}_n$ is *not* efficiently learnable.

Example 9.2. [187] A formula on U_n is said to be a **monomial** if it is a conjunction of some of the $2n$ literals $u_1, \ldots, u_n, \neg u_1, \ldots, \neg u_n$. Let $\mathcal{M}_n$ denote the set of monomials on U_n. It is shown that $\mathcal{M}_n$ is efficiently learnable.

As a first step, let us estimate the cardinality $|\mathcal{M}_n|$. Suppose $f \in \mathcal{M}_n$ is a monomial. Each variable u_i can either be missing completely from f, or be present as u_i, or else be present as $\neg u_i$. Since there are three possibilities for each of the n variables, it follows that $|\mathcal{M}_n| = 3^n$. Hence

$$d_n := \text{VC-dim}(\mathcal{M}_n) \leq n \lg 3.$$

Since d_n is linear in n, this graded concept class is efficiently learnable if it is possible to find an algorithm that produces a consistent hypothesis in polynomial-time. Such an algorithm is described next.

Suppose we are given a data set of length m. A typical element of this set consists of an ordered pair (x_i, b_i) where $x_i \in \{0,1\}^n$ and $b_i \in \{0,1\}$. Now a monomial consistent with the data set can be produced as follows: Begin with the hypothesis

$$h_0 := u_1 \wedge \neg u_1 \wedge \cdots \wedge u_n \wedge \neg u_n,$$

and set the step counter i to zero. At step i, examine if $b_i = 0$ or 1. If $b_i = 0$ (meaning that x_i is a so-called "negative example"), set $h_i = h_{i-1}$; in other words, do nothing. If $b_i = 1$ so that x_i is a "positive example," update h_{i-1} as follows: If the j-th bit of x_i equals 1, then delete $\neg u_j$ from h_{i-1} (if it has not already been deleted earlier); similarly, if j-th bit of x_i is zero, then delete u_j from h_{i-1}. Update in this manner for $j = 1, \ldots, n$ to produce h_i. Now a small technical point needs to be cleared up. Actually h_0 as defined above is *not* a monomial, since a monomial cannot contain *both* u_j and $\neg u_j$. Similarly the various h_i might also not be monomials. To get around this difficulty, let us think of h_i as defined above as mere notational devices. The actual monomial at any step is obtained from the above representation of h_i by deleting any pairs of the form $u_j \wedge \neg u_j$. Thus h_0 is actually a representation of the "empty" monomial.

It is easy to see that the hypothesis h_i is consistent with the data. Moreover, the algorithm is recursive, in the sense that h_i depends only h_{i-1} and the labelled sample (x_i, b_i). One consequence of this recursive nature of the algorithm is that the running time of the algorithm on m data points is just m times the running time of the algorithm on a single data point. In this case, the running time for each step is $O(n)$, so the overall algorithm on m data sets has running time $O(mn)$. Since the algorithm is consistent and polynomial-time, it follows that the concept class is efficiently learnable.

Example 9.3. ([187]) Let $k \geq 3$ be a fixed integer, and let $\mathcal{C}(k, n)$ consist of the set of formulas on U_n in CNF whereby each clause in the CNF is a disjunction of no more than k literals. In shorthand, $\mathcal{C}(k, n)$ is referred to as the set of k-CNF's in n variables. It is shown that $\mathcal{C}(k, n)$ is efficiently learnable.

For this purpose, define $Z(k, n)$ to be the set of all disjunctions of no more than k literals. Thus each element of $Z(k, n)$ is of the form $v_1 \vee \ldots \vee v_j$, where $j \leq k$ and each v_i is a literal. Let us first estimate the cardinality of the set $Z(k, n)$. The number of disjunctions containing *exactly* j literals is at most $\binom{2n}{j}$, which implies that

$$|Z(k,n)| \leq \sum_{j=0}^{k} \binom{2n}{j} \leq \left(\frac{2en}{k}\right)^k \leq (2n)^k \text{ if } k \geq 3,$$

where the upper bound for the summation follows from Lemma 4.3. Now, every k-CNF formula on the original set of variables U_n can be written as an equivalent conjunction of some elements of $Z(k, n)$ (but not their negatives). Hence

$$|\mathcal{C}(k,n)| \leq 2^{|Z(k,n)|} \leq 2^{(2n)^k},$$

and as a consequence

$$\text{VC-dim}(\mathcal{C}(k,n)) \leq (2n)^k.$$

Since the VC-dimension increases polynomially with respect to n (note that the integer k is *independent* of n), it follows that the graded concept class is efficiently learnable if a consistent hypothesis can be found in polynomial-time.

The key observation here is that learning a k-CNF is the same as learning a conjunction of variables over the enlarged set $Z(k, n)$, and we already know how to do this from Example 9.2. Every labelled sample (x_i, b_i) involving the original variables in U_n can be converted into a corresponding labelled sample involving the new variables in $Z(k, n)$, and then the algorithm of Example 9.2 can be run on this labelled sample. From that example we already know that the running time of the algorithm is $O(m|Z(k,n)|) \leq O(mn^k)$. Hence the graded concept class is efficiently learnable.

Example 9.4. Let $k \geq 3$ be a fixed integer, and let $\mathcal{D}(k,n)$ denote the set of k-DNF's; that is, $\mathcal{D}(k,n)$ consists of all formulas over U_n in DNF, with the property that each clause is a conjunction of no more than k literals. It is shown that this class is efficiently learnable.

First, as in Example 9.3 above, the cardinality of $\mathcal{D}(k,n)$ is bounded by

$$|\mathcal{D}(k,n)| \leq 2^{(2n)^k}.$$

Note that, for the purposes of counting, it does not matter whether the formula is a k-CNF or a k-DNF. Thus

$$\text{VC-dim}(\mathcal{D}(k,n)) \leq (2n)^k,$$

so there are no information-theoretic barriers to efficient learnability.

Next, let us show that there exists a polynomial-time consistent algorithm. As shown above, there are no more than $(2n)^k$ monomials that are conjunctions of k or fewer literals. To begin with, let h_0 equal the disjunction of *all* monomials with k or fewer literals. Given the labelled sample (x_i, b_i) at step i, if $b_i = 1$, let $h_i = h_{i-1}$. If $b_i = 0$, then delete from h_{i-1} any monomials that evaluate to 1 on the sample x_i, since it is clear that the unknown target formula cannot contain any such monomial. Since h_i contains at most $(2n)^k$ monomials, the running time of each step is $O(n^k)$, and since the algorithm is iterative, the running time of the algorithm on a data set of length m is $O(mn^k)$.

Example 9.5. ([160], [9], pp. 44-48) Let $k \geq 3$ be a fixed integer, and define $\mathcal{C}_n^k$ to be the set of formulas in n variables in CNF where the number of disjunctive clauses is no larger than k (but there is no limit on the number of literals within each disjunctive clause). Such a formula is said to be in "k-term CNF." The reader is urged to reflect carefully upon the difference between the present set $\mathcal{C}_n^k$ and the set $\mathcal{C}(k,n)$ of Example 9.3. In the present instance, it turns out that the problem of finding a consistent hypothesis is NP-hard. Thus one would expect that the concept class is not efficiently learnable.

Let us first establish that the VC-dimension of $\mathcal{C}_n^k$ is *not* a source of difficulty. As shown in Example 9.2, there are 3^n possible disjunctive clauses in n variables. Thus the number of CNF's with k or fewer clauses is

$$|\mathcal{C}_n^k| \leq \sum_{j=0}^{k} \binom{3^n}{j} \leq \left(\frac{3^n e}{k}\right)^k \leq (3^n)^k = 3^{nk} \text{ if } k \geq 3,$$

where once again Lemma 4.3 is used. Thus

$$\text{VC-dim}(\mathcal{C}_n^k) \leq nk \lg 3,$$

which is linear in n.

Now it is established that the problem of finding a formula in $\mathcal{C}_n^k$ consistent with a given data set is NP-hard. For this purpose, an auxiliary problem known as the "graph colouring problem" is introduced. Consider an (undirected) graph (V, E) where $V = \{1, \ldots, n\}$ is the set of vertices, and $E \subseteq V \times V$ is the set of edges. A k-**colouring** of the graph is a map $r : V \to \{1, \ldots, k\}$ such that if $(i, j) \in E$, then $r(i) \neq r(j)$; in other words, adjacent vertices have distinct colours. Now the k-colouring problem is simply stated: Given a graph with n vertices, together with a fixed integer k independent of n, does there exist a k-colouring of the graph?[1] If the integer n is treated as the size of the problem, then the problem is NP-complete [73].

Now it is shown that the problem of finding a consistent hypothesis in $\mathcal{C}_n^k$ is NP-hard, by mapping the k-colouring problem into it. Thus, if there were to exist a polynomial-time algorithm to solve the consistent hypothesis problem in $\mathcal{C}_n^k$, then there would exist a polynomial-time algorithm for the k-colouring problem. Since the latter problem is known to be NP-complete, it follows that the former problem is NP-hard.

Thus the aim is to show that the k-colouring problem for a graph with n vertices can be mapped into a consistent hypothesis problem in $\mathcal{C}_n^k$ in such a way that the length m of the data set is polynomial in n. This can be done as follows: Given a graph (V, E) with $|V| = n$, generate $m := |V| + |E|/2$ labelled data points, as follows:

1. For each vertex i, $1 \leq i \leq n$, let $x_l \in \{0, 1\}^n$ equal the i-th unit vector $\mathbf{e}_i$, i.e., a vector with a 1 in the i-component and 0's elsewhere; and let the label b_l equal 0.
2. For each edge $(i, j) \in E$, let $x_l \in \{0, 1\}^n$ equal the vector $\mathbf{e}_{ij} \in \{0, 1\}^n$ with 1's in components i and j and 0's elsewhere; and let the label b_l equal 1. Note that, since the graph is undirected, we have that $(i, j) \in E$ if and only if $(j, i) \in E$. Thus the above construction needs to be performed only for those $(i, j) \in E$ with $i < j$.

The above construction results in $m = |V| + |E|/2$ labelled data points. Suppose this set of data points in fed into an algorithm that returns (if possible) a formula in $\mathcal{C}_n^k$ that is consistent with the data, or else states that no such consistent hypothesis exists. It is established that such an algorithm would also answer the question as to whether there exists a k-colouring of the given graph. Specifically, the claim is that there exists a k-colouring of the graph if and only if there exists a formula in $\mathcal{C}_n^k$ that is consistent with the above data set. This claim is established below. But for now, note that $|E| \leq n(n-1)/2$, whence $m \leq n(n+1)/2$. Hence the size of the data set is polynomial in n. Therefore the NP-hardness of the consistent hypothesis problem is proved by establishing the claim.

---

[1] The famous four-colour theorem states that if the graph is *planar*, then an upper bound of $k = 4$ will do irrespective of n; however this is not true for an arbitrary graph.

Accordingly, suppose first that there exists a formula $f \in \mathcal{C}_n^k$ consistent with the data set defined above. This fact is used to define a k-colouring on the graph. Since $f \in \mathcal{C}_n^k$, it is a conjunction of no more than k disjunctive clauses. Let $s \leq k$ denote the number of disjunctive clauses, and suppose

$$f = \bigwedge_{l=1}^{s} f_l(u_1, \ldots, u_n).$$

Since the formula is consistent with the data set, we have that $f(\mathbf{e}_i) = 0$ for each i. Since $f(\mathbf{e}_i) = \wedge_{l=1}^{s} f_l(\mathbf{e}_i)$, it follows that for each i there exists at least one integer l such that $f_l(\mathbf{e}_i) = 0$. Now define a map $r: \{1, \ldots, n\} \to \{1, \ldots, s\}$ as follows: For each $i \in \{1, \ldots, n\}$, let

$$r(i) := \min\{l : f_l(\mathbf{e}_i) = 0\}.$$

It is shown that r defines an s-colouring of the graph.[2] For this purpose, suppose $(i,j) \in E$; it is to be shown that $r(i) \neq r(j)$. Suppose to the contrary that $r(i) = r(j) = l$, say. Thus $f_l(\mathbf{e}_i) = f_l(\mathbf{e}_j) = l$. Since f_l is a disjunction and $f(\mathbf{e}_i) = 0$, it follows that, if at all the variable $u_h, h \neq i$ appears in f_l, it must appear positively, i.e., as u_h and not $\neg u_h$. By the same reasoning, u_i can only appear negatively, if at all. Since $f_l(\mathbf{e}_j) = 0$, the preceding remarks also apply with i replaced by j. Thus, in the disjunctive clause f_l, the variables u_i and u_j must appear negatively (if at all), while $u_h, h \neq i, h \neq j$ must appear positively, if at all. Now let $\mathbf{e}_{ij}$ equal the vector with a 1 in components i and j, and 0's elsewhere. Then the preceding sentence shows that $f_l(\mathbf{e}_{ij}) = 0$. In turn this shows that $f(\mathbf{e}_{ij}) = 0$, since $f(\mathbf{e}_{ij})$ is a disjunction of several clauses, one of which is $f_l(\mathbf{e}_{ij})$. But by construction $f(\mathbf{e}_{ij}) = 1$ whenever $(i,j) \in E$. This contradiction shows that $(i,j) \in E$, $f_l(\mathbf{e}_i) = 0$ and $f_l(\mathbf{e}_j) = 0$ for some l is not possible. Thus $r(i) \neq r(j)$ whenever $(i,j) \in E$, and $r(\cdot)$ defines an s-colouring. This establishes the first half of the claim.

To prove the other half of the claim, suppose $r: \{1, \ldots, n\} \to \{1, \ldots, k\}$ is a k-colouring of the graph. For $1 \leq l \leq k$, define f_l to be the disjunctive clause

$$f_l(\mathbf{u}) := \bigvee_{r(j) \neq l} u_j,$$

and define

$$f(\mathbf{u}) := \bigwedge_{l=1}^{k} f_l(\mathbf{u}).$$

It is shown that f is consistent with the data set defined previously. (It is obvious that $f \in \mathcal{C}_n^k$.) First, let $\mathbf{u} = \mathbf{e}_i$ for some $i \in \{1, \ldots, n\}$, and let $l := r(i)$. Then the variable u_i does not appear in the clause f_l. Since all variables appear positively in f_l, and since $u_j = 0$ for all $j \neq i$, it follows that

---

[2] Note that, since $s \leq k$, an s-colouring is also a k-colouring.

$f_l(\mathbf{e}_i) = 0$, whence $f(\mathbf{e}_i) = 0$. Next, suppose $(i,j) \in E$, and let $\mathbf{e}_{ij} \in \{0,1\}^n$ denote the vector with 1's in components i and j and 0's elsewhere. For each index $l \in \{1,\ldots,k\}$, either $r(i) \neq l$ or else $r(j) \neq l$ (or both), since $r(i) \neq r(j)$. Hence every clause contains either u_i or u_j, or both. Consequently, $f_l(\mathbf{e}_i) = 1$ for all l, which in turn implies that $f(\mathbf{e}_{ij}) = 1$.

The above discussion follows [9]; by using a slightly more involved construction, it is shown in [160] that the consistent hypothesis problem is NP-hard also when $k = 2$.

Example 9.6. ([160], [99]) Let $k \geq 3$ be a fixed integer, and define $\mathcal{D}_n^k$ to be the set of formulas in n variables in DNF, where the number of conjunctive clauses is no larger than k (but there is no restriction on the number of literals within each clause). Such formulas are referred to as "k-term DNF's." Proceeding exactly as in Example 9.5 above, one can show that

$$|\mathcal{D}_n^k| \leq 3^{nk}, \text{ and VC-dim}(\mathcal{D}_n^k) \leq nk \lg 3.$$

At the same time, the problem of finding a formula in $\mathcal{D}_n^k$ consistent with a given data set is NP-hard. This can be shown by reducing the k-colouring problem to the above problem. The reduction follows the same lines as in Example 9.5 above, except that each vertex is assigned the value of 1, while each edge is assigned the value of 0; this is the inverse of the assignment in Example 9.5. The details are left to the reader.

Now the importance of *representation* in influencing efficient learnability is discussed very briefly. Note that, by repeated use of the the distributivity laws, every formula in $\mathcal{D}_n^k$ can be written equivalently as another formula in $\mathcal{C}(k,n)$ (see Example 9.3 above). Now it is known from Example 9.3 that $\mathcal{C}(k,n)$ *is* efficiently learnable. Thus we have an apparently paradoxical situation whereby a set $\mathcal{C}(k,n)$ *is* efficiently learnable, whereas a subset of it (namely $\mathcal{D}_n^k$) is *not* efficiently learnable. Note that this phenomenon has no analog in *statistical* learning theory. A perusal of the definition of $r(m,\epsilon)$ in (3.2.1) reveals that if a concept class $\mathcal{C}$ is PAC learnable, then so is every subset of $\mathcal{C}$.

The resolution of this apparent paradox lies in the issue of *representation*. Given a data set of length m, it is NP-hard to find a formula in $\mathcal{D}_n^k$ consistent with the data, but it is quite efficient to find a formula in $\mathcal{C}(k,n)$ consistent with the data. This is so even if it is known ahead of time that the data is in fact being generated by a target formula in $\mathcal{D}_n^k$. Thus, in *computational* learning theory, it is natural to embed a *concept class* to be learnt within a larger *hypothesis class*, just to make it possible to find a consistent hypothesis in polynomial-time.

Example 9.7. At last we come to an example that does not involve Boolean formulas. For each integer $n \geq 1$, let $X_n = \mathbb{R}^n$, $\mathcal{S}_n =$ the Borel σ-algebra on $\mathbb{R}^n$, $\mathcal{C}_n =$ the set of closed half-planes in $\mathbb{R}^n$, denoted by $\mathcal{H}_n$, and as always, $\mathcal{P}_n = \mathcal{P}_n^*$, the set of all probability measures on $(X_n, \mathcal{S}_n)$. Then, as shown in

Example 4.3, VC-dim($\mathcal{H}_n$) = $n + 1$. Since the VC-dimension of this graded concept class grows linearly with respect to n, the concept class is efficiently learnable if the problem of finding a consistent hypothesis class can be solved in polynomial-time. On the basis of recent results in linear programming [94], we know that the consistency problem can be solved in polynomial-time if the data is represented to *finite* precision, e.g., each component of every vector in $X_n = \mathbb{R}^n$ is represented to no more than b bits of accuracy, where b is independent of n. Thus $\{\mathcal{H}_n\}$ is efficiently learnable.

Next, let $k \geq 2$ be a fixed integer, and define $\mathcal{H}_n^k$ to be the collection of intersections of k or fewer closed half-planes in $\mathbb{R}^n$. Then it is known from Theorem 4.5 that

$$\text{VC-dim}(\mathcal{H}_n^k) \leq 2k \lg(ek)\,(n+1).$$

Since this bound is linear in n, this graded collection is efficiently learnable if the consistent hypothesis problem can solved in polynomial-time. Unfortunately, even if $k = 2$, finding an element in $\mathcal{H}_n^2$ consistent with a given data set is an NP-complete problem [31]; this is so even if all the data vectors are restricted to be binary. See also Example 10.1 for further discussion of this example.

As in Example 9.6 above, we have here a situation that is peculiar to *computational* learning theory, and has no analog in *statistical* learning theory. A consequence of Theorem 4.5 is that, if a graded concept class $\mathcal{C}_n$ has VC-dimension that grows polynomially with n, then so does the collection of sets $\mathcal{C}_n^k$ obtained by performing an arbitrary but fixed Boolean operation on sets in $\mathcal{C}_n$. Thus the number of samples required by a consistent algorithm grows only polynomially with respect to n. But, as soon as one begins to consider the issue of finding an *efficient* consistent algorithm, the process of performing Boolean operations on sets can destroy efficient learnability, as is the case in the present example.

9.2 Active Learning

In this section, we study the problem of "active" learning, whereby the learner is able to "query" the oracle with a view towards obtaining some information about the unknown target concept. This is in contrast to "passive" learning, in which the learner has access only to randomly generated labelled samples. The hope of course is that by suitably choosing the queries to the oracle, the learner would be able to accelerate the learning process. In this connection, one can ask two types of questions: (i) Are there instances of concept classes that are *not* learnable when the learning is passive, but become learnable when the learning is active? In essence, this question asks whether the passive nature of the learning process *by itself* sometimes poses an information-theoretic barrier to learnability. It is shown that the answer to this question is "No." Even if the learner is permitted to make an *arbitrary* binary query at each step, the only concept classes that are learnable are those that are also passively learnable. (ii) Given a concept class that is passively learnable, does the *sample complexity* of learning an unknown target concept decrease significantly if the learning is permitted to be active? The answer to this question is both "Yes" and "No," depending on the type of queries that the learner is permitted to make. If the learner is permitted to make *arbitrary* binary queries (which are the most powerful type of query), then the answer is "Yes" – there can be a substantial reduction in the number of samples needed to learn an unknown target concept to a specified level of accuracy. On the other hand, suppose the learner is only permitted to make so-called "membership" queries; that is, the learner can select an element $x \in X$, and an oracle returns the value $I_T(x)$, where T is the unknown target concept. In this case, it is possible to construct examples where the sample complexity of active learning is roughly of the same order as that of passive learning.

Recall that in passive learning, the efficacy of an algorithm is measured by the quantity

$$r(m, \epsilon) := \sup_{P \in \mathcal{P}} \sup_{T \in \mathcal{C}} P^m \{\mathbf{x} \in X^m : d_P[T, H_m(T; \mathbf{x})] > \epsilon\}.$$

As pointed out in Subsection 3.2.4, in the above expression the probability measure P plays two distinct roles. The first role is in $P^m\{\cdot\}$, namely that of measuring the probability that the learning algorithm will be given a bad multisample. This particular role of P is no longer meaningful in active learning, since the multisamples are directly generated by the learner; thus, if the multisamples generated lead to a poor hypothesis, the "fault" can only be attributed to the algorithm and not to chance. Even if the learner were to use a randomized algorithm to generate the training samples, the probability space underlying the source of randomness would not be $(X, \mathcal{S}, P)$ but something else. The second role of P is in $d_P(T, H_m)$, namely to quantify the disparity between the current hypothesis and the unknown target concept. This role continues to be meaningful even in active learning. However, there is no reason to require that the metric on $\mathcal{C}$ that measures the distance between H_m and T must be induced by a probability measure.[3] In fact, one can choose *any* metric ρ on $\mathcal{C}$ from a prespecified family of metrics $\mathcal{R}$ on $\mathcal{C}$; then the learner attempts to approximate T closely with respect to the metric ρ, by using some queries and some algorithm.

The basic ingredients of active learning are:

– A set X,
– A collection $\mathcal{C}$ of subsets of X,
– A family $\mathcal{R}$ of metrics on $\mathcal{C}$, and

---

[3] The main reason for imposing this requirement is to interpret $d_P(T, H_m)$ as the probability of misclassifying a testing input selected at random according to P.

– A query model.

The first three items are self-explanatory, but the last item requires an explanation. Several query models are used in the literature, but only two are studied here in the interests of brevity, namely the membership query and the arbitrary binary query. In the membership query, the learner is permitted to input an element $x \in X$, and an oracle returns the value $I_T(x)$, where T is the unknown target concept. Thus the membership query is equivalent to selecting an $x \in X$ and asking the question: Does a chosen x belong to the unknown target concept T? In the arbitrary binary query, the learner is permitted to input an arbitrary subset S of $\mathcal{C} \times \mathcal{R}$, and an oracle returns the value $I_S(T, \rho)$, where T is the target concept and ρ is the metric used to measure the proximity of the hypothesis to the target concept. Thus the arbitrary binary query is equivalent to selecting an arbitrary subset S of $\mathcal{C} \times \mathcal{R}$, and asking the question: Does the pair (T, ρ) belong to S?

Learning takes place as follows: A target concept $T \in \mathcal{C}$ is fixed. An "algorithm" in active learning consists of two separate indexed families of maps. The first of these is the "query generator," which at the m-th step maps the first $(m - 1)$ queries and the most recent hypothesis H_{m-1} into the m-th query. Thus, in the membership query model, the query generator maps $X^{m-1} \times \mathcal{C}$ into X, whereas in the arbitrary query model, the m-th query generator maps $(2^{\mathcal{C} \times \mathcal{R}})^{m-1} \times \mathcal{C}$ into $2^{\mathcal{C} \times \mathcal{R}}$, where $2^{\mathcal{C} \times \mathcal{R}}$ denotes the power set of $\mathcal{C} \times \mathcal{R}$, that is, the collection of all subsets of $\mathcal{C} \times \mathcal{R}$. The second indexed family is the "hypothesis generator." At the m-th step, it maps the first m queries and their outcomes into an element of $\mathcal{C}$, which is denoted by H_m. Some authors permit both generators to use some sort of randomization; but this case is not studied here in the interests of simplicity; the reader is referred to [111] for a discussion of this more general case. The algorithm is said to **actively learn** the concept class $\mathcal{C}$ with respect to the family of metrics $\mathcal{R}$ if

$$\sup_{\rho \in \mathcal{R}} \sup_{T \in \mathcal{C}} \rho(T, H_m) \to 0 \text{ as } m \to \infty.$$

The concept class $\mathcal{C}$ is said to be **actively learnable** with respect to the family of metrics $\mathcal{R}$ if there exists a suitable algorithm that actively learns it. As in passive learning, the metric ρ used to measure the disparity between the target concept T and the hypothesis H_m can either be known to the learner, or completely unknown (except for the information that ρ belongs to the family $\mathcal{R}$, which is known to the learner). For reasons that will become clear, we also examine a third possibility, namely: the metric ρ is known to the learner, and the learner is permitted to use the knowledge of ρ explicitly in the algorithm; however, the convergence of the error measure $\rho(T, H_m)$ to zero must be uniform with respect to ρ (and of course with respect to T) as indicated above.

9.2.1 Fixed-Distribution Learning

Let us first consider the case where $\mathcal{R} = \{\rho\}$, a singleton set; note that if $\rho = d_P$ where P is a probability measure on X, then the problem is an "active" version of the fixed-distribution learning problem studied in Chapter 6. We begin with the case where the learner is permitted to make arbitrary binary queries.

Theorem 9.1. [111] *A concept class $\mathcal{C}$ is actively learnable with respect to the metric ρ using arbitrary binary queries if and only if*

$$N(\epsilon, \mathcal{C}, \rho) < \infty \ \forall \epsilon > 0.$$

Suppose the above condition holds; then $\lceil \lg N(\epsilon, \mathcal{C}, \rho) \rceil$ samples are necessary and sufficient to learn $\mathcal{C}$ to accuracy ϵ. That is:

1. *For each $\epsilon > 0$, there exists a binary query algorithm such that the condition*

$$\sup_{T \in \mathcal{C}} \rho(T, H_m) \leq \epsilon$$

is satisfied for

$$m \geq \lceil \lg N(\epsilon, \mathcal{C}, \rho) \rceil$$

samples.
2. *Any binary query algorithm requires at least $\lceil \lg N(\epsilon, \mathcal{C}, \rho) \rceil$ samples in order to learn $\mathcal{C}$ to accuracy ϵ.*

Proof. Suppose $N(\epsilon, \mathcal{C}, \rho)$ is finite for a fixed ϵ. Construct a minimal ϵ-cover $\{A_1, \ldots, A_N\}$ for $\mathcal{C}$, where N is a shorthand for $N(\epsilon, \mathcal{C}, \rho)$. Then at the first step one can ask the oracle: Is it true that $\rho(T, A_i) \leq \epsilon$ for some $i \leq N/2$? This is equivalent to asking the oracle the value of the indicator function $I_{\mathcal{A}_1}(T)$, where

$$\mathcal{A}_1 := \bigcup_{i=1}^{N/2} \mathcal{B}(\epsilon, A_i, \rho),$$

and

$$\mathcal{B}(\epsilon, A_i, \rho) := \{B \in \mathcal{C} : \rho(A_i, B) \leq \epsilon\}$$

is the ball of radius ϵ centered at A_i. Hence this query is valid under the arbitrary binary query model. Now, if $I_{\mathcal{A}_1}(T) = 1$, then the learner knows that $T \in \mathcal{A}_1$, which has an ϵ-cover of cardinality $N/2$, namely $A_1, \ldots, A_{N/2}$. Similarly, if $I_{\mathcal{A}_1}(T) = 0$, the learner knows that $T \in \mathcal{C} \setminus \mathcal{A}_1$, which also has an ϵ-cover of cardinality $N/2$, namely $A_{1+N/2}, \ldots, A_N$. In either case, T is "localized" to a smaller set which has an ϵ-cover consisting of $N/2$ sets in $\mathcal{C}$. Now the process can be repeated with the smaller set. Thus it is clear that $\lceil \lg N(\epsilon, \mathcal{C}, \rho) \rceil$ queries are enough to identify T within a distance of ϵ.

Conversely, suppose it is desired to approximate T within a distance of ϵ. Since $A_1, \ldots, A_N$ is a *minimal* ϵ-cover, it is clear that the set

330   9. Alternate Models of Learning

$$\mathcal{C} \setminus \bigcup_{j \neq i} \mathcal{B}(\epsilon, A_j, \rho)$$

is nonempty for every $i \in \{1, \ldots, N\}$, and in principle T could belong to any one of these N sets. Hence, irrespective of whatever queries are made, the learner must be able to encode N distinct possibilities (namely, that $T \in \mathcal{B}(\epsilon, A_i, \rho)$ for exactly one value of i), and encoding N possibilities requires at least $\lceil \lg N \rceil$ queries. In particular, if $N(\epsilon, \mathcal{C}, \rho)$ is infinite for some ϵ, then $\mathcal{C}$ cannot be learnt to accuracy ϵ, even with arbitrary binary queries. ∎

To compare the present results on active learning with those in Chapter 6 on passive learning under a fixed probability, suppose $\rho = d_P$ where P is a known probability measure on X. Comparing Theorem 9.1 with Theorem 6.7 shows that the condition

$$N(\epsilon, \mathcal{C}, d_P) < \infty \ \forall \epsilon > 0$$

is necessary and sufficient for $\mathcal{C}$ to be learnable, whether the learning is active or passive. Thus even the ability to ask *arbitrary* binary queries (which is in some sense the strongest possible form of active learning) does not in any way enlarge the set of learnable concept classes. Similarly, Theorem 6.10 shows that, if the above condition is satisfied, then the concept class is learnable even if the oracle occasionally returns an erroneous output. Taken together, these three theorems demonstrate that, in some sense, learnability is an *intrinsic* property of a concept class, in the sense that a concept class is learnable in any one of the three learning models (namely: active learning, passive learning with a perfect oracle, and passive learning with a noisy oracle) if and only if it is learnable in all three models.

On the other hand, while learnability is an intrinsic property, the *sample complexity* can be considerably smaller in the case of active learning. From Theorem 6.3, it is known that in order to learn $\mathcal{C}$ to accuracy ϵ and confidence δ, it is enough to generate

$$m_{\text{passive}} = \frac{32}{\epsilon} \ln \frac{N(\epsilon/2, \mathcal{C}, d_P)}{\delta}$$

i.i.d. samples. If the oracle returns an incorrect classification with a probability of $\alpha < 0.5$, then it is enough to generate

$$m_{\text{noisy}} = \frac{8}{(1-2\alpha)^2 \epsilon^2} \ln \frac{N(\epsilon/2, \mathcal{C}, d_P)}{\delta}$$

samples. In contrast, with active learning using arbitrary binary queries,

$$m_{\text{active}} = \lg N(\epsilon, \mathcal{C}, d_P)$$

queries are enough. To compare the rate of growth of these estimates, suppose the concept class $\mathcal{C}$ has a finite upper metric dimension, call it μ_0. This means that

$$\limsup_{\epsilon \to 0^+} \frac{\ln N(\epsilon, \mathcal{C}, d_P)}{\ln(1/\epsilon)} = \mu_0 < \infty.$$

Hence for every $\mu > \mu_0$, there exist finite constants M and ϵ_0 such that

$$N(\epsilon, \mathcal{C}, d_P) \leq \frac{M}{\epsilon^\mu} \quad \forall \epsilon \leq \epsilon_0.$$

Ignoring constants, we can write $N(\epsilon, \mathcal{C}, d_P) = O[(1/\epsilon)^\mu]$ for ϵ small enough. Therefore

$$m_{\text{passive}} = O\left(\frac{1}{\epsilon}\ln\frac{1}{\epsilon}\right),$$

whereas

$$m_{\text{active}} = O\left(\ln\frac{1}{\epsilon}\right).$$

Finally, from Theorem 6.10, it follows that with a noisy oracle,

$$m_{\text{noisy}} = O\left(\frac{1}{\epsilon^2}\ln\frac{1}{\epsilon}\right)$$

samples are enough. These estimates indicate clearly the effect of the learning model on the sample complexity.

Now let us continue our study of the case where $\mathcal{R} = \{\rho\}$, a singleton set, but the learner is permitted to make only *membership* queries. In this case, the condition

$$N(\epsilon, \mathcal{C}, \rho) < \infty \; \forall \epsilon > 0$$

continues to be necessary and sufficient for active learnability. This can be seen as follows: Since membership queries are less powerful than arbitrary binary queries, the above condition is necessary. Since passive learning is even less powerful than active learning with membership queries, and the above condition is sufficient for passive learnability, it is also sufficient for active learnability with membership queries. But the interesting question is what happens to the sample complexity. If it is possible to divide the "search space" into two at each step by a single membership query, then it is possible to achieve a dramatic reduction in the sample complexity, as in the case of arbitrary binary queries. However, it is also possible to construct examples where membership queries do not significantly reduce the sample complexity. These "extreme" possibilities are illustrated by the next two examples.

Example 9.8. Let $X = \mathbb{R}^2$, and suppose $\mathcal{C}$ equals the set of half-planes where the boundary of the half-plane *passes through the origin*. Suppose that the metric ρ is induced by the uniform probability measure concentrated on the unit circle in $\mathbb{R}^2$. In this case, the concept class is parametrized by a single number θ, which is the angle made by the dividing line with the x-axis. Since it is necessary to specify *which side* of the line is the target concept, the angle θ varies over $[0, 2\pi)$, even though rotating a line by π results in exactly the same line. Thus the unknown target concept can be thought of as an unknown angle $\theta_t \in [0, 2\pi)$. If the target concept corresponds to the angle θ_t and the hypothesis H corresponds to the angle θ_h, then it is easy to see that $\rho(T, H) = |\theta_t - \theta_h|/(2\pi)$. In this special case, it is possible to reduce the uncertainty about the true value of θ_t by a factor of two at each step using membership queries alone. This can be achieved as follows: Select a unit vector $\mathbf{x}_1 \in \mathbb{R}^2$ such that the argument of $\mathbf{x}$ is some fixed number α; for example, if $\alpha = \pi/2$, then choose $\mathbf{x} = [0 \ 1]^t$. If $\mathbf{x} \in T$, then one can conclude that the true value θ_t belongs to $[0, \pi]$; otherwise, $\theta_t \in (\pi, 2\pi)$. Depending on the outcome of this membership query, then one can choose $\mathbf{x}_2$ so as to make the width of the uncertainty about the true value of θ_t equal to $\pi/2$, and so on. In general, after m queries, the value of θ_t will be localized to an interval of width $2\pi/2^m$, and as a result, the distance $\rho(T, H_m)$ will at most be equal to $1/2^m$. Thus the inequality $\rho(T, H_m) \leq \epsilon$ can be satisfied after $m = \lceil \lg(1/\epsilon) \rceil$ membership queries. Note that $1/\epsilon$ is precisely the ϵ-covering number of $\mathcal{C}$ with respect to ρ. Thus, in this example, the sample complexity of active learning using membership queries is the same as that using active binary queries.

Example 9.9. It is shown how one might go about constructing an example in which the sample complexity using membership queries is roughly of the same order as that of passive learning. Suppose that, for some fixed number ϵ, the concept class $\mathcal{C}$ consists of a *disjoint* union of k balls of radius ϵ. Suppose also that a membership query can only localize the unknown target concept to *any one* of these k balls. Then the sample complexity of active learning using membership queries is k, whereas as that of passive learning is $O(k \ln k)$. If arbitrary binary queries are permitted, then as in the proof of Theorem 9.1, it is possible to localize the target concept to the ball containing it using just $\lceil \lg k \rceil$ queries. Thus, in this case, there is no dramatic difference between the sample complexity of active learning with membership queries and that of passive learning. ■

In the computer science community, another form of active learning is popular, whereby the concept class $\mathcal{C}$ is *countable*, and the objective of querying is to determine the unknown target concept *exactly* within a finite number of steps. This type of learning is not discussed here. The interested reader is referred to [5] for further discussion and some pertinent references.

9.2.2 Distribution-Free Learning

Now let us study the case where the metric ρ on $\mathcal{C}$ is not fixed, but is permitted to vary over some specified family $\mathcal{R}$ of metrics. For instance, suppose $\mathcal{R} = \mathcal{R}^* := \{d_P : P \in \mathcal{P}^*\}$, where $\mathcal{P}^*$ denotes the set of *all* probability measures on $(X, \mathcal{S})$ for some σ-algebra $\mathcal{S} \supseteq \mathcal{C}$; this is the active learning analog of distribution-free learning. In such a case, one can make a distinction between two cases: (i) The metric $\rho \in \mathcal{R}$ is known to the learner, and the learner is permitted to make explicit use of ρ in the learning algorithm. (ii) The metric ρ is not known to the learner. At a first glance, it might appear that case (ii) is the correct active analog of distribution-free learning, since in passive distribution-free learning the learner does not know the metric d_P. But this impression is misleading. In fact, since the i.i.d. samples in passive learning are generated according to the unknown probability P, some information about P is actually being transmitted to the learner. For example, if $X = \mathbb{R}$, then the empirical distribution on X formed using the i.i.d. samples converges to the true distribution function of P in a particular topology; this is known as Sanov's theorem. To put it another way, in passive learning, even though the probability P is not *explicitly* known to the learner, the fact that the i.i.d. samples are generated according to P means that the learner is given relatively more samples in that part of X where P is concentrated. In contrast, in the active learning case where ρ is unknown, the learner is obliged to generate queries without having any idea of the criterion used to assess the quality of the hypothesis generated by the learning algorithm. To focus the discussion, suppose $\mathcal{R} = \{d_P : P \in \mathcal{P}\}$, where $\mathcal{P} \subseteq \mathcal{P}^*$ is some family of probability measures. In this case, if the concept class $\mathcal{C}$ has a finite ϵ-cover with respect to the metric $\bar{d}_\mathcal{P}$ defined in Chapter 8, then it is easy to modify Theorem 9.1 to learn using arbitrary binary queries. However, in the remarks following Theorem 8.4, it is shown that if $\mathcal{P} = \mathcal{P}^*$, then a concept class is totally bounded with respect to $\bar{d}_\mathcal{P}$ if and only if $\mathcal{C}$ is finite. Not surprisingly, an analogous result can be proved for the case of active learning as well.

Theorem 9.2. [111] *Suppose $(X, \mathcal{S})$ is a measurable space and let $\mathcal{R} = \mathcal{R}^* = \{d_P : P \in \mathcal{P}^*\}$. Suppose $\mathcal{C} \subseteq \mathcal{S}$. Then $\mathcal{C}$ is actively learnable when the learner does not know ρ if and only if $\mathcal{C}$ is finite.*

The proof of Theorem 9.2 is based on the following preliminary lemma.

Lemma 9.2. *Suppose $(X, \mathcal{S})$ is a measurable space, and that $\mathcal{C} \subseteq \mathcal{S}$ is an infinite set of concepts. Let $\{C_1, \ldots, C_n\}$ be any finite subset of $\mathcal{C}$; then there exists a set $C_{n+1} \in \mathcal{C}$ and a probability measure P on $(X, \mathcal{S})$ such that*

$$d_P(C_{n+1}, C_i) = \frac{1}{2} \text{ for } i = 1, \ldots, n.$$

Proof. **of the lemma** Consider all sets of the form $\cap_{i=1}^n A_i$ where each A_i is either C_i or its complement C_i^c. Then there are at most 2^n distinct subsets $B_1, \ldots, B_{2^n}$ in $\mathcal{S}$ of this form, and their union is X. Since $\mathcal{C}$ is infinite, there exists a set $C_{n+1} \in \mathcal{C}$ such that for some nonempty subset B_k, $C_{n+1} \cap B_k \neq \emptyset$ and $C_{n+1} \cap B_k \neq B_k$. Hence there exist points $x_1, x_2 \in X$ such that $x_1 \in (C_{n+1} \cap B_k)$ and $x_2 \in B_k \setminus C_{n+1}$. Now let P be the atomic measure concentrated uniformly on x_1, x_2. For each $i = 1, \ldots, n$, either $B_k \cap C_i = \emptyset$ or else $B_k \subseteq C_i$. In either case, $C_{n+1} \Delta C_i$ is either $\{x_1\}$ or $\{x_2\}$, so that $d_P(C_{n+1}, C_i) = 1/2$ for $i = 1, \ldots, n$. ∎

Proof. **of the theorem** Suppose first that $|\mathcal{C}|$ is finite. then $\lceil \lg |\mathcal{C}| \rceil$ binary queries are enough to learn $\mathcal{C}$ exactly (i.e., to determine the unknown target concept T exactly). This proves the "if" part of the theorem.

To prove the "only if" part, suppose $\mathcal{C}$ is infinite. It is shown that, after a finite number of binary queries, however formulated, there are still infinitely many candidate concepts that are at least 1/2-apart under infinitely many probabilities in $\mathcal{P}^*$. By applying Lemma 9.2 repeatedly, it is possible to generate an infinite sequence of concepts $\{C_i\}$ and a corresponding sequence of probability measures $\{P_i\}$ such that

$$d_{P_i}(C_i, C_j) = \frac{1}{2} \text{ for } 1 \leq j \leq i-1, \text{ for all } i \geq 1.$$

Now an arbitrary binary query can only be of the form: "Is $(C, P) \in S$?" where S is some subset of $\mathcal{C} \times \mathcal{P}^*$. Consider the sequence of pairs $\{(C_i, P_i)\}$. Whatever be the set S, either S or its complement S^c (or both) contain an infinite number of pairs (C_i, P_i). Thus, after a finite number of binary queries, an infinite number of pairs (C_i, P_i) still remain as candidates, so that

$$\sup_{P \in \mathcal{P}^*} \sup_{T \in \mathcal{C}} d_P(T, H_m) \geq \frac{1}{2},$$

irrespective of the algorithm used. Hence $\mathcal{C}$ is not actively learnable with respect to $\mathcal{R}^*$ to any accuracy $\epsilon < 0.5$. ∎

To overcome the difficulty described above, let us consider an alternate form of active learning, in which the probability $P \in \mathcal{P}^*$ is made known to the learner, but the sample complexity must be uniformly bounded with respect to P. Then we have the following result:

Theorem 9.3. *Suppose $\mathcal{R} = \mathcal{R}^* = \{d_P : P \in \mathcal{P}^*\}$, and that the learner knows the probability measure P. Then $\mathcal{C}$ is actively distribution-free learnable if and only if $\mathcal{C}$ has finite VC-dimension. Suppose $d := \text{VC-dim}(\mathcal{C})$ is finite. Then $\mathcal{C}$ is actively distribution-free learnable using no more than*

$$m_{\text{active}} = \lceil d \lg \left(\frac{2e}{\epsilon} \ln \frac{2e}{\epsilon} \right) \rceil$$

binary queries.

Proof. It follows from Theorem 9.1 that, in the present model, $\mathcal{C}$ is actively distribution-free learnable if and only if

$$\sup_{P \in \mathcal{P}^*} N(\epsilon, \mathcal{C}, d_P) < \infty \text{ for all } \epsilon > 0.$$

From Corollary 7.1, this condition holds if and only if $\mathcal{C}$ has finite VC-dimension.

Next, suppose $d := \text{VC-dim}(\mathcal{C})$ is finite. Then it follows from Corollary 4.2 that

$$N(\epsilon, \mathcal{C}, d_P) \leq \left(\frac{2e}{\epsilon} \ln \frac{2e}{\epsilon}\right)^d \forall P \in \mathcal{P}^*.$$

The bound on the number of binary queries now follows from Theorem 9.1. ∎

9.3 Learning with Prior Information: Necessary and Sufficient Conditions

Recall the definition of the quantity that measures the efficacy of an algorithm, namely (cf. (3.2.1)):

$$\bar{r}(m, \epsilon, \mathcal{P}) := \sup_{T \in \mathcal{C}} \sup_{P \in \mathcal{P}} P^m\{\mathbf{x} \in X^m : d_P[T, H_m(T; \mathbf{x})] > \epsilon\},$$

where $H_m(T; \mathbf{x})$ denotes the hypothesis produced by the algorithm when the target concept is T and the multisample is $\mathbf{x}$. Note that, in order to facilitate the subsequent discussion, the definition of $\bar{r}(m, \epsilon, \mathcal{P})$ is slightly modified from (3.2.1), in that the order of taking the two suprema is interchanged; however, this interchange does not affect the value of $\bar{r}(m, \epsilon, \mathcal{P})$. As it stands, the above definition of $\bar{r}(m, \epsilon, \mathcal{P})$ is *worst-case* with respect to both the probability measure P that generates the learning samples as well as the unknown target concept T. It is perhaps reasonable to take the supremum with respect to $P \in \mathcal{P}$, since we would like the algorithm to produce uniformly good hypotheses irrespective of the probability that is generating the learning samples. However, there might be a case for permitting an algorithm to work for "most" target concepts, even if it might fail occasionally. This consideration motivates the present section. First, a notion of learnability with prior information is given that formalizes the above notion of an algorithm that works for "most" target concepts. Then some simple sufficient conditions are given. Finally, necessary and sufficient conditions for learnability with prior information are given in terms of a notion called "dispersability."

9.3.1 Definition of Learnability with Prior Information

In this subsection, a formal definition is given of PAC learnability with prior nformation.

As before, it is assumed that one is given a measurable space $(X, \mathcal{S})$, a function class $\mathcal{F} \subseteq [0, 1]^X$, and a family $\mathcal{P}$ of probability measures on $(X, \mathcal{S})$. The additional feature now is that one is also given a family $\mathcal{Q}$ of probability measures *on the function class* $\mathcal{F}$. As before, a passive learner receives i.i.d. samples $x_1, x_2, \ldots$ from X generated in accordance with a fixed but

336   9. Alternate Models of Learning

unknown $P \in \mathcal{P}$, together with the values of an unknown "target" function $f(x_1), f(x_2), \ldots$. Again as before, an "algorithm" is an indexed family of mappings $\{A_m\}_{m \geq 1}$, where A_m maps $(X \times [0,1])^m$ into $\mathcal{F}$. The major departure lies in the definition of a "figure of merit" to assess the efficacy of the algorithm. Instead of (and in contrast to) the definition of $\bar{r}(m, \epsilon)$ given above, one defines

$$\bar{u}(m, \epsilon) := \sup_{Q \in \mathcal{Q}} \sup_{P \in \mathcal{P}} (Q \times P^m)\{(f, \mathbf{x}) \in \mathcal{F} \times X^m : d_P[f, h_m(f; \mathbf{x})] > \epsilon\},$$

where as always $h_m(f; \mathbf{x})$ denotes the hypothesis generated by the algorithm.

Definition 9.1. *The algorithm is said to* **PAC learn with prior information (WPI)** *if $\bar{u}(m, \epsilon) \to 0$ as $m \to \infty$ for each ϵ, and the triplet $(\mathcal{F}, \mathcal{P}, \mathcal{Q})$ is said to be* **PAC learnable WPI** *if there exists a suitable algorithm that PAC learns WPI.*

Next, let us interpret the property of PAC learning WPI in terms of the convergence of a stochastic process, which in turn permits us to give an alternate and equivalent definition. In the interests of simplicity, suppose $\mathcal{P} = \{P\}$, $\mathcal{Q} = \{Q\}$ are singleton sets. First define the function $\phi : \mathbb{R} \to \mathbb{R}$ by

$$\phi(\sigma) = \begin{cases} 1, & \text{if } \sigma > 0, \text{ and} \\ 0, & \text{if } \sigma \leq 0. \end{cases}$$

Thus $\phi(\cdot)$ is the same as the step function $\eta(\cdot)$ defined in Chapter 3, except that $\phi(0) = 0$ whereas $\eta(0) = 1$. Also, for each integer m, each $\epsilon > 0$, and each $f \in \mathcal{F}$, define

$$\beta_m(f, \epsilon) := P^m\{\mathbf{x} \in X^m : d_P[f, h_m(f; \mathbf{x})] > \epsilon\}.$$

Thus $\beta_m(f, \epsilon)$ is the measure of the set of "bad" multisamples for the target function f, where a multisample $\mathbf{x} \in X^m$ is deemed to be "bad" if $d_P[f, h_m(f; \mathbf{x})] > \epsilon$. Equivalently,

$$\beta_m(f, \epsilon) = E_{P^m}[\phi(d_P[f, h_m(f; \mathbf{x})] - \epsilon)].$$

Then, since $Q \times P^m$ is a product measure on $\mathcal{F} \times X^m$, it follows that[4]

$$\begin{aligned} u(m, \epsilon) &= (Q \times P^m)\{(f, \mathbf{x}) \in \mathcal{F} \times X^m : d_P[f, h_m(f; \mathbf{x})] > \epsilon\} \\ &= E_Q E_{P^m}[\phi(d_P[f, h_m(g; \mathbf{x})] - \epsilon)]. \end{aligned} \quad (9.3.1)$$

Now let us interpret $\{\beta_m(f, \epsilon)\}$ for a fixed ϵ as a stochastic process on $\mathcal{F}$. Then the algorithm PAC learns WPI if $E_Q[\beta_m(f, \epsilon)] \to 0$ as $m \to \infty$ for each $\epsilon > 0$. Further, since $\beta_m(f, \epsilon) \in [0, 1]$ for each m, ϵ, T, it is easy to see that the

---
[4] Note that we use $u(m, \epsilon)$ instead of $\bar{u}(m, \epsilon)$ since both $\mathcal{Q}$ and $\mathcal{P}$ are singleton sets.

9.3 Learning with Prior Information: Necessary and Sufficient Conditions

condition above is equivalent to the requirement that the stochastic process $\{\beta_m(f,\epsilon)\}$ *converges to zero in probability* as $m \to \infty$ for each $\epsilon > 0$, i.e., to the requirement that for each $\epsilon, \delta, \alpha > 0$, there exists an $m_0 = m_0(\epsilon, \delta, \alpha)$ such that

$$Q\{f \in \mathcal{F} : \beta_m(f, \epsilon) > \delta\} \leq \alpha, \ \forall m \geq m_0.$$

Let us expand the above relationship by substituting for $\beta_m(f, \epsilon)$. This gives

$$Q\{f \in \mathcal{F} : P^m\{\mathbf{x} \in X^m : d_P[f, h_m(f; \mathbf{x})] > \epsilon\} > \delta\} \leq \alpha, \ \forall m \geq m_0. \tag{9.3.2}$$

In other words, it can be said that the algorithm PAC learns to accuracy ϵ and confidence δ for all functions in $\mathcal{F}$ except for those belonging to a set of measure α. By drawing more and more samples, all of the numbers ϵ, δ, α can be made arbitrarily small.

Now let us modify the above interpretation to the situation where Q and P are no longer fixed. Then the question arises as to whether the algorithm is permitted to use the probability measure $Q \in \mathcal{Q}$ explicitly. It is perhaps consistent with the spirit of "prior information" if the algorithm were to have Q available to it, since Q is just the prior distribution of target concepts. Moreover, if the algorithm *cannot* use Q explicitly, and if $\mathcal{Q}$ equals $\mathcal{Q}^*$, the set of *all* probabilities on $\mathcal{F}$, then it is easy to see that this type of learning with prior information reduces to the standard PAC problem formulation. Thus it can be assumed that Q is known and available to the learner, in which case the problem "decouples" into a *collection* of problems, one for each fixed Q. The final conclusion therefore is that Q can be assumed to be fixed throughout. As for P, if P is known only to belong to a family of probability measures $\mathcal{P}$, then the definition of the stochastic process $\beta_m(f, \epsilon)$ can be modified to

$$\beta_m(f, \epsilon) := \sup_{P \in \mathcal{P}} P^m\{\mathbf{x} \in X^m : d_P[f, h_m(f; \mathbf{x})] > \epsilon\}.$$

Thus an algorithm PAC learns WPI if, for each $\epsilon, \delta, \alpha > 0$, there exists an $m_0 = m_0(\epsilon, \delta, \alpha)$ such that

$$Q\{f \in \mathcal{F} : \sup_{P \subset \mathcal{P}} P^m\{\mathbf{x} \in X^m : d_P[f, h_m(f; \mathbf{x})] > \epsilon\} > \delta\} \leq \alpha, \ \forall m \geq m_0. \tag{9.3.3}$$

9.3.2 Some Simple Sufficient Conditions

Now a few simple *sufficient* conditions for PAC learnability WPI are given. The problem of deriving *necessary* conditions is studied in the next section. In view of the comments above, it is assumed that Q is a known fixed probability measure on $\mathcal{F}$.

338   9. Alternate Models of Learning

Theorem 9.4. *Suppose $\mathcal{P} = \{P\}$, a singleton set. Then the triple $(\mathcal{F}, P, Q)$ is PAC learnable WPI if for each $\alpha > 0$ there exists an "exceptional set" $\mathcal{E}_\alpha \subseteq \mathcal{F}$ such that $Q(\mathcal{E}_\alpha) \leq \alpha$, and*

$$N(\epsilon, \mathcal{F} \setminus \mathcal{E}_\alpha, d_P) < \infty, \ \forall \epsilon > 0.$$

In particular, if $\mathcal{F}$ is countable, then $(\mathcal{F}, P, Q)$ is PAC learnable WPI for every P, Q.

Proof. First, suppose $\mathcal{F}, P, Q$ satisfy the hypothesis of the theorem. Given any $\epsilon, \delta, \alpha > 0$, choose a set $\mathcal{E}_\alpha \subseteq \mathcal{F}$ such that $Q(\mathcal{E}_\alpha) \leq \alpha$, and such that the complement $\mathcal{F} \setminus \mathcal{E}_\alpha$ satisfies the finite metric entropy condition. For each *fixed* $\epsilon > 0$, let $N := N(\epsilon/2, \mathcal{F} \setminus \mathcal{E}_\alpha, d_P)$, and choose a minimal $\epsilon/2$-cover $g_1, \ldots, g_N$ for $\mathcal{F} \setminus \mathcal{E}_\alpha$. Define the algorithm to be the minimal empirical risk algorithm applied to $\{g_1, \ldots, g_N\}$. Then from Theorem 6.2, we have that

$$\beta(f, \epsilon) \leq N e^{-m\epsilon^2/8}, \text{ if } f \in \mathcal{F} \setminus \mathcal{E}_\alpha.$$

Hence if we choose

$$m \geq \frac{8}{\epsilon^2} \ln \frac{N}{\delta},$$

then the condition (9.3.2) is satisfied.

To prove the last sentence, suppose $\mathcal{F}$ is countable, say $\mathcal{F} = \{f_1, f_2, \ldots\}$, and let Q be an arbitrary probability measure on $\mathcal{F}$. Then Q is completely characterized by the sequence of weights $q_i := Q(\{f_i\})$. Note that $\sum_{i=1}^{\infty} q_i = 1$. Hence, given any $\alpha > 0$, there exists an integer M such that $\sum_{i=M+1}^{\infty} q_i \leq \alpha$. Let $\mathcal{E}_\alpha := \{f_i, i \geq M+1\}$. Then the collection $\mathcal{F} \setminus \mathcal{E}_\alpha = \{f_1, \ldots, f_M\}$ satisfies the finite metric entropy condition for every ϵ, since it is itself a finite collection. ∎

Theorem 9.5. *Suppose $\mathcal{F}, \mathcal{P}, Q$ are given. Then $\mathcal{F}$ is PAC learnable WPI if, for each $\alpha > 0$, there exists a set $\mathcal{E}_\alpha \subseteq \mathcal{F}$ such that $Q(\mathcal{E}_\alpha) \leq \alpha$ and the collection $\mathcal{F} \setminus \mathcal{E}_\alpha$ has the property of uniform convergence of empirical probabilities uniformly in probability (UCEPUP) with respect to $\mathcal{P}$. Then the triple $(\mathcal{F}, \mathcal{P}, Q)$ is PAC learnable WPI; in particular, every consistent algorithm is PAC WPI.*

Proof. Recall from Theorem 5.9 that if a collection of sets $\mathcal{G}$ has the UCEPUP property with respect to $\mathcal{P}$, then so does the collection

$$\mathcal{G}\Delta\mathcal{G} := \{f \Delta g : f, g \in \mathcal{G}\}.$$

In particular, given any $\epsilon, \delta > 0$, there exists an m_0 such that

$$\sup_{P \in \mathcal{P}} P^m\{\mathbf{x} \in X^m : \exists f, g \in \mathcal{G} \text{ with } \hat{d}(f, g; \mathbf{x}) = 0 \text{ and } d_P(f, g) > \epsilon\}$$

$$\leq \delta \ \forall m \geq m_0, \quad (9.3.4)$$

9.3 Learning with Prior Information: Necessary and Sufficient Conditions

where $\hat{d}(f, g; \mathbf{x})$ is the empirical distance between f and g based on the multisample $\mathbf{x}$, defined by

$$\hat{d}(f, g; \mathbf{x}) := \frac{1}{m} \sum_{j=1}^{m} |f(x_j) - g(x_j)|.$$

To apply this result to the problem at hand, let $\alpha > 0$ be chosen arbitrarily, and select $\mathcal{E}_\alpha \subseteq \mathcal{F}$ such that $Q(\mathcal{E}_\alpha) \leq \alpha$ and the collection $\mathcal{F} \setminus \mathcal{E}_\alpha$ has the UCEPUP property with respect to $\mathcal{P}$. Given $\epsilon, \delta > 0$, select m_0 as above so that (9.3.4) holds with $\mathcal{G}$ replaced by $\mathcal{F} \setminus \mathcal{E}_\alpha$. Choose any *consistent* algorithm. Then

$$\hat{d}[f, h_m(f; \mathbf{x})] = 0, \ \forall f \in \mathcal{F} \setminus \mathcal{E}_\alpha, \ \forall \mathbf{x} \in X^m, \ \forall m.$$

In particular, for all $m \geq m_0$, we have

$$P^m\{\mathbf{x} \in X^m : d_P[f, h_m(f; \mathbf{x})] > \epsilon\} \leq \delta, \ \forall m \geq m_0, \ \forall f \in \mathcal{F} \setminus \mathcal{E}_\alpha, \ \forall P \in \mathcal{P}.$$

Finally, (9.3.3) is satisfied with m_0 chosen as above. ∎

Theorem 9.6. *Suppose $\mathcal{P} = \mathcal{P}^*$, the set of all probability measures on $(X, \mathcal{S})$. Then the triple $(\mathcal{F}, \mathcal{P}^*, Q)$ is PAC learnable WPI if, for each $\alpha > 0$, there exists an $\mathcal{E}_\alpha \subseteq \mathcal{F}$ such that $Q(\mathcal{E}_\alpha) \leq \alpha$, and F-dim$_\gamma(\mathcal{F} \setminus \mathcal{E}_\alpha) < \infty$, $\forall \gamma > 0$. In particular, every consistent algorithm is PAC WPI.*

The proof is almost identical to that of Theorem 9.5 and is thus left to the reader.

Corollary 9.1. *Suppose $\mathcal{C}$ is a nested collection of the form*

$$\mathcal{C} = \bigcup_{i=1}^{\infty} \mathcal{C}_i, \ where \ \mathcal{C}_i \subseteq \mathcal{C}_{i+1},$$

or else a disjoint union of the form

$$\mathcal{C} = \bigcup_{i=1}^{\infty} \mathcal{A}_i, \ where \ \mathcal{A}_i \cap \mathcal{A}_j = \emptyset \ if \ i \neq j.$$

Suppose Q is an arbitrary probability measure on $\mathcal{C}$ such that each $\mathcal{C}_i$ is measurable, or each $\mathcal{A}_i$ is measurable, as appropriate. Suppose VC-dim$(\mathcal{A}_i) < \infty$ for all i, or that VC-dim$(\mathcal{C}_i) < \infty$ for all i, as appropriate. Then the triple $(\mathcal{C}, \mathcal{P}^*, Q)$ is PAC learnable WPI. In particular, every consistent algorithm is PAC WPI.

Proof. Observe first that both hypotheses are equivalent, since a disjoint union can be turned into a nested union by defining

$$\mathcal{C}_i = \bigcup_{j=1}^{i} \mathcal{A}_j, \ \forall i.$$

Similarly, a nested union can be turned into a disjoint union by defining

$$\mathcal{A}_1 = \mathcal{C}_1, \text{ and } \mathcal{A}_i = \mathcal{C}_i \setminus \mathcal{C}_{i-1} \text{ for } i \geq 2.$$

Also, from Lemma 4.11, it follows that

$$\text{VC-dim}(\mathcal{C}_i) \leq i - 1 + \sum_{j=1}^{i} \text{VC-dim}(\mathcal{A}_j), \text{ and}$$

$$\text{VC-dim}(\mathcal{A}_i) \leq \text{VC-dim}(\mathcal{C}_i) \text{ since } \mathcal{A}_i \subseteq \mathcal{C}_i.$$

Finally, the measurability of each $\mathcal{A}_i$ implies that of each $\mathcal{C}_i$, and vice versa. Accordingly, let us examine the case of a disjoint union. Let Q be an arbitrary probability measure on $\mathcal{C}$ such that each $\mathcal{A}_i$ is measurable. Define $q_i = Q(\mathcal{A}_i)$, and observe that $\sum_{i=1}^{\infty} q_i = 1$.

Thus, given any $\alpha > 0$, we can find an integer M such that $\sum_{i=M+1}^{\infty} q_i \leq \alpha$. Now apply Theorem 9.6 with

$$\mathcal{E}_\alpha := \bigcup_{i=M+1}^{\infty} \mathcal{A}_i,$$

and observe that $Q(\mathcal{E}_\alpha) \leq \alpha$. Also, as observed above,

$$\text{VC-dim}(\mathcal{C} \setminus \mathcal{E}_\alpha) \leq i - 1 + \sum_{j=1}^{i} \text{VC-dim}(\mathcal{A}_j) < \infty.$$

This completes the proof. ∎

Example 9.10. Consider the collection of support sets of dyadic functions, as defined in Example 6.10. For each $x \in [0, 1)$, let $b_i(x)$ denote the i-th bit in the binary expansion of x in the form

$$x = \sum_{i=1}^{\infty} b_i(x) 2^{-i}.$$

Finally, define

$$C_i = \{x \in X : b_i(x) = 1\}, \ i = 1, 2, \ldots.$$

It is shown in Example 6.10 that if P is the uniform probability measure on $[0, 1)$, then the countable concept class $\mathcal{C} := \{C_i\}$ is *not* PAC learnable, since it fails to satisfy the finite metric entropy condition of Theorem 6.3. In contrast, Theorem 9.5 implies that, for *any* probability measure Q on $\mathcal{C}$, the triple $(\mathcal{C}, Q, \mathcal{P}^*)$ is (distribution-free) PAC learnable WPI. Thus permitting an algorithm to fail on a few target concepts has a dramatic positive effect on learnability.

9.3 Learning with Prior Information: Necessary and Sufficient Conditions

Example 9.11. Let $X = [0,1]^n$ where n is a fixed integer, and let $\mathcal{C}$ equal the set of closed convex polygons in X. Then, by a straight-forward adaptation of Example 4.5, it follows that VC-dim($\mathcal{C}$) is infinite, so that $\mathcal{C}$ by itself is *not* distribution-free PAC learnable. Now let $\mathcal{A}_k$ denote the set of closed convex polygons in X with exactly k faces, i.e., those closed convex polygons obtained by a nontrivial intersection of exactly k closed half-planes in $\mathbb{R}^n$. Then it is easy to see that $\mathcal{C}$ is a disjoint union of $\mathcal{A}_k, k \geq 1$. Also, it follows from Example 4.3 and Theorem 4.5 that

$$\text{VC-dim}(\mathcal{A}_k) \leq 2k(n+1)\lg(ek) < \infty, \ \forall k.$$

Now suppose Q is any probability measure on $\mathcal{C}$ such that each $\mathcal{A}_k$ is measurable. Then from Theorem 9.6 it follows that the triple $(\mathcal{C}, \mathcal{P}^*, Q)$ is PAC learnable WPI.

9.3.3 Dispersability of Function Classes

In the remainder of this section, we derive some necessary and sufficient conditions for a function family to be learnable WPI. Specifically, in this subsection we introduce a notion called "dispersability," which turns out to be sufficient for learnability WPI in all cases, and necessary for learnability WPI in the case of concept classes. Thus dispersability plays the same role in the problem of learnability WPI as does the finite metric entropy property in the case of conventional PAC learnability with a fixed distribution.

Definition 9.2. *(Dispersion under a partition) Consider a partition π of the function class $\mathcal{F}$, i.e. a collection $\{\mathcal{F}_i \in \mathcal{F}\}_{i=1}^r$ such that $\cup_{i=1}^r \mathcal{F}_i = \mathcal{F}$ and $\mathcal{F}_i \cap \mathcal{F}_j = \emptyset, i \neq j$. The **dispersion** of the class F under the partition π is defined as*

$$\text{disp}(\pi) := \sum_{i=1}^r \inf_{f \in F} \int_{\mathcal{F}_i} d_P(g, f) \, Q(dg).$$

The expression $\inf_{f \in \mathcal{F}} \int_{\mathcal{F}_i} d_P(g, f) \, Q(dg)$ is a measure of the dispersion of the set $\mathcal{F}_i$ (the i-th element of the partition π) where each function $g \in \mathcal{F}_i$ is given a weight according to the probability Q. Therefore $\text{disp}(\pi)$ quantifies the dispersion of a function class once it has been split into the subclasses forming the partition.

Suppose now one is allowed to select a partition π of given cardinality r so as to minimize the dispersion. The resulting dispersion is the so-called minimal dispersion:

Definition 9.3. *The **minimal dispersion** of $\mathcal{F}$ under a partition of cardinality r is defined as*

$$\text{disp}(r) := \inf_{\pi : |\pi| = r} \text{disp}(\pi).$$

342   9. Alternate Models of Learning

A partition π is said to be "optimal" when its dispersion is minimal, that is, $\mathrm{disp}(\pi) = \mathrm{disp}(r)$. Note that an optimal dispersion need not exist in general. However, there will always exist a partition π of cardinality r such that $\mathrm{disp}(\pi)$ is arbitrarily close to $\mathrm{disp}(r)$. In the proofs of the various theorems below, it is always assumed that an optimal partition exists. This is done purely to reduce notational clutter, and the proofs can be readily amended to cater to the case where optimal partitions do not exist.

Finally we come to the notion of dispersability itself.

Definition 9.4. *The function class F is said to be* **dispersable** *if*

$$\lim_{r \to \infty} \mathrm{disp}(r) = 0.$$

Thus a function class is dispersable if its dispersion can be made arbitrarily small by considering partitions of larger and larger cardinality.

Next, it is shown that finite metric entropy implies dispersability, and then it is shown that the converse is not true in general.

Lemma 9.3. *Suppose a function class $\mathcal{F}$ satisfies the finite metric entropy condition with respect to d_P, and let Q be an arbitrary probability measure on $\mathcal{F}$. Then $\mathcal{F}$ is dispersable.*

Proof. Let $N(\epsilon, \mathcal{F}, d_P)$ denote the minimum number of balls of radius ϵ needed to cover $\mathcal{F}$. The proof consists of showing that $\mathrm{disp}(r) \leq \epsilon \ \forall r \geq N(\epsilon) = N(\epsilon, \mathcal{F}, d_P)$, from which it follows that $\lim_{r \to \infty} \mathrm{disp}(r) = 0$, i.e., that $\mathcal{F}$ is dispersable. Consider a collection of $N(\epsilon)$ closed balls B_i centered at f_i, for $i = 1, \ldots, N(\epsilon)$, such that $\cup_{i=1}^{N(\epsilon)} B_i = \mathcal{F}$. Define $\mathcal{F}_i = B_i \setminus \cup_{j=1}^{i-1} B_j$, $i = 1, \ldots, N(\epsilon)$. Then

$$\begin{aligned}
\mathrm{disp}(N(\epsilon)) &= \inf_{\pi : |\pi| = N(\epsilon)} \mathrm{disp}(\pi) \\
&\leq \sum_{i=1}^{N(\epsilon)} \inf_{f \in \mathcal{F}} \int_{\mathcal{F}_i} d_P(g, f) \, Q(dg) \\
&\leq \sum_{i=1}^{N(\epsilon)} \int_{\mathcal{F}_i} d_P(g, f_i) \, Q(dg) \\
&\leq \sum_{i=1}^{N(\epsilon)} \epsilon \, Q(\mathcal{F}_i) \\
&= \epsilon.
\end{aligned} \qquad (9.3.5)$$

On the other hand, dispersability is a milder property than the finite metric entropy property, as shown next.

Lemma 9.4. *Suppose (Y, ρ) is a separable metric space such that $\rho(y, y') \leq 1$, for every $y, y' \in Y$, and let $\mathbf{Y}$ denote the corresponding Borel σ-algebra on Y. Suppose Q is a probability measure on $(Y, \mathbf{Y})$. Then Y is dispersable.*

9.3 Learning with Prior Information: Necessary and Sufficient Conditions

Proof. Given $\epsilon > 0$, select a countable set $\{y_i \in Y\}$ such that, with B_i equal to the closed ball of radius $\epsilon/2$ centered at y_i, we have that $\cup_i B_i = Y$. Such a countable set exists since Y is separable. Set $Y_n := \cup_{i=1}^{n-1} B_i$, and note that $Q(Y_n) \uparrow 1$. Choose $n(\epsilon)$ such that $Q(Y_{n(\epsilon)}) \geq 1 - \epsilon/2$. Define $\mathcal{F}_i := Y_i \setminus Y_{i-1} = B_i \setminus \cup_{j=1}^{i-1} B_j$. Then

$$\begin{aligned}
\mathrm{disp}(n(\epsilon)) &\leq \sum_{i=1}^{n(\epsilon)-1} \int_{\mathcal{F}_i} \rho(y, y_i)\, Q(dy) + Q(Y \setminus Y_{n(\epsilon)}) \\
&\leq \frac{\epsilon}{2} Q(Y_{n(\epsilon)}) + Q(Y \setminus Y_{n(\epsilon)}) \\
&\leq \frac{\epsilon}{2} + \frac{\epsilon}{2} \\
&= \epsilon.
\end{aligned}$$

Since ϵ is arbitrary, this implies that Y is dispersable.

In particular, Lemma 9.4 implies that every countable set is dispersable under every bounded metric, because a countable set is always separable. On the other hand, it is easy to construct examples of countable sets with a bounded metric that do not satisfy the finite metric entropy condition, such as the dyadic functions of Example 6.10. This shows that dispersability is in general a weaker property than finite metric entropy.

The next result shows that dispersability is perhaps "too weak" a property, in that every subset of a separable metric space is dispersable. This result is therefore applicable to practically all examples in the learning literature.

Lemma 9.5. *Suppose X is a separable metric space and let $\mathcal{S}$ denote the associated Borel σ-algebra. Let $\mathcal{F}$ denote the family of all measurable functions from X to $[0,1]$. Let P be any probability measure on $(X, \mathcal{S})$, and let d_P denote the corresponding pseudometric on $\mathcal{F}$. Finally, let Q be any probability measure on the Borel σ-algebra of the metric space $(\mathcal{F}, d_P)$. Then, the function class $\mathcal{F}$ is dispersable.*

Proof. The theorem is proven by showing that $(\mathcal{F}, d_P)$ is a separable metric space. Once the separability of $(\mathcal{F}, d_P)$ is established, its dispersability follows from Lemma 9.4.

Note first that the Borel σ-algebra on a separable matric space X is countably generated (by all the balls with rational radius centered on a dense countable subset of X). Thus, $\mathcal{S}$ is countably generated. Next, apply [26], Theorem 19.2, which states that the space $L^p(X)$, $1 \leq p < \infty$, where X is a set with σ-finite measure, is separable provided that the σ-algebra on X is countably generated. This leads to the conclusion that the space $L^1(X)$ of summable functions on $(X, \mathcal{S})$ is separable. Finally the conclusion follows on observing that $(\mathcal{F}, d_p)$ is is a subset of $L^1(X)$. ∎

Even though the notion of dispersability is very weak in general, it is still possible to find examples of function classes that are not dispersable.

344   9. Alternate Models of Learning

Example 9.12. Let $X = R^{[0,1]}$ be the set of real functions defined on the interval $[0, 1]$. The variable $t \in [0, 1]$ is interpreted as time and an element $x \in X$ is a trajectory of a stochastic process. Endow X with a σ-algebra and a probability by means of the standard procedure based on Kolmogorov's existence theorem ([26], Theorem 36.1). Specifically, given any finite set of time instants $t_1, \ldots, t_k \in [0, 1]$, define the finite dimensional distribution corresponding to $t_1, \ldots, t_k$ as the uniform distribution in the hypercube $[0, 1]^k$. This completely defines the probability of cylinder sets, that is, sets of form $\{x \in X : (x(t_1), \ldots, x(t_k)) \in H\}$, where $x(t_j)$ represents the value of the trajectory x at t_j and H is a Borel set in R^k. This system of finite dimensional distributions is "consistent" in the sense stated in Section 36 of [26]. Then by Kolmogorov's existence theorem, it follows that there exists a probability P defined over the σ-algebra generated by the cylinder sets whose finite dimensional distributions coincide with the given uniform distributions. This completes the definition of the probability space $(X, \mathcal{S}, P)$.

Now define the concept $C(t) := \{x \in X : x(t) \in [0, 0.5]\}$, where $t \in [0, 1]$. Thus the concept $C(t)$ consists of all trajectories $x(\cdot)$ that assume a value in the interval $[0, 0.5]$ at time t. Let $\mathcal{C}$ the corresponding concept class. Thus $\mathcal{C} := \{C(t), t \in [0, 1]\}$. Note that the concept class $\mathcal{C}$ is uncountable. By Lemma 9.4, any countable set is dispersable. Observe that $d_P(C_t, C_\tau) = 0.5$ if $t \neq \tau$, because

$$\begin{aligned} d_P(C_t, C_\tau) &= P(C_t \Delta C_\tau) \\ &= P(\{x \in X : x(t) \in [0, 0.5] \text{ and } x(\tau) \in (0.5, 1]\} \\ &\quad \bigcup \{x \in X : x(\tau) \in (0.5, 1] \text{ and } x(\tau) \in [0, 0.5]\}) \\ &= 0.25 + 0.25 = 0.5. \end{aligned}$$

The collection $\mathcal{C}$ can be placed in a one-to-one correspondence with the unit interval $[0, 1]$ by the association $t \leftrightarrow c(t)$. Let $\mathcal{S}(\mathcal{C})$ denote the σ-algebra on $\mathcal{C}$ corresponding to the Borel subsets of the unit interval $[0, 1]$. Now it is claimed that $\mathcal{C}$ is not dispersable. This can be verified by noting that, for any partition $\pi = \{C_i\}_{i=1}^r$, we have

$$\begin{aligned} \mathrm{disp}(\pi) &= \sum_{i=1}^r \inf_{B \in \mathcal{C}} \int_{\mathcal{C}_i} d_P(A, B)\, Q(dA) \\ &= \sum_{i=1}^r 0.5 Q(\mathcal{C}_i) \\ &= 0.5. \end{aligned}$$

In other words, $\mathrm{disp}(\pi) = 0.5$ for *every* partition π. Thus $\mathcal{C}$ is not dispersable.

9.3.4 Connections Between Dispersability and Learnability WPI

In this subsection, it is shown that if a function family is dispersable, then a minimum empirical risk algorithm applied to a suitably selected partition

9.3 Learning with Prior Information: Necessary and Sufficient Conditions

of the function class is PAC WPI. An analysis of the complexity of this algorithm in the present setting is carried out in the next subsection. It is also shown that, in the case of *concept* classes, the dispersability condition is also *necessary* for PAC learnability WPI. Since this latter result has a very short proof, it is proved first.

In the sequel, a set of concepts $\mathcal{C}\{C\}$ is identified with the corresponding function class $\mathcal{F} := \{I_C, C \in \mathcal{C}\}$. In particular, we say that "a concept class $\mathcal{C}$ is learnable (or dispersable)" if $\mathcal{F} = \{I_C, C \in \mathcal{C}\}$ is learnable (or dispersable).

Theorem 9.7. *A concept class $\mathcal{C}$ is PAC learnable WPI only if it is dispersable.*

Proof. Consider an algorithm that PAC learns $\mathcal{C}$ WPI, and denote by $\{H_m(T;\mathbf{x})\}$ the corresponding random sequence of hypotheses corresponding to the target concept T and multisample $\mathbf{x}$. The probability space $(\mathcal{C} \times X^m, Q \times P^m)$ in which $H_m(T;\mathbf{x})$ resides can be embedded in the larger probability space $(\mathcal{C} \times X^\infty, Q \times P^\infty)$, so as to eliminate an explicit dependence on the number of samples m. In this space $(\mathcal{C} \times X^\infty, Q \times P^\infty)$, the PAC learnability WPI assumption implies that as $m \to \infty$, the sequence $\{d_P(T, H_m(T;\mathbf{x}))\}_{m=1}^\infty$ converges to zero in probability with respect to the measure $Q \times P^\infty$. Therefore, from the sequence $\{d_P(T, H_m(T;\mathbf{x}))\}_{m=1}^\infty$ it is possible to extract a subsequence $\{d_P(T, H_{m_n}(T;\mathbf{x}))\}_{n=1}^\infty$ that converges to zero $Q \times P^\infty$ almost surely (see, e.g., [114]). This implies that $\forall \rho > 0$, there exists a $\mathcal{C}(\rho) \subset \mathcal{C}$ such that

1. $Q(\mathcal{C}(\rho)) \geq 1 - \rho$.
2. $\lim_{n \to \infty} \sup_{C \in \mathcal{C}(\rho)} P^\infty \{d_P(T, H_{m_n}(T;\mathbf{x})) > \epsilon\} = 0$, $\forall \epsilon > 0$.

By virtue of Theorem 6.3, Condition 2 implies that $\mathcal{C}(\rho)$ satisfies the finite metric entropy condition, and is therefore dispersable by Lemma 9.3. Now select a partition π of $\mathcal{C}(\rho)$ such that the dispersion of $\mathcal{C}(\rho)$ is less than or equal to ρ; then the partition $\pi \cup (\mathcal{C} \setminus \mathcal{C}(\rho))$ of $\mathcal{C}$ has a dispersion not greater than 2ρ. Since ρ is arbitrary, this proves that $\mathcal{C}$ is dispersable. ∎

The remainder of the subsection is devoted to showing that dispersability is a *sufficient* condition for PAC learnability WPI for a function class, by constructing a suitable learning algorithm.

Suppose the function family $\mathcal{F}$ is dispersable. Consider a partition $\pi = \{\mathcal{F}_i\}_{i=1}^r$. In the interests of simplifying the notation, it is assumed throughout the sequel that there exist functions f_i, $i = 1, \ldots, r$, minimizing the dispersion of each element $\mathcal{F}_i \in \pi$. In other words, it is assumed that for each i there exists an $f_i \in \mathcal{F}_i$ such that

$$\int_{\mathcal{F}_i} d_P(g, f_i)\, Q(dg) = \inf_{f \in \mathcal{F}} \int_{\mathcal{F}_i} d_P(g, f)\, Q(dg). \tag{9.3.6}$$

In case this condition is not satisfied, suitable approximations could be used in place of the f_i's.

346   9. Alternate Models of Learning

The following algorithm is simply a minimal empirical error algorithm applied to a partition with minimum dispersion.

Algorithm 9.1. Select an increasing integer-valued function $r(m) \uparrow \infty$. After m samples, do the following:

1. Determine an optimal partition π_m of cardinality $r(m)$. Thus $\text{disp}(\pi_m) = \text{disp}(r(m))$.
2. Let $r = r(m)$. Determine functions $\{f_i\}_{i=1}^r$ such that (9.3.6) holds.
3. Compute the empirical error of each function f_i:

$$\hat{d}_{P,m}(g, f_i) := \frac{1}{m} \sum_{j=1}^{m} |g(x_j) - f_i(x_j)|, \quad i = 1, \ldots, r.$$

4. Select h_m to be the minimizer of the empirical distance $\hat{d}P, m(g, f_i)$; thus

$$h_m := \arg \min_{f_i, i=1,\ldots,r} \hat{d}_{P,m}(g, f_i).$$

The behaviour of the above algorithm is captured in the next result.

Theorem 9.8. *If $\mathcal{F}$ is dispersable and $r(m) = \exp(o(m))$, then Algorithm 1 PAC learns class $\mathcal{F}$ WPI.*

Proof. To show that the above algorithm PAC learns WPI, we compute the probability in the product probability space $\mathcal{F} \times X^m$ that $d_P(g, h_m(g; \mathbf{x}))$ exceeds a given value $\epsilon > 0$. It is claimed that

$$\Pr\{d_P(g, h_m(g; \mathbf{x})) > \epsilon\} \leq (r+1) \exp(-m\epsilon^2/8) + \frac{2}{\epsilon} \text{disp}(\pi), \quad \forall \epsilon > 0. \quad (9.3.7)$$

Suppose for the time being that (9.3.7) is established. Since $r(m) = \exp(o(m))$, the right side of (9.3.7) tends to zero for every $\epsilon > 0$. Hence the algorithm PAC learns WPI. Thus the proof is complete once (9.3.7) is established.

For this purpose, fix $g \in \mathcal{F}$, and choose an $f^0 \in \{f_1, \ldots, f_r\}$ such that

$$d_P(g, f^0) = \min_{1 \leq i \leq r} d_P(g, f_i).$$

Thus, while h_m is the minimizer of the *empirical* distance between the target function g and the f_i's, f^0 is the minimizer of the *true* distance between g and the f_i's. Let h_m be a shorthand for $h_m(g; \mathbf{x})$. Note that by definition of $\text{disp}(\pi)$ we have

$$\int_{\mathcal{F}} d_P(g, f^0) Q(dg) \leq \text{disp}(\pi). \quad (9.3.8)$$

Now let us compute the probability that $d_P(g, h_m) - d_P(g, f^0)$ exceeds $\epsilon/2$. Note that if

$$d_P(g, f_i) - \hat{d}_{P,m}(g, f_i) \leq \frac{\epsilon}{4} \text{ for } i = 1, \ldots, r, \text{ and} \quad (9.3.9)$$

9.3 Learning with Prior Information: Necessary and Sufficient Conditions

$$\hat{d}_{P,m}(g, f^0) - d_P(g, f^0) \leq \frac{\epsilon}{4}, \qquad (9.3.10)$$

then it follows that

$$d_P(g, h_m) - \hat{d}_{P,m}(g, h_m) \leq \frac{\epsilon}{4} \text{ since } h_m \in \{f_1, \ldots, f_r\},$$

$\hat{d}_{P,m}(g, h_m) - \hat{d}_{P,m}(g, f^0) \leq 0$ by the manner of choosing h_m,

$$\hat{d}_{P,m}(g, f^0) - d_P(g, f^0) \leq \frac{\epsilon}{4}.$$

Adding these three inequalities leads to

$$d_P(g, h_m) - d_P(g, f^0) \leq \frac{\epsilon}{2}.$$

Hence the probability that $d_P(g, h_m) - d_P(g, f^0) > \epsilon/2$ is at most equal to the sum of the probabilities that one of the $r+1$ inequalities in (9.3.9) or (9.3.10) is violated. By Hoeffding's inequality (Lemma 2.7), the probability that any one of these inequalities is violated does not exceed $\exp(-m\epsilon^2/8)$. Hence

$$P^m\{d_P(g, h_m) - d_P(g, f^0) > \epsilon/2\} \leq (r+1)\exp(-m\epsilon^2/8). \qquad (9.3.11)$$

Finally,

$$\begin{aligned}
\Pr\{d_P(g, h_m) > \epsilon\} &= \int_{\mathcal{F}} P^m\{d_P(g, h_m) > \epsilon\} Q(dg) \\
&\leq \int_{\mathcal{F}} P^m \left(\left\{ d_P(g, h_m) - d_P(g, f^0) > \frac{\epsilon}{2} \right\} \right. \\
&\qquad \left. \cup \left\{ d_P(g, f^0) > \frac{\epsilon}{2} \right\} \right) Q(dg) \\
&\leq \int_{\mathcal{F}} P^m \left\{ d_P(g, h_m) - d_P(g, f^0) > \frac{\epsilon}{2} \right\} Q(dg) \\
&\quad + \frac{2}{\epsilon} \int_{\mathcal{F}} d_P(g, f^0) Q(dg) \\
&< (r+1)\exp(-m\epsilon^2/8) + \frac{2}{\epsilon}\mathrm{disp}(\pi),
\end{aligned}$$

where in the last inequality we have used equation (9.3.11) for bounding the first term and equation (9.3.8) for the second one. This establishes the desired inequality (9.3.7) and completes the proof. ∎

Corollary 9.2. *A concept class is PAC learnable WPI if and only if it is dispersable.*

The "only if" part is proven in Theorem 2. The "if" part follows from Theorem 3 which proves the existence of an algorithm that PAC learns class $\mathcal{F}$ WPI.

Theorem 9.9. *Let X be a separable metric space, equipped with the associated Borel σ-algebra. Let $\mathcal{F}$ denote the set of all measurable functions mapping X into $[0,1]$. Finally, let Q be any probability measure on $\mathcal{F}$. Then $\mathcal{F}$ is PAC learnable WPI.*

The proof follows readily from Lemma 9.5 and Theorem 9.8.

Theorem 9.9 shows that in the most widely studied situation where X is a subset of some Euclidean space $\mathbb{R}^n$ for some integer n, learnability WPI is automatic.

We now present an alternative to Algorithm 9.1. In the first step of Algorithm 9.1, one is obliged to determine functions f_i's that are at a minimal average distance from the functions in the i-th element of the partition π. However, determining these functions may be very difficult. It would be much easier simply to select an $f_i \in \mathcal{F}_i$ at random for each i, according to the probability Q. It is shown next that, even with the "optimal" choice of f_i replaced by a "random" choice of f_i, the algorithm is still PAC WPI. Note that there is now an extra element of randomness in this algorithm, since at the first step of this procedure functions f_i's are randomly selected. As a consequence, the hypothesis h_m is now a random element in the probability space $\mathcal{F} \times X^m \times \mathcal{F}_1 \times \cdots \times \mathcal{F}_{r(m)}$. Let us denote by $Q_{\mathcal{F}_i}$ the probability Q restricted to $\mathcal{F}_i$ (i.e. $Q_{\mathcal{F}_i} = Q/Q(\mathcal{F}_i)$), the probability on $\mathcal{F} \times X^t \times \mathcal{F}_1 \times \cdots \times \mathcal{F}_r$ is then given by $Q \times P^t \times Q_{\mathcal{F}_1} \times \cdots \times Q_{\mathcal{F}_r}$. Then we have the following result.

Lemma 9.6. *With h_m generated according to the above algorithm, we have*

$$\Pr\{d_P(g, h_m) > \epsilon\} \leq (r+1)\exp(-m\epsilon^2/8) + \frac{4}{\epsilon}\mathrm{disp}(\pi), \ \forall \epsilon > 0. \quad (9.3.12)$$

Hence, if $\mathcal{F}$ is dispersable and $r(m) = \exp(o(m))$, Algorithm 9.1 is still PAC WPI with a random choice of the f_i.

The proof is analogous to that of Lemma 9.5 and therefore omitted.

By comparing (9.3.7) and 9.3.12), we see that the upper bound for the probability of error with a random choice of the f_i increases by a factor of less than 2 over the upper bound with an optimal choice of the f_i.

9.3.5 Distribution-Free Learning with Prior Information

This subsection is devoted to the problem of learning with prior information in the case in which the probability P is not fixed and it can in fact be *any* probability on $(X, \mathcal{S})$. Define $\mathcal{P}^*$ to be the set of all probabilities on $(X, \mathcal{S})$.

Let $\mathcal{S}(\mathcal{F})$ denote a given σ-algebra on $\mathcal{F}$, and let Q denote a probability measure on $(\mathcal{F}, \mathcal{S}(\mathcal{F}))$. The probability Q constitutes the a priori probability that a function f happens to be the target function, or else the relative importance placed on different target functions. The probability Q is assumed to be known to the learner. According to the philosophy of learning with an

9.3 Learning with Prior Information: Necessary and Sufficient Conditions

arbitrary distribution, given a function $g \in \mathcal{F}$, the probability $P(g)$ according to which the samples x_j are collected is allowed to be any probability in $\mathcal{P}^*$. Moreover, the probability P may be different for different functions g. By the symbol K we denote a kernel of probabilities indexed by $g \in \mathcal{F}$:

$$K := \{P(g), g \in \mathcal{F}\}.$$

In other words, for a given g, $P(g)$ is a probability over X and the probability $P(g, A)$ of a set $A \in \mathcal{S}$ is $\mathcal{S}(\mathcal{F})$-measurable. In the context of distribution-free learning, K plays a role similar to that of P in the fixed distribution setting. Throughout, it is assumed that K is not known and can be any kernel. The set of all kernels is denoted by $\mathcal{K}^*$.

Given a kernel K, the probability Q allows us to define a corresponding probability $\Pr_m$ in the product measurable space $(\mathcal{F} \times X^m, \mathcal{S}(\mathcal{F}) \times \mathcal{S}^m)$ as the unique probability measure which extends the definition $\Pr_m(A \times B) := \int_B P^m(g, A) \, Q(dg)$, $A \in \mathcal{S}^t$, $B \in \mathcal{S}(\mathcal{F})$, to the σ-algebra $\mathcal{S}(\mathcal{F}) \times \mathcal{S}^m$.

The first step in the development of a distribution-free learning theory with prior information is the extension of the previous definitions of learning WPI to the present setting.

Definition 9.5. *An algorithm $\{A_t\}$ is* **distribution-free (d.f.) probably approximately correct (PAC) with prior information (WPI) to accuracy ϵ** *if*

$$\lim_{m \to \infty} \sup_{K \in \mathcal{K}^*} \Pr_m \{d_{P(g)}(g, h_m) > \epsilon\} = 0. \qquad (9.3.13)$$

The algorithm $\{A_m\}$ is **distribution-free PAC WPI** *if it is distribution-free PAC WPI to accuracy ϵ, for every $\epsilon > 0$. The function class $\mathcal{F}$ is* **distribution-free PAC learnable WPI** *if there exists an algorithm that is distribution-free PAC WPI.*

The distinctive feature of Definition 9.5 compared to Definition 9.1 is that in (9.3.13) convergence is required to hold uniformly with respect to $K \in \mathcal{K}^*$; that is, the probability P is allowed to depend on g and this dependence can be arbitrary since $\{P(g)\}$ can be any kernel.

Next we wish to extend the notion of dispersability to the distribution-free setting. There are some intricacies that need to be addressed.

In the fixed distribution setting, the dispersability condition is equivalent to the following requirement: As the cardinality of the partition π approaches infinity, the sum over the elements $\mathcal{F}_i$ (forming the partition π) of the average (with respect to Q) d_P-distance between the functions in $\mathcal{F}_i$ and some representative function f depending on $\mathcal{F}_i$ tends to zero. In mathematical terms, this requirement can be recast in the following statement equivalent to Definition 9.4. Denote by M the set of all maps $f : \mathcal{F} \to \mathcal{F}$ such that $f(g)$ is constant over $\mathcal{F}_i, i = 1, \ldots, r$. Then dispersability is equivalent to requiring that

$$\inf_{f \in M} E_Q\left[d_P(g, f(g))\right]$$

tends to zero when the size r of the partition $\pi = \{\mathcal{F}_i, i = 1,\ldots,r\}$ tends to infinity (compare with Definition 9.4). Extending this idea to a distribution-free setting requires some care. A straightforward, but rather naive, extension would consist in requiring that

$$\inf_{f \in M} \sup_{K \in \mathcal{K}^*} E_Q \left[d_{P(g)}(g, f(g)) \right] \qquad (9.3.14)$$

tends to zero when the partition size r increases. However, a little thought reveals that sending the quantity in (9.3.14) to zero is in general an impossible task. Suppose for instance that we are considering concept learning. Then the integrand $d_{P(g)}(g, f(g))$ can be always made equal to 1 by suitably selecting the probability $P(g)$, whenever $g \neq f(g)$.

The trouble with the above attempt to extend the definition of dispersability comes from the fact that one is asked to determine a partition able to reduce the dispersion, and yet, the metric $d_{P(g)}$ used to measure such a dispersion is unknown. Clearly this is an unfair game. To make the problem formulation more meaningful, the learner must be in a position to form some estimate of $P(g)$ before he is asked to determine the partition. This leads to the notion of *data dependent* partitions.

Consider a multisample $x = (x_1,\ldots,x_s)$. A partition π of cardinality r based on the multisample x is simply a collection of partitions indexed by x:

$$\pi = \{\mathcal{F}_i(x), i = 1,\ldots,r\}.$$

Let M be the set of maps $f : X^s \times \mathcal{F} \to \mathcal{F}$ such that for all $x \in X^s$ and $g \in \mathcal{F}$, $f(x, g)$ is constant over $\mathcal{F}_i(x), i = 1,\ldots,r$. The dispersion of the class $\mathcal{F}$ under partition π is then defined as

$$\mathrm{disp}(\pi) := \inf_{f \in M} \sup_{K \in \mathcal{K}^*} E_{Pr_s} \left[d_{P(g)}(g, f(x, g)) \right], \qquad (9.3.15)$$

where, in analogy with previous notation, Pr_s is defined as the product measure $Q \times [P(g)]^s$. The interpretation of (9.3.15) is as follows. Fix a map $f \in M$. Clearly $d_{P(g)}(g, f(x, g))$ is a random variable that depends on the multisample x and the target function g and it is therefore defined on $\mathcal{F} \times X^s$. Such a random variable depends on the kernel K through $P(g)$. Next, the operator E_{Pr_s} performs integration over $\mathcal{F} \times X^s$, thus returning the average distance of each g from the corresponding $f(x, g)$. The average here is with respect to the target function g and the random multisample x. So, all in all, $E_{Pr_s}\left[d_{P(g)}(g, f(x, g))\right]$ is a deterministic number that measures the average dispersion of g from the corresponding $f(x, g)$; it depends on the map f and the kernel K. Finally, $\mathrm{disp}(\pi)$ is defined as $\inf_f \sup_K E_{Pr_s}\left[d_{P(g)}(g, f(x, g))\right]$ and, therefore, it quantifies how small such an average dispersion can be made in the worst case with respect to K by suitably selecting the map f.

In analogy with (9.3.14), (9.3.15) is worst case owing to the presence of the quantifier $\sup_{K \in \mathcal{K}^*}$. However, in contrast with (9.3.14), in (9.3.15) the

9.3 Learning with Prior Information: Necessary and Sufficient Conditions 351

partition is allowed to depend on $x \in X^s$ and the dispersion is computed as an average over $\mathcal{F} \times X^s$. Such a dependence gives one the possibility of forming some estimate of $P(g)$ before $\mathcal{F}$ is partitioned. Finally, the *minimal dispersion* $\mathrm{disp}(r,s)$ is defined as the infimum of $\mathrm{disp}(\pi)$ when π ranges over the set of all partitions of cardinality r based on the multisample $x \in X^s$.

We are now in a position to define the notion of distribution-free dispersability.

Definition 9.6. *The function class $\mathcal{F}$ is* **distribution-free (d.f.) dispersable** *if*
$$\lim_{r,s \to \infty} \mathrm{disp}(r,s) = 0.$$

Note that $\mathrm{disp}(\cdot, \cdot)$ is a non-increasing function of both arguments and, therefore, the order in which the limit $r, s \to \infty$ is taken in Definition 9.6 is immaterial. The fact that $\mathrm{disp}(\cdot, \cdot)$ is non-increasing can be seen as follows. The function $d_{P(g)}(g, f(x,g))$ defined on $X^s \times \mathcal{F}$ can be embedded in the larger invariant space $X^\infty \times \mathcal{F}$. Then $E_{Pr_s}[d_{P(g)}(g, f(x,g))]$ becomes $E_{Pr_\infty}[d_{P(g)}(g, f(x,g))]$, which exhibits no explicit dependence on s. Now by increasing r and/or s, the set of maps M over which the infimum in (9.3.15) is taken becomes larger. It follows that $\mathrm{disp}(r,s)$ is a non-increasing function of r and s.

Next we introduce an algorithm which generalizes Algorithm 9.1 to a distribution-free framework.

Algorithm 9.2. Select two increasing integer-valued functions $r(m) \uparrow \infty$ and $s(m) \uparrow \infty$ such that $s(m) < m$ for all m. At time m, do the following:

1. Determine an optimal partition π_m of cardinality $r(m)$ based on the multisample $x \in X^{s(m)}$, i.e., a partition π_m such that $\mathrm{disp}(\pi_m) = \mathrm{disp}(r(m), s(m))$;
2. Determine a map f such that
$$\mathrm{disp}(\pi_m) = \sup_{K \in \mathcal{K}^*} E_{Pr_s}[d_{P(g)}(g, f(x,g))];$$
3. Compute the empirical error of each function $f_i(x)$, $i = 1, \ldots, r(m)$, associated with the map f, where $x = (x_1, \ldots, x_{s(m)})$ is the first $s(m)$-dimensional portion of the multisample $x = (x_1, \ldots, x_m)$:
$$\hat{d}_{P(g),t}(g, f_i(x)) := \frac{1}{m - s(m)} \sum_{j=s(m)+1}^{m} |g(x_j) - f_i(x, x_j)|, i = 1, \ldots, r(t);$$
4. Select
$$h_m := \arg \min_{f_i(x), i=1,\ldots,r(m)} \hat{d}_{P(g),m}(g, f_i(x)).$$

In the above algorithm, it is assumed that there exist an optimal partition π_m and a suitable map f. If this is not the case, then one can use a partition and a function that approach the minimum dispersion arbitrarity closely.

Now we summarize the main results in distribution-free learnability WPI. In the interests of brevity, the proofs are omitted and can be found in [43].

Theorem 9.10. *Suppose that the function class $\mathcal{F}$ is d.f. dispersable. If $s(m) = o(m)$ and $r(m) = \exp(o(m - s(m)))$, then Algorithm 9.2 d.f. PAC learns class $\mathcal{F}$ WPI.*

The proof can be found in [43], Theorem 7.

Since a concept class with finite VC-dimension is PAC learnable, it is also PAC learnable WPI. Thus one would expect that such a concept class would be d.f. dispersable. This is indeed the case, as shown next.

Theorem 9.11. *If $VC\text{-}dim(\mathcal{C}) < \infty$, then the concept class $\mathcal{C}$ is d.f. dispersable. If $P\text{-}dim(\mathcal{F}) < \infty$, then the function class $\mathcal{F}$ is d.f. dispersable.*

For the proof, see [43], Theorem 8.

Theorem 9.12. *If $\mathcal{K}^*$ is the set of all families of probabilities $P(g)$ indexed by $g \in \mathcal{F}$, then a concept class $\mathcal{C}$ is d.f. PAC learnable WPI if and only if it is d.f. dispersable.*

For the proof, see [43], Theorem 9.

9.4 Learning with Prior Information: Bounds on Learning Rates

In this section, some upper bounds are obtained on the "rates" at which learning takes place when there is a prior probability Q on the concept class. The notion of a "learning curve" is introduced that quantifies the rate at which an algorithm learns the unknown target concept, in terms of being able to predict the output of the unknown target concept on a given input. Two types of prediction models are used, namely Bayesian and Gibbsian. In the case where the concept class $\mathcal{C}$ has the property that empirical probabilities converge uniformly to their true values (in which case every consistent algorithm is PAC – see Theorem 6.1), it is possible to give *explicit* estimates of the number of errors made by each of the prediction models, both in terms of the "instantaneous" error and the "average" error. The material in this section follows [84] and [83].

To motivate the material presented in this section, suppose first that $\mathbf{x}^* \in X^\infty$, the countably infinite Cartesian product of X with itself; in other words, $\mathbf{x}^*$ is a sequence in X. Suppose $T \in \mathcal{C}$ is a fixed and unknown target concept,

9.4 Learning with Prior Information: Bounds on Learning Rates 353

and that an oracle returns the values of the indicator function $T(x_i), i \geq 1$.[5] Then, after m samples are drawn, the target concept T is "localized" to the set

$$\mathcal{A}_m(T; \mathbf{x}^*) := \{A \in \mathcal{C} : A(x_i) = T(x_i) \text{ for } i = 1, \ldots, m\}.$$

Note that $\mathcal{A}_m(T; \mathbf{x}^*)$ is precisely the collection of concepts that are consistent with the first m measurements. The set $\mathcal{A}_m(T; \mathbf{x}^*)$ is called the m-th **version space**. Suppose as before that the target concept T is known *a priori* to be distributed according to the probability measure Q on $\mathcal{C}$. Then, after m samples and the corresponding oracle outputs are available, the *posterior* distribution of T is obtained by restricting Q to the set $\mathcal{A}_m(T; \mathbf{x}^*)$ and renormalizing. In other words, for every subset $\mathcal{F} \subseteq \mathcal{C}$, the posterior probability measure Q_m is defined by

$$Q_m(\mathcal{F}) := \frac{Q[\mathcal{F} \cap \mathcal{A}_m(T; \mathbf{x}^*)]}{Q[\mathcal{A}_m(T; \mathbf{x}^*)]}.$$

For future reference, define

$$V_m(T; \mathbf{x}^*) := Q[\mathcal{A}_m(T; \mathbf{x}^*)].$$

The number $V_m(T; \mathbf{x}^*)$ quantifies the extent to which the target concept T is localized by the first m measurements – the smaller this number, the faster the learning process.

Now one can ask: How much *additional* information is obtained from the $(m+1)$-st sample, above and beyond that provided by the first m samples? After $m+1$ samples, it becomes known that $T \in \mathcal{A}_{m+1}(T; \mathbf{x}^*)$, which is a subset of $\mathcal{A}_m(T; \mathbf{x}^*)$. The **information gain** at the $(m+1)$-st step is defined as

$$\mathcal{I}_{m+1}(T; \mathbf{x}^*) := -\lg Q\{A \in \mathcal{A}_m(T; \mathbf{x}^*) : A(x_{m+1}) = T(x_{m+1})\}.$$

It is easy to see that

$$\mathcal{I}_{m+1}(T; \mathbf{x}^*) = -\lg \frac{V_{m+1}(T; \mathbf{x}^*)}{V_m(T; \mathbf{x}^*)},$$

where $V_0(T; \mathbf{x}^*)$ is taken as 1. For future reference, define the $(m+1)$-st **volume ratio** $\zeta_{m+1}(T; \mathbf{x}^*)$ by

$$\zeta_{m+1}(T; \mathbf{x}^*) := \frac{V_{m+1}(T; \mathbf{x}^*)}{V_m(T; \mathbf{x}^*)}.$$

Then

$$\mathcal{I}_{m+1}(T; \mathbf{x}^*) = -\lg[\zeta_{m+1}(T; \mathbf{x}^*)].$$

Next, let us consider the learning problem at its basic level, namely: Given the values $T(x_1), \ldots, T(x_m)$, predict the value $T(x_{m+1})$ when x_{m+1} is given.

---
[5] In the interests of simplifying the notation, the indicator function $I_T(\cdot)$ is denoted by $T(\cdot)$; a similar convention is adopted for other sets as well.

Two algorithms are proposed for this purpose. The well-known **Bayesian algorithm** is as follows: Given the m-th version space $\mathcal{A}_m(T;\mathbf{x}^*)$ and an element $x_{m+1} \in X$, divide $\mathcal{A}_m(T;\mathbf{x}^*)$ into two parts:

$$S_m^0 := \{A \in \mathcal{A}_m(T;\mathbf{x}^*) : A(x_{m+1}) = 0\}, \text{ and}$$

$$S_m^1 := \{A \in \mathcal{A}_m(T;\mathbf{x}^*) : A(x_{m+1}) = 1\}.$$

Then the Bayesian algorithm is as follows: If $Q(S_m^1) > Q(S_m^0)$,[6] predict $T(x_{m+1}) = 1$. If $Q(S_m^1) < Q(S_m^0)$, predict $T(x_{m+1}) = 0$. If $Q(S_m^1) = Q(S_m^0)$, predict $T(x_{m+1})$ as the outcome of tossing a fair coin. As stated in this fashion, the algorithm is probabilistic. To make it deterministic, let us arbitrarily predict $T(x_{m+1}) = 1$ if $Q(S_m^1) = Q(S_m^0)$. In other words, the (modified and deterministic) Bayesian algorithm is to predict $T(x_{m+1})$ to be $\eta[Q(S_m^1) - Q(S_m^0)]$, where $\eta(\cdot)$ is the Heaviside function.

The Bayesian algorithm is *optimal*, in the sense that if the target concept T is distributed according to the prior probability Q, then at each step, the probability that the actual value $T(x_{m+1})$ equals the predicted value is maximized. However, there is a potentially serious drawback to the algorithm. Let us fix samples $x_1, \ldots, x_m$, and view the map

$$x_{m+1} \mapsto \eta[Q(S_m^1) - Q(S_m^0)]$$

as the hypothesis generated at the $(m + 1)$-st step. Then in general this map *need not* belong to the given concept class $\mathcal{C}$. Thus, while the Bayesian algorithm produces an optimal hypothesis, it can also in general produce a hypothesis that does not belong to the original concept class.

To get around this difficulty, another algorithm known as the **Gibbsian algorithm** is proposed. It is simply this: Given the sequence of samples $x_1, \ldots, x_m$, construct the version space $\mathcal{A}_m(T;\mathbf{x}^*)$, together with the posterior probability Q_m on $\mathcal{A}_m(T;\mathbf{x}^*)$. Then simply pick an element $H_m \in \mathcal{A}_m(T;\mathbf{x}^*)$ at random according to Q_m, and predict that $T(x_{m+1})$ equals $H_m(x_{m+1})$. Though the Gibbsian algorithm is not optimal as the Bayesian algorithm is, it has the advantage that it always produces a hypothesis that belongs to the original concept class.

In the remainder of the section, an analysis is presented of the error probabilities of the two algorithms. Two types of errors are analyzed. The first is the *instantaneous* error, that is, the probability that $T(x_{m+1})$ is incorrectly predicted, viewed as a function of m. The second is the *average* error, that is, the fraction of incorrect predictions of $T(x_i)$ for $1 \leq i \leq m$, viewed as a function of m. Explicit bounds are derived for each of these quantities in terms of the instantaneous and cumulative information gains.

Consider first the Bayesian algorithm. In this case, the probability that $T(x_{m+1})$ is incorrectly predicted equals 1 if $V_{m+1}(T;\mathbf{x}^*) < V_m(T;\mathbf{x}^*)/2$. To

---

[6] This is equivalent to $Q_m(S_m^1) > Q_m(S_m^0)$, since the only difference between $Q_m(\cdot)$ and $Q(\cdot)$ is a normalization factor.

9.4 Learning with Prior Information: Bounds on Learning Rates 355

see this, note that if $V_{m+1}(T;\mathbf{x}^*) < V_m(T;\mathbf{x}^*)/2$, then T belongs to the "smaller half" of the version space $A_m(T;\mathbf{x}^*)$, whereas the Bayesian algorithm always predicts that T belongs to the "bigger half." Similarly, the probability that $T(x_{m+1})$ is incorrectly predicted equals 0 if $V_{m+1}(T;\mathbf{x}^*) > V_m(T;\mathbf{x}^*)/2$. Finally, if $V_{m+1}(T;\mathbf{x}^*) = V_m(T;\mathbf{x}^*)/2$, then $T(x_{m+1})$ is incorrectly predicted with a probability of 0.5, irrespective of whether the randomized or the modified deterministic Bayesian algorithm is used. Hence the instantaneous error rate of the Bayesian algorithm at the $(m+1)$-st step, denoted by $e^B_{m+1}(T;\mathbf{x}^*)$, is given by

$$e^B_{m+1}(T;\mathbf{x}^*) = E_Q[\theta(0.5 - \xi_{m+1}(T;\mathbf{x}^*))], \qquad (9.4.1)$$

where the expectation is taken with respect to $T \in \mathcal{C}$, and the function $\theta : \mathbb{R} \to \mathbb{R}$ is defined by

$$\theta(y) = \begin{cases} 1 & \text{if } y > 0, \\ 0.5 & \text{if } y = 0, \\ 0 & \text{if } y < 0. \end{cases}$$

Next, let us consider the error rate of the Gibbsian algorithm. In this algorithm, the predicted value of $T(x_{m+1})$ is correct if $H_m \in A_{m+1}(T;\mathbf{x}^*)$. Since H_m is chosen at random from $A_m(T;\mathbf{x}^*)$, the probability that $H_m \in A_{m+1}(T;\mathbf{x}^*)$ is precisely the volume ratio

$$\xi_{m+1}(T;\mathbf{x}^*) = Q_m[A_{m+1}(T;\mathbf{x}^*)] = \frac{Q[A_{m+1}(T;\mathbf{x}^*)]}{Q[A_m(T;\mathbf{x}^*)]}.$$

Hence the instantaneous error rate of the Gibbsian algorithm at the $(m+1)$-st step, denoted by $e^G_{m+1}(T;\mathbf{x}^*)$, is equals

$$e^G_{m+1}(T;\mathbf{x}^*) = E_Q[\xi_{m+1}(T;\mathbf{x}^*)]. \qquad (9.4.2)$$

To proceed further, a useful lemma is stated.

Lemma 9.7. *Suppose* $g : [0,1] \to [0,1]$. *Then*

$$E_Q[g(\xi_{m+1}(T;\mathbf{x}^*))] = E_Q[\xi_{m+1}g(\xi_{m+1}) + (1 - \xi_{m+1})g(1 - \xi_{m+1})],$$

where $\xi_{m+1} = \xi_{m+1}(T;\mathbf{x}^*)$.

Proof. Let us first generalize the notation slightly. Let $\mathbf{v}^* \in \{0,1\}^{\mathbf{N}}$, where $\mathbf{N}$ denotes the set of natural numbers, and define

$$\mathcal{A}_m(\mathbf{v}^*;\mathbf{x}^*) := \{A \in \mathcal{C} : A(x_i) = v_i \text{ for } i = 1, \ldots, m\}.$$

If in particular $v_i = T(x_i)$ for all i, then $\mathcal{A}_m(\mathbf{v}^*;\mathbf{x}^*)$ equals the version space $\mathcal{A}_m(T;\mathbf{x}^*)$. In an analogous fashion, define

$$V_m(\mathbf{v}^*;\mathbf{x}^*) := Q[\mathcal{A}_m(\mathbf{v}^*;\mathbf{x}^*)],$$

and
$$\xi_{m+1}(\mathbf{v}^*; \mathbf{x}^*) := \frac{V_{m+1}(\mathbf{v}^*; \mathbf{x}^*)}{V_m(\mathbf{v}^*; \mathbf{x}^*)}.$$
These symbols are obvious generalizations of corresponding earlier symbols. Now note that
$$\xi_{m+1}(\mathbf{v}^*; \mathbf{x}^*) = \frac{Q\{A \in \mathcal{A}_m(\mathbf{v}^*; \mathbf{x}^*) : A(x_{m+1}) = v_{m+1}\}}{V_m(\mathbf{v}^*; \mathbf{x}^*)}.$$
Hence
$$1 - \xi_{m+1}(\mathbf{v}^*; \mathbf{x}^*) = \frac{Q\{A \in \mathcal{A}_m(\mathbf{v}^*; \mathbf{x}^*) : A(x_{m+1}) = 1 - v_{m+1}\}}{V_m(\mathbf{v}^*; \mathbf{x}^*)}.$$
Now let $\mathbf{v}_m \in \{0,1\}^m$ denote the first m components of $\mathbf{v}^*$, and observe that $\xi_{m+1}(\mathbf{v}^*; \mathbf{x}^*)$ depends only on $\mathbf{v}_{m+1}$. Hence we may write $\xi_{m+1}(\mathbf{v}_{m+1}; \mathbf{x}^*)$ instead. Similarly one can write $V_{m+1}(\mathbf{v}_{m+1}; \mathbf{x}^*)$ instead of $V_{m+1}(\mathbf{v}^*; \mathbf{x}^*)$. Now, whatever be $T \in \mathcal{C}$, the vector $[T(x_1) \ldots T(x_m)]^t$ belongs to $\{0,1\}^m$. Hence
$$E_Q[g(\xi_{m+1}(T; \mathbf{x}^*))] = \sum_{\mathbf{v}_{m+1} \in \{0,1\}^{m+1}} g(\xi_{m+1}(\mathbf{v}_{m+1}; \mathbf{x}^*)) \cdot V_{m+1}(\mathbf{v}^*; \mathbf{x}^*).$$
Since v_{m+1} has only two possible values (namely 0 and 1), one can write
$$E_Q[g(\xi_{m+1}(T; \mathbf{x}^*))] = \sum_{\mathbf{v}_m \in \{0,1\}^m} f_m,$$
where
$$f_m := g(\xi_{m+1}(\mathbf{v}_m, 0; \mathbf{x}^*)) \cdot V_{m+1}(\mathbf{v}_m, 0; \mathbf{x}^*)$$
$$+ g(\xi_{m+1}(\mathbf{v}_m, 1; \mathbf{x}^*)) \cdot V_{m+1}(\mathbf{v}_m, 1; \mathbf{x}^*).$$
Note that the summation in the previous equation is over $\mathbf{v}_m$ and not $\mathbf{v}_{m+1}$. Next, using the definition of V_m, one observes that
$$V_{m+1}(\mathbf{v}_m, 0; \mathbf{x}^*) = V_m(\mathbf{v}_m; \mathbf{x}^*) - V_{m+1}(\mathbf{v}_m, 1; \mathbf{x}^*),$$
and similarly
$$\xi_{m+1}(\mathbf{v}_m, 0; \mathbf{x}^*) = 1 - \xi_{m+1}(\mathbf{v}_m, 1; \mathbf{x}^*).$$
Hence
$$E_Q[g(\xi_{m+1}(T; \mathbf{x}^*))] = \sum_{\mathbf{v}_{m+1} \in \{0,1\}^{m+1}} f_{m+1} V_{m+1}(\mathbf{v}_{m+1}; \mathbf{x}^*),$$
where
$$f_{m+1} := \xi_{m+1}(\mathbf{v}_{m+1}; \mathbf{x}^*) \cdot g(\xi_{m+1}(\mathbf{v}_{m+1}; \mathbf{x}^*))$$
$$+ [1 - \xi_{m+1}(\mathbf{v}_{m+1}; \mathbf{x}^*)] \cdot g[1 - \xi_{m+1}(\mathbf{v}_{m+1}; \mathbf{x}^*)].$$
But it is clear that the last summation is just
$$E_Q[\xi_{m+1} g(\xi_{m+1}) + (1 - \xi_{m+1}) g(1 - \xi_{m+1})].$$
This completes the proof. ■

9.4 Learning with Prior Information: Bounds on Learning Rates

Now we are in a position to start deriving some bounds on the error rates. For this purpose, let us define the Shannon entropy map $\eta : [0,1] \to [0,1]$ by

$$\eta(p) := p \lg p - (1-p) \lg(1-p).$$

It is well-known, and in any case easy to verify, that the function $\eta(\cdot)$ has the following properties: $\eta(p) = \eta(1-p)$, $\eta(\cdot)$ is concave, and $\eta(\cdot)$ is strictly increasing for $p \in [0, 0.5]$, and strictly decreasing for $p \in [0.5, 1]$. Accordingly, define $\eta^{-1} : [0,1] \to [0, 0.5]$ by setting $\eta^{-1}(y)$ equal to the unique $p \in [0, 0.5]$ such that $\eta(p) = y$. Then $\eta^{-1}(\cdot)$ is well-defined and *convex*. Moreover, there exists a constant c_0 such that

$$\eta^{-1}(y) \geq \frac{c_0 y}{\lg(2/y)}. \tag{9.4.3}$$

The following useful inequalities about $\eta(\cdot)$ are easily proven:

$$p(1-p) \leq \min\{p, 1-p\} \leq 2p(1-p) \leq \frac{\eta(p)}{2}. \tag{9.4.4}$$

Now we state the main results concerning the instantaneous error rates of both the Bayesian and the Gibbsian algorithms. Both upper and lower bounds for the error rates are given, as well as a comparison of the two algorithms.

Theorem 9.13. *The following chains of inequalities hold for each $\mathbf{x}^* \in X^\infty$:*

$$E_Q[e^G_{m+1}(T; \mathbf{x}^*)] \geq E_Q[e^B_{m+1}(T; \mathbf{x}^*)] \geq \frac{c_0 E_Q[\mathcal{I}_{m+1}(T; \mathbf{x}^*)]}{\lg(2/E_Q[\mathcal{I}_{m+1}(T; \mathbf{x}^*)])}. \tag{9.4.5}$$

$$E_Q[e^B_{m+1}(T; \mathbf{x}^*)] \leq E_Q[e^G_{m+1}(T; \mathbf{x}^*)] \leq \frac{1}{2} E_Q[\mathcal{I}_{m+1}(T; \mathbf{x}^*)]. \tag{9.4.6}$$

$$E_Q[e^B_{m+1}(T; \mathbf{x}^*)] \leq E_Q[e^G_{m+1}(T; \mathbf{x}^*)] \leq 2\, E_Q[e^B_{m+1}(T; \mathbf{x}^*)]. \tag{9.4.7}$$

Remarks: The set of inequalities (9.4.5) gives *lower* bounds on the instantaneous error of the two algorithms, while the set of inequalities (9.4.6) gives *upper* bounds; finally, the set of inequalities (9.4.7) "brackets" the performance of the Gibbsian algorithm in terms of that of the Bayesian algorithm.

Proof. Let us begin by applying Lemma 9.7 to each of the functions

$$g_1(\xi) = -\lg \xi, \quad g_2(\xi) = \theta(0.5 - \xi), \quad g_3(\xi) = 1 - \xi.$$

This leads succesively to the following estimates: Note that throughout $\xi_{m+1}(T; \mathbf{x}^*)$ is abbreviated as ξ_{m+1} so as to make the formulae more compact. First,

$$E_Q[\mathcal{I}_{m+1}(T; \mathbf{x}^*)] = E_Q[-\lg \xi_{m+1}]$$

$$= E_Q[-\xi_{m+1}\lg\xi_{m+1} - (1-\xi_{m+1})\lg(1-\xi_{m+1})]$$
$$= E_Q[\eta(\xi_{m+1})]. \tag{9.4.8}$$

Second,
$$E_Q[e^B_{m+1}(T;\mathbf{x}^*)] = E_Q[\theta(0.5-\xi)]$$
$$= E_Q[\xi_{m+1}\,\theta(0.5-\xi) + (1-\xi_{m+1})\theta(-0.5+\xi_{m+1})]$$
$$= E_Q[\min\{\xi_{m+1}, 1-\xi_{m+1}\}], \tag{9.4.9}$$

after observing that
$$y\,\theta(0.5-y) + (1-y)\,\theta(-0.5+y) = \min\{y, 1-y\}.$$

Third,
$$E_Q[e^G_{m+1}(T;\mathbf{x}^*)] = E_Q[1-\xi] = E_Q[\xi_{m+1}(1-\xi_{m+1}) + (1-\xi_{m+1})\xi_{m+1}]$$
$$= 2E_Q[\xi_{m+1}(1-\xi_{m+1})]. \tag{9.4.10}$$

Now (9.4.4) implies that
$$\xi_{m+1}(1-\xi_{m+1}) \le \min\{\xi_{m+1}, 1-\xi_{m+1}\} \le 2\xi_{m+1}(1-\xi_{m+1}).$$

Combining this observation with the expressions (9.4.9) and (9.4.10) establishes the set of inequalities (9.4.7). To prove (9.4.5), observe that the left inequality in (9.4.5) is already contained in (9.4.7). To prove the right inequality, observe that
$$(\eta^{-1}\cdot\eta)(p) = \min\{p, 1-p\}, \quad \forall p \in [0,1]. \tag{9.4.11}$$

Moreover, since $\eta^{-1}(\cdot)$ is convex, it follows from Jensen's inequality (see e.g., [35], p. 80) and the above relationships that
$$\eta^{-1}(E_Q[\mathcal{I}_{m+1}(T;\mathbf{x}^*)]) = \eta^{-1}(E_Q[\eta(\xi_{m+1})]) \text{ by } (9.4.8)$$
$$\le E_Q[(\eta^{-1}\cdot\eta)(\xi_{m+1})] \text{ by Jensen's inequality}$$
$$= E_Q[\min\{\xi_{m+1}, 1-\xi_{m+1}\}] \text{ by } (9.4.11)$$
$$= E_Q[e^B_{m+1}(T;\mathbf{x}^*)] \text{ by } (9.4.9).$$

On the other hand, from (9.4.3) we get
$$\eta^{-1}(E_Q[\mathcal{I}_{m+1}(T;\mathbf{x}^*)]) \ge \frac{c_0 E_Q[\mathcal{I}_{m+1}(T;\mathbf{x}^*)]}{\lg(2/E_Q[\mathcal{I}_{m+1}(T;\mathbf{x}^*)])}.$$

Combining these two inequalities establishes (9.4.5). Finally, (9.4.4) implies that
$$2\xi_{m+1}(1-\xi_{m+1}) \le \frac{\eta(\xi_{m+1})}{2}.$$

9.4 Learning with Prior Information: Bounds on Learning Rates

Combining this observation with (9.4.8) and (9.4.10) shows that

$$E_Q[e_{m+1}^G(T;\mathbf{x}^*)] \leq \frac{1}{2} E_Q[\mathcal{I}_{m+1}(T;\mathbf{x}^*)],$$

which is the right inequality in (9.4.6). The left inequality in (9.4.6) is the same as that in (9.4.7). ∎

Theorem 9.13 applies to the "instantaneous" probability that $T(x_{m+1})$ is incorrectly predicted after m samples, as a function of m. But it is also of interest to estimate the "average" error probability after m samples, i.e., the fraction of the first m samples $I(x_i)$, $1 \leq i \leq m$ that are misclassified, as a function of m. Except for a factor of m, this is the same as the "cumulative" error studied in [84]. Moreover, the various bounds in Theorem 9.13 pertain to a *fixed* sequence $\mathbf{x}^* \in X^\infty$. It is desirable to take the *expectation* of the average error probability with respect to $\mathbf{x}^* \in X^\infty$, or equivalently, with respect to $\mathbf{x}_m \in X^m$ (since the error probability does not depend on x_i for $i > m$). Let us denote the average error probability for the Bayesian algorithm by $a^B(m)$, and that for the Gibbsian algorithm by $a^G(m)$. Thus, from (9.4.3),

$$a^B(m) = \frac{1}{m} E_{P^\infty} \sum_{i=0}^{m-1} E_Q[\theta(0.5 - \xi_{m+1}(T;\mathbf{x}^*))].$$

Since the summation is finite and $\theta(\cdot)$ is bounded, one can use Fubini's theorem to interchange the order of taking the expectations and move the summation inside the expectations. This gives

$$a^B(m) = E_Q E_{P^\infty}\left[\frac{1}{m}\sum_{i=0}^{m-1}\theta(0.5 - \xi_{m+1}(T;\mathbf{x}^*))\right].$$

This identity has an appealing interpretation: $a^B(m)$ is the expected value (with respect to $T \in \mathcal{C}$) of the average number of classification errors, averaged with respect to the training sequence $\mathbf{x}^*$. Similarly, for the Gibbsian algorithm, we have

$$a^G(m) = \frac{1}{m} E_{P^\infty}\left[\sum_{i=0}^{m-1} E_Q[\xi_{m+1}(T;\mathbf{x}^*)]\right] = E_Q E_{P^\infty}\left[\frac{1}{m}\sum_{i=0}^{m-1}\xi_{m+1}(T;\mathbf{x}^*)\right].$$

It is difficult to obtain estimates directly for these two quantities. However, by using Theorem 9.13 together with the *additive* property of entropy, it is possible to obtain some useful bounds. In order to state these bounds, a little notation is recalled from Chapters 4 and 5. Suppose $\mathbf{x}^* \in X^\infty$, and let $\mathbf{x}_m \in X^m$ denote the first m components of $\mathbf{x}^*$. Then $\pi(\mathbf{x}_m;\mathcal{C})$ denotes the number of distinct vectors of the form

$$[A(x_1)\ldots A(x_m)]^t \in \{0,1\}^m$$

that can be generated by varying A over C. The symbol $d(\mathbf{x}_m)$ denotes the VC-dimension of C restricted to $\{x_1, \ldots, x_m\}$; in other words $d(\mathbf{x}_m)$ denotes the largest integer n such that there exists a set $\{x_{i_1}, \ldots, x_{i_n}\} \subseteq \{x_1, \ldots, x_m\}$ that is shattered by C. Note that $d(\mathbf{x}_m) \leq m$ for all m. Hence, from Theorem 4.1, it follows that

$$\pi(\mathbf{x}_m; C) \leq \left[\frac{em}{d(\mathbf{x}_m)}\right]^{d(\mathbf{x}_m)}.$$

Also, recall from Theorem 5.4 that C has the property of uniform convergence of empirical probabilities (UCEP) with respect to the probability measure P if and only if

$$\lim_{m \to \infty} E_{P^m} \left[\frac{d(\mathbf{x}_m)}{m}\right] = 0.$$

For future reference, define

$$\alpha_m := E_{P^m}\left[\frac{d(\mathbf{x}_m)}{m}\right]. \tag{9.4.12}$$

Thus C has the UCEP property with respect to P if and only if $\alpha_m \to 0$ as $m \to \infty$.

Theorem 9.14 below gives an explicit *upper* bound on the average error rate as a function of the constant α_m defined in (9.4.12). In the case where the concept class C has the UCEP property with respect to P, the same theorem also provides an upper bound of the *rate* at which the average error rate approaches zero as $m \to \infty$.

Theorem 9.14. *With all symbols defined as above, we have*

$$a^B(m) \leq a^G(m) \leq \frac{e}{2}\eta(\alpha_m/e), \ \forall m \geq 1. \tag{9.4.13}$$

In particular, if C has finite VC-dimension, say $d(C)$, then

$$a^B(m) \leq a^G(m) \leq \frac{e}{2}\eta\left[\frac{d(C)}{em}\right], \ \forall m \geq 1. \tag{9.4.14}$$

Proof. Since the instantaneous error probabilities satisfy

$$e^B_{m+1}(T; \mathbf{x}^*) \leq e^G_{m+1}(T; \mathbf{x}^*), \ \forall T \in C, \ \forall \mathbf{x}^* \in X^\infty,$$

it is obvious that $a^B(m) \leq a^G(m)$ for all m, and it only remains to prove the upper bounds for $a^G(m)$. For this purpose, we invoke the right inequality in (9.4.6). After taking expectations, this inequality implies that

$$a^G(m) \leq \frac{1}{2m} E_{P^\infty} E_Q \left[\sum_{i=0}^{m-1} \mathcal{I}_{i+1}(T; \mathbf{x}^*)\right]. \tag{9.4.15}$$

Now note that

9.4 Learning with Prior Information: Bounds on Learning Rates   361

$$\mathcal{I}_{i+1}(T;\mathbf{x}^*) = -\lg \xi_{i+1}(T;\mathbf{x}^*) = -\lg V_{i+1}(T;\mathbf{x}^*) + \lg V_i(T;\mathbf{x}^*).$$

Hence

$$\sum_{i=0}^{m-1} \mathcal{I}_{i+1}(T;\mathbf{x}^*) = \sum_{i=0}^{m-1} [-\lg V_{i+1}(T;\mathbf{x}^*) + \lg V_i(T;\mathbf{x}^*)] = -\lg V_m(T;\mathbf{x}^*),$$

since $V_0(T;\mathbf{x}^*) = 1$. Therefore

$$E_Q\left[\sum_{i=0}^{m-1} \mathcal{I}_{i+1}(T;\mathbf{x}^*)\right] = E_Q[-\lg V_m(T;\mathbf{x}^*)].$$

To evaluate the expected value on the right side, we proceed as follows: For each fixed $\mathbf{x}^* \in X^\infty$ and each integer m, define an equivalence relationship $\sim$ on $\mathcal{C}$ by setting $A \sim B$ if $A(x_i) = B(x_i)$ for $i = 1, \ldots, m$. Then $\mathcal{C}/\sim$ consists of precisely $\pi(\mathbf{x}_m;\mathcal{C})$ equivalence classes. For the moment use π as a shorthand for $\pi(\mathbf{x}_m;\mathcal{C})$, and write $\mathcal{C} = \cup_{j=1}^\pi \mathcal{A}_j$, where $\mathcal{A}_1, \ldots, \mathcal{A}_\pi$ are the disjoint equivalence classes. Then $V_m(T;\mathbf{x}^*) = Q(\mathcal{A}_j)$, where $\mathcal{A}_j$ is the equivalence class to which T belongs. Moreover, the probability that $T \in \mathcal{A}_j$ equals $Q(\mathcal{A}_j)$. Therefore

$$E_Q[-\lg V_m(T;\mathbf{x}^*)] = \sum_{j=1}^\pi -Q(\mathcal{A}_j)\lg[Q(\mathcal{A}_j)].$$

This quantity has a natural interpretation in terms of entropy. The vector $\mathbf{x}_m \in X^m$ induces a partitioning of $\mathcal{C}$ into $\pi(\mathbf{x}_m;\mathcal{C})$ disjoint collections, and assigns the probability $r_j := Q(\mathcal{A}_j)$ to each collection $\mathcal{A}_j$. Now

$$E_Q[-\lg V_m(T;\mathbf{x}^*)] = \sum_{j=1}^\pi -r_j \lg r_j$$

is just the entropy of this partitioning. It is well-known, and in any case easy to demonstrate, that the entropy of a partitioning into π classes is maximized when each class is assigned an equal weight, and that the maximum equals $\lg[\pi(\mathbf{x}_m;\mathcal{C})]$. Thus it has been established that

$$E_Q\left[\sum_{i=0}^{m-1} \mathcal{I}_{i+1}(T;\mathbf{x}^*)\right] \leq \lg[\pi(\mathbf{x}_m;\mathcal{C})]. \qquad (9.4.16)$$

To complete the proof, recall that

$$\pi(\mathbf{x}_m;\mathcal{C}) \leq \left[\frac{em}{d(\mathbf{x}_m)}\right]^{d(\mathbf{x}_m)},$$

where $d(\mathbf{x}_m)$ is the restricted VC-dimension defined above. Combining this bound with (9.4.15) and (9.4.16) leads to

$$a^G(m) \le \frac{1}{2} E_{P^m}\left[\frac{d(\mathbf{x}_m)}{m} \lg \frac{em}{d(\mathbf{x}_m)}\right]$$

$$= \frac{e}{2} E_{P^m}\left[-\frac{d(\mathbf{x}_m)}{em} \lg \frac{d(\mathbf{x}_m)}{em}\right]$$

$$= \frac{e}{2} E_{P^m}\left[\eta\left(\frac{d(\mathbf{x}_m)}{em}\right)\right].$$

Note that E_{P^∞} has been replaced by E_{P^m}, since the argument of $E_{P^\infty}(\cdot)$ does not depend on x_i for $i > m$. Since $\eta(\cdot)$ is a concave function, Jensen's inequality ([35], p. 80) implies that

$$E_{P^m}\left[\eta\left(\frac{d(\mathbf{x}_m)}{em}\right)\right] \le \eta\left(E_{P^m}\left[\frac{d(\mathbf{x}_m)}{em}\right]\right) = \eta(\alpha_m/e),$$

where α_m is defined in (9.4.12). Combining the previous two inequalities establishes (9.4.13).

Finally, if $\mathcal{C}$ has finite VC-dimension, say $d(\mathcal{C})$, then $d(\mathbf{x}_m) \le d(\mathcal{C})$ for all m and all $\mathbf{x}^* \in X^\infty$. Therefore

$$\alpha_m = E_{P^m}\left[\frac{d(\mathbf{x}_m)}{m}\right] \le \frac{d(\mathcal{C})}{m}.$$

Since $\eta(\cdot)$ is nondecreasing, it follows that

$$\eta(\alpha_m/e) \le \eta\left[\frac{d(\mathcal{C})}{m}\right].$$

This establishes (9.4.14). ■

Notes and References The material in Section 9.1 is standard in the computational learning theory literature. The reader is directed to the texts [147], [9] or [99] for a thorough discussion of various topics that are only touched upon here. Lemma 9.1 is an adaptation of a more general result due to Haussler et al. [81]. Example 9.2 on learning monomials and Example 9.3 on learning k-CNF's are essentially from Valiant's seminal paper [187], which can be said to have launched the field of computational learning theory. Example 9.5 on the difficulty of learning k-term DNF's is given by Pitt and Valiant [160], but the treatment here follows [9]. Example 9.7 on the difficulty of learning the class of intersections of two or more half-planes is due to Blum and Rivest [31]. However, it is shown by Blum and Kannan [30] that learning the class of intersections of a bounded number of half-planes under a *uniform* distribution is a tractable problem. The material in Section 9.2 is taken from the paper by Kulkarni et al. [111]. Another form of active learning is very popular in the computational learning theory community, whereby the concept class is countable, and the learner is expected to determine the target

9.4 Learning with Prior Information: Bounds on Learning Rates

concept *exactly* in a finite number of steps; see the survey paper by Angluin [5] for a fairly comprehensive discussion of the results available up to that date. Nowadays there are several negative results showing that membership queries sometimes do not help very much; see the paper by Angluin and Kharitonov [6] for a recent set of results. The use of cryptographic methods to provide lower bounds on the complexity of learning problems is a recent trend; see Kharitonov [102] for some recent results.

The necessary and sufficient conditions in Section 9.3 on learning with prior information are taken from [43]. The treatment of learning curves in Section 9.4 is taken from Haussler *et al.* [84], with Theorem 9.14 presenting a slight refinement over the corresponding result in [84]; see also [155] and [83] for related discussion.

10. Applications to Neural Networks

In this chapter, we study some applications of the results derived thus far to the problem of designing and/or training neural networks. The past decade has witnessed a tremendous surge of interest in the application of neural networks for a variety of purposes. In essence, almost all of these applications can be summarized as follows: Given a set of randomly generated data, and a family of neural networks all sharing a common "architecture," construct a neural network from this family that best approximates the data, with high probability. Stated in this form, the problem is one of nonlinear curve-fitting or nonlinear regression. What distinguishes the use of neural networks for this purpose, as opposed to many other standard techniques, is the widespread belief that "neural networks can generalize." In other words, it is believed that, after a neural network has been "trained" on a sufficiently large number of input-output pairs, it can then correctly predict all future input-output pairs, even for those inputs that the network has not seen previously. In the absence of the generalization ability, there is no reason to use a neural network merely to reproduce *known* input-output pairs – a simple table look-up scheme would serve just as well. It is shown in Section 3.2, specifically Example 3.6, that *perfect* generalization by a neural network is an impossibility. Rather, all that one can aspire to is that, after a sufficient amount of training, the trained neural network can predict the correct output *with high probability* on a randomly selected test input. It is also shown in Section 3.2 that the PAC learning problem formulation is a very natural way of mathematically modelling the notion of "generalization with high probability." Moreover, the model-free learning problem formulation of Section 3.3 captures the idea of finding a "nearly best" fit to randomly generated data using a neural network from a specified class.

One of the challenging problems in neural network design is that of choosing a suitable architecture. Ideally, one would like to choose an architecture that is both easy to train and has good predictive power. Unfortunately, the two objectives are in conflict. A network containing a large number of neurons is more "expressive" than one containing fewer neurons, in the sense that the family of input-output mappings that can be realized by a network (by suitably adjusting the various parameters in the network) becomes richer as the number of neurons is increased. At the same time, using a network with too many neurons leads to the problem of "over-fitting". As optimal trade-off can be achieved using an approach called "structural risk minimization," which is described in Section 10.4.

A perusal of the contents of this chapter would reveal that the main technique used throughout is to estimate the VC-dimension of a collection of subsets of some Euclidean space $\mathbb{R}^k$, or the P-dimension of a collection of functions mapping some Euclidean space $\mathbb{R}^k$ into $[0, 1]$. Thus, though the title of the chapter mentions neural networks, both the methods used as well as the specific results obtained here can be used in a wide variety of applications.

10.1 What is a Neural Network?

Neural networks are interconnections of artificial "neurons" that are greatly simplified versions of the biological neurons found in the human brain. Strictly speaking, the networks described here should be termed "artificial" neural networks, so as to distinguish them from the biological neural networks occurring in the the brains of humans and other living organisms. However, as there is no danger of confusion, the prefix "artificial" is dropped hereafter.

The basic building block of a neural network is the "neuron," which computes a mapping g from $\mathbb{R}^k$ into $\mathbb{R}$, where k denotes the number of inputs to the neuron. Sometimes the number of inputs is also called the "fan-in." The function g is a composition of two other functions: a **shaping function** $\psi : \mathbb{R}^k \to \mathbb{R}$, which is a polynomial, and an **activation function** $\alpha : \mathbb{R} \to \{0,1\}$ or $\alpha : \mathbb{R} \to [0,1]$. Thus the input-output relationship of a neuron is of the form

$$g(x_1, \ldots, x_k) = \alpha(z), \; z = \psi(u_1, \ldots, u_k).$$

The degree of the polynomial ψ is called the **order** of the neuron. Note that the output of a neuron is always limited to the range $[0,1]$. Of course, there is nothing special about the values 0 and 1; the key point is that the output of each neuron has *bounded* range, in which case the limits of the range can be normalized to 0 and 1.

One of the simplest neurons is the **perceptron**, introduced in Section 3.2. The input-output relationship of a perceptron is given by

$$y = \eta(z), \; z = \sum_{i=1}^{k} w_i x_i - \theta, \tag{10.1.1}$$

where $y \in \{0,1\}$ is the output, $\mathbf{x} := [x_1 \ldots x_k]^t \in \mathbb{R}^k$ is the input, and the step function $\eta(\cdot)$ is defined by

$$\eta(z) = \begin{cases} 1, & \text{if } z \geq 0, \\ 0, & \text{if } z < 0. \end{cases}$$

The numbers $w_1, \ldots, w_k$ are called the "weights" of the perceptron, and the number θ is called the "threshold." It is easy to see that the perceptron as defined above has a fan-in of k. It also has $k+1$ adjustable parameters, namely $\theta, w_1, \ldots, w_k$. Some authors refer to the step function as the "Heaviside" function, after the British mathematician O. Heaviside. In this chapter, the symbol $\eta(\cdot)$ is used to denote the step function, and a perceptron.

A perceptron is an example of a *binary*, or *two-state* neuron, since its output can assume only one of two possible values (which can be taken to be 0 and 1 without loss of generality). Another commonly used neural characteristic is the **standard sigmoid**

$$y = \frac{1}{1 + \exp(-z)}, \quad z = \sum_{i=1}^{k} w_i x_i - \theta. \tag{10.1.2}$$

In this chapter, the symbol $\sigma(\cdot)$ is used for the above function. One of the appealing features of the standard sigmoid function is that

$$\sigma'(z) = \sigma(z) - [\sigma(z)]^2, \quad \forall z.$$

The significance of this property can be found in Section 10.3.7.

More generally, one can let

$$y = \alpha(z), \quad z = \sum_{i=1}^{k} w_i x_i - \theta,$$

where $\alpha : \mathbb{R} \to [0, 1]$ is continuous, continuously differentiable almost everywhere, nondecreasing, and satisfies

$$\lim_{z \to \infty} \alpha(z) = 1, \text{ and } \lim_{z \to -\infty} \alpha(z) = 0.$$

Such a function is referred to as being "sigmoidal." This nomenclature is not to be confused with *the* standard sigmoid defined above.[1] An example of such a sigmoidal characteristic is the hard-limiter defined by

$$y = \lambda(z) := \begin{cases} 1, & \text{if } z > 1, \\ z, & \text{if } 0 \leq z \leq 1, \\ 0, & \text{if } z < 0. \end{cases} \tag{10.1.3}$$

Note that the standard perceptron, the standard sigmoid, and the hard limiter are all examples of first-order neurons, though the term is rarely used.

A **neural network** is a labelled acyclic directed graph, with the following features:

---

[1] Note that the above terminology is not altogether standard. Some authors refer to the function $\sigma(z) = 1/(1 + e^{-z})$ as just the "sigmoid" and omit the prefix "standard."

368   10. Applications to Neural Networks

– The graph has at least one node with fan-in zero. The set of nodes with fan-in zero is called the set of **input nodes** of the network. Similarly, the graph has at least one node with fan-out zero. The set of nodes with zero fan-out is called the set of **output nodes** of the network. For each node, the length of the longest path from an input node to that node is called the **level** of the node. The length of the longest path from an input node to an output node is called the **depth** of the network.
– Every node of the graph, except for the input nodes, is labelled with a suitable input-output mapping. Thus, if node i has fan-in r_i, then the label of node i is a mapping $g_i : \mathbb{R}^{r_i} \to \mathbb{R}$, chosen from some prescribed *family* of mappings $\mathcal{G}_i$.

The functioning of the neural network is as follows: Let k denote the number of input nodes, and let $X \subseteq \mathbb{R}^k$ be a given set, called the **input set**. Whenever a vector $\mathbf{x} := [x_1 \ldots x_k]^t \in X^k$ is assigned to the input nodes, every subsequent node i applies the mapping g_i to its inputs. Since the graph is acyclic, this procedure is well-defined. By proceeding in this sequential fashion, one can eventually assign values to each node of the network, and in particular, to each of the output nodes. In this way, the neural network defines a map $h : X \to \mathbb{R}^n$, where n is the number of output nodes.

Throughout the chapter, the following symbols are invariably used:

– k = the number of inputs to the network
– l = the number of adjustable parameters in the network
– t = the depth of the network.

Thus, in the present formulation, a neural network consists of: (i) an acyclic directed graph, (ii) an input set $X \subseteq \mathbb{R}^k$, and (iii) a label g_i for each node i other than the input nodes, where g_i is chosen from a prespecified family $\mathcal{G}_i$. The **architecture** of the neural network consists of its graph, its input set X, and the *families* of maps $\mathcal{G}_i$. Each choice of g_i from $\mathcal{G}_i$ leads to a mapping h from X to $\mathbb{R}^n$. By *varying* each g_i over the corresponding set $\mathcal{G}_i$, one generates a *family* $\mathcal{H}$ of mappings from X into $\mathbb{R}^n$. In the context of learning theory, an input-output mapping h is referred to as a "hypothesis function," while the family $\mathcal{H}$ is referred to as the "hypothesis class." Now all of the theory developed in the preceding chapters can be applied to analyze various properties of the family $\mathcal{H}$.

As an illustration of this abstract definition, consider a multi-layer perceptron network with a single output node. For each assignment $\mathbf{w}$ of the weights and thresholds to the various perceptrons, the network computes a map $h_\mathbf{w}$ from $\mathbb{R}^k$ into $\{0, 1\}$. As $\mathbf{w}$ is varied over $\mathbb{R}^l$, where l denotes the total number of adjustable parameters in the network, the corresponding collection of binary-valued maps $\{h_\mathbf{w}\}$ defines the family $\mathcal{H}$.

Several simplifying assumptions have been made in the above model of neural computation, of which only two are highlighted. The first simplifying assumption is that each node in the graph computes a function g_i from $\mathbb{R}^{r_i}$

into $\mathbb{R}$. Thus, in the language of system theory, each node computes a *memoryless* function of its inputs. In particular, there is no explicit notion of *time*. However, in applications such as speech recognition and control systems, the dynamical nature of the underlying signals plays an essential role, and in such applications, it is common to use so-called "timed" neural networks. In such networks, the inputs belong to $X^{\mathbf{N}}$, that is, they are *sequences* in X, indexed by the natural numbers. Each computational node in a timed neural network has associated with it an integer q_i, called the "delay" of the node, and an associated mapping $g_i : \mathbb{R}^{q_i r_i} \to \mathbb{R}$, where r_i is the fan-in of the node. Thus the output of the node depends not only on the values of the r_i inputs at the *current* instant of time, but also on their values during the preceding $q_i - 1$ instants of time. With more elaborate notation, the theory presented here can also be made applicable to timed neural networks, provided there is a prior bound on the maximum delay of any node. The second, and more serious, assumption is that the graph underlying the neural network is acyclic. This assumption rules out so-called "recurrent" neural networks such as those of the Hopfield type. Removing this assumption is not easy, since much of the theory presented below depends in a crucial manner upon the assumption that the graph underlying the network is acyclic.

10.2 Learning in Neural Networks

In this section we formally state various problems in neural network learning, and recapitulate the relevant results from the preceding chapters.

10.2.1 Problem Formulation

Essentially all problems in neural computation can be divided into one of two categories. In the first category of problems, which can be termed as the *identification* of an unknown neural network based on input-output measurements, it is assumed that the data is being generated by an unknown "target" neural network belonging to a known family of neural networks. If the output of the unknown target neural network is available to the learner without any measurement error, then the problem is the same as the standard PAC learning problem introduced in Section 3.2. If the output of the target neural network is available to the learner only after being corrupted by measurement noise, then the problem can be formulated as a model-free learning problem as in Section 3.3. In either case, the objective is to "train" a hypothesized neural network h_m (where m denotes the number of training inputs) in such a way that h_m is a good approximation to the unknown target network f.[2] As in Chapter 3, the objective is to ensure that, if a randomly selected testing

---

[2] Here the same symbol is used both for the neural network as well as the associated input-output mapping.

input $\mathbf{x} \in X$ is applied to both the trained neural network h_m and the true network f, then the expected value of the difference $|h_m(\mathbf{x}) - f(\mathbf{x})|$ is small, with high confidence. In case both networks have binary outputs, this is the same as saying that the trained network misclassifies the testing input with a small probability of error, with high confidence. Thus, to summarize, the aim of the identification problem is one of *achieving near-perfect generalization with high probability*. The second category of problems can be referred to as *data-fitting* using a family of neural networks. In this case, one is given a stream of data generated at random (and perhaps by a probability distribution that is itself unknown), together with a family of neural networks that can be used to fit the data. The objective is to select one network (call it h_m) from this family of neural networks (call it $\mathcal{H}$) such that, with high probability, h_m nearly achieves the best possible fit to the data by any network in the family $\mathcal{H}$. In this setting there is no interpretation of "generalization," since there is no longer an assumption that the data is being generated by a "true but unknown" network. Note that some authors refer to this type of learning as "agnostic" learning; but the term "model-free" learning is preferred here.

From a purely mathematical standpoint, the model-free learning problem includes as special cases both the other problems (namely: the data being generated by an unknown target network, either with perfect measurements, or with noisy measurements). Thus, for the purposes of analysis, one can focus exclusively on the model-free learning problem. But this should not obscure the fact that the identification problem is philosophically quite different from the data-fitting problem. Ideally one would like to be able to prove that the rate at which the trained neural network h_m approaches the optimal network depends only on the richness of the model family $\mathcal{H}$ and not on the properties of the data. Indeed, we will succeed in proving essentially this.

Let us recollect the salient features of the model-free learning problem as formulated in Section 3.3 and studied in subsequent chapters, as specialized to neural networks. Let k, n denote the number of inputs and outputs of a neural network, respectively. Let $X = \mathbb{R}^k$, and $Y = U = [0,1]^n$. It is reasonable to assume that, while the input space of the network could be all of $\mathbb{R}^k$, the components of the outputs vector are restricted to lie in the interval $[0,1]$, because the various types of commonly used neural characteristics all have this property. Let $\ell : Y \times U \to [0,1]$ denote the loss function. Typically ℓ is chosen as a pseudometric on Y, such as $\ell(y, u) = \|y - u\|$ where $\|\cdot\|$ is some norm on $\mathbb{R}^n$; however, this assumption is not necessary in what follows. Let $\bar{\mathcal{P}}$ be a given family of probability measures on $X \times Y$, and let $\mathcal{H}$ denote a collection of maps from X into U. Typically $\mathcal{H}$ consists of all maps from X into U that can be realized by a neural network of a given fixed architecture by varying all adjustable parameters over their respective ranges. Sometimes, however, $\mathcal{H}$ consists of a *union* of the form $\mathcal{H} = \cup_{i=1}^{\infty} \mathcal{H}_i$, where $\mathcal{H}_i \subseteq \mathcal{H}_{i+1}$; in such a case, the family is said to be "nested," and $\mathcal{H}$ is said to be a "graded" hypothesis space.

10.2 Learning in Neural Networks

Suppose a probability measure $\bar{P} \in \bar{\mathcal{P}}$ is fixed but unknown, and i.i.d. samples $(x_1, y_1), \ldots, (x_m, y_m)$ are drawn in accordance with $\bar{P}$. For each map $h \in \mathcal{H}$, define the associated "risk" function

$$J(h, \bar{P}) := \int_{X \times Y} \ell(y, h(x)) \, \bar{P}(dx, dy).$$

The minimum achievable risk is denoted by $J^*(\bar{P})$; thus

$$J^*(\bar{P}) := \inf_{h \in \mathcal{H}} J(h, \bar{P}).$$

The objective of model-free learning is to construct an "algorithm" that produces a sequence of hypotheses $\{h_m\}$, based on the data $\{(x_1, y_1), \ldots, (x_m, y_m)\}$, such that h_m becomes nearly optimal with high probability. More precisely, suppose the algorithm consists of an indexed family of mappings $\{A_m\}$, where

$$A_m : (X \times Y)^m \to \mathcal{H}.$$

Now let

$$h_m(\mathbf{x}, \mathbf{y}) := A_m[(x_1, y_1), \ldots, (x_m, y_m)]$$

denote the hypothesis produced by the algorithm after m steps, and define

$$r_{\mathrm{mf}}(m, \epsilon) := \sup_{\bar{P} \in \bar{\mathcal{P}}} \bar{P}^m \{ (\mathbf{x}, \mathbf{y}) \in X^m \times Y^m : J(h_m, \bar{P}) > J^*(\bar{P}) + \epsilon \}.$$

Thus $r_{\mathrm{mf}}(m, \epsilon)$ is the measure of the set of multisamples that result in a hypothesis that is more than ϵ-worse compared to the optimum performance. Equivalently, $r_{\mathrm{mf}}(m, \epsilon)$ is the probability that, after m random samples have been drawn, the hypothesis produced by the algorithm is more than ϵ-worse compared to the optimum achievable performance. The algorithm $\{A_m\}$ is said to be **probably approximately correct** if $r_{\mathrm{mf}}(m, \epsilon) \to 0$ as $m \to \infty$, for each $\epsilon > 0$. The triplet $(\mathcal{H}, \bar{P}, \ell)$ is said to be **model-free learnable** if there exists an algorithm that is PAC.

Now suppose that a particular algorithm is indeed PAC. Then by assumption, for each $\epsilon, \delta > 0$ there exists an integer $m_0 = m_0(\epsilon, \delta)$ such that

$$r_{\mathrm{mf}}(m, \epsilon) \leq \delta \; \forall m \geq m_0.$$

The integer $m_0(\epsilon, \delta)$ is referred to as the "sample complexity" of the algorithm. Note that it is usually impossible to compute the sample complexity exactly, but it is normally enough to obtain upper bounds for the sample complexity. Thus, if it is known that $m \geq m_0(\epsilon, \delta)$ for some ϵ, δ, then one can state with confidence $1 - \delta$ that the hypothesis h_m produced by the algorithm applied to m randomly drawn samples performs within ϵ of the optimal achievable performance. For this reason, it is customary to refer to ϵ as the accuracy parameter, and to δ as the confidence parameter.

10.2.2 Reprise of Sample Complexity Estimates

A neat *sufficient* condition for the model-free learning problem to have a solution is given by Theorem 3.2, which can briefly be summarized as follows: Suppose a family $\mathcal{L}_\mathcal{H}$ associated with the learning problem has a property known as UCEMUP (uniform convergence of empirical means uniformly in probability); then every algorithm that nearly minimizes empirical risk is PAC. Let us now recall in detail what all these terms mean.

For each $h \in \mathcal{H}$, define the associated function $\ell_h : X \times Y \to [0, 1]$ by

$$\ell_h(x, y) := \ell(y, h(x)), \ \forall x, y,$$

and let $\mathcal{L}_\mathcal{H}$ denote the collection of functions ℓ_h as h varies over $\mathcal{H}$. Given a multisample $\mathbf{z} := (\mathbf{x}, \mathbf{y}) \in X^m \times Y^m$ and a function $h \in \mathcal{H}$, define

$$\hat{J}(h; \mathbf{z}) := \frac{1}{m} \sum_{i=1}^{m} \ell[y_i, h(x_i)].$$

Thus $\hat{J}(h; \mathbf{z})$ is the *empirical estimate* of $J(h, \bar{P})$ based on the multisample $\mathbf{z}$; sometimes $\hat{J}(h; \mathbf{z})$ is referred to as the "empirical risk." In analogy with (3.1.4), define

$$\bar{q}(m, \epsilon, \mathcal{L}_\mathcal{H}) := \sup_{\bar{P} \in \bar{\mathcal{P}}} \bar{P}^m \{\mathbf{z} \in X^m \times Y^m : \sup_{h \in \mathcal{H}} |J(h, \bar{P}) - \hat{J}(h; \mathbf{z})| > \epsilon\}.$$

The family $\mathcal{L}_\mathcal{H}$ is said to **have the UCEMUP property** if $\bar{q}(m, \epsilon, \mathcal{L}_\mathcal{H}) \to 0$ as $m \to \infty$, for each $\epsilon > 0$. This property means that the empirical estimate $\hat{J}(h; \mathbf{z})$ converges to the true value $J(h, \bar{P})$ *uniformly* with respect to $h \in \mathcal{H}$ and $\bar{P} \in \bar{\mathcal{P}}$; see Section 3.1 for a detailed discussion of the UCEMUP property.

Now let us turn to the properties of the algorithm itself. Given a multisample $\mathbf{z} := (\mathbf{x}, \mathbf{y}) \in X^m \times Y^m$ and a function $h \in \mathcal{H}$, define $\hat{J}(h; \mathbf{z})$ as above. Now let

$$\hat{J}^*(\mathbf{z}) := \inf_{h \in \mathcal{H}} \hat{J}(h; \mathbf{z}),$$

and note that $\hat{J}^*(\mathbf{z})$ is the minimum achievable *empirical* risk based on the multisample $\mathbf{z}$. For each algorithm, define the quantity

$$t(m, \epsilon) := \bar{P}^m \{\mathbf{z} \in Z^m : \hat{J}[h_m(\mathbf{z}); \mathbf{z}] > \hat{J}^*(\mathbf{z}) + \epsilon\}.$$

Thus $t(m, \epsilon)$ is the probability that, after m random samples are drawn, the empirical risk $\hat{J}[h_m(\mathbf{z}); \mathbf{z}]$ of the hypothesis $h_m(\mathbf{z})$ generated by the algorithm is more than ϵ-worse compared to the minimum achievable value $\hat{J}^*(\mathbf{z})$. Then the algorithm is said to "nearly minimize empirical risk with high probability (NMER)" if $t(m, \epsilon) \to 0$ as $m \to \infty$. Now Theorem 3.2 states that if the family $\mathcal{L}_\mathcal{H}$ has the UCEMUP property, then every NMER algorithm is PAC. More precisely, Theorem 3.2 establishes the following: Given an accuracy parameter $\epsilon > 0$ and a confidence parameter $\delta > 0$, choose a number $m_0 = m_0(\epsilon, \delta)$ such that

$$\bar{q}(m, \epsilon/4, \mathcal{L}_\mathcal{H}) \leq \delta/2, \text{ and } t(m, \epsilon/4) \leq \delta/2, \forall m \geq m_0. \tag{10.2.1}$$

Then
$$r_{\mathrm{mf}}(m, \epsilon) \leq \delta \; \forall m \geq m_0.$$

The preceding discussion suggests that, in order to solve the problem of model-free learning of neural networks, one can adopt the following two-pronged strategy: (i) Derive conditions under which the family $\mathcal{L}_\mathcal{H}$ has the UCEMUP property. (ii) Develop algorithms that nearly minimize empirical risk with high probability. Roughly speaking, the first task falls within the domain of "statistical" learning theory, whereas the second task falls within the domain of "computational" learning theory. Note that Theorem 3.2 and its proof not only provide a means of separating the model-free learning problem into the above two constituent subproblems, but also give *explicit estimates*, as in (10.2.1), of the number of samples that are sufficient to produce a hypothesis that is accurate to ϵ with confidence $1 - \delta$.

The remainder of this subsection is devoted to a discussion of some sufficient conditions for ensuring that the family $\mathcal{L}_\mathcal{H}$ has the UCEMUP property, while the next subsection is addressed to the problem of choosing an algorithm that nearly minimizes empirical risk with high probability.

Next, a brief review is given of the notions of VC-dimension, P-dimension, and their application in obtaining sample complexity estimates for learning problems. In the interests of notational simplicity, it is assumed that the number of outputs of the neural network equals one; this assumption can be removed at the expense of more cumbersome notation and formulae. Throughout, the symbol $\mathcal{H}$ is used to denote the family of input-output mappings that can be realized by neural networks of a given architecture, by varying all the adjustable parameters over their respective ranges. Throughout, the symbol k is used to denote the number of inputs of the network. Thus every function in $\mathcal{H}$ is a mapping from (some subset of) $\mathbb{R}^k$ into either $\{0, 1\}$ or $[0, 1]$, depending on the class of networks under study.

Let us begin with the case in which every function in $\mathcal{H}$ is binary-valued.

Definition A set $S = \{x_1, \ldots, x_n\}$ is said to be **shattered** by the family of functions $\mathcal{H}$ if each of the 2^n possible functions $f : S \to \{0, 1\}$ is the restriction to S of some function in $\mathcal{H}$. The **Vapnik-Chervonenkis (VC-) dimension** of $\mathcal{H}$, denoted by VC-dim($\mathcal{H}$), is the largest integer n such that there exists a set S of cardinality n that is shattered by $\mathcal{H}$.

Section 4.1 contains several examples of the computation of the VC-dimension of some families of sets. By identifying a binary-valued function with its support set (and conversely, by identifying each set with its indicator function), it is possible to convert each of these examples into another corresponding example that computes the VC-dimension of a family of binary-valued functions.

374   10. Applications to Neural Networks

In the case where the functions in $\mathcal{H}$ map $\mathbb{R}^k$ into $[0, 1]$, the notion of the VC-dimension is replaced by a more general notion.

Definition A set $S = \{x_1, \ldots, x_n\}$ is said to be **P-shattered** by the family of functions $\mathcal{H}$ if there exists a vector $\mathbf{c} \in [0, 1]^n$ such that, for every binary vector $\mathbf{e} \in \{0, 1\}^n$, there exists a corresponding function $f_\mathbf{e} \in \mathcal{H}$ such that

$$f_\mathbf{e}(x_i) \geq c_i \text{ if } e_i = 1, \text{ and } f_\mathbf{e}(x_i) < c_i \text{ if } e_i = 0.$$

The **P-dimension** of $\mathcal{H}$, denoted by P-dim$(\mathcal{H})$, is the the largest integer n such that there exists a set S of cardinality n that is P-shattered by $\mathcal{H}$.

The concept of P-shattering by a family of real-valued functions can be understood with reference to Figure 10.1, which is the same as Figure 4.5. Fix a real vector $\mathbf{c} \in [0, 1]^n$. At each point $x_i \in S$ and for each function

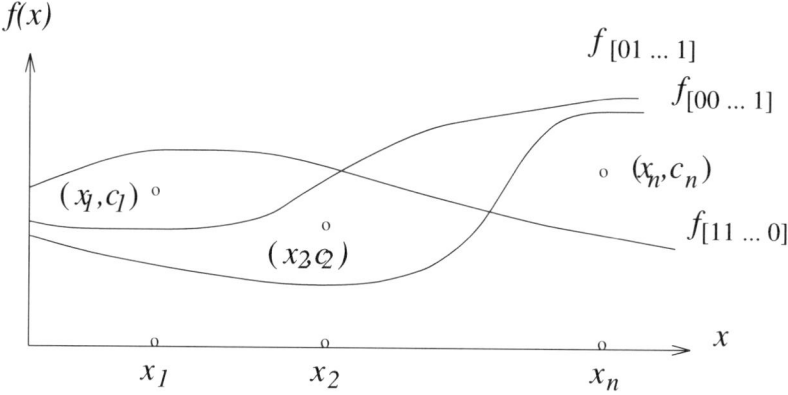

Fig. 10.1. Illustration of P-Shattering

$f \in \mathcal{H}$, the graph of $f(x_i)$ can either pass *above* (or through) c_i, or else *below* c_i. Thus there are 2^n possible different behaviours as f varies over $\mathcal{H}$. The set S is P-shattered by $\mathcal{H}$ if each of these 2^n possible behaviours is realized by some function $f \in \mathcal{H}$.

In the case of neural networks, it is possible to find a useful relationship between the P-dimension of a family of *real-valued functions*, and the VC-dimension of an associated family of *binary-valued functions*. Suppose $\mathcal{H}$ is a family of $[0, 1]$-valued functions realizable by a particular architecture, as shown in Figure 10.2. Let us now modify the architecture by adding one more input $c \in [0, 1]$, and then passing the difference $y - c$ (between the output of the original network and the new input) through a Heaviside function or perceptron, indicated in the figure by the symbol $\eta(\cdot)$. Let us denote the output of the modified network by y', and observe that $y' \in \{0, 1\}$. Now let $\mathcal{H}'$ denote the family of *binary-valued* mappings that can be realized by varying the *original* neural network over $\mathcal{H}$.

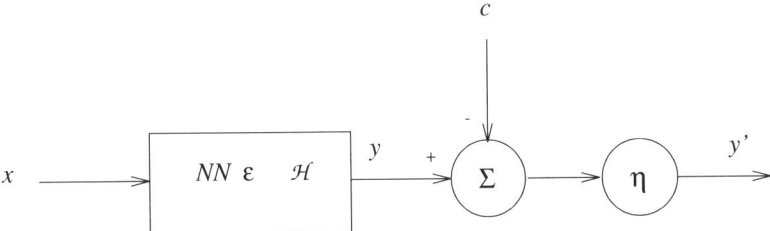

Fig. 10.2. Relating the VC-Dimension and P-Dimension of Neural Networks

Lemma 10.1. *With $\mathcal{H}$ and $\mathcal{H}'$ defined above, we have*

$$P\text{-}dim(\mathcal{H}) = VC\text{-}dim(\mathcal{H}').$$

Proof. Suppose a set $S = \{x_1, \ldots, x_n\} \subseteq \mathbb{R}^k$ is P-shattered by $\mathcal{H}$. Then, by definition, there exists a vector $\mathbf{c} \in [0,1]^n$ such that the set

$$S' := \{(x_1, c_1), \ldots, (x_n, c_n)\}$$

is shattered by $\mathcal{H}'$. Thus P-dim($\mathcal{H}$) $\leq$ VC-dim($\mathcal{H}'$). To prove the opposite inequality, suppose a set $S' := \{\mathbf{z}_1, \ldots, \mathbf{z}_n\} \subseteq \mathbb{R}^k \times [0,1]$ is shattered by $\mathcal{H}'$, and partition each $\mathbf{z}_i$ as (x_i, c_i) where $x_i \in \mathbb{R}^k$ and $c_i \in [0,1]$. Then it is clear that each of the x_i's must be distinct, and that the set $S = \{x_1, \ldots, x_n\} \subseteq \mathbb{R}^k$ is P-shattered by $\mathcal{H}$. Thus VC-dim($\mathcal{H}'$) $\leq$ P-dim($\mathcal{H}$). ∎

Having defined the VC-dimension for families of binary-valued functions, and the P-dimension for families of $[0,1]$-valued functions, we now give *explicit upper bounds* for the quantity $\bar{q}(m, \epsilon, \mathcal{L}_\mathcal{H})$. In turn, these upper bounds lead to corresponding estimates of the number m of samples that are sufficient to ensure that $\bar{q}(m, \epsilon, \mathcal{L}_\mathcal{H}) \leq \delta$. Actually, the bounds that follow are *distribution-free*, in the sense that they hold when $\bar{\mathcal{P}}$ equals $\bar{\mathcal{P}}^*$, the set of *all* probability measures on the set $X \times Y$. To highlight this fact, we introduce the quantity $q^*(m, \epsilon, \mathcal{L}_\mathcal{H})$ defined as follows:

$$q^*(m, \epsilon, \mathcal{L}_\mathcal{H}) := \sup_{\bar{P} \in \bar{\mathcal{P}}^*} \bar{P}^m \{\mathbf{z} \in X^m \times Y^m : \sup_{h \in \mathcal{H}} |J(h, \bar{P}) - \hat{J}(h; \mathbf{z})| > \epsilon\}.$$

The two theorems below give upper bounds for this quantity. Using these estimates, it is possible to derive bounds on the number of samples that are sufficient to ensure that $q^*(m, \epsilon, \mathcal{L}_\mathcal{H}) \leq \delta$.

Theorem 10.1. *Suppose the family $\mathcal{H}$ has finite P-dimension, say d, and that the loss function ℓ satisfies the uniform Lipschitz condition*

$$|\ell(y, u_1) - \ell(y, u_2)| \leq \mu |u_1 - u_2|, \; \forall y, u_1, u_2 \in [0,1],$$

for some constant μ. Then the family $\mathcal{L}_\mathcal{H}$ has the property of distribution-free uniform convergence of empirical means. Moreover,

376   10. Applications to Neural Networks

$$q^*(m,\epsilon,\mathcal{L}_\mathcal{H}) \leq 8\left(\frac{16e\mu}{\epsilon}\ln\frac{16e\mu}{\epsilon}\right)^d \exp(-m\epsilon^2/32), \; \forall m,\epsilon.$$

The inequality $q^*(m,\epsilon,\mathcal{L}_\mathcal{H}) \leq \delta$ is satisfied provided at least

$$m \geq \frac{32}{\epsilon^2}\ln\left[\frac{8}{\delta} + d\left(\ln\frac{16e\mu}{\epsilon} + \ln\ln\frac{16e\mu}{\epsilon}\right)\right]$$

samples are drawn.

Proof. The bound on $q^*(m,\epsilon,\mathcal{L}_\mathcal{H})$ is a direct consequence of Theorem 7.5. The estimate for m follows readily by setting the right side of the inequality less than or equal to δ and solving for m. ∎

Theorem 10.2. *Suppose $Y = U = \{0,1\}$, $\ell(y,u) = |y-u|$, and that $\mathcal{H}$ has finite VC-dimension, say d. Then the family of loss functions $\mathcal{L}_\mathcal{H}$ has the property of distribution-free uniform convergence of empirical means. Moreover,*

$$q^*(m,\epsilon,\mathcal{L}_\mathcal{H}) \leq 4\left(\frac{2em}{d}\right)^d \exp(-m\epsilon^2/8), \; \forall m,\epsilon.$$

The inequality $q^*(m,\epsilon,\mathcal{L}_\mathcal{H}) \leq \delta$ is satisfied provided at least

$$m \geq \max\left\{\frac{16}{\epsilon^2}\ln\frac{4}{\delta}, \frac{32d}{\epsilon^2}\ln\frac{32e}{\epsilon^2}\right\}$$

samples are drawn.

Proof. The bound on $q^*(m,\epsilon,\mathcal{L}_\mathcal{H})$ follows readily from Theorem 7.6. To prove the estimate for m, we use a procedure similar to that in the proof of Theorem 7.8. Given $\epsilon, \delta > 0$, set

$$4\left(\frac{2em}{d}\right)^d \exp(-m\epsilon^2/8) \leq \delta,$$

or equivalently,

$$m \geq \frac{8}{\epsilon^2}\left(\ln\frac{4}{\delta} + d\ln\frac{2em}{d}\right).$$

This inequality is satisfied if

$$\frac{m}{2} \geq \frac{8}{\epsilon^2}\ln\frac{4}{\delta}, \text{ and } \frac{m}{2} \geq \frac{8d}{\epsilon^2}\ln\frac{2em}{d}.$$

The first of these inequalities is satisfied if

$$m \geq \frac{16}{\epsilon^2}\ln\frac{4}{\delta}.$$

By applying Lemma 4.6 with $\alpha = 16d/\epsilon^2$, $\beta = 2e/d$, and $\gamma = 0$, it can be seen that the second inequality is satisfied if

$$m \geq \frac{32d}{\epsilon^2}\ln\frac{32e}{\epsilon^2}.$$

This completes the proof. ∎

10.2.3 Complexity-Theoretic Limits to Learnability

The results of the previous subsection can be interpreted to mean that, for "reasonable" families $\mathcal{H}$ and loss functions $\mathcal{L}_\mathcal{H}$, the family $\mathcal{L}_\mathcal{H}$ has the UCEMUP property. Thus, in most cases, there are no *information-theoretic* barriers to model-free learning. In order to develop *efficient* learning algorithms for model-free learning of neural networks, all that remains is to determine computationally efficient algorithms that nearly minimize empirical risk with high probability, and do so within a polynomial number of time steps. The problem of determining whether or not a given neural network architecture is capable of realizing a given set of input-output pairs is known in the computer science community as the "loading problem." Unfortunately, in many very innocent-looking situations, the problem of constructing a hypothesis that minimizes empirical risk turns out to be NP-complete or even NP-hard. This means that there are no known polynomial-time algorithms to sovle such problems, and it is widely believed that no polynomial-time algorithms exist in such problems. Thus it could be said that there exist very serious *complexity-theoretic* barriers to efficient model-free learning. The situation is illustrated by a couple of examples.

Example 10.1. ([31], [50]) Consider the following problem: One is given an integer k, together with a simple network consisting of just three perceptrons, as shown in Figure 10.3. The k inputs are restricted to be Boolean, so that

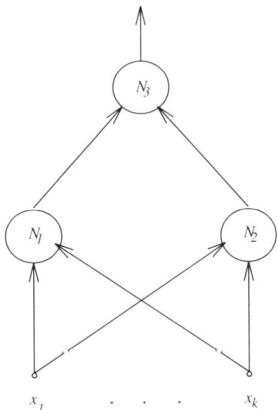

Fig. 10.3. A Neural Network for which Loading is NP-Complete

every input to the network belongs to $\{0,1\}^k$. However, the weights and thresholds of the perceptrons need not be Boolean.

It is easy to see that the total number of adjustable parameters in the network is $2k+5$ ($k+1$ for each of the two hidden-layer perceptrons, and 3 for the output-layer perceptron). Hence, by Theorem 10.3 below, it follows that the collection of maps $\mathcal{H}$ from $\{0,1\}^k$ into $\{0,1\}$ that can be realized by this class of networks has VC-dimension $O(k \lg k)$. Suppose we use the natural loss function

$$\ell(y, u) := |y - u|.$$

Then it follows from Theorem 5.12 that the family $\mathcal{L}_\mathcal{H}$ has the same VC-dimension as $\mathcal{H}$; in other words, VC-dim($\mathcal{L}_\mathcal{H}$) = $O(k \lg k)$. By the discussion above, we conclude that if the learning algorithm consists of simply minimizing the empirical risk, then the number of samples needed to achieve an accuracy ϵ and confidence δ is polynomial in $1/\epsilon$ and $\ln(1/\delta)$. Therefore, if it were possible to determine a function h_m in $\mathcal{H}$ (i.e., an assignment of the various parameters in the network) that (nearly) minimizes the empirical risk $\hat{J}(h_m; \mathbf{z})$ in a polynomial number of operations, then this class of networks would be *effectively* learnable. Unfortunately, this is *impossible*, as shown in [31].

The specific result proved in [31] is the following: Suppose one is given a collection of points $(x_1, y_1), \ldots, (x_m, y_m)$ from $\{0,1\}^{k+1}$, where $m = O(k)$; it is desired to know whether or not this collection of input-output pairs can be realized by the network in Figure 10.3 by a suitable adjustment of the various parameters. This decision problem is NP-complete. Thus it is NP-complete merely to decide whether the minimum achievable empirical risk $\hat{J}^*(\mathbf{z})$ equals zero or not, let alone actually determining what this minimum achievable value is.

Only a sketch of the proof is given here, and the reader is referred to [31] for complete details. One can think of $x_1, \ldots, x_m$ as being the vertices of a hypercube in $\mathbb{R}^k$. Each vertex is labelled as a positive ($y_i = 1$) or negative ($y_i = 0$) example. Now, the zero set of each of the two functions computed by the hidden-layer perceptrons is a $(k-1)$-dimensional hyperplane in $\mathbb{R}^k$. These two hyperplanes divide $\mathbb{R}^k$ into four quadrants (or fewer, in degenerate cases). Since the output node receives only the outputs of the two hidden-layer nodes, it follows that every x_i in the same quadrant produces the same output. In other words, the network cannot distinguish between points in the same quadrant. Moreover, the output node cannot output a 1 when the inputs are $(1,1)$, $(0,0)$, and 0 when the inputs are $(0,1)$, $(1,0)$ – this is just the XOR counterexample. Thus the question can be reformulated as follows: Given $O(k)$ points in $\{0,1\}^k$, each labelled '+' (for $y_i = 1$) or '−' (for $y_i = 0$), does there exist either (i) a single hyperplane that separates the '+' points from the '−' points, or else (ii) two hyperplanes such that one quadrant contains all the '+' points and no '−' points, or vice versa. It is shown in [31] that this decision problem is NP-complete. ■

Let us discuss briefly the implications of this finding. It means that, as the dimension of the input space becomes larger and larger, there is no known polynomial-time (in k) algorithm for minimizing the empirical risk. (Note that it is an NP-complete problem simply to determine whether or not the minimal empirical risk is zero, i.e., whether or not the data can be correctly fitted by a suitably chosen network of this family.) Thus, even though the theory tells us that there is no *information-theoretic* barrier to learning, there is definitely a *complexity-theoretic* barrier to *efficient* learning.

Now suppose the nature of the computing nodes is altered slightly, so that the two hidden-layer nodes are no longer preceptrons, but are hard limiters of the form defined in (10.1.3). In this case, the proof in [31] is no longer applicable. However, it is shown in [50] that, even in this case, the problem remains as difficult. Specifically, suppose one is given $O(k)$ points in $\{0,1\}^{k+1}$, and it is desired to determine whether this set of k input-output pairs can be realized by the network of Figure 10.3, with the hidden-layer nodes now changed to hard limiters. This problem is also NP-complete.

Finally, if the computing nodes are changed yet again to standard sigmoids instead of hard limiters, it is not known whether the loading problem still remains NP-complete. See Problem 12.12.

The present example is crucially dependent on the *input dimension* approaching infinity. Suppose the problem is altered slightly so that the input dimension is fixed, but the number of hidden-layer perceptrons is increased, as in Figure 10.4. In this case it can be shown that, given a finite set of

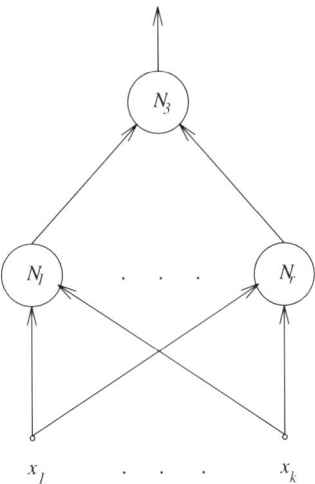

Fig. 10.4. A Neural Network for which Loading is Tractable

points $(x_1, y_1), \ldots, (x_m, y_m)$, each belonging to $[0,1]^k \times \{0,1\}$, there exists a polynomial-time algorithm (in r) to determine whether or not these input-output pairs can be realized by a network of the form shown in Figure 10.4. The proof is based on the work of Meggido [131] and is found in [50]. Note that the number of adjustable parameters in this architecture is $O(r)$. So by Theorem 10.3 below, the VC-dimension of the family $\mathcal{H}$ of functions realizable by this architecture is $O(r \lg r)$. Hence, so far as the VC-dimension goes, both architectures in Figures 10.3 and 10.4 are comparable. However, from a complexity-theoretic standpoint, the problem of finding the minimum achievable empirical risk has entirely different behaviour in the two cases.

The purpose of the next example is to demonstrate that, from a *complexity-theoretic* point of view, there can be a substantial difference between concept (or function) learning and model-free learning, even when the underlying hypothesis class is the same in both cases. This is a counterpoint to Theorems 5.12 and 7.6, which imply that the *information-theoretic* complexity of model-free learning is dependent only on the VC-dimension of the hypothesis class $\mathcal{H}$, and is not affected by whether the learning problem is one of concept learning, or model-free learning. More generally, Theorems 5.11 and 7.5 imply that, if a hypothesis class is learnable with an oracle that returns perfect measurements, then *any* model-free learning problem with the same hypothesis class is also learnable, provided only that the loss function satisfies a very reasonable equicontinuity assumption; thus, from an *information-theoretic* standpoint, learnability is pretty much a property of the hypothesis class alone.

Example 10.2. Let $X = \mathbb{R}^k$, and let $\mathcal{H}$ consist of all half-planes in X. Then the VC-dimension of $\mathcal{H}$ is $k + 1$; see Example 4.3. Now let us consider in succession three learning problems:

1. The PAC learning problem of Section 3.2 when the concept class is $\mathcal{H}$, and the oracle returns the membership function $I_T(\cdot)$, where $T \in \mathcal{H}$ is the unknown target function.
2. Same as Problem 1, except that the oracle occasionally returns erroneous measurements of $I_T(\cdot)$. Thus, given an input $x \in \mathbb{R}^k$, the oracle returns $I_T(x)$ with probability $1 - \alpha$, and $1 - I_T(x)$ with probability α, where $\alpha \in (0, 0.5)$ is the error rate.
3. Random i.i.d. elements of $X \times \{0, 1\}$ are generated according to an unknown probability $\bar{P}$, and the objective is to find the (nearly) best approximation to the data by an element of $\mathcal{H}$.

Note that Problems 2 and 3 fall into the category of model-free learning as defined in Section 3.3, with the loss function $\ell(y, u) := |y - u|$. Hence the family $\mathcal{L}_\mathcal{H}$ has exactly the same VC-dimension as $\mathcal{H}$, namely $k + 1$. As a consequence, from the information-theoretic standpoint, all three problems have exactly the same sample complexity.

Now let us examine the algorithmic side. In Problem 1, it is a consequence of Theorem 7.8 that *every* consistent algorithm is PAC. Suppose we are given data $(x_1, y_1), \ldots, (x_m, y_m)$ from $\mathbb{R}^k \times \{0, 1\}$. Then it is known beforehand that *the data is consistent with some target concept.* In other words, *there exists* a vector $[w_0 \; w_1 \ldots w_k]^t \in \mathbb{R}^{k+1}$ such that the set of m inequalities

$$w_0 + \sum_{i=1}^k w_j x_{ij} \geq 0 \text{ if } y_i = 1, \text{ and } < 0 \text{ if } y_i = 0$$

has a solution. Such data is said to be "linearly separable." Once it is known that the data is linearly separable so that the above set of inequalities is solvable, a feasible solution for $\mathbf{w} \in \mathbb{R}^{k+1}$ can be found using linear programming. Moreover, if one makes the natural assumption that each of the x_i's is represented to a finite number of bits specified *a priori*, then the running time of the algorithm is $O(k^3)$ [94].

Consider next Problem 2. Given the data $(x_1, y_1), \ldots, (x_m, y_m)$ from $\mathbb{R}^k \times \{0, 1\}$, it is known that if an "average" of αm of the x_i's are "neglected," then the data becomes linearly separable. Using this idea, it is possible to develop a polynomial-time algorithm to learn this problem [29].

Finally, let us discuss Problem 3. In this case, one is simply given data $(x_1, y_1), \ldots, (x_m, y_m)$ from $\mathbb{R}^k \times \{0, 1\}$, without any idea of how close the data is to being linearly separable. Now, Theorem 3.2 implies that any algorithm that nearly minimizes empirical risk is PAC. Thus one might be tempted to try to find a hyperplane that misclassifies the fewest number of points. Unfortunately, it is shown in [87] that the following problem is NP-hard with respect to k, the dimension of x: Given a subset $S \subseteq \{0, 1\}^k$, a classification $c : S \to \{0, 1\}$, and an integer $l \geq 0$, determine whether there exists a halfspace in $\mathbb{R}^k$ that misclassifies at most l points in S. Thus, even though the family of induced loss functions has finite VC-dimension, there is no known polynomial-time algorithm for actually computing a PAC hypothesis.

The moral of the above two examples is this: The metric entropy and VC-dimension-based results of Chapters 6 through 8 address *only a part* of the learning problem. But it should be emphasized that both the preceding examples show NP-completeness or NP-hardness with respect to the integer k, the number of inputs to the neural network. One could argue that, in practice, the number of inputs rarely increases without bound, and that one is more concerned about how the computational complexity of finding a PAC hypothesis increases with respect to the *number of neurons*, as opposed to the input dimension. Not very much is known about this case. However, Maass [125] argues that, in a number of practically important cases, the model-free learning problem *is* tractable as the number of neurons becomes larger and larger.

10.3 Estimates of VC-Dimensions of Families of Networks

In view of Theorems 10.1 and 10.2, it is clearly desirable to derive upper bounds for the VC-dimension and the P-dimension of the set of input-output mappings realizable by various neural network architectures. Actually, it is enough to focus on the VC-dimension of neural network architectures with a binary output, since every VC-dimension bound can be translated into a corresponding P-dimension bound using Lemma 10.1. Over the years, various researchers have evolved several interesting approaches to this problem, most of which are presented here.

It is relatively easy to prove that a neural network architecture has finite VC-dimension provided (i) the input vector is restricted to belong to a *compact* subset of $\mathbb{R}^k$, (ii) the vector of adjustable parameters is restricted to belong to a *compact* subset of $\mathbb{R}^l$, and (iii) each neural activation function is Lipschitz-continuous; see for example [80], Theorem 11. However, the bounds thus obtained approach infinity if either the maximum norm of the parameter vector or the maximum Lipschitz constant of the activation functions approaches infinity. Thus it is clear that such bounds are rather conservative. In particular, the finite VC-dimension results derived in this section *cannot* be derived using the results of [80], Theorem 11.

It turns out that all the estimates derived here depend on only three things: the number of adjustable parameters in the neural network, the type of activation function of the neurons (e.g., perceptron, piecewise-polynomial, standard sigmoid, etc.), and the depth of the network.

Except for the results on multi-layer perceptron networks given next, the estimates derived in this section involve the use of extremely advanced methods drawn from several diverse branches of mathematics such as algebraic geometry and mathematical logic. The interesting aspect, however, is that while the proof techniques might be accessible only to specialists, the bounds themselves are very easy to understand and to apply. Thus a reader who is unfamiliar with or uninterested in the various mathematical arguments will nevertheless be able to use the theorems directly, without having to go through the proofs. To follow such a program, the reader should read through Subsection 10.3.3, and understand the general philosophy of considering a neural network architecture as evaluating a family of formulas. In the remainder of the section, only the theorem statements can be read, and the proofs can be omitted.

10.3.1 Multi-Layer Perceptron Networks

In this subsection, it is shown that for a neural network consisting exclusively of perceptrons of the form (10.1.1), the VC-dimension is $O(l \lg l)$, where l is the number of adjustable parameters in the network. Moreover, this bound is the best possible, in that for arbitrarily large values of l there exist networks whose VC-dimension is $\Omega(l \lg l)$.

Theorem 10.3. *([47], [19]) Suppose a neural network contains only perceptrons, and let l denote the total number of parameters in the network (i.e., the total number of weights and thresholds of all perceptrons). Let $\mathcal{H}$ denote the family of mappings from $\mathbb{R}^k$ into $\{0,1\}$ that can be realized by varying this set of parameters over $\mathbb{R}^l$. Then*

$$VC\text{-}dim(\mathcal{H}) \leq 2l \lg(el).$$

10.3 Estimates of VC-Dimensions of Families of Networks

Proof. From Example 4.3, it is known that the VC-dimension of the set of mappings achievable by a perceptron with r inputs is $r + 1$, which happens to be the same as the number of adjustable parameters (r weights and one threshold). Hence it follows from Theorem 4.1 that, given a subset S of $\mathbb{R}^r$ of cardinality v, there are at most $\phi(v, r+1)$ different ways of partitioning it using a half-space, where the function $\phi(\cdot, \cdot)$ is defined in (4.2.2). Now suppose $S \subseteq \mathbb{R}^k$ has cardinality n. The first-level perceptrons all compute various maps from S into $\{0, 1\}$, and the outputs of these maps are in turn fed into the second level, and so on. Thus the total number of maps from S into $\{0, 1\}$ that can be generated by the network can be bounded by simply multiplying together the numbers for each peceptron. Let a_i denote the number of adjustable parameters of the i-th perceptron; thus $a_i = r_i + 1$, where r_i is the number of inputs of the i-th perceptron. Let g denote the total number of perceptrons in the network. Suppose without loss of generality that $n \geq a_i$ for each i. Then by repeated application of Theorem 4.1, it follows that the total number of maps from S into $\{0, 1\}$ in $\mathcal{H}$ is at most

$$\prod_{i=1}^{g} \phi(n, a_i) \leq \prod_{i=1}^{g} \left(\frac{en}{a_i}\right)^{a_i} \leq \prod_{i=1}^{g} (en)^{a_i} = (en)^l,$$

where $l = \sum_{i=1}^{g} a_i$ denotes the total number of weights and thresholds in the network. Now $(en)^l$ grows only *polynomially* with respect to n. In particular, we have from Lemma 4.6 that

$$n \geq 2l \lg(el) \implies n > l \lg(en) \implies 2^n > (en)^l.$$

Hence a set of cardinality $n \geq 2l \lg(el)$ cannot be shattered by $\mathcal{H}$, since the number of distinct maps from S into $\{0, 1\}$ in $\mathcal{H}$ is less than 2^n. ∎

Theorem 10.3 states that the VC-dimension of a perceptron network with l adjustable parameters is $O(l \lg l)$. One can ask whether this bound is the best possible. It is shown below that the answer is "yes" by constructing for arbitrarily large l a family of neural networks whose VC-dimension is $\Omega(l \lg l)$.

Example 10.3. [122], [123] Suppose d is an arbitrary power of 2. A neural network is constructed with $2d + \lg d$ inputs and at most $17d^2$ edges such that its VC-dimension is at least $d^2 \lg d$. By noting that the number l of adjustable parameters in a perceptron network is at most equal to twice the number of edges, we see that the VC-dimension of the network is $\Omega(l \lg l)$.

Suppose $d = 2^m$ for some integer $m \geq 1$. We construct a network that shatters the set

$$S = \{\mathbf{e}_p \, \mathbf{e}_q \, \bar{\mathbf{e}}_r, 1 \leq p, q \leq d, 1 \leq r \leq m\},$$

where $\mathbf{e}_i$ denotes a unit vector of length d with a one in the i-th position, whereas $\bar{\mathbf{e}}_j$ denotes a unit vector of length m with a one in the j-th position.

It is easy to see that $|S| = d^2 m = d^2 \lg d$. Now we construct a perceptron network of *fixed* architecture such that, by suitable adjustment of the various parameters, the network can realize *any* map $f : S \to \{0,1\}$. This shows that the corresponding family of input-output mappings shatters S, whence VC-dim$(\mathcal{H}) \geq d^2 \lg d$.

The construction is rather elaborate, and proceeds in several steps. As a first step, let $f : S \to \{0,1\}$ be an arbitrary map, and define

$$g(p,q) := [f(\mathbf{e}_p \mathbf{e}_q \bar{\mathbf{e}}_1) \ldots f(\mathbf{e}_p \mathbf{e}_q \bar{\mathbf{e}}_m)]^t \in \{0,1\}^m.$$

Thus g can be thought of as a map from $\{1,\ldots,d\}^2$ into $\{0,1\}^m$. Now it is known ([120], Lemma 4) that, given *any* function $g : \{1,\ldots,d\}^2 \to \{0,1\}^m$, there exist four functions $g_1, g_2, g_3, g_4 : \{1,\ldots,d\}^2 \to \{0,1\}^m$ such that

$$g(p,q) = \begin{cases} g_1(p,q) \oplus g_2(p,q), & \text{if } p \leq d/2, \text{ and} \\ g_3(p,q) \oplus g_4(p,q), & \text{if } p > d/2, \end{cases} \quad (10.3.1)$$

where $\oplus$ denotes the componentwise exclusive-or mapping, and in addition, each of the maps g_j has the property that $g_j(\cdot, q) : \{1,\ldots,d\} \to \{0,1\}^m$ is one-to-one and onto for each $q \in \{1,\ldots,d\}$ (and of course for each j). Suppose these four functions have been constructed.

In the next step, suppose that a function $g : \{1,\ldots,d\}^2 \to \{0,1\}^m$ is given, with the additional property that $g(\cdot, q) : \{1,\ldots,d\} \to \{0,1\}^m$ is one-to-one and onto. (Essentially this g can be any one of the g_j's above; the subscript j is dropped for convenience.) We now construct a perceptron network with $2d$ inputs and m outputs such that, when the input to the network equals $\mathbf{e}_p \mathbf{e}_q$, then the b-th output equals one if and only if there exists an index $i \in \{0,\ldots,d-1\}$ such that the b-th bit of the binary expansion of i is one, and in addition $g(p,q)$ equals the binary expansion of i. For $i \in \{0,\ldots,d-1\}$, let bin(i) denote the binary expansion of i, and note that bin$(i) \in \{0,1\}^m$. Define a weight set $w_{i,q}$ for each $q \in \{1,\ldots,d\}$ and each $i \in \{0,\ldots,d-1\}$ as follows:

$$w_{i,q} = p \Leftrightarrow g(p,q) = \text{bin}(i).$$

By assumption, for each q, i, there exists a unique p such that $g(p,q) = \text{bin}(i)$, so that $w_{i,q}$ is well-defined. At the first level, the network contains $2d$ perceptrons, labelled as η_i^+ and η_i^- for $i = 0,\ldots,d-1$. The structure of these perceptrons is shown in Figures 10.5 and 10.6, respectively. Thus

$$\eta_i^+(\mathbf{x}) = 1 \Leftrightarrow \sum_{s=1}^{d} s x_s \geq \sum_{s=1}^{d} w_{i,s} x_{s+d}, \text{ and}$$

$$\eta_i^-(\mathbf{x}) = 1 \Leftrightarrow \sum_{s=1}^{d} s x_s \leq \sum_{s=1}^{d} w_{i,s} x_{s+d}.$$

10.3 Estimates of VC-Dimensions of Families of Networks

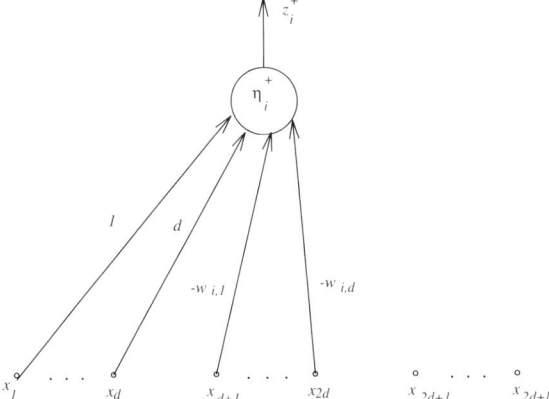

Fig. 10.5. Architecture of the Perceptron η_i^+

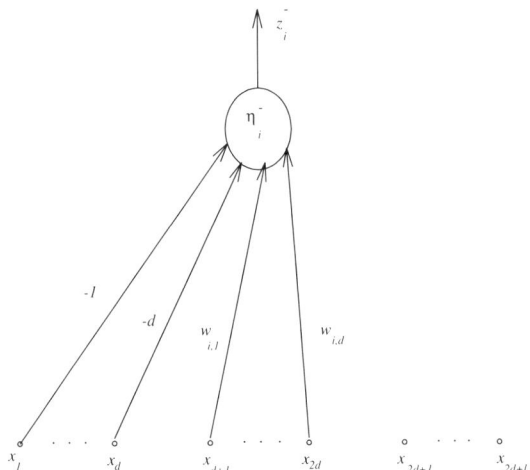

Fig. 10.6. Architecture of the Perceptron η_i^-

At the second level, the network consists of $m = \lg d$ perceptrons, as shown in Figure 10.7. Note that the perceptron labelled π_b $(1 \le b \le m)$ has the input-output relationship

$$y_b = \eta \left[\sum_{[\text{bin}(i)]_b = 1} (z_i^+ + z_i^-) - \left(\frac{d}{2} + 1 \right) \right].$$

In other words, only the outputs of those η_i^+ and η_i^- such that the b-th bit of $\text{bin}(i) = 1$ are connected to π_b. The weight of each input to π_b is 1, and the threshold is $1 + d/2$.

Now suppose the input $\mathbf{e}_p \mathbf{e}_q$ is applied to this network, and the weights at the first level are set to $w_{i,s}$ as shown in Figures 10.5 and 10.6. It is clear

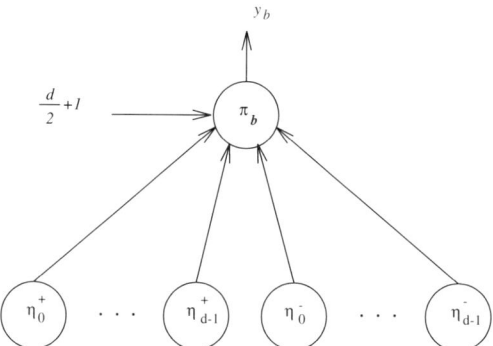

Fig. 10.7. Architecture of Second-Level Neurons

that, whatever be the weights and the input vector, the output of *at least one* of the two perceptrons η_i^+ and η_i^- equals one. Moreover, *both* perceptrons have an output of 1 if and only if

$$\sum_{s=1}^{d} sx_s = \sum_{s=1}^{d} w_{i,s} x_{s+d}.$$

If $\mathbf{x} = \mathbf{e}_p \mathbf{e}_q$, then

$$\sum_{s=1}^{d} sx_s = p, \text{ and } \sum_{s=1}^{d} w_{i,s} x_{s+d} = w_{i,q}.$$

Thus the outputs z_i^+ and z_i^- (see Figures 10.5 and 10.6) both equal 1 if and only if $w_{i,q} = p$, or equivalently (from the definition of $w_{i,q}$), $g(p,q) = \text{bin}(i)$. Now consider the next level perceptron π_b. Clearly its output y_b equals 1 if and only if there exists an index i such that $[\text{bin}(i)]_b = 1$ and $z_i^+ = z_i^- = 1$. Equivalently, $y_b = 1$ if and only if there exists an index i such that $[\text{bin}(i)]_b = 1$ and $g(p,q) = \text{bin}(i)$.

Next, suppose that neural networks $\mathcal{N}_1$ through $\mathcal{N}_4$ have been constructed, corresponding to the functions g_1 through g_4, and let ϕ_i denote the input-output map of $\mathcal{N}_i$. Thus $\phi_i : \mathbb{R}^{2d} \to \{0,1\}^m$. Now construct another network with $2d$ inputs and m binary outputs, as follows (see Figure 10.8): Let $x_1, \ldots, x_{2d}$ denote the inputs to the overall network. Then connect x_1 through $x_{d/2}$, and again x_{d+1} through x_{2d}, but not $x_{1+d/2}$ through x_d, to networks $\mathcal{N}_1$ and $\mathcal{N}_2$. This is equivalent to assigning weights 1 to the connected inputs, and weights zero to the unconnected inputs. Similarly, connect inputs $x_{1+d/2}$ through x_{2d} to $\mathcal{N}_3$ and $\mathcal{N}_4$. Take the outputs of the four networks, and perform two sets of component-wise exclusive-or operations (which can be achieved using perceptrons). Finally, add the outputs of these two exclusive-or networks. Let $\phi : \mathbb{R}^{2d} \to \{0,1\}^l$ denote the input-output map of this network. Now suppose the input $\mathbf{x}$ equals $\mathbf{e}_p \mathbf{e}_q$. Then there are two

10.3 Estimates of VC-Dimensions of Families of Networks

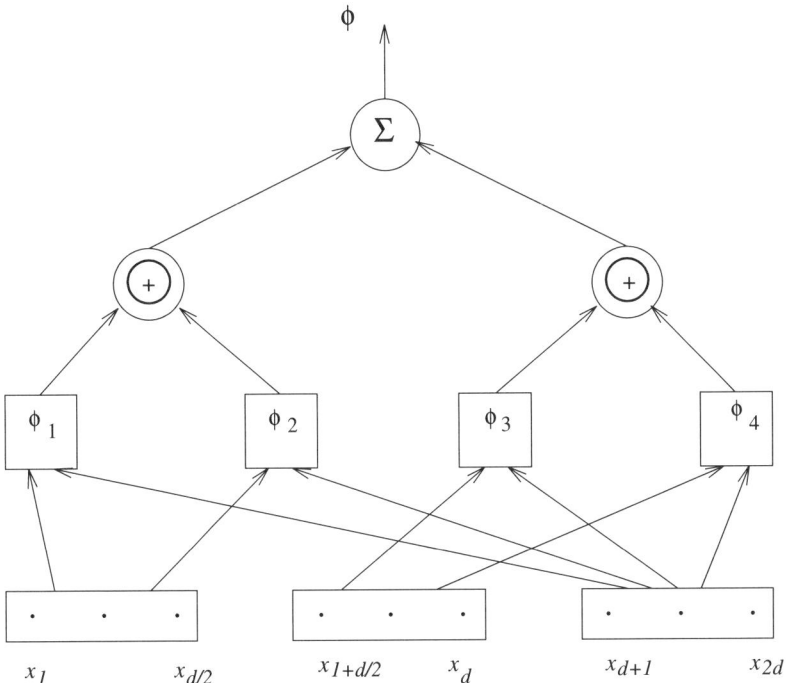

Fig. 10.8. Realization of the Map $\phi(\cdot)$

cases to consider, namely: $p \leq d/2$, and $p > d/2$. If $p \leq d/2$, then effectively both $\mathcal{N}_1$ and $\mathcal{N}_2$ see the input $\mathbf{e}_p\mathbf{e}_q$, whereas effectively both $\mathcal{N}_3$ and $\mathcal{N}_4$ see the input $(\mathbf{0}, \mathbf{e}_q)$. With this input, both the networks $\mathcal{N}_3$ and $\mathcal{N}_4$ produce the zero output. This can be seen as follows: With this input, we have

$$\sum_{s=1}^{d} w_{i,s} x_{s+d} = w_{i,q}, \text{ and } \sum_{s=1}^{d} s x_s = 0.$$

Since $w_{i,q} = p > 0$ for each i, it follows that the condition $z_i^+ = z_i^- = 1$ is never satisfied, and as a result $y_b = 0$ for every b. Thus the overall output equals $\phi_1(\mathbf{x}) \oplus \phi_2(\mathbf{x})$. The situation is reversed if $p > d/2$.

To complete the argument, the network of Figure 10.8 is embedded into a larger network as shown in Figure 10.9. The first $2d$ inputs are fed into the realization of $\phi(\cdot)$, while the last m inputs are fed into a perceptron along with the m outputs of $\phi(\cdot)$. Thus the output of the overall network is

$$h(\mathbf{x}) = \eta \left[\sum_{b=1}^{m} a_{b,1} \phi_b(\mathbf{x}) + a_{b,2} x_{2d+b} - 1.5 \right].$$

Now suppose the input to the network is $\mathbf{x} = [\mathbf{e}_p \mathbf{e}_q \bar{\mathbf{e}}_r]$. By choosing the weights

388   10. Applications to Neural Networks

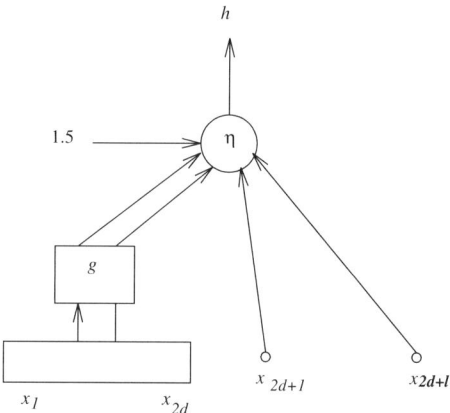

Fig. 10.9. Realization of Overall Network

$$a_{b,1} = a_{b,2} = \begin{cases} 1, & \text{if } b = r, \text{ and} \\ 0, & \text{if } b \neq r, \end{cases}$$

it can be ensured that $h(\mathbf{x}) = 1$ if and only if $\phi_r(\mathbf{e}_p\mathbf{e}_q) = 1$. From the manner in which the map $\phi(\cdot)$ is realized in Figure 10.8, we see that

$$\phi(\mathbf{e}_p\mathbf{e}_q) = \begin{cases} \phi_1(\mathbf{e}_p\mathbf{e}_q) \oplus \phi_2(\mathbf{e}_p\mathbf{e}_q) & \text{if } p \leq d/2, \text{ and} \\ \phi_3(\mathbf{e}_p\mathbf{e}_q) \oplus \phi_4(\mathbf{e}_p\mathbf{e}_q), & \text{if } p > d/2. \end{cases}$$

Now recall that each ϕ_j is a "realization" of the function $g_j : \{1, \ldots, d\}^2 \to \{0,1\}^m$, in the sense that the b-th bit of $\phi_j(\mathbf{e}_p\mathbf{e}_q)$ equals 1 if and only if there exists an index i such that $g_j(p, q) = \text{bin}(i)$ and the b-th bit of the binary expansion of i equals 1. Recall also that

$$g(p,q) = [f(\mathbf{e}_p\mathbf{e}_q\bar{\mathbf{e}}_1) \ldots f(\mathbf{e}_p\mathbf{e}_q\bar{\mathbf{e}}_m)]^t \in \{0,1\}^m.$$

Hence the r-th bit of $\phi(\mathbf{e}_p\mathbf{e}_q)$ equals 1 if and only if there exists an index i such that the r-th bit of the binary expansion of i equals 1, and $g(p,q) = \text{bin}(i)$. This is equivalent to the requirement that $f(\mathbf{e}_p\mathbf{e}_q\bar{\mathbf{e}}_r) = 1$. In this way, an arbitrary function mapping S (defined at the beginning of the example) into $\{0,1\}$ can be realized.

10.3.2 A Network with Infinite VC-Dimension

The upper bound derived in the preceding subsection on the VC-dimension of multi-layer perceptron networks depends crucially on the fact that the outputs of even the *intermediate-level* perceptrons are all binary, and not just the output of the highest-level neuron. If we consider instead a network in which the intermediate-level neurons are not perceptrons but have continuous activation functions, then it is easy to construct very simple neural networks that have *infinite* VC-dimension, as shown next.

10.3 Estimates of VC-Dimensions of Families of Networks

Example 10.4. [175] Consider the neural network shown in Figure 10.10, where $\phi(\cdot)$ is the function defined by

$$\phi(x) := \frac{1}{\pi} \tan^{-1} x + \frac{1}{2} + \frac{\cos x}{\alpha(1+x^2)},$$

and the output-layer perceptron has both input weights equal to one and a threshold of one. Thus the input-output relationship of the network is given

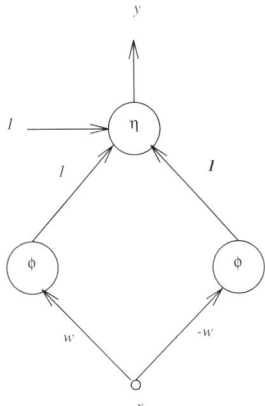

Fig. 10.10. A Neural Network with Infinite VC-Dimension

by

$$y = \eta[\rho(x)],$$

where

$$\rho(x) := \phi(wx) + \phi(-wx) - 1 = \frac{2\cos wx}{\alpha(1+w^2x^2)}.$$

Note that the function $\phi(\cdot)$ satisfies

$$\lim_{x \to \infty} \phi(x) = 1, \text{ and } \lim_{x \to -\infty} \phi(x) = 0.$$

Moreover, if $\alpha > 2\pi$, then $\phi'(x) > 0$ for all x. Hence the function $\phi(\cdot)$ is sigmoidal whenever $\alpha > 2\pi$.

Now it is claimed that the VC-dimension of this network is infinite. Actually, a much stronger property is established, namely: Let $n \geq 2$ be an *arbitrary* integer; then the set of n-tuples $(x_1, \ldots, x_n)$ that is shattered by the network is *dense* in $\mathbb{R}^n$.[3] To see this, choose numbers $x_1, \ldots, x_n \in \mathbb{R}$ that are rationally independent; this means that there do not exist any rational numbers $g_1, \ldots, g_n$ such that $\sum_{i=1}^{n} g_i x_i = 0$. Then it follows (see e.g. Lemma

---

[3] By a slight abuse of notation, we say "shattered by the network" to mean "shattered by the family $\mathcal{H}$ of input-output mappings of networks of this architecture."

2.7 of [127]) that the set of n-tuples $(wx_1, \ldots, wx_n)$ modulo 2π generated by varying w over the positive integers is dense in $[0, 2\pi]^n$. Consequently the set of vectors of the form

$$[\cos(wx_1) \ldots \cos(wx_n)]^t, \ w \in \mathbf{N}$$

is a dense subset of $[-1, 1]^n$. Thus the vector

$$[\rho(wx_1) \ldots \rho(wx_n)]^t, \ w \in \mathbf{N}$$

can be made to achieve *any* desired set of signs in $\{-1, 1\}^n$ by picking a suitable weight w. As a result, the family of input-output maps of this network shatters the set $\{x_1, \ldots, x_n\}$. Moreover, it is well-known that the set of rationally independent n-tuples is dense in $\mathbb{R}^n$. Thus it has been shown that the VC-dimension of this network is infinite. Moreover, for every n, the set of $(x_1, \ldots, x_n)$ that is shattered by the network is *dense* in $\mathbb{R}^n$. ■

The preceding example shows that, unless some restrictions are placed on the nature of the individual neurons, a network where the neural activation functions are continuous can have infinite VC-dimension. A topic of active research is to determine "reasonable" conditions on the neural activation functions that lead to the network having finite VC-dimension. Several such activation functions have been identified, and are discussed later in this section.

10.3.3 Neural Networks as Verifiers of Formulas

A fruitful approach to analyzing the VC-dimension of neural network architectures is to think of a neural network with binary output as evaluating the truth or falsity of a "formula" over the set of real numbers. For instance, a perceptron with l inputs, weights $w_1, \ldots, w_l$, and threshold θ evaluates the formula

$$w_1 x_1 + \ldots + w_l x_l - \theta \geq 0.$$

If the input l-tuple $(x_1, \ldots, x_l)$ satisfies this formula, then the perceptron outputs a 1; otherwise it outputs a 0. To make this intuitive notion precise, it is necessary to define what a "formula" is. This leads us to a deep and difficult subject known as the "first-order theory of the real numbers," which is barely touched upon here. Strictly speaking, one should say *a*, and not *the*, first-order theory of the reals, since it is in fact possible to define several such theories.

A quick review is now given of first-order predicate logic, with particular reference to the real numbers. Note that in predicate logic, it is customary to make a distinction between a *language* and a *structure*. Roughly speaking, a structure is obtained when the abstract symbols in a language are given specific interpretations in a specific setting. However, in the present setting,

10.3 Estimates of VC-Dimensions of Families of Networks

there is no reason to engage in such abstraction, since we are only interested in the real number system. One begins with an infinite set of constant symbols, an infinite set of variable symbols, and a collection (finite or infinite) of functions. Each function has associated with it an "arity," which is nothing more than the number of its arguments. Thus if f is an n-ary function, then f maps $\mathbb{R}^n$ into $\mathbb{R}$. It is assumed that the standard binary functions $+, -, \cdot$ are among the set of functions. But usually there are several other functions as well. Indeed, it is precisely the set of functions that distinguishes one model of the real number system from another. Next there are the binary relations $>, <, =$ with the usual meanings. Then there are the standard logical symbols $\wedge$ (and), $\vee$ (or), and $\neg$ (not). Finally there are the quantifiers $\exists$ (exists) and $\forall$ (for all). Now we are in a position to define a **term**, which is a prelude to defining a formula. The definition of a term is as follows:

- Every constant symbol and every variable is a term.
- If f is an n-ary function and $t_1, \ldots, t_n$ are terms, then $f(t_1, \ldots, t_n)$ is a term.
- No string of symbols is a term unless it can be proven to be a term by repeated application of the above two rules.

Finally we can define a **formula**.

- If t_1, t_2 are terms, then $t_1 = t_2$, $t_1 > t_2$, $t_1 < t_2$ are formulas.
- If ϕ_1, ϕ_2 are formulas, then $\phi_1 \wedge \phi_2$, $\phi_1 \vee \phi_2$, and $\neg \phi_1$ are formulas.
- If ϕ is a formula and z is a variable, then $\forall z \phi$ and $\exists z \phi$ are formulas.
- No string of symbols is a formula unless it can be proven to be so by repeated application of the above three rules.

The reader is referred to any standard text on first-order logic for a more thorough discussion.

Note that, since both $+$ and $\cdot$ are always included in every model of real computation, all polynomials can be defined in every model of real computation. However, by enlarging the set of functions beyond these two standard functions, one obtains a richer model of computation.

As an illustration of the above abstract definition, suppose exp and cos are (1-ary) functions in our model of the real number system, and x, y, z are variables. Then

$$\exists z (\exp(x \cdot z) > \cos(y \cdot z))$$

is a formula. The "meaning" of the above formula is obvious. Strictly speaking, we have been sloppy in not using parentheses liberally, as required by the rules of logic; but where there is no danger of confusion, the number of parentheses is kept to a minimum. We also resort to standard abbreviations such as ab for $a \cdot b$, and $t \geq 0$ for $(t > 0) \vee (t = 0)$, and so on. The basic idea is that, unless specified otherwise, all symbols have their normal everyday meaning.

In predicate logic, one makes a distinction between "free" and "bound" variables. Roughly speaking, a variable is "free" unless it follows one of the quantification symbols $\forall$ or $\exists$. Thus in the formula above, x and y are free variables, whereas z is a bound variable. For the most part, we deal in this chapter with so-called "quantifier-free" formulas, which means that the symbols $\forall$ and $\exists$ do not appear in the formula.

Suppose ϕ is a formula containing the free variables $x_1, \ldots, x_n$. If specific values $a_1, \ldots, a_n$ are substituted for these variables, then we get a "valuation" $\phi(a_1, \ldots, a_n)$ which is either "true" or "false" in the usual meaning of the words. If $\phi(a_1, \ldots, a_n)$ evaluates to "true," it is customary in logic to write $\mathbb{R} \models \phi(a_1, \ldots, a_n)$. Thus, with each formula ϕ, one can associate the set

$$S(\phi) := \{(a_1, \ldots, a_n) \in \mathbb{R}^n : \mathbb{R} \models \phi(a_1, \ldots, a_n)\}.$$

In other words, $S(\phi)$ is the set of all n-tuples $(a_1, \ldots, a_n) \in \mathbb{R}^n$ that cause $\phi(a_1, \ldots, a_n)$ to evaluate to "true." Less formally, one can think of $S(\phi)$ as the set of all n-tuples $(a_1, \ldots, a_n) \in \mathbb{R}^n$ that "satisfy" the formula ϕ.

For example, suppose $n = 2$, and suppose that ϕ is the formula

$$\phi := 2x_1 - 3x_2 + 0.5 \geq 0.$$

Here 2, 3, and 0.5 are constants, and $\geq$ is used as a convenient shorthand to avoid writing

$$\phi := (2x_1 - 3x_2 + 0.5 > 0) \vee (2x_1 - 3x_2 + 0.5 = 0).$$

Then the set $S(\phi)$ is given by

$$S(\phi) := \{(x_1, x_2) \in \mathbb{R}^2 : 2x_1 - 3x_2 + 0.5 \geq 0\}.$$

It is intuitively clear that every neural network with a binary output "evaluates" a formula in a model of the real number system, provided the activation functions of the various neurons all belong to the set of functions included in the model of the real number system. For example, suppose both the standard sigmoid function and the linear function are in the function set, and consider the network shown in Figure 10.11. Let us write down the formula computed by the network, i.e., a formula $\phi(x_1, x_2)$ that evaluates to "true" if and only if the output of the network equals one. Now there are two approaches that one can adopt. In the first approach, one can introduce auxiliary symbols z_1, z_2 for the outputs of the two hidden-layer nodes; in this case, one obtains a formula containing the existential quantifier, as follows: Note that the symbols v_1 and v_2 are merely used as a shorthand for the strings $w_{11}x_1 + w_{22}x_2 - \theta_1$ and $w_{21}x_1 + w_{22}x_2 - \theta_2$ respectively, so as to make the formula readable; in particular, v_1 and v_2 are *not* variable names.

$$\phi(x_1, x_2) := \exists z_1 \exists z_2 (a_1 z_1 + a_2 z_2 - b \geq 0) \wedge (z_1 = \sigma(v_1)) \wedge$$
$$\{[(v_2 < 0) \wedge (z_2 = 0)] \vee [(0 \leq v_2) \wedge (v_2 \leq 1) \wedge (z_2 = v_2)] \vee [(1 < v_2) \wedge (z_2 = 1)]\}.$$

10.3 Estimates of VC-Dimensions of Families of Networks

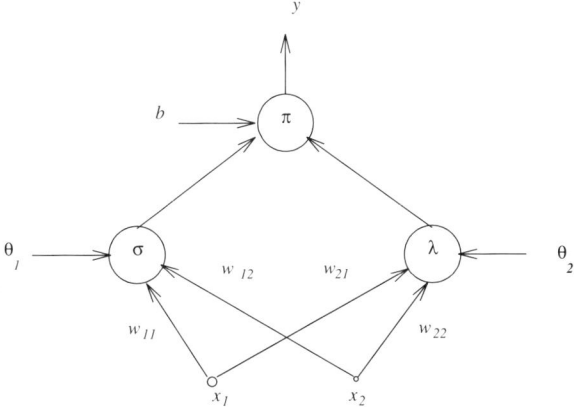

Fig. 10.11. A Neural Network that Verifies a Formula

The reader is urged to go through the above formula carefully and become persuaded that the formula is indeed being computed by the network under study. The main drawback of the above formula is that it contains the existential quantifier and a couple of bound variables. To avoid this difficulty, one could try an alternate approach whereby one simply applies the distributive law repeatedly, and in effect "eliminates" the bound variables z_1 and z_2. In general, converting a formula containing the existential or universal quantifiers into an equivalent quantifier-free formula is a tricky business. However, in the special types of formulas that arise in connection with neural networks, this is quite easy (at least in principle). For example, for the network in Figure 10.11, the formula evaluated by the network can be written as follows:

$$\phi(x_1, x_2) := [(v_2 < 0) \wedge (a_1 \sigma(v_1)) - b \geq 0)]$$
$$\vee [(0 \leq v_2) \wedge (v_2 \leq 1) \wedge (a_1 \sigma(v_1)) + a_2 v_2 - b \geq 0)]$$
$$\vee [(1 < v_2) \wedge (a_1 \sigma(v_1)) + a_2 - b \geq 0)].$$

The key point to note is that, since the activation function $\lambda(\cdot)$ of the hard limiter is defined by dividing $\mathbb{R}$ into three regions, we have written down one formula each corresponding to the possibility that the input to the hard-limiter lies in one of these regions. In general, if there are n neurons, and if the activation function of the i-th neuron is defined in a "piecewise" fashion by dividing the input space $\mathbb{R}$ into r_i regions, then one can enumerate a total of $\prod_{i=1}^{n} r_i$ possible combinations of input regions for the totality of neurons. Of course, not all of these possible combinations might make sense. But this number represents an *upper bound* on the total number of combinations.

The above example also makes it clear that the distinction between "constants" and "variables" in a formula is not always clear-cut, especially in the context of neural networks. In the formula above, there are only two "genuine" constants, namely 0 and 1. The remaining symbols, such as the

394   10. Applications to Neural Networks

w_{ij}, θ_i, a_i, b can either be thought of as variables or as constants. To make this idea more precise, suppose a neural network architecture has k inputs and l adjustable parameters. With this *architecture*, one can associate a *family* of formulas $\phi(\mathbf{x}; \mathbf{w})$, where each choice of the parameter vector $\mathbf{w}$ corresponds to a specific neural network belonging to the architecture. With each $\mathbf{w}$ one associates the set

$$S_\mathbf{w}(\phi) := \{\mathbf{x} \in \mathbb{R}^k : \mathbb{R} \models \phi(\mathbf{x}; \mathbf{w})\}.$$

One can also associate the function $h_\mathbf{w}$ defined by

$$h_\mathbf{w}(\mathbf{x}) = 1 \text{ if } \mathbf{x} \in S_\mathbf{w}(\phi), \ 0 \text{ otherwise}.$$

The totality of the sets $\{S_\mathbf{w}(\phi) : \mathbf{w} \in \mathbb{R}^l\}$, or equivalently, the collection of binary-valued functions $\{h_\mathbf{w}(\cdot) : \mathbf{w} \in \mathbb{R}^l\}$, is the hypothesis class associated with the neural network architecture. Thus, in order to apply the theory developed thus far to the problem of generalization by neural networks, it is desirable to be able to compute (or at least estimate) the VC-dimension of a collection of sets of the form $\{S_\mathbf{w}(\phi) : \mathbf{w} \in \mathbb{R}^l\}$.

In this connection, it is sometimes useful to define the so-called "dual" VC-dimension. To define this notion, it is not necessary to restrict oneself to first-order predicate logic. More generally, let X, W be arbitrary sets, and suppose $\phi : X \times W \to \{0, 1\}$ is some function. Then one can define the sets

$$S_w := \{x \in X : \phi(x, w) = 1\} \subseteq X, \ \forall w \in W,$$

$$\mathcal{H}_x := \{S_w : w \in W\} \subseteq 2^X,$$

and define d_x to be the VC-dimension of the collection of sets $\mathcal{H}_x$. In a dual fashion, one can also define

$$S_x := \{w \in W : \phi(x, w) = 1\} \subseteq W, \ \forall x \in X,$$

$$\mathcal{H}_w := \{S_x : x \in X\} \subseteq 2^W,$$

and define d_w to be the VC-dimension of the collection of sets $\mathcal{H}_w$. The number d_w is called the **dual VC-dimension**. There is a useful relationship between d_x and d_w, as brought out next.

Theorem 10.4. *With the notation as above, we have*

$$d_w \geq \lfloor \lg d_x \rfloor, \text{ and } d_x \geq \lfloor \lg d_w \rfloor.$$

Proof. It is enough to show that $d_w \geq \lfloor \lg d_x \rfloor$, since the second inequality follows by symmetry between x and w. For this purpose, suppose $d_x \geq 2^n$ for some integer n. It is shown that $d_w \geq n$.

By assumption, there exists a set $S = \{x_0, \ldots, x_{2^n-1}\}$ of cardinality 2^n that is shattered by $\mathcal{H}_x$. In some natural fashion, establish a one-to-one

10.3 Estimates of VC-Dimensions of Families of Networks

correspondence between the numbers $0, \ldots, 2^n - 1$ and the 2^n subsets of $\{0, \ldots, n-1\}$. For example, given a subset $L \subseteq \{0, \ldots, n-1\}$, one could define $b(L) := \sum_{i \in L} 2^i$. With this relabelling, the assumption is that the set $S = \{x_L : L \subseteq \{0, \ldots, n-1\}\}$ is shattered by $\mathcal{H}_x$. In other words, given any collection $\mathcal{A}$ of subsets of $\{0, \ldots, n-1\}$, there exists an element $w_{\mathcal{A}} \in W$ such that
$$\phi(x_L, w_{\mathcal{A}}) = 1 \Leftrightarrow L \in \mathcal{A}.$$

Next, for each integer $i \in \{0, \ldots, n-1\}$, define
$$\mathcal{A}_i := \{L \subseteq \{0, \ldots, n-1\} : i \in L\}.$$

In words, $\mathcal{A}_i$ consists of all subsets of $\{0, \ldots, n-1\}$ that contain i. Now denote $w_i := w_{\mathcal{A}_i}$ as above. Thus
$$\phi(x_L, w_i) = 1 \Leftrightarrow i \in L \ (\Leftrightarrow L \in \mathcal{A}_i). \tag{10.3.2}$$

Now it is claimed that the set $\{w_0, \ldots, w_{n-1}\}$ is shattered by $\mathcal{H}_w$. To see this, let B be an arbitrary subset of $\{w_0, \ldots, w_{n-1}\}$, and define L to be the corresponding subset of indices; in other words, if $B = \{w_{i_1}, \ldots, w_{i_d}\}$, then $L = \{i_1, \ldots, i_d\}$. Then by (10.3.2) above, we have that
$$\phi(x_L, w_i) = 1 \Leftrightarrow i \in L \Leftrightarrow w_i \in B.$$

In other words, $S_{x_L} = B$. This shows that $\{w_0, \ldots, w_{n-1}\}$ is shattered by $\mathcal{H}_w$, and thus establishes that $d_w \geq n$. ∎

Corollary 10.1. *With all symbols as above, we have*
$$d_x \leq 2^{d_w+1}, \text{ and } d_w \leq 2^{d_x+1}.$$

Proof. Once again, it is enough to prove the first inequality, since the second follows by symmetry. The first inequality follows readily by observing that
$$d_w \geq \lfloor \lg d_x \rfloor \geq \lg d_x - 1.$$

A slight rearrangement yields the desired result. ∎

In a certain narrow sense (see [113], p. 380), the bounds in Theorem 10.4 are "sharp." But in general these bounds might not be very good. Consider for example the collection of perceptrons with zero threshold for which
$$\phi(x, w) := \eta \left(\sum_{i=1}^{n} w_i x_i \right).$$

In this case, it is easy to show that $d_w = d_x = n$. Thus the bound given by Theorem 10.4 is not very good. If one imposes a little "structure" on the set W by assuming that it is a Cartesian product of smaller sets, then it is possible to prove sharper bounds; see [52], Theorem 2.

Now it is possible to state the general philosophy behind the various bounds on the VC-dimension of neural networks presented from here onwards. Suppose a neural network architecture is specified, as in Figure 10.11, for example. As shown above, with each architecture it is possible to associate (i) a first-order model of the real number system that incorporates the various activation functions used in the network, and (ii) a formula $\phi(\mathbf{x}, \mathbf{w})$, where $\mathbf{x} \in \mathbb{R}^k$ denotes the input vector to the network, and $\mathbf{w} \in \mathbb{R}^l$ denotes the set of adjustable parameters. Now one can think of $\phi(\mathbf{x}, \mathbf{w})$ as being generated by a Boolean operation on a set of **atomic formulas**. For the present purposes, it is enough to define a set of atomic formulas as follows: If $\phi_1(\mathbf{x}, \mathbf{w}), \ldots, \phi_s(\mathbf{x}, \mathbf{w})$ are formulas and if

$$\phi(\mathbf{x}, \mathbf{w}) = u[\phi_1(\mathbf{x}, \mathbf{w}), \ldots, \phi_s(\mathbf{x}, \mathbf{w})],$$

where $u : \{0, 1\}^s \to \{0, 1\}$ is a Boolean map, then $\phi_1(\mathbf{x}, \mathbf{w}), \ldots, \phi_s(\mathbf{x}, \mathbf{w})$ are atomic formulas of $\phi(\mathbf{x}, \mathbf{w})$. The above definition does not uniquely specify what a set of atomic formulas is, and indeed, there need not exist a unique set of atomic formulas corresponding to a given formula. However, in the specific types of neural network architectures studied here, we shall see that there usually exists a natural choice of atomic formulas.

Suppose now that a set $S = \{\mathbf{x}_1, \ldots, \mathbf{x}_v\}$ is shattered by the family of formulas $\{\phi(\cdot, \mathbf{w}), \mathbf{w} \in \mathbb{R}^l\}$. This means that, for each subset $L \subseteq \{1, \ldots, v\}$, there exists a $\mathbf{w}_L \in \mathbb{R}^l$ such that

$$\mathbb{R} \models \phi(\mathbf{x}_i, \mathbf{w}_L) \Leftrightarrow i \in L.$$

Now look at the set of atomic formulas $\phi_j(\mathbf{x}_i, \mathbf{w})$ for $1 \le i \le v$ and $1 \le j \le s$. More precisely, examine the vector

$$\mathbf{a}(\mathbf{w}) := [\phi_j(\mathbf{x}_i, \mathbf{w}), 1 \le i \le v, 1 \le j \le s]^t \in \{0, 1\}^{vs}. \tag{10.3.3}$$

If $\mathbf{a}(\mathbf{w}) = \mathbf{a}(\mathbf{w}')$, then $\phi_j(\mathbf{x}_i, \mathbf{w}) = \phi_j(\mathbf{x}_i, \mathbf{w}')$ for all i, j; consequently, $\phi(\mathbf{x}_i, \mathbf{w}) = \phi(\mathbf{x}_i, \mathbf{w}')$, and as a result,

$$\{\mathbf{x}_i \in S : \mathbb{R} \models \phi(\mathbf{x}_i, \mathbf{w})\} = \{\mathbf{x}_i \in S : \mathbb{R} \models \phi(\mathbf{x}_i, \mathbf{w}')\}.$$

Thus, in order for a set of cardinality v to be shattered, the vector $\mathbf{a}(\mathbf{w})$ must assume at least 2^v distinct values in $\{0, 1\}^{vs}$ as $\mathbf{w}$ varies over $\mathbb{R}^l$. Now it turns out that, in a variety of important situations, the number of distinct vectors $\mathbf{a}(\mathbf{w})$ generated by fixing a set S and varying $\mathbf{w}$ is bounded by a *polynomial* in $v = |S|$, and thus must be less than 2^v for v sufficiently large. In this way we can derive an upper bound on the VC-dimension of a neural network architecture for some useful special cases.

10.3.4 Neural Networks with Piecewise-Polynomial Activation Functions

The general philosophy described above is applied in this subsection to neural networks where the activation function of each neuron is a piecewise-polynomial. The results presented in this section depend on some results from

10.3 Estimates of VC-Dimensions of Families of Networks

algebraic geometry. However, if the central results are "taken on faith," then the derivation of the upper bounds for the VC-dimension is quite straightforward.

We begin with a basic lemma.

Lemma 10.2. ([209], Theorem 2) *Suppose $f_1, \ldots, f_m$ are polynomials of degree at most d in $n \leq m$ variables. Let $S_i \subseteq \mathbb{R}^n$ denote the zero set of f_i, and let $S := \cup_{i=1}^{n} S_i$. Then the complement of S contains at most $(4emd/n)^n$ connected components.*

Proof. As shown in [209], Theorem 2, the number of connected components of the complement of S is bounded (for every n, m) by

$$\sum_{k=0}^{n} 2\,(2d)^n\, 2^k \binom{m}{k} =: \psi(n, m, d),$$

with the convention that $\binom{m}{k} = 0$ if $m < k$. Now suppose $m \geq n$. Then

$$\psi(n, m, d) \leq (4emd/n)^n$$

as shown in the proof of [209], Theorem 3. ∎

A corollary of Lemma 10.2 is given next. Given polynomials $f_1, \ldots, f_m$ as in the lemma, for each $\mathbf{x} \in \mathbb{R}^n - S$ define the sign vector

$$\mathbf{s}(\mathbf{x}) := [\mathrm{sign} f_1(\mathbf{x}) \ldots \mathrm{sign} f_m(\mathbf{x})]^t \in \{-1, 1\}^m.$$

Note that if $\mathbf{x} \notin S$, then $f_i(\mathbf{x}) \neq 0$ for *every* i, so that the sign vector is well-defined.

Lemma 10.3. ([209], Theorem 3) *With all symbols as in Lemma 10.2, the number of distinct sign vectors that can be generated by varying $\mathbf{x}$ is at most $(4emd/n)^n$.*

Proof. This follows from the observation that $\mathbf{s}(\mathbf{x})$ is constant on each connected component of $\mathbb{R}^n - S$. Hence the number of connected components provides an upper bound on the number of distinct sign vectors. ∎

Now a slight modification is made of the estimate in Lemma 10.2 in order to make it applicable to neural networks. With all symbols as above, define

$$\mathbf{b}(\mathbf{x}) := [\mathrm{sign} f_1(\mathbf{x}) \ldots \mathrm{sign} f_m(\mathbf{x})]^t \in \{-1, 0, 1\}^m,$$

where the sign of zero is taken as zero. The difference between $\mathbf{s}(\mathbf{x})$ and $\mathbf{b}(\mathbf{x})$ is that the latter is defined for *all* n-tuples $\mathbf{x}$, even those belonging to the zero set of some f_i, and not just for those $\mathbf{x}$ in $\mathbb{R}^n - S$.

Lemma 10.4. ([76], Corollary 2.1) *Let all symbols be as in Lemma 10.3 above. Then the number of distinct vectors* $\mathbf{b}(\mathbf{x})$ *that can be generated by varying* $\mathbf{x}$ *over* $\mathbb{R}^n$ *is at most* $(8edm/n)^n$.

Proof. Let

$$\mathcal{F} := \{f_1, \ldots, f_m\}, \text{ and } \mathcal{F}' := \{f_1 + \epsilon, f_1 - \epsilon, \ldots, f_m + \epsilon, f_m - \epsilon\},$$

whre ϵ is to be specified later. We use the symbols $\mathbf{b}(\mathbf{x}; \mathcal{F})$ and $\mathbf{s}(\mathbf{x}; \mathcal{F}')$ to make clear which family of functions is under consideration. The claim is that, for ϵ sufficiently small, every sign vector $\mathbf{b}(\mathbf{x}; \mathcal{F})$ corresponds to a unique sign vector $\mathbf{s}(\mathbf{x}; \mathcal{F}')$. Accordingly, suppose $\mathbf{v} \in \{-1, 0, 1\}^m$ equals $\mathbf{b}(\mathbf{x}; \mathcal{F})$ for some $\mathbf{x} \in \mathbb{R}^n$. Let ϵ be smaller than the absolute value of all the *nonzero* values among $f_1(\mathbf{x}), \ldots, f_m(\mathbf{x})$. Then $f_i(\mathbf{x}) \pm \epsilon \neq 0$ for *all* i, so that the sign vector $\mathbf{s}(\mathbf{x}; \mathcal{F}')$ is well-defined. Since $\mathcal{F}'$ contains $2m$ polynomials, applying Lemma 10.3 to $\mathcal{F}'$ leads to the desired estimate. ∎

We are now in a position to derive some useful upper bounds on the VC-dimension of neural networks.

Theorem 10.5. [76] *Suppose a class of neural networks evaluates a formula* $\phi(\mathbf{x}, \mathbf{w})$, *which is a Boolean formula containing up to s atomic formulas of the form*

$$\phi_j(\mathbf{x}, \mathbf{w}) := \{f_j(\mathbf{x}, \mathbf{w}) > 0, \text{ or } = 0, \text{ or } < 0\}, \ 1 \leq j \leq s,$$

where each $f_j(\mathbf{x}, \mathbf{w})$ is a polynomial of degree no larger than d in $\mathbf{w} \in \mathbb{R}^l$ for each $\mathbf{x} \in \mathbb{R}^k$. *Let $\mathcal{H}$ be the family of input-output mappings of the network obtained by varying $\mathbf{w}$ over $\mathbb{R}^l$. Then*

$$\text{VC-dim}(\mathcal{H}) < 2l \lg(8esd). \tag{10.3.4}$$

Proof. The proof is based on the general philosophy described at the end of the preceding subsection. Suppose a set $S = \{\mathbf{x}_1, \ldots, \mathbf{x}_v\}$ is shattered by $\mathcal{H}$. Then the vector $\mathbf{a}(\mathbf{w}) \in \{0, 1\}^{vs}$ defined by (10.3.3) is the set of truth assignments to a set of vs polynomial inequalities (or equalities) in the variables $\mathbf{w} \in \mathbb{R}^l$. Moreover, each polynomial (in)equality is of degree at most d in $\mathbf{w}$. Hence the number of distinct vectors $\mathbf{a}(\mathbf{w})$ that can be generated by varying $\mathbf{w}$ over $\mathbb{R}^l$ is bounded, from Lemma 10.4, by $(8evsd/l)^l$. If a set of cardinality v is shattered by $\mathcal{H}$, then at least 2^v distinct vectors $\mathbf{a}(\mathbf{w})$ must be generated; that is,

$$2^v \leq (8evsd/l)^l, \text{ or equivalently } v \leq l \lg \frac{8evsd}{l}.$$

By Lemma 4.6, this inequality implies that

$$v < 2l \lg(8esd).$$

This completes the proof. ∎

10.3 Estimates of VC-Dimensions of Families of Networks

In the preceding theorem, it is very important to note that d is the degree of the atomic formula *with respect to the weight vector* **w**, and not with respect to the input vector **x**. To illustrate this point, suppose one of the formulas corresponds to a higher-order perceptron of the form

$$\phi_j(\mathbf{x}, \mathbf{w}) := \sum_{i=1}^{k} \sum_{l=1}^{k} w_{in}^{j} x_i x_n \geq 0.$$

In this case, the index d equals one, not two, because $\phi(\mathbf{x}, \mathbf{w})$ is linear in the components of **w**, even though it is quadratic in the components of x.

Theorem 10.5 is stated in terms of the number of atomic formulas and the maximum degree of the polynomial (in)equalities that define these atomic formulas. It would be desirable to recast this bound in a manner analogous to Theorem 10.3, that is, in terms of the number of adjustable parameters in the network. This is done next. The theorem as stated below is a slight improvement over [125], Theorem 2.5.

Theorem 10.6. *Suppose a neural network consists of an interconnection of neurons whose activation functions are piecewise-polynomial. Specifically, suppose each activation function consists of no more than q pieces (that is, the input space is partitioned into no more than q disjoint regions), and that over each region, the output is a polynomial of degree no more than d in the various adjustable parameters. Finally, suppose each neuron is order no more than r, i.e., suppose the shaping (polynomial) function of each neuron has degree no more than r. Let l denote the total number of adjustable parameters in the network, and let t denote the depth of the network. Then the VC-dimension of the family $\mathcal{H}$ of input-output mappings realizable by such a network is bounded by*

$$\text{VC-dim}(\mathcal{H}) \leq 2l \lg(8eq^l(dr)^t) = 2l[\lg(8e) + l \lg q + t(\lg d + \lg r)]. \quad (10.3.5)$$

Proof. The proof is based on Theorem 10.5. Since the input space of each neuron is partitioned into at most q regions, the total number of distinct atomic formulas in the formula evaluated by the network is at most q^l. Moreover, since the longest path from the input to the output contains t edges, the degree of each atomic formula is no more than $(dr)^t$. This is because, in each partitioned region of $\mathbb{R}$, the input-output mapping of each neuron is a composition of a shaping polynomial of degree $\leq r$ and an activation function (polynomial) of degree $\leq d$, and is thus a polynomial of degree $\leq dr$. The longest path from any input node to the output node has length t, so the degree of the overall polynomial (in each partitioned region of the input space) is no larger than $(dr)^t$. Applying the bound (10.3.4) to the problem at hand by substituting $s = q^l$ and replacing d by $(dr)^t$ leads to the estimate

$$\text{VC-dim}(\mathcal{H}) \leq 2l \lg(8eq^l(dr)^t) = 2l[\lg(8e) + l \lg q + t(\lg d + \lg r)],$$

which is the desired result. ∎

The estimate (10.3.5) shows that, if the number of neurons is fixed, as well as the integers q, d, r, then the upper bound is lower for a network of smaller depth. Thus, all other things being equal, an architecture having smaller depth (i.e., fewer hidden layers) can be expected to have a smaller VC-dimension as well.

As in the case of Theorem 10.3, it is natural to ask whether the present bound is the best possible, in terms of the rate of growth of the bound as a function of l, the number of adjustable parameters. As shown next, the answer is "yes" – there exist neural networks whose VC-dimension does indeed grow *quadratically* with respect to the number of adjustable parameters.

Example 10.5. [106] It is shown that, for every integer $n \geq 1$, there exists a neural network with $w = O(n)$ adjustable parameters, consisting of linear and threshold neurons, with VC-dimension n^2. This is achieved as follows: Let n be specified, and define

$$T := \{w = \sum_{i=1}^{n} a_i 2^{-i} : a_i \in \{0,1\}\}.$$

In other words, T consists of all real numbers in $[0, 1)$ whose binary expansion terminates after at most n terms. The weights of the network will be chosen from the set T, so as to shatter the set $S = \{1, \ldots, n\}^2$.

Given a set of weights $\mathbf{w} = (w_1, \ldots, w_n) \in T^n$, let us design a network that computes the function $f_\mathbf{w} : S \to \{0, 1\}$ defined as follows: If $(x, y) \in S$, then $f_\mathbf{w}(x, y)$ is the x-th bit of the number w_y. Clearly, for each Boolean function $f : S \to \{0, 1\}$, there exists a unique n-tuple $\mathbf{w} \in T^n$ such that $f = f_\mathbf{w}$. At the first level of the neural network, define

$$f_\mathbf{w}^1(y) := w_1 + \sum_{z=2}^{n} (w_z - w_{z-1})\, \eta(y - z + 0.5),$$

where $\eta(\cdot)$ is the Heaviside (or step) function. It can be easily verified that $f_\mathbf{w}^1(y) = w_y$. Thus the first-level function $f_\mathbf{w}^1$ "extracts" the weight w_y from the n-tuple $\mathbf{w} = (w_1, \ldots, w_n)$. The first-layer function can be realized using one linear neuron, $(n-1)$ perceptrons, and $3(n-1) + 1$ weights. Next we would like the network to extract the x-th bit of w_y. As a prelude to this step, let us define a network that maps a real number $w \in T$ into its binary representation $(a_1, \ldots, a_n) =: f^2(w)$. Thus the second layer of the network has a single input and n binary outputs. It is well-known that the a_i's can be computed recursively. That is,

$$a_i = \eta \left[2^{i-1} \left(w - \sum_{j=1}^{i-1} a_j 2^{-j} \right) - 0.5 \right].$$

It is clear that the above relationship can be realized by a network consisting of n linear neurons and n perceptrons with $4n$ weights, as shown in Figure

10.3 Estimates of VC-Dimensions of Families of Networks

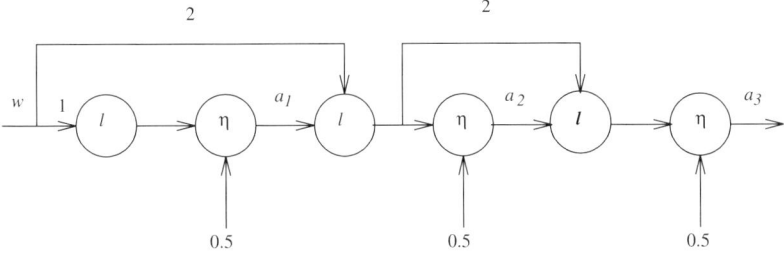

Fig. 10.12. A Neural Network that Computes Binary Coefficients

10.12. Finally, at the third level, given an $x \in \{1, \ldots, n\}$ and the binary representation $w = (a_1, \ldots, a_n)$ as inputs, the network should output a_x. As a first attempt, let us try

$$f^3(x, w) := a_1 + \sum_{z=2}^{n} [a_z\, \eta(x - z + 0.5) - a_{z-1}\, \eta(x - z + 0.5)].$$

However, this is not legal, since the inputs $a_1, \ldots, a_n$ cannot *multiply* the output of an $\eta(\cdot)$ – they can only form an *argument* of an $\eta(\cdot)$. To overcome this difficulty, note that

$$uv = \eta(u + v - 1.5) \text{ for } u, v \in \{0, 1\}.$$

Using this identity, one can define the third-level function as follows:

$$f^3(x, w) := a_1 + \sum_{z=2}^{n} \{\eta[a_z + \eta(x-z+0.5) - 1.5] - \eta[a_{z-1} + \eta(x-z+0.5) - 1.5]\}.$$

This function can be computed by a network with one linear neuron, $4(n-1)$ perceptrons, and $12(n-1) + n$ weights. The overall network is thus defined by

$$f(x, y) := f^3(x, f^2(f^1_{\mathbf{w}}(y))).$$

The realization of this network as given above consists of $n+2$ linear neurons, $n - 1 + n + 4(n-1) = 6n - 5$ perceptrons, and $3n - 2 + 4n + 12n - 11 = 19n - 13$ weights. Clearly the number of weights is $O(n)$ while the VC-dimension is n^2. ∎

The paper [106] contains several other interesting examples of neural networks whose VC-dimension is of quadratic order as a function of the number of weights. In particular, instead of using two different types of neurons (namely: linear neurons and perceptrons), it is possible to use a *single fixed* activation function $\phi(\cdot)$. The idea is that, if $\phi(\cdot)$ is sigmoidal and continuously differentiable with a nonzero derivative at some point, then $\phi(\lambda \cdot)$ looks like a perceptron as $\lambda \to \infty$, and looks like a linear neuron when $\lambda \to 0$.

10.3.5 A General Approach

In this subsection, we present a very general approach due to [95], [96]. This approach involves the application of yet another theorem due to Warren (not the same as Lemma 10.2). Using this approach, it is possible to make a slight refinement of Theorem 10.5. More important, it is possible to provide an *explicit* upper bound for the VC-dimension of neural networks containing the standard sigmoid $\sigma(\cdot)$ defined in (10.1.2). These results improve those in [126], wherein it is shown only that such networks have finite VC-dimension, but no explicit bound is given. It turns out that the standard sigmoid function is an instance of a so-called "Pfaffian" function, and the results presented here apply to any network architecture where the neuron activation functions are so-called "Pfaffian" functions. The general approach is presented in this subsection, while the application of the general result to the case of Pfaffian activation functions is presented in the next subsection.

The set-up is as follows: As before, let k denote the number of inputs to the neural network, and let l denote the number of adjustable parameters. Let $\tau_1, \ldots, \tau_s$ be given C^∞ (infinitely differentiable) functions mapping $\mathbb{R}^{k+l}$ into $\mathbb{R}$, and let $\phi_1, \ldots, \phi_s$ be corresponding atomic formulas, such that $\phi_j(\mathbf{x}, \mathbf{w})$ is of the form $\tau_j(\mathbf{x}, \mathbf{w}) > 0$, or $\tau_j(\mathbf{x}, \mathbf{w}) = 0$, or $\tau_j(\mathbf{x}, \mathbf{w}) < 0$. Suppose the formula computed by the neural network architecture, denoted by $\phi(\mathbf{x}, \mathbf{w})$, is a Boolean formula involving the atomic formulas $\phi_1(\mathbf{x}, \mathbf{w}), \ldots, \phi_s(\mathbf{x}, \mathbf{w})$. For each $\mathbf{w} \in \mathbb{R}^l$, define

$$C_{\mathbf{w}} := \{\mathbf{x} \in \mathbb{R}^k : \mathbb{R} \models \phi(\mathbf{x}, \mathbf{w})\},$$

and let

$$\mathcal{C} := \{C_{\mathbf{w}} : \mathbf{w} \in \mathbb{R}^l\}$$

denote the collection of sets generated by the neural network architecture by varying all the adjustable parameters. Thus the family of hypothesis functions computed by the network consists of the characteristic functions of the various sets $C_{\mathbf{w}}$ as $\mathbf{w}$ varies over $\mathbb{R}^l$. The objective is to derive an upper bound on the VC-dimension of the collection of sets $\mathcal{C}$ in terms of the behaviour of the C^∞ functions $\tau_1, \ldots, \tau_s$.

The central assumption in the subsequent derivation is now stated. In order to state this assumption, the notion of a regular value is introduced first. Suppose r, l are given integers, and that $\mathbf{f} : \mathbb{R}^l \to \mathbb{R}^r$ is a C^∞ mapping. (This means that every component of $\mathbf{f}$ is infinitely differentiable, or equivalently, that every element of the Jacobian matrix ∇f is infinitely differentiable.) Now a vector $\mathbf{y} \in \mathbb{R}^r$ is said to be a **regular value** of the mapping $\mathbf{f}$ if:

- Either $r \leq l$, and $\mathbf{f}^{-1}(\mathbf{y})$ is either empty, or else the matrix $\nabla f(\mathbf{x})$ has full row rank at each $\mathbf{x}$ in the preimage $\mathbf{f}^{-1}(\mathbf{y})$; the latter requirement is sufficient to ensure that $\mathbf{f}^{-1}(\mathbf{y})$ is an $(l-r)$-dimensional submanifold of $\mathbb{R}^r$.

10.3 Estimates of VC-Dimensions of Families of Networks

– Or else $r > l$, and $\mathbf{f}^{-1}(\mathbf{y})$ is empty.

If $\mathbf{y} \in \mathbb{R}^r$ is *not* a regular value of $\mathbf{f}$, then it is said to be a **critical value**. A famous and remarkable theorem due to Sard [170] states very simply that the set of critical values of $\mathbf{f}$ has zero measure in $\mathbb{R}^r$, whatever be the mapping $\mathbf{f}$. In particular, suppose $r \leq l$. Thus Sard's theorem implies that for all $\mathbf{y} \in \mathbb{R}^r$ except for those belonging to a set of measure zero, the preimage $\mathbf{f}^{-1}(\mathbf{y})$ is either empty or or else an $(l-r)$-dimensional submanifold of $\mathbb{R}^r$. Now we are in a position to state:

The Fundamental Assumption: Let $\mathbf{x}_1, \ldots, \mathbf{x}_v \in \mathbb{R}^k$. From the sv functions $\tau_j(\mathbf{x}_i, \cdot) : \mathbb{R}^l \to \mathbb{R}$, choose $r \leq l$ functions, and label them as $\theta_1, \ldots, \theta_r$. Define $\mathbf{f} : \mathbb{R}^l \to \mathbb{R}^r$ by

$$\mathbf{f}(\mathbf{w}) := [\theta_1(\mathbf{w}) \ldots \theta_r(\mathbf{w})]^t \in \mathbb{R}^r.$$

Suppose $\mathbf{y} \in \mathbb{R}^r$ is a regular value of $\mathbf{f}$. (By Sard's theorem, almost all $\mathbf{y}$ are regular values of $\mathbf{f}$.) The assumption is as follows: There is a number B such that, if $\mathbf{f}^{-1}(\mathbf{y})$ is an $(l-r)$-dimensional submanifold of $\mathbb{R}^r$, then $\mathbf{f}^{-1}(\mathbf{y})$ contains no more than B connected components. The key point to note is that the upper bound B is independent of the $\mathbf{x}_i$'s, as well as the integer $r \leq l$ and the choice of the functions $\theta_1, \ldots, \theta_r$ from among the $\tau_j(\mathbf{x}_i, \cdot)$.

Before proceeding further, let us persuade ourselves that the above assumption holds in at least one practically relevant case. Suppose each of the $\tau_j(\mathbf{x}_i, \cdot)$ is a polynomial of degree no larger than d. Then by a theorem of Milnor ([136], Theorem 2), it follows that the number of connected components of $\mathbf{f}^{-1}(\mathbf{y})$ is no larger than $d(2d-1)^l$.[4] This number can be taken as B. However, for convenience in later calculations, let us take $B = (2d)^l$, which is larger than $d(2d-1)^l$.

Now we state the main result of this subsection, from which several specific bounds can be derived. Note that the theorem is a slight refinement of [96], Theorem 2.

Theorem 10.7. [96] *With all symbols as above, we have*

$$\text{VC-dim}(\mathcal{C}) \leq 2 \lg B + 2l \lg(2es).$$

Corollary 10.2. *Suppose each $\tau_j(\mathbf{x}, \mathbf{w})$ is a polynomial in $\mathbf{w}$ of degree no larger than d. Then*

$$\text{VC-dim}(\mathcal{C}) \leq 2l \lg(4eds).$$

The corollary follows at once from the theorem by substituting $B = (2d)^l$. Note that the above bound is slightly better than that given in Theorem 10.5, in that the term $8esd$ is replaced by $4eds$.

The proof of Theorem 10.7 depends on the following result, also due to Warren.

---

[4] Actually, Milnor states the result when $\mathbf{y} = \mathbf{0}$, but the result holds whenever $\mathbf{y}$ is a regular value of the mapping $\mathbf{f}$.

404   10. Applications to Neural Networks

Lemma 10.5. ([209], **Theorem 1**) *Suppose $\mathcal{M}$ is a connected l-dimensional topological manifold, and let $\mathcal{M}_1, \ldots, \mathcal{M}_r$ be connected $(l-1)$-dimensional submanifolds of $\mathcal{M}$ such that the following conditions are satisfied:*

1. *Each $\mathcal{M}_i$ is closed in $\mathcal{M}$.*
2. *The intersection of any $m \leq l$ of the $\mathcal{M}_i$ is either empty, or else is an $(l-m)$-dimensional submanifold of the intersection of any $m-1$ of the m manifolds.*
3. *Any intersection of more than l of the $\mathcal{M}_i$ is empty.*

Let $b_m, 0 \leq m \leq l$, denote the total number of connected components of all possible (nonempty) intersections of any m of the $\mathcal{M}_i$. Then the set $\mathcal{M} - \cup_{i=1}^{r} \mathcal{M}_i$ has at most $\sum_{i=1}^{m} b_m$ connected components.

Proof. (Of the Theorem) Suppose a set $S = \{\mathbf{x}_1, \ldots, \mathbf{x}_v\}$ is shattered by the collection of sets $\mathcal{C}$. Then, for each subset E of $\{1, \ldots, v\}$, there exists a corresponding vector $\mathbf{w}_E \in \mathbb{R}^l$ such that

$$\mathbb{R} \models \phi(\mathbf{x}_i, \mathbf{w}_E) \Leftrightarrow i \in E.$$

There are 2^v such vectors $\mathbf{w}_E$. Now, for a given $\mathbf{w} \in \mathbb{R}^l$, the set $\{\mathbf{x}_i \in S : \mathbb{R} \models \phi(\mathbf{x}_i, \mathbf{w})\}$ depends only on the *signs* of the sv functions $\tau_j(\mathbf{x}_i, \mathbf{w})$. Thus the vector

$$[\mathrm{sign}\,\tau_j(\mathbf{x}_i, \mathbf{w})]^t \in \{-1, 0, 1\}^{sv}$$

must achieve at least 2^v different sign vectors as $\mathbf{w}$ varies over $\mathbb{R}^l$. Thus the proof consists of finding an upper bound for the number of different sign vectors, and showing that it is less than 2^v for large enough v.

As a first step, choose an $\epsilon > 0$ small enough that, if any $\tau_j(\mathbf{x}_i, \mathbf{w}_E) \neq 0$, then $|\tau_j(\mathbf{x}_i, \mathbf{w}_E)| > \epsilon$. Next, choose numbers $\epsilon_{ij} \in (0, \epsilon)$ for $1 \leq i \leq v$, $1 \leq j \leq s$. Let $\mathcal{M} = \mathbb{R}^l$, and choose the $\mathcal{M}_i$ as the zero sets of the functions $\tau_j(\mathbf{x}_i, \cdot) + \epsilon_{ij}$, $\tau_j(\mathbf{x}_i, \cdot) - \epsilon_{ij}$. Thus there are $2sv$ functions, and $2sv$ zero sets in all. Let $\mathbf{e} := [\epsilon_{ij}] \in (0, \epsilon)^{sv}$. Then, by Sard's theorem, one can conclude that for almost all vectors $\mathbf{e}$ except for those belonging to a set of measure zero, the following statements are true:

(i) The intersection of $m \leq l$ of these zero sets is either empty, or else is an $(l-m)$-dimensional submanifold of $\mathbb{R}^l$.
(ii) The intersection of more than l of these zero sets is empty.

For an elaboration of this argument, see [96]. Of course, we cannot directly apply Warren's theorem to these zero sets, since the zero sets need not be connected. Instead, we can apply Lemma 10.5 by letting the $\mathcal{M}_i$ equal the *connected components* of the $2sv$ zero sets.

Finally, in order to apply Warren's theorem, it is necessary to estimate the integers b_m for $1 \leq m \leq l$. (Clearly $b_0 = 1$.) There are $2sv$ zero sets. Moreover, any intersection of sets of the form

$$\{\mathbf{w} \in \mathbb{R}^l : \tau_j(\mathbf{x}_i, \mathbf{w}) = \epsilon_{ij}\} \cap \{\mathbf{w} \in \mathbb{R}^l : \tau_j(\mathbf{x}_i, \mathbf{w}) = -\epsilon_{ij}\}$$

10.3 Estimates of VC-Dimensions of Families of Networks

equals the empty set since $\epsilon_{ij} > 0$. Thus the only way to obtain a nonempty intersection is to choose some m pairs (i, j), then assign arbitrarily plus or minus signs to the corresponding ϵ_{ij}, and then take the corresponding intersections. (The claim is only that *all other* intersections are empty – not that every intersection of the above type is nonempty.) There are $\binom{sv}{m}$ different ways of choosing the m pairs (i, j), and for each choice of these m pairs, there are 2^m different ways of assigning plus or minus signs to the selected ϵ_{ij}. Also, by the fundamental assumption, the intersection of any m of these zero sets contains no more than B connected components. Hence

$$b_m \leq 2^m \binom{sv}{m} B, \text{ for } 1 \leq m \leq l.$$

Since $b_0 = 1$, the inequality also holds for $m = 0$. Now

$$\sum_{m=0}^{l} b_m \leq \sum_{m=0}^{l} 2^m \binom{sv}{m} B \leq 2^l B \sum_{m=0}^{l} \binom{sv}{m}$$

$$\leq 2^l B \left(\frac{esv}{l}\right)^l = B \left(\frac{2esv}{l}\right)^l,$$

where the last inequality follows from Sauer's lemma (Theorem 4.1). Finally, by Warren's theorem, the complement of the union of the various zero sets contains no more than $B(2esv/l)^l$ connected components. Since the sign vector

$$[\tau_j(\mathbf{x}_i, \mathbf{w}_E) + \epsilon_{ij}, \tau_j(\mathbf{x}_i, \mathbf{w}_E) - \epsilon_{ij}]^t \in \{-1, 1\}^{2sv}$$

is constant in each connected component of the complement of the union of these zero sets (since $\epsilon_{ij} < \epsilon$ for all i, j), it follows that the number of different sign vectors achievable by varying E over all subsets of $\{1, \ldots, v\}$ is no larger than $B(2esv/l)^l$. Since the set $\{\mathbf{x}_1, \ldots, \mathbf{x}_v\}$ of cardinality v is shattered by $\mathcal{C}$, it follows that

$$2^v \leq B \left(\frac{2esv}{l}\right)^l,$$

or equivalently,

$$v \leq \lg B + l \lg \frac{2esv}{l}.$$

Now apply Lemma 4.6 with $\alpha = l$, $\beta = 2es/l$, and $\gamma = \lg B$. This leads to the upper bound

$$v < 2 \lg B + 2l \lg(2es).$$

This completes the proof. ∎

10.3.6 An Improved Bound

In deriving Corollary 10.2 from Theorem 10.7, we had used a bound due to Milnor [136] on the number of connected components of a polynomial variety.[5] Actually, we used a slight variation of Milnor's result, whereby the integer B is chosen as $(2d)^n$. This bound, while easy to use, is also quite conservative, since it makes use of only the number of variables and the maximum degree of the various polynomials, but does not use any more detailed information about the structure of the polynomials. In this subsection, we present an alternative upper bound on the number of connected components of a polynomial variety, which takes into account some more detailed information about the polynomials. This improved bound is due to Rojas [166]. The new result sometimes gives an improved bound for the integer B. By virtue of Theorem 10.7, the improved bound on the integer B directly translates into an improved bound on the VC-dimension of a family of neural networks. The improved bound is, *in all cases*, less conservative than the earlier bound of Goldberg and Jerrum [75]. Moreover, it is intuitively appealing, as the improvement can be quantified as the relative entropy of two probability vectors, whose dimension equals the number of layers in the neural network. An example is given to illustrate how the improved bound works.

To state this improved result, a little notation is introduced. Let S_d^n denote the n-dimensional simplex with side d. That is,

$$S_d^n := \{(x_1, \ldots, x_n) \mathbb{R}^n : x_i \geq 0 \, \forall i, \sum_{i=1}^n x_i \leq d\}.$$

Let $\mu_n(\cdot)$ denote the uniform measure on S_1^n, normalized so that $\mu_n(S_1^n) = 1$. It is easy to see that, if $L_n(\cdot)$ denotes the Lebesgue measure on $\mathbb{R}^n$, then $\mu_n(\cdot) = n! L_n(\cdot)$. This is because $L_n(S_1^n) = 1/n!$. This last assertion can be easily proven by induction on n, starting with the observation that

$$L_n(S_1^n) = \int_0^1 dx_n \int_0^{1-x_n} dx_{n-1} \int_0^{1-x_n-x_{n-1}} dx_{n-2} \ldots \int_0^{1-x_n-\ldots-x_2} dx_1.$$

Now the following result is proved in [166].

Theorem 10.8. *Suppose $\tau_1(\mathbf{w}), \ldots, \tau_r(\mathbf{w})$ are polynomials in the k-dimensional vector $\mathbf{w}$. Let $\mathbf{e}_1, \ldots, \mathbf{e}_k$ denote the standard elementary unit vectors in $\mathbb{R}^k$. Further, let V denote the convex hull of the l vectors $\mathbf{e}_1, \ldots, \mathbf{e}_k$, together with the set of all k-tuples $\mathbf{i} = (i_1, \ldots, i_k)$ with the property that $w_1^{i_1} \ldots w_k^{i_k}$ is a monomial of one of the $\theta_j(\cdot)$. Then*

$$B \leq 2^k \mu_k(V) = 2^k \, k! \, L_k(V). \tag{10.3.6}$$

---

[5] Actually, bounds essentially equivalent to those of Milnor were known in the literature prior to the publication of [136]; see for example [154]. See also [182] for related work.

10.3 Estimates of VC-Dimensions of Families of Networks

By blindly assuming that *every* k-tuple with $\sum_{j=1}^{k} i_j \leq d$ occurs in V, we recover the (adjusted) Milnor bound $(2d)^k$.

Next, it is shown that using the bound of Theorem 10.8 leads to improved VC-dimension bounds for a class of neural networks with polynomial activation functions. Let us begin recalling the description of the class of neural networks under study. It is assumed that the network has k real inputs denoted by $x_1, \ldots, x_k$, where the value of k is not important. There are t levels in the network, and at level i there are q_i output neurons; however, at the output layer (level t) there is only a single neuron (see below). Let l_i denote the number of adjustable parameters, or "weights," at level i, and let $l = \sum_{i=1}^{t} l_i$ denote the total number of adjustable parameters. Let $\mathbf{w}_i := (w_{i,1}, \ldots, w_{i,l_i})$ denote the weight vector at level i, and $\mathbf{w} = (\mathbf{w}_1 \ldots \mathbf{w}_l)$ denote the total weight vector. The input-output relationship of each neuron at level i is of the form

$$y_{i,j} = \tau_{i,j}(\mathbf{w}_i, y_{i-1,1}, \ldots, y_{i-1,q_{i-1}}), \; j = 1, \ldots, q_i.$$

where $y_{i,j}$ is the output of neuron j at level i, and $\tau_{i,j}$ is a polynomial of degree no larger than α_i in the components of the weight vector $\mathbf{w}_i$, and no larger β_i in the components of the vectors $y_{i-1,j}$. At the final layer, there is a simple perceptron device following the polynomial activation function.

With this class of neural networks, it is clear that the output will equal one if and only if a polynomial inequality of the form

$$y_t(\mathbf{w}, \mathbf{x}) \geq 0,$$

is satisfied, where $\mathbf{w}$ is the weight vector and $\mathbf{x} = (x_1 \ldots x_k)$ is the input vector. Thus we can apply Theorem 10.7 with $s = 1$. The issue now is to determine the number of connected components B of the polynomial variety defined by $y_t(\mathbf{w}, \mathbf{x}) = \mathbf{y}$.

Now we are in a position to state the main result. To facilitate theorem statement, we introduce some more notation. Define

$$d_t = \alpha_t, \; d_{t-1} = \alpha_{t-1}\beta_t, \ldots, d_i = \alpha_i \prod_{j=i+1}^{t} \beta_j, \; i = 1, \ldots, t-1.$$

Recall that l_i denotes the number of adjustable parameters at level i, and that l denotes the total number of adjustable parameters. Define the probability vectors

$$\mathbf{v} := (l_1/l \ldots l_t/l), \; \mathbf{u} := (d_1/d \ldots, d_t/d),$$

and define the "binary" relative entropy $H(\mathbf{v}|\mathbf{u})$ as

$$H(\mathbf{v}|\mathbf{u}) := \sum_{i=1}^{l} v_i \lg(v_i/u_i).$$

Note that the above is the same as the conventional relative entropy of two probability vectors, except that we use binary logarithms instead of natural logarithms. Following standard convention, we take $0 \lg(0/0) = 0$.

Theorem 10.9. *With the above notation, we have*

$$B \leq 2^k k! \prod_{i=1}^{l} \frac{d_i^{k_i}}{k_i!} \quad (10.3.7)$$

$$\approx (2d)^k 2^{-kH(\mathbf{v}|\mathbf{u})}. \quad (10.3.8)$$

Consequently, the VC-dimension of the neural network architecture is bounded by

$$VC\text{-}dim(\Phi) \leq 2k \lg(4ed) - 2kH(\mathbf{v}|\mathbf{u}). \quad (10.3.9)$$

Remark: The above theorem shows that the reduction in the VC-dimension estimate over that of Theorem 10.5 is precisely $2k$ times the (binary) relative entropy of the two probability vectors (k_i/k) and (d_i/d). Thus if $k_i/k = d_i/d$ for all i, there will not be any reduction at all. In general, the "percentage" reduction equals the ratio $H(\mathbf{v}|\mathbf{u})/(\lg(4ed))$.

Proof. The proof depends on a careful book-keeping of the degree of $y_t(\mathbf{w}, \mathbf{x})$ with respect to the various components of $\mathbf{w}$. From the architecture of the neural network, it is clear that at the first level, each of the $y_{1,j}$ is a polynomial in the components of $\mathbf{w}_1$ of degree no larger than α_1. At the second level, each of the $y_{2,j}$ is a polynomial, whose monomials are of (combined) degree no larger than α_2 in the components of $\mathbf{w}_2$, and of (combined) degree no larger than $\beta_2 \alpha_1$ in the components of $\mathbf{w}_1$. Thus, while each $y_{2,j}$ could have a total degree of $\alpha_2 + \beta_2 \alpha_1$ in the components of $\mathbf{w}_1$ and $\mathbf{w}_2$, the total degree of the monomial terms involving the components of $\mathbf{w}_1$ does not exceed $\beta_2 \alpha_1$, while the total degree of the monomial terms involving the components of $\mathbf{w}_2$ does not exceed α_2. Extrapolating this argument, we see that at the output layer (level t), the single output y_t is a polynomial whose monomials have total degree no larger than $d_t = \alpha_t$ in the components of $\mathbf{w}_t$, no larger than $d_{t-1} = \beta_t \alpha_{t-1}$ in the components of $\mathbf{w}_{t-1}$, and so on. With the d_i's defined as above, the components of each $\mathbf{w}_i$ appear with total degree no larger than d_i. Thus the total degree of y could be as large as $d = \sum_{i=1}^{t} d_i$, but the monomial terms involving the components of $\mathbf{w}_i$ have total degree no larger than d_i. So the set V defined in Theorem 10.8 satisfies the following containment:

$$V \subseteq \prod_{i=1}^{t} S_{d_i}^{k_i}.$$

Because of this containment, it follows that the (Lebesgue) volume of V satisfies

$$L_k(V) \leq \prod_{i=1}^{t} \frac{d_i^{k_i}}{k_i!}.$$

Thus

10.3 Estimates of VC-Dimensions of Families of Networks

$$\mu_k(V) \leq k! \prod_{i=1}^{l} \frac{d_i^{k_i}}{k_i!}.$$

Combining this with the bound (10.3.6) establishes the first estimate (10.3.7).

To prove the second estimate, we use Stirling's approximation. Strictly speaking, the argument below is valid only for "large" values of k_i and d_i. However, the inequality itself is valid for *all* k_i, d_i, and a proof for this general case can be derived, for example, by mimicking that in [54], p. 14. Let us define

$$C := k! \prod_{i=1}^{l} \frac{d_i^{k_i}}{k_i!}$$

Then, by Stirling's approximation, we have

$$\begin{aligned}
\ln C &\approx k \ln k - k + \sum_{i=1}^{l}(-k_i \ln k_i + k_i + k_i \ln d_i) \\
&= \sum_{i=1}^{l} k_i(\ln k - \ln k_i + \ln d_i) \text{ since } \sum_{i=1}^{l} k_i = k \\
&= k \ln d + \sum_{i=1}^{l} k_i(\ln k - \ln k_i + \ln d_i - \ln d) \text{ since } \sum_{i=1}^{l} k_i = k \\
&= k \ln d - k \sum_{i=1}^{l} \frac{k_i}{k} \ln\left(\frac{k_i/k}{d_i/d}\right) \\
&= k \ln d - k \frac{H(\mathbf{v}|\mathbf{u})}{\lg(e)}. \quad (10.3.10)
\end{aligned}$$

Exponentiating both sides shows that

$$C \approx d^k \cdot 2^{-kH(\mathbf{v}|\mathbf{u})}.$$

Now the estimate for the number B of connected components, as derived from Theorem 10.8, becomes

$$B \leq (2d)^k \cdot 2^{-kH(\mathbf{v}|\mathbf{u})}.$$

The VC-dimension estimate (10.3.9) now follows readily from Theorem 10.7. ■

Example 10.6. Consider a network with four inputs, five hidden-layer neurons at the first level and an output-layer neuron. As is common, let us suppose that $\alpha_i = 1$ for all i. This means that all the adjustable parameters *enter linearly* into the corresponding activation function. Suppose $\beta_1 = 2, \beta_2 = 3$. This means that the hidden-layer neurons have quadratic activation functions, whereas the output-layer neuron has a cubic activation function. It remains to specify the integers k_1 and k_2, representing the number of adjustable parameters. Let us assume that practically all of the monomial terms are present in each neural characteristic. Thus it is reasonable to assume $k_1 = 50, k_2 = 20$. Finally, $d_1 = 3, d + 2 = 1$. With these figures, one has

$$\mathbf{v} = [5/7 \ 2/7], \ \mathbf{u} = [0.25 \ 0.75],$$

$$H(\mathbf{v}|\mathbf{u}) \approx 0.684033, \ \lg(4ed) \approx 5.4427, \ \frac{H(\mathbf{v}|\mathbf{u})}{\lg(4ed)} \approx 0.12567.$$

Thus, in this case, the improved bound is roughly 12.5% less conservative.

10.3.7 Networks with Pfaffian Activation Functions

In this subsection, we specialize the general result of Theorem 10.7 to the case where each neural activation function is a so-called "Pfaffian" function. Commonly used activation functions such as the standard sigmoid $1/(1+e^{-z})$ and the inverse tangent function

$$y = \frac{1}{\pi} \tan^{-1} z + \frac{1}{2}$$

are both of this type.

Suppose $f_1, \ldots, f_q : \mathbb{R}^m \to \mathbb{R}$ are continuously differentiable. Then the sequence $\{f_1, \ldots, f_q\}$ is said to be a **Pfaffian chain of length q and degree D** if, for each $j \in \{1, \ldots, q\}$, each of the partial derivatives $\partial f_j / \partial x_i$, $1 \leq i \leq m$, can be expressed as a polynomial of degree no larger than D in the functions $f_1, \ldots, f_q$, and in $x_1, \ldots, x_m$. This abstract definition can be illustrated by a couple of examples. The standard sigmoid

$$\sigma(x) := \frac{1}{1 + \exp(-x)}$$

is a Pfaffian "chain" of length $q = 1$ and degree $D = 2$, since

$$\sigma'(x) = \sigma(x) - \sigma^2(x).$$

Similarly, the chain

$$f_1(x) := \frac{1}{1+x^2}, \ f_2(x) := \frac{1}{\pi} \tan^{-1} x + \frac{1}{2}$$

is a Pfaffian chain of length $q = 2$ and degree $D = 3$, since

$$f_1'(x) = -\frac{2x}{(1+x^2)^2} = -2x f_1^2(x), \text{ and}$$

$$f_2'(x) = \frac{1}{\pi(1+x^2)} = \frac{f_1(x)}{\pi}.$$

10.3 Estimates of VC-Dimensions of Families of Networks

One of the useful features of Pfaffian chains is that, if the functions $\tau_1, \ldots, \tau_s$ described in the preceding subsection are polynomials in $\mathbf{x}, \mathbf{w}$, as well as members of a Pfaffian chain, then the fundamental assumption is satisfied, and moreover, it is possible to find an explicit upper bound B, so that Theorem 10.7 can be applied. The specific result is stated below, and is taken from [96], Section 1.4; in turn, it is based on [103], Example 3.

Lemma 10.6. *Suppose $f_1, \ldots, f_q$ is a Pfaffian chain of degree D in the variables $\mathbf{x}, \mathbf{w}$, and that each τ_j is a polynomial of degree no larger than d in $\mathbf{x}, \mathbf{w}$ and $f_1, \ldots, f_q$. Then the fundamental assumption of the preceding subsection is satisfied with*

$$B = 2^{lq(lq-1)/2}\, d^l\, [l(d+D)]^l\, [l^2(d+D)]^{lq} = [2^{q(lq-1)/2}\, d\, l^{2q+1}\, (d+D)^{q+1}]^l. \tag{10.3.11}$$

Now we can combine the above estimate for B with the results of Theorem 10.7 to obtain explicit estimates for the VC-dimension of a neural network whose activation functions belong to a Pfaffian chain. The network is assumed to have k input nodes and only one output node. Let m denote the total number of nodes in the network. Each node n has associated with it an input-shaping polynomial ψ_n and an activation function α_n. The input-output relationship of the n-th neuron is given by

$$y_n = \alpha_n(z_n),\ z_n = \psi_n(\mathbf{x}; \mathbf{w}; y_r, r \in R_n).$$

Thus the input z_n to the n-th neuron depends on the external inputs $\mathbf{x}$, the adjustable parameters $\mathbf{w}$, and *some* of the other neural outputs $y_r, r \in R_n$. Since the graph describing the network is acyclic, the neurons can always be numbered in such a way that the set R_n (the neurons whose outputs become inputs to neuron n) is a subset of $\{1, \ldots, n-1\}$. At the output node alone, the quantity y is passed through a step (or Heaviside) function, so that the output y' of the network is binary.

All this notation is absolutely standard. Now the additional assumption is that for each n, the activation function α_n equals one of the functions $f_{s(n)}$, where $\{f_1, \ldots, f_q\}$ is a Pfaffian chain of length q and degree D. As a matter of practicality, one normally uses the *same* activation function throughout a network, so it will normally be the case that all α_n will equal f_q, the *last* function in the chain. However, the theory itself does not exclude the possibility of using "intermediate" functions in the chain as activation functions at different neurons. The above assumption is satisfied in the practically important cases where the neuron activation function is the standard sigmoid, or the inverse tangent function. Now let d denote the largest degree of the polynomials ψ (i.e., the maximum "order" of any neuron). As always, let t denote the depth of the network. Finally, let $\mathcal{H}$ denote the collection of input-output relationships of such a network obtained by varying $\mathbf{w}$ over $\mathbb{R}^l$.

Theorem 10.10. *With all symbols as above, we have*

$$VC\text{-}dim(\mathcal{H}) \leq 2l[l(q+1)^2 m^2/2 + \lg d + (2(q+1)m + 1)\lg l$$
$$+ ((q+1)m + 1)\lg((t+1)(d+D)) + \lg(2e)]. \qquad (10.3.12)$$

Remarks Before proceeding to the proof of the theorem, a few remarks may help to put the above bound in perspective. The integers q, d, D are all characteristic of the type of activation function(s) used. The depth t cannot exceed the number m of nodes. Thus the above result shows that

$$VC\text{-}dim(\mathcal{H}) = O(l^2 m^2).$$

In other words, our upper bound is quadratic in the product of the number of adjustable parameters and the number of nodes. Usually l is much larger than m, so the dominant term is l^2 and not m^2. However, since in principle one could construct a network where $m = O(l)$, in the worst case the above upper bound can be $O(l^4)$.

Proof. As stated above, the nodes of the network can be numbered in such a way that z_n depends only y_i for $i < n$ (and of course on $\mathbf{x}, \mathbf{w}$); assume this is done. Now it is shown that the sequence of functions $\{z_n, f_1(z_n), \ldots, f_q(z_n), 1 \leq n \leq m\}$ is a Pfaffian chain of length $(q+1)m$; the degree of this chain is estimated by and by. This claim can be established recursively with respect to the level of the node. At the first level, each z_n depends only on $\mathbf{x}, \mathbf{w}$ and not on any y_r. So

$$\frac{\partial z_n}{\partial x_i} = \frac{\partial \psi}{\partial x_i}, \frac{\partial z_n}{\partial w_i} = \frac{\partial \psi}{\partial w_i}$$

are polynomials in $\mathbf{x}, \mathbf{w}$ of degree no more than $d-1$. Next, since $\{f_1, \ldots, f_q\}$ is a Pfaffian chain of degree D, it follows that

$$\frac{\partial f_s(z_n)}{\partial x_i} = \frac{\partial f_s}{\partial z_n} \cdot \frac{\partial z_n}{\partial x_i}$$

is a polynomial of degree no greater than $d + D - 1$ in $\mathbf{x}, \mathbf{w}, z_n$ and $f_1(z_n), \ldots, f_s(z_n)$. Similar remarks apply to $\partial f_s(z_n)/\partial w_i$. At the λ-th level in the network, we have

$$\frac{\partial z_n}{\partial x_i} = \frac{\partial \psi_n}{\partial x_i} + \sum_{r \in R_n} \frac{\partial \psi_n}{\partial y_r} \cdot \frac{\partial y_r}{\partial x_i}.$$

Since each y_r equals $f_{s(r)}(z_r)$ for some $s(r) \in \{1, \ldots, q\}$, it follows that $\partial z_n/\partial x_i$ is a polynomial of degree no larger than $(\lambda - 1)(d + D - 1)$ in $\mathbf{x}, \mathbf{w}$, and some $f_s(z_r)$ for $r < n$. Similar remarks apply to $\partial z_n/\partial w_i$. In the same way, at level λ,

$$\frac{\partial f_s(z_n)}{\partial x_i} = \frac{\partial f_s(z_n)}{\partial z_n} \cdot \frac{\partial z_n}{\partial x_i}$$

is a polynomial of degree no larger than $D+(\lambda-1)(d+D-1)$ in $\mathbf{x}, \mathbf{w}$, and some $f_t(z_r)$ for $t \leq s$ and $r \leq n$. Thus we conclude that $\{z_n, f_1(z_n), \ldots, f_q(z_n), 1 \leq n \leq m\}$ is a Pfaffian chain of length $\bar{q} := (q+1)m$ and degree

$$\bar{D} := D + (t-1)(d+D-1) \leq t(d+D).$$

Now the number B can be estimated by replacing q and D by $\bar{q}$ and $\bar{D}$ respectively in (10.3.11). This gives, after replacing some terms by larger terms so as to make the formula simpler,

$$\lg B \leq l[l(q+1)^2m^2/2 + \lg d + (2(q+1)m+1)\lg l$$
$$+ ((q+1)m+1)\lg((t+1)(d+D))].$$

(Note that $\bar{q}(l\bar{q}-1)$ has been replaced by $l\bar{q}^2 = l(q+1)^2m^2$, and $d+\bar{D}$ has been replaced by $(t+1)(d+D)$.) To estimate VC-dim($\mathcal{H}$), one observes that the network evaluates a single formula of the type $\tau(\mathbf{x}, \mathbf{w}) \geq 0$, where $\tau(\mathbf{x}, \mathbf{w})$ equals y, the input to the final Heaviside function. It is easy to see that $\tau(\mathbf{x}, \mathbf{w})$ is one of the functions in the Pfaffian chain. Thus the VC-dimension of the collection of sets generaetd by this formula by varying $\mathbf{w}$ over $\mathbb{R}^l$ can be obtained from Theorem 10.7 with $s = 1$, and $\lg B$ bounded as above. This leads to the desired estimate in (10.3.12). ∎

10.3.8 Results Based on Order-Minimality

Up to now, the emphasis has been on deriving *explicit* upper bounds on the VC-dimension of a given collection of sets. In this last subsection, we present an alternative approach that, for the moment at least, can be used only to prove that a given collection of sets *has finite VC-dimension*, without providing an explicit upper bound. Thus, from the standpoint of leading to estimates of the sample complexity in a learning problem, the approach presented in this subsection is not too useful at the moment. On the other hand, the approach is extremely powerful. As new discoveries are made in model theory, it is entirely possible that the approach can also be used to obtain useful explicit upper bounds. Actually, in principle explicit upper bounds can be obtained even with the existing theory, but these bounds tend to be too large to be of any practical use.

Recall the description of first-order predicate logic given in Section 10.3.3. In this setting, what distinguishes one model of the real number system from another is the *set of functions* included in the model. In principle, one also has to the freedom to choose several *relations* on $\mathbb{R}$, but it is customary to restrict these to $=$ and $<$ (and their derivatives such as $\leq$, $>$, etc.).

Definition A model of the real number system is said to be **order-minimal**, or **o-minimal** if, for every integer $l \geq 1$, every $\mathbf{w} \in \mathbb{R}^l$, and every formula $\phi(x, \mathbf{w})$, the set

$$S_\mathbf{w} := \{x \in \mathbb{R} : \mathbb{R} \models \phi(x, \mathbf{w})\}$$

is a finite union of points and intervals with end points in $\mathbb{R} \cup \{-\infty, \infty\}$.

The point to note here is that, while $\mathbf{w} \in \mathbb{R}^l$ can be a vector, $x \in \mathbb{R}$ is a scalar. Thus a model of the real number system is o-minimal if, whatever be the formula $\phi(x, \mathbf{w})$ and the "adjustable parameter vector" $\mathbf{w} \in \mathbb{R}^l$, the set $S_\mathbf{w}$ of x such that $\phi(x, \mathbf{w})$ is true is a finite union of points and intervals, some of which could be semi-infinite intervals.

The importance of o-minimal models arises from the following very general result due to Laskowski; see [113], Corollary 2.5.

Theorem 10.11. *Suppose a model of the real number system is o-minimal. Let $k, l \geq 1$ be arbitrary integers, and let $\phi(\mathbf{x}, \mathbf{w})$ be an arbitrary formula, where $\mathbf{x} \in \mathbb{R}^k$ and $\mathbf{w} \in \mathbb{R}^l$. Let*

$$C_\mathbf{w} := \{\mathbf{x} \in \mathbb{R}^k : \mathbb{R} \models \phi(\mathbf{x}, \mathbf{w})\}, \text{ and } \mathcal{C} := \{C_\mathbf{w} : \mathbf{w} \in \mathbb{R}^l\}.$$

Then $\mathcal{C}$ has finite VC-dimension.

Roughly speaking, Theorem 10.11 states the following: Suppose a model of computation over the real numbers has the feature that, with just a single "input variable" x but an arbitrary number of "adjustable parameter variables" $\mathbf{w}$, the set of x that satisfy any one formula $\phi(x, \mathbf{w})$ is a finite union of points and intervals. Then increasing the number of input variables beyond one still leads to a collection of sets with finite VC-dimension.

In view of the above theorem, there is a clear interest in demonstrating that a model of real computation incorporating several "natural" functions is in fact o-minimal. Suppose that we begin with the bare minimum model, consisting of just two binary functions, namely addition and multiplication. Then it is clear that every polynomial can be defined using just these two functions. Now classical Tarski-Seidenberg elimination theory states that such a model of real computation is o-minimal. In other words, if $\phi(x, \mathbf{w})$ is a polynomial, then the set $S_\mathbf{w}$ satisfies the feature of being a finite union of points and intervals. Now Theorem 10.11 enables us to conclude that a neural network where all activation functions are piecewise polynomials has finite VC-dimension. This is the approach taken in [180]. Of course, Theorem 10.6 leads to an explicit estimate, but that is another matter. Similarly, using recent results in model theory in conjunction with Theorem 10.11, it is possible to show that various other classes of neural networks have finite VC-dimension. Some useful results on model theory are presented in [189], [188]; see also [126], Section 4.2. To state these results, we introduce the notion of "restricted analytic" functions. Suppose $n \geq 1$ is a given integer. A function $f : \mathbb{R}^n \to \mathbb{R}$ is said to be **restricted analytic (RA)** if there exists a number $M < \infty$ and another function $g : \mathbb{R}^n \to \mathbb{R}$ such that (i) g is (real) analytic over some neighbourhood of the hypercube $[-M, M]^n$, and (ii) $f(\mathbf{y}) = g(\mathbf{y})$ if $\mathbf{y} \in [-M, M]^n$, and $f(\mathbf{y}) = 0$ otherwise. Then we have the following result [189], [188]:

Lemma 10.7. *Any model of real computation involving only the exponential function and a set of RA functions is o-minimal.*

Combining Lemma 10.7 and Theorem 10.11 leads to the following observation: Suppose a neural network has the feature that every activation function is definable using only the exponential and some RA functions; this means that the input vector and the output scalar of each neuron satisfy a formula in this model of computation. Then such a network has finite VC-dimension.

Note that a wide variety of activation functions can be defined in the above model. For instance, the standard sigmoid $y = 1/(1 + e^{-z})$ can be defined in this model of computation, since its input and output satisfy the formula

$$(1 + \exp(-z)) \cdot y = 1,$$

which is a valid formula in this model.[6] Thus, using Theorem 10.11, one can conclude that a network where all activation functions are standard sigmoids has finite VC-dimension. Of course, this result is subsumed by Theorem 10.10.

More interestingly, suppose we define the RA functions SIN and COS, which are the same as the familiar sine and cosine functions, but restricted to the interval $[-\pi, \pi]$. Thus

$$\text{SIN}(x) := \begin{cases} \sin x & \text{if } \pi \leq x \leq \pi, \\ 0 & \text{otherwise} \end{cases}$$

and similarly for $\text{COS}(x)$. Then the inverse tangent function is RA-definable, since

$$y = \tan^{-1}(x) \Leftrightarrow y \in [-\pi/2, \pi/2] \text{ and } \text{SIN}(y) = x \cdot \text{COS}(y).$$

Thus any neural network where the activation functions are definable using SIN, COS, exp, and $\tan^{-1}$ has finite VC-dimension. This observation provides an interesting counterpoint to Example 10.4, which shows that if SIN and COS are replaced by their "unrestricted" versions sin and cos, then the VC-dimension can be infinite.

10.4 Structural Risk Minimization

Up to now we have studied the learning problem with a *fixed* neural network architecture. In the present section, we change the problem slightly by focusing on a *family* of architectures $\{\mathcal{H}_n\}$. It is usually the case that the family is "nested," in the sense that $\mathcal{H}_n \subseteq \mathcal{H}_{n+1}$; in other words, every input-output mapping that can be realized by the n-th architecture can also be realized by the $(n+1)$-st architecture. As an illustration, suppose a network has k real inputs and one $[0, 1]$-valued output. Thus every input-output relationship maps $\mathbb{R}^k$ into $[0, 1]$. Suppose now that we define the n-th architecture as

---

[6] Note that $y = \sigma(z)$ is *not* a valid formula in this model, since σ is neither an exponential nor an RA function.

follows: Each of the k inputs is connected to each of n hidden-layer neurons; in turn, the output of each hidden-layer neuron is fed into the output neuron. Let us fix an input-shaping function and an activation function for the hidden-layer and output neurons, for example, the standard sigmoid, with the input-shaping polynomial being just a weighted sum of the inputs minus a threshold. Finally, let $\mathcal{H}_n$ denote the set of input-output mappings achievable by such a network using n hidden-layer neurons. Then it is easy to see that $\mathcal{H}_n \subseteq \mathcal{H}_{n+1}$, and the family is nested. Strictly speaking, the theory presented below does not require a nested family of hypothesis classes; but the underlying issues become a little clearer if the family is nested.

A serious problem often encountered in neural network design is that of "over-fitting." Roughly speaking, the problem can be described as follows: Suppose one is given a data set of fixed finite length. In the language of model-learning, suppose one has available a data set $(x_1, y_1), \ldots, (x_m, y_m)$. The aim is to choose an input-output mapping $h \in \mathcal{H}_n$ in such a way that, with high probability, h is nearly the best possible fit to the data in $\mathcal{H}_n$. In the case where the integer n (and thus the hypothesis class $\mathcal{H}_n$) is *fixed*, this is the standard model-learning problem discussed in Section 10.2. However, it often happens in neural network design that the class of hypothesis functions is *not* fixed, and that the network designer has the freedom to *change* (usually, to *enlarge*) the class of hypothesis functions. Invariably, what happens is this: In the notation of Section 10.2, let

$$\mathbf{x} := [x_1 \ldots x_m]^t, \ \mathbf{y} := [y_1 \ldots y_m]^t, \ \mathbf{z} := (\mathbf{x}, \mathbf{y})$$

denote the data set. Let $\ell(y, u)$ denote the loss function, e.g., $\ell(y, u) := |y - u|$. Finally, for each $h \in \mathcal{H}_n$, let

$$\hat{J}(h; \mathbf{z}) := \frac{1}{m} \sum_{i=1}^{m} \ell(y_i, h(x_i))$$

denote the empirical risk, and let

$$\hat{J}_n^*(\mathbf{z}) := \inf_{h \in \mathcal{H}_n} \hat{J}(h; \mathbf{z})$$

denote the minimum achievable empirical risk. Note that a subscript "n" is added to $\hat{J}^*(\mathbf{z})$ to remind us that the infimum is over the class $\mathcal{H}_n$. As shown in Section 10.2, it is a good strategy to choose h so that $\hat{J}(h; \mathbf{z})$ is close to the infimum $\hat{J}_n^*(\mathbf{z})$. The process of choosing such an h can be called "empirical risk minimization," and various standard methods of neural network design (e.g., the popular backpropagation method [168], [169]) aim to achieve precisely this. In other words, most currently popular methods of neural network design aim to achieve *the best possible fit to the data set*, using a member of the specified hypothesis class. To avoid unnecessary technicalities, suppose that the infimum above is in fact a minimum, and that h is chosen such that $\hat{J}(h; \mathbf{z})$ equals $\hat{J}_n^*(\mathbf{z})$.

10.4 Structural Risk Minimization

Now suppose that, for a given value of the integer n, the network designer deems that the number $\hat{J}_n^*(\mathbf{z})$ is unacceptably large. In other words, the hypothesis class $\mathcal{H}_n$ is not rich enough to provide a reasonable fit to the data set. In such a case, it is natural for the designer to enlarge the hypothesis class so as to reduce the error in fitting the data. For instance, in the example discussed above, if n hidden-layer neurons are not sufficient to provide a good enough fit to the data, it is but natural to introduce one more hidden-layer neuron, and retrain the network. If $\mathcal{H}_{n+1} \supseteq \mathcal{H}_n$, then the minimum empirical error $\hat{J}_{n+1}^*(\mathbf{z})$ is at worst equal to $\hat{J}_n^*(\mathbf{z})$, and is usually less. Thus, on the *training* data, one is sure to obtain better and better performance by making the hypothesis class richer and richer. However, it is often noted in practice that, on a randomly chosen *test input*, beyond a certain point the performance of the trained network often *deteriorates* as the hypothesis class is made richer and richer. Ultimately one reaches a situation where the trained network can reproduce the training data nearly perfectly, but fails miserably on test inputs. In effect, the network merely "memorizes" the data. This phenomenon is known as "overfitting."

It is possible to give an intuitive mathematical explanation of this phenomenon. To focus the discussion, suppose the integer n refers to some quantity of physical significance, such as the number of hidden-layer neurons in the example above. Then, as n is increased, the number of network parameters to be adjusted becomes larger and larger. With a data set of a fixed size, the estimate that one gets of this ever-increasing set of parameters becomes poorer and poorer.

Thus it is natural to ask whether there exists an "optimal" choice of the integer n, and if so, how it is possible to make an optimal choice in a *systematic* manner. The method of structural risk minimization, described in [190], [191], provides a means of doing this.

Suppose an integer m is given, as well as a random data set $(x_1, y_1), \ldots, (x_m, y_m)$ of length m. Fix a confidence level δ to which the method of structural risk minimization is to be applied. Suppose each hypothesis class $\mathcal{H}_n$ has finite P-dimension, and let $d_n \geq \text{P-dim}(\mathcal{H}_n)$; thus d_n can be just an *upper bound* for P-dim($\mathcal{H}_n$). Finally, suppose there exists a finite constant μ such that ℓ satisfies the uniform Lipschitz condition

$$|\ell(y, u_1) - \ell(y, u_2)| \leq \mu |u_1 - u_2|, \ \forall y, u_1, u_2.$$

Now we are in a position to describe the method. Let ϵ_n denote the largest solution of the equation

$$8 \left(\frac{16e}{\epsilon} \ln \frac{16e}{\epsilon} \right)^{d_n} \exp(-m\epsilon^2/32) = \delta,$$

where δ is the selected confidence level. Now Theorem 10.1 implies that

$$|J(h, P) - \hat{J}(h; \mathbf{z})| \leq \epsilon_n \ \forall h \in \mathcal{H}_n, \text{ and } \forall \mathbf{z} \text{ with confidence } \geq 1 - \delta.$$

In particular,
$$J(h, P) \le \hat{J}(h; \mathbf{z}) + \epsilon_n, \ \forall h \in \mathcal{H}_n, \ \text{and} \ \forall \mathbf{z} \ \text{w.p.} \ge 1 - \delta.$$

Now suppose h_n is selected such that
$$\hat{J}(h_n; \mathbf{z}) = \hat{J}_n^*(\mathbf{z}) := \inf_{h \in \mathcal{H}_n} \hat{J}(h; \mathbf{z}).$$

As mentioned above, we are simplifying the problem somewhat by assuming that the infimum on the right side is actually achieved; this assumption can be removed at the expense of more cumbersome notation. Then, with confidence at least $1 - \delta$, it can be assumed that
$$J_n^*(P) := \inf_{h \in \mathcal{H}_n} J(h, P) \le J(h_n, P) \le \hat{J}(h_n; \mathbf{z}) + \epsilon_n = \hat{J}_n^*(\mathbf{z}) + \epsilon_n.$$

This inequality means the following: Given a random data set $\mathbf{z}$ of length m, suppose we compute the minimum empirical error $\hat{J}_n^*(\mathbf{z})$. Then we can assert with confidence at least $1 - \delta$ that the unknown quantity $J_n^*(P)$ is no larger than $\hat{J}_n^*(\mathbf{z}) + \epsilon_n$. Ideally we would like to choose the integer n so as to minimize $J_n^*(P)$; unfortunately this is not possible, since $J_n^*(P)$ is unknown. As a next best option, we can minimize a *known upper bound* for $J_n^*(P)$; that is, the "optimal" choice of n is the one that minimizes the sum $\hat{J}_n^*(\mathbf{z}) + \epsilon_n$. This is the method of structural risk minimization.

The method as described above brings out clearly the trade-off between the richness of the hypothesis class and the achievable performance. In the natural case of nested hypothesis classes, we have that $\mathcal{H}_n \subseteq \mathcal{H}_{n+1}$ for all n, which usually implies that $d_{n+1} \ge d_n$ (this would be the case if d_n were the *exact* P-dimension, rather than just an upper bound). As a result, $\epsilon_{n+1} \ge \epsilon_n$ for all n. Thus the uncertainty in our estimate of $J_n^*(P)$ *increases* with n. On the other hand, the minimal empirical risk $\hat{J}_n^*(\mathbf{z})$ *decreases* with n, because the infimum is taken over larger and larger hypothesis classes. Hence, by minimizing the sum $\hat{J}_n^*(\mathbf{z}) + \epsilon_n$ with respect to n, it is possible to select an "optimal" choice of n. Note that the optimal choice of n depends on the length of the data set m, as well as the confidence level δ. If δ is kept fixed, then the optimal value of n usually increases as m becomes larger, but more slowly than m.

See [133] for a case study in structural risk minimization.

Notes and References As soon as the connection between the VC-dimension of a neural network architecture and its generalization ability became well-understood, several researchers interested in neural networks started investigating bounds on the VC-dimension. Theorem 10.3 was derived as far back as 1968, as an upper bound on the "capacity" of a neural network. It is interesting that the same bound proves useful in the context of assessing the generalization ability of a neural network. Maass [122] shows that the bound in this theorem is the best possible in terms of the rate of growth.

The problem of estimating the VC-dimension of subsets of a Euclidean space $\mathbb{R}^k$ is also well-established in statistics, and several researchers have tackled the problem using several different approaches. Interestingly, the same problem also appears in model theories of the real number system, in a different guise. Specifically, a first-order formula $\phi(\mathbf{x}, \mathbf{w})$ over the reals is said to have the "independence property" if, in the present terminology, the associated collection of sets $\{S_\mathbf{w}\}$ has infinite VC-dimension. The paper of Shelah [173] cited by Sauer [171] actually addresses the independence property. Laskowski [113] makes a vital connection between order-minimality of a particular model of the real number system and the finiteness of the VC-dimension of collections of sets defined in such a model. Stengle and Yukich [180] use classical Tarski-Seidenberg elimination theory together with a result of Dudley [62] to show that a collection of sets defined by a finite set of polynomial inequalities has finite VC-dimension; but their bounds grow rather too quickly to be of practical use. Goldberg and Jerrum [75], [76] derive less conservative bounds than Stengle and Yukich. The application of the results of Goldberg and Jerrum to derive bounds on the VC-dimension of networks with polynomial activation functions is given by Maass [125]. Koiran and Sontag [106] show that the bound given by Maass is the best possible in terms of the rate of growth. The general approach to bounding the VC-dimension in terms of the connected components of a "generic" preimage is due to Karpinski and Macintyre [95], [96]. In specific situations, they use bounds on the number of connected components derived by Warren [209] and Khovanski [103]. Since the problem of bounding the number of connected components of a set is well-studied in algebraic topology, it seems likely that as time goes on more and more such connections will be made. Similarly, now that the connection to model theory of real numbers has been made, it is reasonable to expect that many more papers will appear that exploit this connection. See for example the recent papers by Sontag [177], [178].

The fact that the "loading problem" for neural networks is sometimes intractable, with all its attendant implications to the learning of neural networks, is brought out by Judd [89] and Blum-Rivest [31]. These results apply only to the case where the number of inputs to the network approaches infinity. It is not yet clear how tractable the loading problem is when the number of neurons (as opposed to the number of inputs) becomes unbounded.

The technique of structural risk minimization is introduced by Vapnik in [190].

There are several recent books that address the topic of neural network learning from the perspective of statistical learning theory. The interested reader is referred to [7, 69] for treatments that are quite similar to the present discussion.

Finally, one of the most important advances in the field of neural network learning is the introduction of support vector machines (SVMs). This topic is not discussed here due to reasons of space. The interested reader is referred to [46, 49, 172].

11. Applications to Control Systems

In this chapter, we examine the applications of learning theory to two specific problems in control system theory. First, it is shown that the methods of the preceding chapters can be used to derive "efficient" (i.e., polynomial-time) randomized algorithms for solving various problems in robust control whose exact solution is NP-hard. Second, it is shown that the problem of identifying an unknown system on the basis of input-output measurements can be recast as a problem in learning theory. Using the results of the previous chapters, quantitatively precise estimates are derived for the rate at which the estimated model converges to the true system or to its best approximation within the given model class.

11.1 Randomized Algorithms for Robustness Analysis

In recent years, there has been considerable interest in studying the computational complexity of various problems in robust controller synthesis. Many of the known results are negative, in the sense that several problems of interest are known to be NP-complete or NP-hard. Thus it is intractable to find an *exact* solution to these problems. In the face of these negative results, one is forced to change one's definition of a "solution," and search for methods that nearly work, most of the time. The approach of using a *randomized* algorithm is a natural mathematical formulation of this idea. It turns out that, whenever the performance conditions that need to be satisfied can be formulated in terms of a finite number of polynomial inequalities, there exists an efficient (i.e., polynomial-time) randomized algorithm for that particular problem. Since many of the commonly studied problems in robust controller synthesis can be formulated in terms of a finite number of polynomial constraints, the conclusion is that the randomized approach is well-suited to a wide class of problems in robust control.

11.1.1 Introduction to Robust Control

One of the central problems in control theory is that of designing a fixed controller for a given *family* of plants, in such a way that the fixed controller works reasonably well for the entire family of plants. This problem is usually referred to as "robust controller synthesis." In this subsection, a brief review is given of this problem, and some of the known results that are germane to the present treatment.

At an abstract level, the problem can be stated as follows: Suppose one is given a family of plants, denoted by $\{G(x), x \in X\}$, and a family of controllers, denoted by $\{K(y), y \in Y\}$.[1] Here x denotes a set of plant parameters, such as the pole-zero locations, coefficients of the numerator and denominator of the plant transfer matrix, elements of the state-space description of the plant, and the like. The symbol X denotes the set of values assumed by the plant parameter vector x. Similarly, y denotes the vector of controller parameters, and Y denotes the set of possible controller parameters. If the plants are continuous-time, then $G(x)$ is actually $G(x, s)$ where s is the Laplacian variable, and if the plant is discrete-time, then $G(x)$ is actually $G(x, z)$ where z is the transform of the unit delay. As with $G(x)$, so too is $K(y)$ either $K(y, s)$ or $K(y, z)$ depending on the situation.

Of the many demands that can be made on a controller, two are of particular interest here, namely:

– Robust stabilization, and
– Robust performance

The robust stabilization problem can be stated as follows: Given the families $\{G(x) : x \in X\}$ and $\{K(y) : y \in Y\}$, find if possible a controller $K(y_0)$ that stabilizes *every* plant $G(x), x \in X$. The notions of stability and stabilization are not defined here, as they are well-known in control theory. The interested reader is referred to any standard source, e.g., [91], [198], [70], or [13]. Now the above problem can be further divided into two parts, namely: controller *analysis*, and controller *synthesis*. In the analysis problem, one is given a "candidate" controller $K(y_0)$, however generated, and the objective is to determine whether $K(y_0)$ does indeed stabilize $G(x)$ for every $x \in X$. The objective of the synthesis problem is either to find a controller that achieves robust stabilization, or else to demonstrate conclusively that *no* controller in the given family $\{K(y) : y \in Y\}$ can achieve robust stabilization.

For some models of plant uncertainty, both the analysis and the synthesis problems are tractable. For example, consider the case of additive sector uncertainty. To introduce this and other uncertainty models, a little notation is introduced first. In the interests of simplicity, all discussion is confined to continuous-time systems. The Hardy space H_∞ consists of all functions $f(s)$ such that (i) $f(\cdot)$ is analytic over the open right half-plane $\{s : \text{Re } s > 0\}$, and (ii) $f(\cdot)$ is essentially bounded over the closed right half-plane $\{s : \text{Re } s \geq 0\}$. The Hardy space H_2 consists of the Laplace transforms of all functions in

---

[1] In the control theory literature, the symbols P for the plant and C for the controller are used nearly universally. However, in the present book, both of these symbols are used to represent other entities. Thus G is used here for the plant and K for the controller.

$L_2[0,\infty)$; equivalently, H_2 consists of all functions $f(s)$ such that (i) $f(\cdot)$ is analytic over the open right half-plane $\{s : \text{Re } s > 0\}$, and (ii)

$$\sup_{\sigma > 0} \int_{-\infty}^{\infty} |f(\sigma + j\omega)|^2 \, d\omega < \infty.$$

The norms on H_∞ and H_2 are defined as follows:

$$\| f \|_\infty := \operatorname*{ess.\,sup}_{\omega \in \mathbb{R}} |f(j\omega)|,$$

or equivalently,

$$\| f \|_\infty := \operatorname*{ess.\,sup}_{\text{Re } s \geq 0} |f(s)|.$$

Next,

$$\| f \|_2 := \left[\frac{1}{2\pi} \int_{-\infty}^{\infty} |f(j\omega)|^2 \, d\omega \right]^{1/2}.$$

Now let us extend these norms to vector- and matrix-valued functions. First, if A is a matrix whose entries are complex numbers, then $\bar{\sigma}(A)$ denotes the largest singular value of A. Next, if $A(\cdot)$ is a matrix-valued function such that each element of $A(\cdot)$ belongs to H_∞, then

$$\| A \|_\infty := \operatorname*{ess.\,sup}_{\omega \in \mathbb{R}} \bar{\sigma}[A(j\omega)].$$

Along similar lines, if $\mathbf{f}(\cdot)$ is a vector-valued function such that each component of $\mathbf{f}(\cdot)$ belongs to H_2, then

$$\| \mathbf{f} \|_2 := \left[\frac{1}{2\pi} \int_{-\infty}^{\infty} [\mathbf{f}^*(j\omega)\mathbf{f}(j\omega)] \, d\omega \right]^{1/2},$$

where $\mathbf{f}^*(j\omega)$ denotes the conjugate transpose of $\mathbf{f}(j\omega)$. Note that if $A(\cdot)$ and $\mathbf{f}(\cdot)$ have compatible dimensions, then

$$\| A\mathbf{f} \|_2 \leq \| A \|_\infty \cdot \| \mathbf{f} \|_2.$$

See [198] and [70] for further background material and additional information.

Now let us return to the tractability of the robust stabilization problem. Suppose $G_0(s)$ represents a nominal plant, which could be multi-input and/or multi-output. It is assumed that $G_0(\cdot)$ is a rational matrix, that is, a matrix whose elements are all rational functions of s. It is *not* assumed that $G_0(\cdot)$ is stable, but it *is* assumed that $G_0(\cdot)$ does not have any poles on the $j\omega$-axis. Let $r(\cdot) \in H_\infty$ be a given rational function. Define the family of plants $\{G(x) : x \in X\}$ to consist of all rational matrices $G(\cdot)$ of the same dimensions as $G_0(\cdot)$ such that (i) $G(\cdot)$ has the same number of unstable poles as $G_0(\cdot)$, and (ii)

$$\bar{\sigma}[G(j\omega) - G_0(j\omega)] \leq |r(j\omega)|, \; \forall \omega \in \mathbb{R}.$$

This is referred to as the "additive uncertainty model." In this instance, X denotes the set of all permissible additive perturbations.

Now suppose $K(\cdot)$ is a rational matrix whose dimensions are complementary to those of $G_0(\cdot)$; this means K has the same dimensions as G_0^t. Then it is known that $K(\cdot)$ stabilizes every plant $G(\cdot)$ in the additive uncertainty model if and only if (i) $K(\cdot)$ stabilizes $G_0(\cdot)$, and (ii)

$$\| [I + KG_0]^{-1} Kr \|_\infty < 1.$$

See [59] or [198], p. 273 for the proof. Moreover, given a rational matrix $A(\cdot)$ whose elements all belong to H_∞, it is a routine matter to verify whether or not $\| A \|_\infty < 1$ using the algorithm of [33]. Hence the *analysis* problem of robust stabilization is quite tractable. As for the synthesis problem, it is known ([208], [198], p. 285) that a robustly stabilizing controller exists if and only if

$$\inf_{R \in M(H_\infty)} \| N(X + R\tilde{D})r \|_\infty < 1, \qquad (11.1.1)$$

where $M(H_\infty)$ denotes the set of matrices with elements in H_∞ of appropriate dimensions; (N, D) and $(\tilde{D}, \tilde{N})$ are respectively a right-coprime factorization and a left-coprime factorization over H_∞ of the nominal plant transfer matrix $G_0(\cdot)$; and $X, Y \in M(H_\infty)$ satisfy $XN + YD = I$. Moreover, using the methods of [58], it is now feasible to determine (i) whether or not (11.1.1) is satisfied, and (ii) if so, to find an $R \in M(H_\infty)$ such that $\| N(X + R\tilde{D})r \|_\infty < 1$. Once such an R is found, $K := (Y - R\tilde{N})^{-1}(X + R\tilde{D})$ is a robustly stabilizing controller. Hence the *synthesis* problem of robust stabilization is also tractable, in the case of the additive uncertainty model. Similar statements apply to other types of uncertainty models, such as the multiplicative and the stable-factor uncertainty models; the reader is referred to [198], Chapter 7 or [70] for further details.

11.1.2 Some NP-Hard Problems in Robust Control

In the preceding subsection we have seen that both the analysis problem and the synthesis problem of robust stabilization are tractable if the uncertainty is additive or multiplicative. The situation is substantially different if the nature of the plant uncertainty is changed to the so-called "structured perturbation" model described next. Consider the feedback system shown in Figure 11.1, where G_0 is the nominal plant, K is the controller, and Δ is the perturbation. By incorporating the controller into the nominal plant, one can recast the system as the feedback system shown in Figure 11.2, where $M = G_0(I + KG_0)^{-1}$. The allowable perturbations consist of three types: real structured, complex structured, and complex unstructured. Specifically, define

$$\mathcal{D} := \{\mathrm{Diag}[a_1 I_{n_1}, \ldots, a_r I_{n_r}, b_1 I_{l_1}, \ldots, b_c I_{l_c}, A_1, \ldots, A_n]\},$$

where $a_i \in \mathbb{R}$ for $1 \leq i \leq r$, $b_j \in \mathbb{C}$ for $1 \leq j \leq c$, and $A_1, \ldots, A_n$ are unrestricted complex matrices. For each real number $\gamma > 0$, define

11.1 Randomized Algorithms for Robustness Analysis

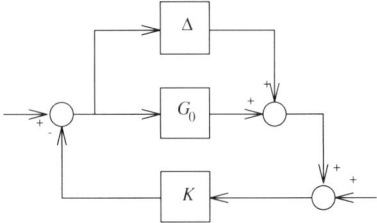

Fig. 11.1. A Perturbed Feedback System

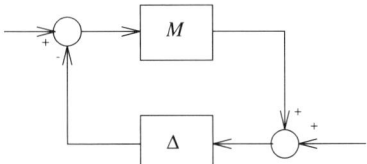

Fig. 11.2. Equivalent Perturbed Feedback System

$$\mathcal{D}_\gamma := \{\Delta \in \mathcal{D} : \bar{\sigma}(\Delta) \leq \gamma\}.$$

A *robustness measure* γ_{opt} for the closed-loop system can be defined as follows: γ_{opt} is the supremum of all γ such that the feedback system is stable for all $\Delta \in \mathcal{D}_\gamma$. Equivalently, γ_{opt} is the smallest value of γ for which there exists a $\Delta \in \mathcal{D}_\gamma$ such that the feedback system is unstable. In principle, γ_{opt} can be found by computing the corresponding structured singular value [57], [156]. Unfortunately, computing the structured singular value *exactly* is often NP-hard [34]. To be more precise, let us restate the problem as a decision problem: Given a number $\gamma \in \mathbb{R}$, is it true that $\gamma_{\text{opt}} \leq \gamma$? Finding the answer to this question is NP-hard if the overall dimension of the matrix Δ is taken as the size parameter. Thus, finding the *exact* value of γ_{opt} is NP-hard as the size of the system increases. This is true for both real and complex structured singular values; see [34, 184]. Moreover, it is NP-hard even to compute an approximation for γ_{opt}, in the following sense: Suppose a "tolerance level" ϵ is fixed, and a constant γ is given. It is NP-hard to determine whether or not $\gamma \leq (1+\epsilon)\gamma_{\text{opt}}$; see for example [48]. Now there are some upper bounds available for the structured singular value. However, the gap between these upper bounds and the actual structured singular value grows without bound as the size of the matrix increases; see [132, 185].

Note that the difficulty is caused solely by the "structured" nature of the perturbations Δ. If $r = c = 0$ and $n = 1$, so that $\mathcal{D}_\gamma$ consists of *all* complex matrices Δ with $\bar{\sigma}(\Delta) \leq \gamma$, then $\gamma_{\text{opt}} = [\|\, M\, \|_\infty]^{-1}$, which can be computed to arbitrarily small accuracy. Moreover, determining whether or not there exists a controller K such that $\|\, M\, \|_\infty \leq \alpha$ for a given α is a standard problem in H_∞-optimization theory. Another NP-hard problem is the following: Given a controller K, compute the maximum real part of the closed-loop poles as Δ varies over a given set $\mathcal{D}_\gamma$. Specifically, given $\gamma > 0$, define $\lambda_{\max}(\gamma)$ to be the smallest number x_0 such that, for every $\Delta \in \mathcal{D}_\gamma$, the poles of the closed-loop system all have real parts less than or equal to x_0. For a discussion of these and other problems, see [161] and [150].

Several more NP-hard problems arise out of so-called "interval matrices." Given an integer n, suppose one is given rational numbers $\alpha_{ij}, \beta_{ij}, 1 \leq i, j \leq n$, such that $\alpha_{ij} \leq \beta_{ij}$ for all i, j. For simplicity, let $\mathbf{z}$ denote the $2n^2$-tuple $(\alpha_{ij}, \beta_{ij}), 1 \leq i, j \leq n$. Then the **interval matrix** corresponding to the parameter vector $\mathbf{z}$ is denoted by $\mathbf{A_z}$ and is defined by

$$\mathbf{A_z} := \{A \in \mathcal{Q}^{n \times n} : \alpha_{ij} \leq a_{ij} \leq \beta_{ij}, \; \forall i, j\},$$

where $\mathcal{Q}$ denotes the set of rational numbers. Thus the interval matrix $\mathbf{A_z}$ is just the set of rational matrices whose ij-th element lies in the interval $[\alpha_{ij}, \beta_i]$. The set of all symmetric matrices in $\mathbf{A_z}$ is denoted by $\mathbf{A}_{s,\mathbf{z}}$.

With the above definitions, the following problems are NP-hard.

1. Determine whether every matrix $A \in \mathbf{A_z}$ is stable (i.e., whether all eigenvalues of every $A \in \mathbf{A_z}$ have negative real parts).
2. Determine whether every matrix in $\mathbf{A_z}$ has norm bounded by a given number (which can be taken as one, without loss of generality).
3. Determine whether every $A \in \mathbf{A_z}$ is nonsingular.
4. Determine whether every $A \in \mathbf{A}_{s,\mathbf{z}}$ is positive definite.

11.1.3 Randomized Algorithms for Robustness Analysis

In view of the difficulty of computing γ_{opt} or $\lambda_{\max}(\gamma)$ either exactly or approximately to within a specified tolerance, a probabilistic approach to such problems has recently gained popularity [165], [128], [181], [101]. Most of these probabilistic methods can be viewed as a variant of the following abstract search problem, which can be termed "estimating the supremum of a function": Suppose $(X, \mathcal{S}, P)$ is a probability space, and that $f : X \to \mathbb{R}$ is a random variable (i.e., a measurable function on $(X, \mathcal{S})$). Let $x_1, \ldots, x_m \in X$ be i.i.d. samples drawn according to P, and define

$$\bar{f}(\mathbf{x}) := \max_{1 \leq i \leq m} f(x_i).$$

Note that $\bar{f}(\cdot)$ is a random variable defined on X^m. One can think of $\bar{f}(\mathbf{x})$ as an "empirical estimate" of the supremum of f. However, it is a slightly different type of empirical estimate from those in Chapter 3 for example, in that no claim is made that $\bar{f}(\mathbf{x})$ is anywhere close to the real supremum of $f(\cdot)$; indeed, $f(\cdot)$ can even be an unbounded function. Rather, the claim is that the set $\{y \in X : f(y) > \bar{f}(\mathbf{x})\}$ has small measure, with high probability. In other words, $f(\cdot)$ is bounded by $\bar{f}(\mathbf{x})$ over "most" of X, with high probability. The next lemma, which makes the result precise, is the basis for the remainder of this subsection.

11.1 Randomized Algorithms for Robustness Analysis

Lemma 11.1. *Suppose $(X, \mathcal{S}, P)$ is a probability space, and that $f : X \to \mathbb{R}$ is a random variable. Let $x_1, \ldots, x_m \in X$ be i.i.d. samples drawn according to P, and define*

$$\bar{f}(\mathbf{x}) := \max_{1 \leq i \leq m} f(x_i).$$

Then

$$P^m \{\mathbf{x} \in X^m : P\{y \in X : f(y) > \bar{f}(\mathbf{x})\} > \epsilon\} \leq (1 - \epsilon)^m. \qquad (11.1.2)$$

Proof. Define the distribution function of the random variable f in the familiar manner, namely: For each $a \in \mathbb{R}$, let

$$r(a) := P\{y \in X : f(y) \leq a\}.$$

Then the function $r(\cdot)$ is right-continuous; see, e.g., [74], p. 162. Given $\epsilon > 0$, define

$$a_\epsilon := \inf\{a : r(a) \geq 1 - \epsilon\}.$$

By the right-continuity of $r(\cdot)$, it follows that $r(a_\epsilon) \geq 1-\epsilon$. Also, by definition, $r(a) < 1 - \epsilon$ if $a < a_\epsilon$. Now suppose $\bar{f}(\mathbf{x}) \geq a_\epsilon$; then

$$P\{y \in X : f(y) \leq \bar{f}(\mathbf{x})\} = r[\bar{f}(\mathbf{x})] \geq 1 - \epsilon.$$

As a result,

$$P\{y \in X : f(y) > \bar{f}(\mathbf{x})\} = 1 - r[\bar{f}(\mathbf{x})] \leq \epsilon.$$

Taking the contrapositive shows that

$$P\{y \in X : f(y) > \bar{f}(\mathbf{x})\} > \epsilon \Rightarrow \bar{f}(\mathbf{x}) < a_\epsilon \Rightarrow r[\bar{f}(\mathbf{x})] < 1 - \epsilon.$$

Now $\bar{f}(\mathbf{x}) < a_\epsilon$ if and only if $f(x_i) < a_\epsilon$ for $i = 1, \ldots, m$. Each of these m events is independent of the rest, and the probability of each event is no larger than $1 - \epsilon$. ∎

On the basis of this lemma, several randomized algorithms can be proposed.

Algorithm 11.1. **Estimating γ_{opt}.** Pick a number γ, which is a "candidate" for γ_{opt}. Choose a probability measure P on $\mathcal{D}_\gamma$, and generate matrices $\Delta_1, \ldots, \Delta_m \in \mathcal{D}_\gamma$ at random. If the feedback system is unstable for some Δ_i, then declare that $\gamma_{\text{opt}} < \gamma$. If the feedback system is stable for each Δ_i, declare that, with confidence at least $1 - (1-\epsilon)^m$, the feedback system is stable for all $\Delta \in \mathcal{D}_\gamma$ except possibly for those Δ belonging to a set of measure $\leq \epsilon$ (with respect to P).

The justification for this algorithm comes readily from Lemma 11.1.

Algorithm 11.2. **Estimating the probability that a matrix in $\mathcal{D}_\gamma$ destabilizes the feedback system.** Generate i.i.d. matrices $\Delta_1, \ldots, \Delta_m \in \mathcal{D}_\gamma$ at random according to P. Suppose the feedback system is unstable for $k \leq m$ of these matrices, and stable for the remaining $m-k$ matrices. Then declare with confidence $\geq 1 - e^{-2m\epsilon^2}$ that the feedback system is stable for all $\Delta \in \mathcal{D}_\gamma$, except possibly for those Δ belonging to a set of measure $\leq \epsilon + (k/m)$.

The justification for this algorithm comes from Hoeffding's inequality. Let $\mathcal{U} \subseteq \mathcal{D}_\gamma$ denote the set of $\Delta \in \mathcal{D}_\gamma$ for which the feedback system is unstable. Then k/m is the empirical measure of $\mathcal{U}$ based on the multisample $\Delta_1, \ldots, \Delta_m$; call $\hat{P}(\mathcal{U})$. Now Hoeffding's inequality implies that

$$P^m\{(\Delta_1, \ldots, \Delta_m) \in \mathcal{D}_\gamma^m : P(\mathcal{U}) > \hat{P}(\mathcal{U}) + \epsilon\} \leq \exp(-2m\epsilon^2),$$

from which the desired conclusion follows readily.

Algorithm 11.3. **Estimating $\lambda_{\max}(\gamma)$.** As above, generate i.i.d. matrices $\Delta_1, \ldots, \Delta_m \in \mathcal{D}_\gamma$ at random according to P. Let $\lambda_M(\Delta)$ denote the maximum real part of the closed-loop poles when the perturbation matrix is Δ, and define

$$\bar{\lambda} := \max_{1 \leq i \leq m} \lambda_M(\Delta_i).$$

Then, for each $\epsilon > 0$, it can be stated with confidence $\geq 1 - (1-\epsilon)^m$ that $\lambda_M(\Delta) \leq \bar{\lambda}$ for all $\Delta \in \mathcal{D}_\gamma$ except possibly for those Δ belonging to a set of measure $\leq \epsilon$.

The justification for this algorithm again comes from Lemma 11.1.

Thus far we have studied the controller *analysis* problem, where the issue is to determine whether a given candidate controller, however generated, can meet various performance requirements. Now let us examine the dual problem of controller *synthesis*. If the plant is allowed to vary over the set $\{G(x), x \in X\}$, then this problem is very difficult. So let us examine the simpler problem where the plant is *fixed*, call it G, and the issue is to determine whether or not there exists a controller within the specified class $\{K(y), y \in Y\}$ that stabilizes G. It is possible to propose a probabilistic algorithm for this purpose along familiar lines. The case where the plant itself varies over a set $\{G(x), x \in X\}$ is studied in the next section.

Algorithm 11.4. **Testing the existence of a stabilizing controller within a given family.** Given the family $\{K(y), y \in Y\}$, postulate a probability measure P on Y. Generate i.i.d. elements $y_1, \ldots, y_m$ from Y according to P. If any of the corresponding controllers $K(y_i)$ stabilizes G, then declare that there exists a controller in the given family that stabilizes G. If not, then declare with confidence $\geq 1 - (1-\epsilon)^m$ that the measure of the controllers in $\{K(y), y \in Y\}$ that stabilize G, if any, is less than ϵ.

This algorithm can also be justified on the basis of Lemma 11.1.

It is easily seen that the above discussion of probabilistic methods does not make use of any of the deep results from the preceding chapters. In contrast, the results in the next section depend in an essential way on the UCEM theory developed in the preceding chapters.

11.2 Randomized Algorithms for Robust Controller Synthesis: General Approach

In the preceding section, we have developed various randomized algorithms for robustness *analysis*. In contrast, the present section is devoted to the problem of *synthesizing* a robust controller.

11.2.1 Paradigm of Robust Controller Synthesis Problem

Suppose one is given a family of plants $\{G(x), x \in X\}$ parametrized by x, and a family of controllers $\{K(y), y \in Y\}$ parametrized by y. The objective of robust controller synthesis is to find a *a single fixed controller* $K(y_0), y_0 \in Y$ that performs reasonably well for almost all plants $G(x)$. By choosing an appropriate performance index, many problems in controller synthesis can be covered by the above statement. The objective of this subsection is to put forward an abstract problem formulation that makes the above statement quite precise, and which forms the "universe of discourse" for the remainder of the section. In particular, it is argued that, to avoid overly conservative designs, the performance of a controller should be taken as its *average* performance as the plant varies over a prespecified family, and not its *worst-case* performance.

Suppose $\psi(\cdot,\cdot)$ is a given cost function. Thus $\psi(G,K)$ is a measure of the performance of the system when the plant is G and the controller is K. The phrase "cost function" implies that lower values of ψ are preferred. For instance, if the objective is merely to choose a stabilizing controller, then one could define

$$\psi(G,K) := \begin{cases} 1, & \text{if the pair } (G,K) \text{ is unstable,} \\ 0, & \text{if the pair } (G,K) \text{ is stable.} \end{cases} \quad (11.2.1)$$

As a second example, in filtering problems, one could choose

$$\psi(G,K) := \begin{cases} 1, & \text{if the pair } (G,K) \text{ is unstable,} \\ J(G,K)/[1+J(G,K)], & \text{if the pair } (G,K) \text{ is stable,} \end{cases} \quad (11.2.2)$$

where

$$J(G,K) = \| W(I+GK)^{-1} \|_2,$$

$\|\cdot\|_2$ denotes the H_2-norm, and W is a given weighting matrix. Two points should be noted in the above definition: (i) The usual weighted H_2-norm denoted by $J(G, K)$ takes values in $[0, \infty)$. However, since all of the UCEM theory in the previous chapters is developed for the case of function families assuming values in $[0, 1]$, this cost function is rescaled by defining $\psi = J/(1 + J)$, so that $\psi(G, K)$ takes values in $[0, 1]$. (ii) To guard against the possibility that $W(I + GK)^{-1}$ belongs to H_2 even though the pair (G, K) is unstable,[2] the cost function $\psi(G, K)$ is explicitly defined to be 1 (corresponding to $J = \infty$), if the pair (G, K) is unstable. As a third and last example, if the objective is to achieve uniform rejection of L_2-norm bounded disturbances, then one can use the same cost function $\psi(G, K)$ as above, but with the modification that

$$J(G, K) := \| W(I + GK)^{-1} \|_\infty .$$

The preceding discussion pertains only to quantifying the performance of a *single plant-controller pair*. However, in problems of robust stabilization and robust performance, the cost function should reflect the performance of a *fixed* controller for a *variety* of plants. Since $G = G(x)$ and $K = K(y)$, let us define

$$g(x, y) := \psi[G(x), K(y)].$$

Note that g depends on both the plant parameter $x \in X$ and the controller parameter $y \in Y$. As such, g maps $X \times Y$ into $[0, 1]$. The aim is to define an objective function of y alone that quantifies the performance of the controller $K(y)$, so that by minimizing this objective function with respect to y one could find an "optimal" controller.

As a first attempt, one could choose

$$h(y) := \sup_{x \in X} g(x, y) = \sup_{x \in X} \psi[G(x), K(y)]. \tag{11.2.3}$$

Thus $h(y)$ measures the *worst-case* performance of a controller $K(y)$ as the plant varies over $\{G(x), x \in X\}$. For instance, if one chooses $\psi(\cdot, \cdot)$ as in (11.2.1), then $h(y) = 0$ if and only if the controller $K(y)$ stabilizes *every single* plant in $\{G(x), x \in X\}$. If $K(y)$ fails to stabilize even a single plant, then $h(y) = 1$. Thus minimizing the present choice of $h(\cdot)$ corresponds to solving the robust (or simultaneous) stabilization problem. Similarly, if $\psi(G, K)$ is chosen as in (11.2.2), then minimizing the associated $h(\cdot)$ corresponds to achieving the best possible *guaranteed* performance with robust stabilization.

It is widely believed that methods such as H_∞-norm minimization for achieving robust stabilization, and μ-synthesis for achieving guaranteed performance and robust stabilization, lead to overly conservative designs. Much of the conservatism of the designs can be attributed to the worst-case nature

---

[2] For example, consider the case where $W = 1/(s+1)^2$ and $(1 + GK)^{-1}$ is improper and behaves as $O(s)$ as $s \to \infty$, but has all of its poles in the open left half-plane.

11.2 Randomized Algorithms for Robust Controller Synthesis: General Approach

of the associated cost function. To illustrate this point, consider the situation depicted in Figure 11.3, where the function $g(\cdot, y)$ is plotted for two different values of y. From this stylized diagram, it can be seen that if the objective function $h(y)$ is defined as in (11.2.3), then the controller $K(y_1)$ would be preferred to the controller $K(y_2)$, even though the latter controller seems to outperform the former for "most" plants $G(x)$. Thus, if a worst-case objective function of the type (11.2.3) is used, then the control system designer is forced to focus all the effort and attention to bring down the value of $g(x, y)$ *for the value of x at which the supremum in (11.2.3) occurs*, which could be rather an unrepresentative plant.

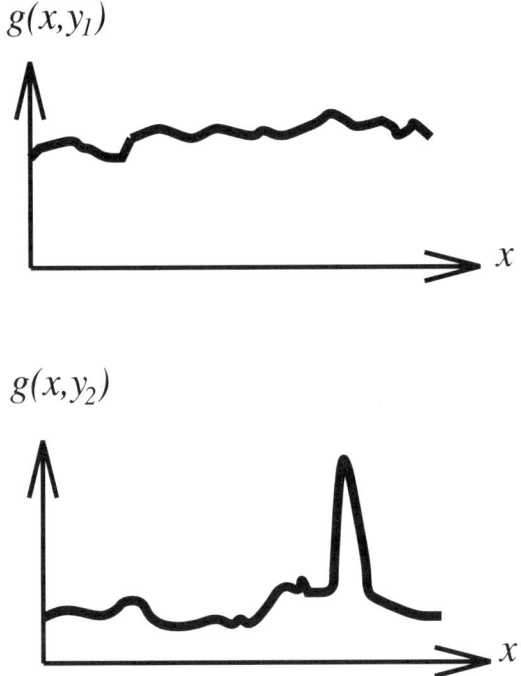

Fig. 11.3. Conservatism of the Worst-Case Objective Function

It seems much more reasonable to settle for controllers that work satisfactorily "most of the time." One way to capture this intuitive idea in a mathematical framework is to introduce a probability measure P_X on the set X, that reflects one's prior belief on the way that the "true" plant $G(x)$ is distributed in the set of possible plants $\{G(x), x \in X\}$. For instance, in a problem of robust stabilization, G_0 can be a nominal, or most likely, plant model, and the probability measure P_X can be "peaked" around G_0. The more confident one is about the nominal plant model G_0, the more sharply peaked the probability measure P_X can be. Once the probability measure P_X

is chosen, the objective function to be minimized can be defined as

$$f(y) := E_{P_X}[g(x,y)] = E_{P_X}[\psi(G(x), K(y))]. \quad (11.2.4)$$

Thus $f(y)$ is the *expected* or *average* performance of the controller $K(y)$ when the plant is distributed according to the probability measure P_X. For instance, if P_X is chosen as the uniform measure in Figure 11.3, then the controller $K(y_2)$ would be preferred to $K(y_1)$ since the area under the curve of $g(x, y_2)$ is less than that of $g(x, y_1)$. Thus the expected value type of objective function captures the intuitive idea that a controller can occasionally be permitted to perform poorly for some plant conditions, provided these plant conditions are not too likely to occur. Of course, the use of the expected value objective function does not rule out the possibility that the controller chosen performs poorly for the "true" plant, however unlikely that might be. The only way to preclude such an eventuality is to use the worst-case objective function of (11.2.3). However, note that, even with the expected value objective function, it is possible to ensure that the optimal controller always stabilizes the true plant, whatever it is. But the performance of the controller might be far from the optimal value of $f(\cdot)$.

While the worst-case objective function defined in (11.2.3) is easy to understand and to interpret, the interpretation of the expected-value type of objective function defined in (11.2.4) needs a little elaboration. Suppose $\psi(\cdot, \cdot)$ is defined as in (11.2.1). Then $f(y)$ is the measure (or "volume" with respect to the measure P_X) of the subset of $\{G(x), x \in X\}$ that *fails* to be stabilized by the controller $K(y)$. Alternatively, one can assert with confidence $1 - f(y)$ that the controller $K(y)$ stabilizes a plant $G(x)$ selected at random from $\{G(x), x \in X\}$ according to the probability measure P_X. More generally, it follows from Markov's inequality that

$$P_X\{x \in X : \psi[G(x), K(y)] > \gamma\} \leq f(y)/\gamma.$$

11.2.2 Various Types of "Near" Minima

In the previous subsection, an abstract formulation of the robust controller synthesis problem was given, which ultimately involves the minimization of a function $f : Y \to \mathbb{R}$, where Y is the set of controller parameter vectors. Thus the synthesis problem is to find a $y^* \in Y$ such that

$$f(y^*) = \min_{y \in Y} f(y).$$

There are many problems, such as those mentioned in Section 11.1, in which finding the *exact* minimum value

$$f^* := \inf_{y \in Y} f(y)$$

is NP-hard. More precisely, given a number f_0, it is NP-hard to determine whether or not $f_0 \geq f^*$. In such cases, one has to be content with "nearly"

11.2 Randomized Algorithms for Robust Controller Synthesis: General Approach

minimizing $f(\cdot)$. The objective of this section is to introduce three different definitions of "near minima," and to discuss their significance.

Definition 11.1. *Suppose $f : Y \to \mathbb{R}$ and that $\epsilon > 0$ is a given number. A number $f_0 \in \mathbb{R}$ is said to be a* **Type 1 near minimum of $f(\cdot)$ to accuracy ϵ**, *or an* **approximate near minimum of $f(\cdot)$ to accuracy ϵ**, *if*

$$\inf_{y \in Y} f(y) - \epsilon \leq f_0 \leq \inf_{y \in Y} f(y) + \epsilon, \tag{11.2.5}$$

or equivalently

$$\left| f_0 - \inf_{y \in Y} f(y) \right| \leq \epsilon.$$

An approximate near minimum perhaps corresponds most closely to what we normaly think of as a "near" minimum. Figure 11.4 depicts the notion of an approximate near minimum.

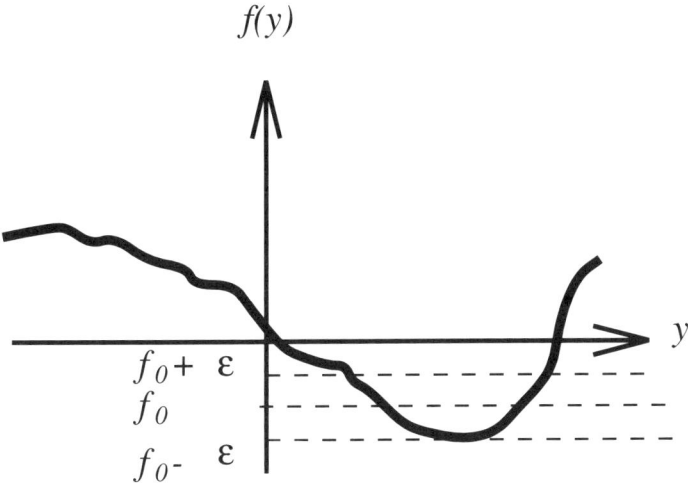

Fig. 11.4. Approximate Near Minimum to Accuracy ϵ

Unfortunately, in some robust stability problems, it is NP-hard to compute even an *approximation* to f^*; see [48] for a result along this direction. In other words, there exist problems in which it is NP-hard to determine not only the exact minimum value f^*, but even a Type 1 (or approximate) near minimum. Thus it is necessary to look for other notions of a near minimum. One such notion is provided in the next definition.

Definition 11.2. *Suppose $f : Y \to \mathbb{R}$, that P_Y is a given probability measure on Y, and that $\alpha > 0$ is a given number. A number $f_0 \in \mathbb{R}$ is said to be a* **Type 2 near minimum of $f(\cdot)$ to level α**, *or a* **probable near minimum of $f(\cdot)$ to level α**, *if $f_0 \geq f^*$, and in addition*

434   11. Applications to Control Systems

$$P_Y\{y \in Y : f(y) < f_0\} \leq \alpha.$$

The notion of a probable near minimum can be interpreted as follows: f_0 is a probable near minimum of $f(\cdot)$ to level α if there is an "exceptional set" S with $P_Y(S) \leq \alpha$ such that

$$\inf_{y \in Y} f(y) \leq f_0 \leq \inf_{y \in Y \setminus S} f(y). \tag{11.2.6}$$

In other words, f_0 is bracketed by the infimum of $f(\cdot)$ over all of Y, and the infimum of $f(\cdot)$ over "nearly" all of Y. This is depicted in Figure 11.5.

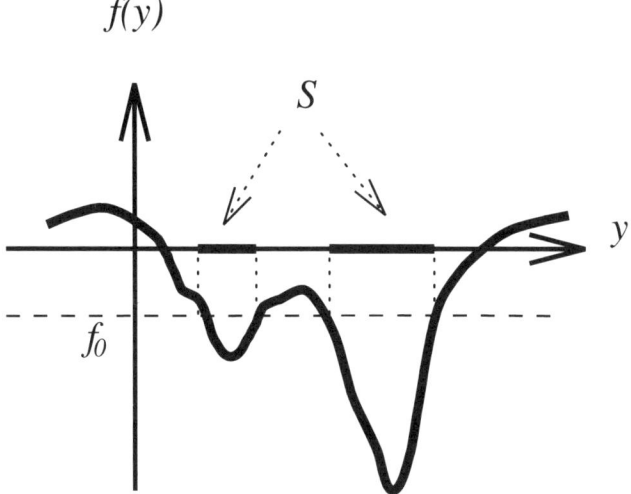

Fig. 11.5. Probable Near Minimum to Level α

Note that the "empirical" maximum (or minimum) described in Lemma 11.1 is in fact a Type 2 near maximum (or minimum). However, the approach described therein requires that the function value $f(y)$ be computable *exactly* for any given $y \in Y$. To cater to situations in which this is not possible, we introduce yet one more type of near minimum.

Definition 11.3. *Suppose $f : Y \to \mathbb{R}$, that P_Y is a given probability measure on Y, and that $\epsilon, \alpha > 0$ are given numbers. A number $f_0 \in \mathbb{R}$ is said to be a* **Type 3 near minimum of $f(\cdot)$ to accuracy ϵ and level α**, *or a* **probably approximate near minimum of $f(\cdot)$ to accuracy ϵ and level α**, *if $f_0 \geq f^* - \epsilon$, and in addition*

$$P_Y\{y \in Y : f(y) < f_0 - \epsilon\} \leq \alpha.$$

Another way of saying this is that there exists an "exceptional set" $S \subseteq Y$ with $P_Y(S) \leq \alpha$ such that

11.2 Randomized Algorithms for Robust Controller Synthesis: General Approach

$$\inf_{y \in Y} f(y) - \epsilon \leq f_0 \leq \inf_{y \in Y \setminus S} f(y) + \epsilon. \qquad (11.2.7)$$

A comparison of (11.2.5), (11.2.6) and (11.2.7) brings out clearly the relationships between the various types of near minima.

11.2.3 A General Approach to Randomized Algorithms

Recall that the objective of robust controller synthesis is to minimize an objective function of the type defined in (11.2.4), namely

$$f(y) = E_{P_X}[g(x, y)].$$

Let us introduce the notation

$$g_y(x) := g(x, y), \; \forall x \in X, \; \forall y \in Y,$$

and note that $f(y)$ is precisely the expected value of $g(\cdot, y)$. Thus for each $y \in Y$, the function $g_y(\cdot)$ maps X into $[0, 1]$. Now define the associated family of functions $\mathcal{G} := \{g_y(\cdot), \; y \in Y\}$.

Ideally we would like to be able to compute $f(y)$ for a given y, so that $f(\cdot)$ can be minimized using, for example, gradient-based methods or even random search as in Lemma 11.1. However, except in very simple situations, it is rather difficult to compute the expected value $E_{P_X}[g(x, y)]$ *exactly*. Suppose now that $\mathbf{x} := [x_1 \ldots x_m]^t \in X^m$ is a collection of i.i.d. samples. For each function $g_y(\cdot) \in \mathcal{G}$, one can define its empirical mean based on the multisample $\mathbf{x}$ as

$$\hat{E}(g_y; \mathbf{x}) := \frac{1}{m} \sum_{j=1}^{m} g_y(x_j) = \frac{1}{m} \sum_{j=1}^{m} g(x_j, y), \; y \in Y.$$

In other words, the *actual* performance $f(y)$ of a controller $K(y)$ is approximated by its *average* performance on the randomly generated plants $G(x_1), \ldots, G(x_m)$. Now let

$$q(m, \epsilon; \mathcal{G}) := P^m \{\mathbf{x} \in X^m : \sup_{g_y \in \mathcal{G}} |\hat{E}(g_y, \mathbf{x}) - E_{P_X}(g_y)| > \epsilon\}. \qquad (11.2.8)$$

Observe that an equivalent way of writing $q(m, \epsilon; \mathcal{G})$ is as follows:

$$q(m, \epsilon; \mathcal{G}) := P^m \{\mathbf{x} \in X^m : \sup_{y \in Y} |\hat{E}(g_y; \mathbf{x}) - f(y)| > \epsilon\}.$$

The family $\mathcal{G}$ has the UCEM property if and only if $q(m, \epsilon; \mathcal{G}) \to 0$ as $m \to \infty$ for each $\epsilon > 0$.

Suppose the family $\mathcal{G}$ does indeed have the UCEM property. Let $\epsilon, \delta \in (0, 1)$ be specified accuracy and confidence parameters, respectively. Choose

m large enough that $q(m, \epsilon; \mathcal{G}) < \delta$. Then it can be said with confidence $1 - \delta$ that
$$|f(y) - \hat{E}(g_y; \mathbf{x})| \leq \epsilon, \ \forall y \in Y.$$
In other words, the function $\hat{E}(g; \mathbf{x})$ is a *uniformly close* approximation to the original objective function $f(\cdot)$. Hence it readily follows that an *exact* minimizer of $\hat{E}(g; \mathbf{x})$ is also an *approximate* near minimizer of $f(\cdot)$ to accuracy ϵ. Moreover, it might be simpler to minimize the empirical mean value $\hat{E}(g_y; \mathbf{x})$ with respect to y, as the latter quantity is easier to compute. However, the optimization method used to minimize $\hat{E}(g_y; \mathbf{x})$ with respect to y is not specified, and is at the discretion of the designer.

11.2.4 Two Algorithms for Finding Probably Approximate Near Minima

The ideas in the preceding subsection can be combined to produce two distinct randomized algorithms for finding a *probably approximate* (or Type 3) near minimum of an objective function $f(\cdot)$ of the form (11.2.4). The first algorithm is "universal," while the second algorithm is applicable only to situations where an associated family of functions has the UCEM property. The sample complexity estimates for the first "universal" algorithm are the best possible, whereas there is considerable scope for improving the sample complexity estimates of the second algorithm.

Suppose real parameters $\epsilon, \alpha, \delta > 0$ are given; the objective is to develop a randomized algorithm that constructs a probably approximate (Type 3) near minimum of
$$f(y) := E_{P_X}[g(x, y)]$$
to accuracy ϵ and level α, with confidence $1-\delta$. In other words, the probability that the randomized algorithms fails to find a probably approximate near minimum to accuracy ϵ and level α must be at most δ.

Algorithm 11.5. Choose integers
$$n \geq \frac{\lg(2/\delta)}{\lg[1/(1-\alpha)]}, \text{ and } m \geq \frac{1}{2\epsilon^2} \ln \frac{4n}{\delta}. \tag{11.2.9}$$
Generate i.i.d. samples $y_1, \ldots, y_n \in Y$ according to P_Y and $x_1, \ldots, x_m \in X$ according to P_X. Define
$$\hat{f}_i := \frac{1}{m} \sum_{j=1}^{m} g(x_j, y_i), \ i = 1, \ldots, n, \text{ and}$$
$$\hat{f}_0 := \min_{1 \leq i \leq n} \hat{f}_i.$$
Then with confidence $1 - \delta$, it can be said that $\hat{f}_0$ is a probably approximate (Type 3) near minimum of $f(\cdot)$ to accuracy ϵ and level α.

11.2 Randomized Algorithms for Robust Controller Synthesis: General Approach

The proof of the claim in Algorithm 11.5 is easy. Once the i.i.d. samples $y_1, \ldots, y_n$ are generated where n satisfies (11.2.9), one can define

$$\bar{f} := \min_{1 \leq i \leq n} f(y_i).$$

Then it follows from Lemma 11.1 that, with confidence $1 - \delta/2$ (*not* $1 - \delta$ — compare (11.2.9) with (11.1.2)), the number $\bar{f}$ is a probable near minimum of $f(\cdot)$ to level α. Now consider the *finite* family of functions $\mathcal{A} := \{g(\cdot, y_i), i = 1, \ldots, n\}$, and note that $2ne^{-2m\epsilon^2} \leq \delta/2$ in view of (11.2.9). Hence it follows from (11.3.5) that with confidence $1 - \delta/2$, we have

$$|f(y_i) - \hat{f}_i| \leq \epsilon, \text{ for } i = 1, \ldots, n.$$

In particular, it follows that

$$|\hat{f}_0 - \bar{f}| \leq \epsilon. \tag{11.2.10}$$

Combining the two statements shows that, with confidence $1 - \delta$, $\hat{f}_0$ is a probably approximate (Type 3) near minimum of $f(\cdot)$ to accuracy ϵ and level α.

While Algorithm 11.5 is "universal" in the sense that it requires no assumptions about the nature of the function $g(\cdot, \cdot)$, it has the drawback that the number m of x-samples is dependent on n, the number of y samples. In particular, as the level parameter α approaches zero, *both* integers m and n need to be increased in tandem. It is now shown that, if the associated family of functions $\mathcal{G}$ defined previously has the UCEM property, then it is possible to make the integer m *independent* of the level parameter α. Moreover, unlike in the case of Algorithm 11.5, there is considerable scope for improving the estimates for the integer m. It is shown in the remainder of the section that the UCEM property does hold in a wide variety of control problems. Thus, by sacrificing a little generality, it is possible to develop a possibly more efficient algorithm.

Algorithm 11.6. Select integers n, m such that

$$n \geq \frac{\lg(2/\delta)}{\lg[1/(1-\alpha)]}, \text{ and } q(m, \epsilon; \mathcal{G}) \leq \delta/2. \tag{11.2.11}$$

Generate i.i.d. samples $y_1, \ldots, y_n \in Y$ according to P_Y and $x_1, \ldots, x_m \in X$ according to P_X. Define

$$\hat{f}_i := \frac{1}{m} \sum_{j=1}^{m} f(x_j, y_i), \ i = 1, \ldots, n, \text{ and}$$

$$\hat{f}_0 := \min_{1 \leq i \leq n} \hat{f}_i.$$

Then with confidence $1 - \delta$, it can be said that $\hat{f}_0$ is a probably approximate (Type 3) near minimum of $f(\cdot)$ to accuracy ϵ and level α.

It can be seen by comparing (11.2.9) and (11.2.11) that the only difference between Algorithms 11.5 and 11.6 is in the number m of x-samples. The key point to note is that m is *independent* of the integer n, which in turn depends on the level parameter α.

11.3 VC-Dimension Estimates for Problems in Robust Controller Synthesis

In the previous section, a general approach was presented for developing randomized algorithms to synthesize robust controllers. Two distinct randomized algorithms were presented, of which one requires that the class of functions $\mathcal{G}$ have the UCEM property. In the present section, it is shown that the UCEM property does indeed hold in a wide variety of situations. In fact, Theorem 11.1 below makes it clear that in *any* controller synthesis problem where the satisfaction of a performance constraint can be expressed in terms of a finite number of polynomial inequalities, the UCEM property holds.

11.3.1 A General Result

Theorem 11.1. *Suppose the controller parameter set $Y \subseteq \mathbb{R}^l$ for some integer l, the plant parameter set $X \subseteq \mathbb{R}^k$ for some integer k, and that for each constant $c \in [0,1]$, the inequality*

$$\psi[G(x), K(y)] - c < 0 \Leftrightarrow g(x,y) - c < 0$$

can be expressed as a Boolean formula involving s polynomial inequalities of the form $\tau_i(x,y) > 0, i = 1, \ldots, s$. Finally, suppose that the degree of each $\tau_i(x,y)$ in y is no larger than d. Then the family of functions $\mathcal{G}$ defined by

$$\mathcal{G} := \{g_y : y \in Y\}$$

has finite P-dimension. Moreover,

$$\text{P-dim}(\mathcal{G}) \leq 2l \lg(4eds).$$

The proof follows readily from Corollary 10.2.

11.3.2 Robust Stabilization

In this subsection, we study whether the UCEM property holds for the family $\mathcal{G}$ that arises in connection with the problem of robust stabilization. In this case, we have

$$g_y(x) := g(x,y) = \begin{cases} 1, & \text{if the pair } (G(x), K(y)) \text{ is unstable, and} \\ 0, & \text{if the pair } (G(x), K(y)) \text{ is stable.} \end{cases}$$

11.3 VC-Dimension Estimates for Problems in Robust Controller Synthesis

We study only the case where the plant parameter x and the controller parameter y appear in the *input-output* description. The case where these parameters enter into the *state-space* description can be studied in an entirely analogous fashion. In the interests of notational simplicity, it is assumed for the moment that both the plant and the controller are single-input, single-output. The extension to the MIMO case is straight-forward and is indicated near the end of the subsection. Suppose each plant $G(x,s)$ is strictly proper and is of the form

$$G(x,s) = \frac{n_G(x,s)}{d_G(x,s)}, \quad \forall x \in X,$$

where n_G, d_G are *polynomials* in x, s. Next, suppose all the plants have McMillan degree α_s. In other words, it is assumed that the degree of d_G with respect to s is α_s for every $x \in X$, and that the degree of n_G with respect to s is no larger than α_s for every $x \in X$. The assumptions about $K(y,s)$ are entirely analogous. It is assumed that

$$K(y,s) = \frac{n_K(y,s)}{d_K(y,s)}, \quad \forall y \in Y$$

is a proper rational function of s, with McMillan degree β_s. Also, it is assumed that $n_K(y,s)$, $d_K(y,s)$ are polynomials in y of degree no larger than β_y. Finally, it is assumed that $X \subseteq \mathbb{R}^k, Y \subseteq \mathbb{R}^l$ for suitable integers k, l.

Note that the above assumptions include the case of "interval coefficients." The above assumptions are made solely in the interests of simplicity of exposition, and can be relaxed considerably. For example, it is possible to treat the case of discrete-time systems instead of continuous-time systems in a manner entirely analogous to that given below. It is also possible to extend the arguments below to the case of MIMO systems instead of SISO systems. Finally, it is possible to permit the coefficients of the various polynomials to be *rational functions* of x and y respectively, rather than polynomials, provided the denominator polynomials have constant sign for all (x, y).

Theorem 11.2. *Under the above assumptions, the family of binary-valued functions* $\mathcal{G} := \{g_y : y \in Y\}$ *has finite VC-dimension. In particular,*

$$VC\text{-}dim(\mathcal{G}) \leq 2l \lg[4e(\alpha_s + \beta_s)^2 \beta_y]. \tag{11.3.1}$$

Remarks: It is important to note that the quantity $\alpha_s + \beta_s$, which is the order of the closed-loop system, appears *inside* the $\lg(\cdot)$ function. Thus the bounds given above will be rather small even for high-order control systems. However, the integer l, representing the number of "degrees of freedom" in the controller parameter y, appears linearly in the upper bound.

Proof. The proof consists of writing down the conditions for the closed-loop system to be stable as a Boolean formula entailing several polynomial inequalities in x and y, and then appealing to Theorem 11.1. For a fixed $x \in X, y \in Y$, the pair $[G(x,s), K(y,s)]$ is stable if and only if two conditions hold: (i) the closed-loop transfer function is proper, and (ii) the characteristic polynomial of the closed-loop system is Hurwitz. Since every plant $G(x,s)$ is *strictly proper* and every controller $K(y,s)$ is proper, the first condition is automatically satisfied and we are left with only the second condition. Next, the closed-loop characteristic polynomial equals

$$\theta(x,y,s) := n_G(x,s)\, n_K(y,s) + d_G(x,s)\, d_K(y,s). \qquad (11.3.2)$$

This is a polynomial of degree $\alpha_s + \beta_s$ in s, where each coefficient of θ is a polynomial in x, y; moreover, the degree of each coefficient with respect to y is no larger than β_y. The stability of the polynomial θ can be tested by forming its Hurwitz determinants as in [72], pp. 190 ff. Let $\delta_c := \alpha_s + \beta_s$ denote the degree of θ with respect to s, and write

$$\theta(x,y,s) = \sum_{i=0}^{\delta_c} a_{\delta_c - i}(x,y)\, s^i.$$

Let H_i denote the i-th Hurwitz determinant of θ; then H_i is of the form

$$H_i(x,y) = \begin{vmatrix} a_1 & a_3 & a_5 & \cdots & a_{2i-1} \\ a_0 & a_2 & a_4 & \cdots & a_{2i-2} \\ 0 & a_1 & a_3 & \cdots & a_{2i-3} \\ \vdots & \vdots & \vdots & \cdots & \vdots \\ 0 & 0 & 0 & \cdots & a_i \end{vmatrix},$$

where $a_i(x,y)$ is written as a_i in the interests of clarity. Note that a_i is taken as zero if $i > \delta_c$. Now θ is a Hurwitz polynomial if and only if $H_i(x,y) > 0$ for $i = 1, \ldots, \delta_c$. Hence the pair $[G(x,s), K(y,s)]$ is stable, and $\psi(x,y) = 0$, if and only if

$$[H_1(x,y) > 0] \wedge \ldots \wedge [H_{\delta_c}(x,y) > 0]. \qquad (11.3.3)$$

This is a Boolean formula. Moreover, $\psi(x,y) = 1$ is just the *negation* of the above Boolean formula, and is thus another Boolean formula with just the same atomic polynomial inequalities.

To apply Theorem 11.1, it is necessary to count the number of atomic polynomial inequalities, and their maximum degree with respect to y. First, the number of atomic inequalities is $\delta_c = \alpha_s + \beta_s$. Thus we take

$$t = \alpha_s + \beta_s.$$

Next, let us examine the degree of each of the atomic polynomial inequalities with respect to y. Each Hurwitz determinant $H_i(x,y)$ is the determinant of an $i \times i$ matrix, each of whose entries is a polynomial in y of degree β_y or less. Therefore the degree of H_i with respect to y is at most $\beta_y i$. Hence the maximum degree of the atomic polynomial inequalities with respect to y can be taken as

11.3 VC-Dimension Estimates for Problems in Robust Controller Synthesis

$$r = (\alpha_s + \beta_s)\beta_y.$$

Finally, applying Theorem 11.1 leads to the estimate

$$\text{VC-dim}(\mathcal{G}) \leq 2l\lg(4ert) \leq 2l\lg[4e(\alpha_s + \beta_s)^2\beta_y].$$

This completes the proof. ∎

The reader might wonder why, in the above proof, the Hurwitz test for stability is used instead of the more widely known Routh test (which is almost always mislabelled as the "Routh-Hurwitz test"). The reason is economy in estimating the integer r. The polynomial $\theta(x, y, s)$ is Hurwitz if and only if the elements in the first column of the Routh array all have the same sign (usually positive). Now it is well-known (see, e.g., [72]) that the first element in the i-th row of the Routh array (call it R_i) is just the product of the first i Hurwitz determinants. As we have seen above, the degree of H_i with respect to y is at most $\beta_y i$. Hence the degree of R_i with respect to y is at most $\beta_y i(i+1)/2$. Thus, the stability condition that R_i is positive for all i leads to exactly $\alpha_s + \beta_s$ inequalities; hence the estimate for the number t of atomic polynomial inequalities is unchanged if the Routh test is used instead of the Hurwitz test. However, the estimate for r now jumps to $(\alpha_s + \beta_s)(\alpha_s + \beta_s + 1)\beta_y/2$.

The above argument can be readily extended to the case where the coefficients of $n_G(x,s)$, $d_G(x,s)$, $n_K(y,s)$, and $d_K(y,s)$ are all *rational functions* of x, y respectively, rather than polynomials, and also to MIMO systems.

11.3.3 Weighted H_∞-Norm Minimization

By using arguments entirely analogous to those in the preceding subsection, it is possible to estimate the P-dimension of the family $\mathcal{G}$ in the case where the objective is to minimize the weighted H_∞-norm of the weighted H_2-norm of the closed-loop transfer function.

We begin with the problem of minimizing the weighted H_∞-norm of the closed-loop transfer function. The problem of minimizing the weighted H_2-norm is studied in the next subsection. Let $W(s)$ be a given weighting function, and define the performance measure to be

$$\psi(G,K) := \begin{cases} 1, & \text{if the pair } (G,K) \text{ is unstable,} \\ J(G,K)/[1+J(G,K)], & \text{if the pair } (G,K) \text{ is stable,} \end{cases}$$

where

$$J(G,K) = \| W(1+GK)^{-1} \|_\infty$$

and $\|\cdot\|_\infty$ denotes the H_∞-norm. Let $G = G(x,s)$, $K = K(y,s)$ be as before, wherein $G = n_G/d_G$, $K = n_K/d_K$, the coefficients of n_G, d_G are polynomials in x, and the coefficients of n_K, d_K are polynomials in y. As before, let β_y denote the maximum degree of any coefficient of n_K, d_K with respect to y. For convenience, let $\delta_c := \alpha_s + \beta_s$ denote the McMillan degree of the closed-loop transfer function, and let δ_w denote the McMillan degree of the *weighted* closed-loop transfer function $W(1+GK)^{-1}$. As before, let $g(x,y) := \psi[G(x),K(y)]$. For each $y \in Y$ define the corresponding function $g_y : X \to [0,1]$ by $g_y(x) := g(x,y)$, and define $\mathcal{G} := \{g_y : y \in Y\}$.

Theorem 11.3. *With all symbols as above, we have*

$$P\text{-}dim(\mathcal{G}) \leq 2l \lg[8e\beta_y \delta_w (2\delta_c + 2\delta_w + 1)]. \tag{11.3.4}$$

Remarks: As in the case of the bounds in (11.3.1) given in Theorem 11.2, the quantities δ_c which is the order of the closed-loop system and δ_w which is the order of the weighted closed-loop transfer function, both appear *inside* the $\lg(\cdot)$ function. Thus the bounds given above will be rather small even for high-order control systems. However, the integer l, representing the number of "degrees of freedom" in the parameter y, appears linearly in the upper bound.

Proof. Let $G = N_G/d_G, K = n_K/d_K, W = n_W/d_W$ denote *polynomial* coprime factorizations with respect to s of G, K, W respectively. Then the closed-loop system is stable if and only if the characteristic polynomial θ defined in (11.3.2) is Hurwitz. Moreover, the weighted sensitivity function can be expressed as

$$\frac{W}{1+KG} = \frac{n_W n_G n_K}{d_W (n_G n_K + d_G d_K)} =: \frac{n_M(x,y,s)}{d_M(x,y,s)}.$$

By assumption, each element of n_K, d_K is a polynomial in y of degree at most β_y. Hence it follows that each element of d_M is also a polynomial in y of degree at most β_y. Moreover, the degree of d_M with respect to s is δ_w.

For each $y \in Y$, define the associated function $\bar{g}_y : X \times [0,1] \to \{0,1\}$ by

$$\bar{g}_y(x,c) = \eta[g_y(x) - c],$$

where $\eta(\cdot)$ is the Heaviside function. Let $\bar{\mathcal{G}} := \{\bar{g}_y : y \in Y\}$. Then it follows from Lemma 10.1 that

$$P\text{-}dim(\mathcal{G}) = VC\text{-}dim(\bar{\mathcal{G}}).$$

Hence we concentrate on estimating $VC\text{-}dim(\bar{\mathcal{G}})$.

For this purpose, we write

$$\eta[g_y(x) - c] = 0$$

as a Boolean formula involving several polynomial inequalities in x, y, c and then apply Theorem 11.1. Note that

$$\eta[g_y(x) - c] = 0 \Leftrightarrow g(x,y) - c < 0.$$

11.3 VC-Dimension Estimates for Problems in Robust Controller Synthesis

Let us write J as a shorthand for $J[G(x), K(y)]$. Then[3]

$$g(x, y) - c < 0 \Leftrightarrow [(c = 1) \wedge (J < \infty)] \vee [(c < 1) \wedge (J < c/(1-c))].$$

Now $J < \infty$ is just closed-loop stability. The results of the preceding subsection imply that this condition can be written in terms of δ_c polynomial inequalities, each of which has degree at most $\beta_y \delta_c$ with respect to y. Next, the condition $J < c/(1-c)$ says that the weighted transfer function $F := W(1 + GK)^{-1}$ has H_∞-norm less than $b := c/(1-c)$. This is the case if and only if

$$(G, K) \text{ is stable, } b > F(\infty), \text{ and } b^2 - F^*(jw)F(jw) > 0 \; \forall \omega.$$

Again, the first requirement involves δ_c polynomial inequalities of degree at most $\beta_y \delta_c$ in y. Next, since $F(s)$ is a *stable* rational function, we can express the condition

$$b^2 - F^*(jw)F(jw) > 0 \; \forall \omega$$

equivalently as a *polynomial* inequality in ω of the form

$$p(\omega) > 0 \; \forall \omega,$$

where the coefficients of $p(\cdot)$ are rational functions of x, y, c. Now $p(\omega)$ is an *even* polynomial in ω of degree $2\delta_w$. Since $b > F(\infty)$, the polynomial $p(\cdot)$ is positive for sufficiently large ω. Hence the condition $p(\omega) > 0 \; \forall \omega$ is equivalent to the requirement that $p(\cdot)$ does not have any positive real zeros. By [90], Theorem 5.3, p. 154, this requirement can be expressed as follows: (i) Both the highest and smallest coefficient of $p(\cdot)$ are positive, and (ii) the number of sign changes in the Routh table for $p(\cdot)$ is $\delta_w - 1$. Now it is known (see, e.g., [72], p. 198) that there is a sign change in the first column of the i-th row in the Routh table if and only if the i-th Hurwitz determinant is negative. Thus Theorem 5.3 of [90] requires that, out of the $2\delta_w$ Hurwitz determinants of $p(\cdot)$, exactly $\delta_w - 1$ are negative. It is easy to see that this requirement can be written as a (very messy) Boolean formula involving the $2\delta_w$ atomic inequalities $H_1(p) > 0$ through $H_{2\delta_w}(p) > 0$, where $H_i(p)$ denotes the i-th Hurwitz determinant of $p(\cdot)$. As in the case of the Hurwitz determinants for closed-loop stability, each of these determinants $H_i(p)$ is a polynomial in x, y, c, and the maximum degree of any determinant with respect to y is $2\beta_y \delta_w$. Since $\delta_w \geq \delta_c$, we can now apply Theorem 11.1 with

$$t = \text{No. of inequalities} = 2\delta_c + 2\delta_w + 1,$$

$$r = \text{Max. degree w.r.t. } y = 2\beta_y \delta_w.$$

This completes the proof. ∎

---

[3] Strictly speaking, we should write $\neg(c < 1)$ instead of $c = 1$.

11.3.4 Weighted H_2-Norm Minimization

In this subsection, an upper bound is derived for the P-dimension of the family $\mathcal{G}$ in the case where the objective is to minimize the weighted H_2-norm of the closed-loop transfer function. As before, let $G(x,s)$ be a strictly proper SISO plant, $K(y,s)$ a proper SISO controller, and $W(s)$ a strictly proper weighting function. Define the performance measure $\psi(G,K)$ to be

$$\psi(G,K) := \begin{cases} 1, & \text{if the pair } (G,K) \text{ is unstable,} \\ J(G,K)/[1+J(G,K)] & \text{if the pair } (G,K) \text{ is stable,} \end{cases}$$

where the objective function $J(G,K)$ is now defined as

$$J(G,K) := \| W(1+GK)^{-1} \|_2,$$

and $\| \cdot \|_2$ denotes the H_2-norm.

Theorem 11.4. *With all symbols as above, we have*

$$P\text{-}dim(\mathcal{G}) \le 2l \lg[8e(\delta_c+1)(\delta_w+1)\beta_y].$$

Proof. Given the plant parameter vector x, the controller parameter y, and the real parameter $c \in [0,1]$, we express the inequality

$$\eta[g_y(x) - c] = 0$$

as a Boolean formula involving polynomial inequalities in x, y, c. As before, we have

$$\eta[g_y(x) - c] = 0 \Leftrightarrow [(c=1) \wedge (J < \infty)] \vee [(c<1) \wedge (J < c/(1-c))].$$

Again as before, the condition $J < \infty$ is the closed-loop stability requirement, and as such, it can be expressed in terms of δ_c polynomial inequalities of degree at most $\delta_c \beta_y$ in y. The only thing that is different in the present instance is the treatment of the condition

$$J < \frac{c}{1-c} \Leftrightarrow J^2 < \left[\frac{c}{1-c}\right]^2.$$

Now it is shown that $J^2[G(x),K(y)]$ is a rational function of x,y, whose degree with respect to y is no larger than $(\delta_w+1)\beta_y$. For this purpose, let

$$H(x,y,s) := \frac{W(s)}{1+G(x,s)K(y,s)}$$

denote the weighted closed-loop sensitivity function, and note that $J^2 = \| H \|_2^2$. Next, we use the results in [151], Appendix E, to compute J^2. Write

11.3 VC-Dimension Estimates for Problems in Robust Controller Synthesis

$$H(x,y,s) := \frac{\sum_{i=0}^{\delta_w-1} c_i s^i}{\sum_{i=0}^{\delta_w} d_i s^i}.$$

This is the same notation as in [151], except that δ_w plays the role of the symbol n. Then (cf. Equation (E.1-7) of [151]),

$$\| H \|_2^2 = \frac{a_{\delta_w-1}}{d_{\delta_w}},$$

where the vector $A := [a_0 \; a_1 \; \ldots \; a_{\delta_w-1}]$ satisfies a linear matrix equation (cf. Equation (E.1-19))

$$DA = C.$$

Now the solution for a_{δ_w-1} can be written using Cramer's rule in the form

$$a_{\delta_w-1} = \frac{\det [\bar{D}|C]}{\det D},$$

where $\bar{D}$ consists of all but the last column of D. Also (cf. (E.1-22) and (E.1-23)), the entries of D are just the coefficients d_i, whereas the entries of C (cf. (E.1.15)) are sums of products of two of the c_i's. We have already seen in the proof of Theorem 11.3 that each of the c_i, d_i is a polynomial in y of degree at most β_y. Hence each element of C is a polynomial in y of degree at most $2\beta_y$. Also, $\det D$ is degree at most $\delta_w \beta_y$ whereas $\det [\bar{D}|C]$ has degree at most $(\delta_w + 1)\beta_y$ with respect to y. This shows that a_{δ_w-1} is a rational function in y of degree at most $(2\delta_w + 1)\beta_y$, and $\| H \|_2^2$ is a rational function in y of degree at most $2(\delta_w + 1)\beta_y$. Now we can apply Theorem 11.1 with the following values:

$$t = \delta_c + 1, \; r = 2(\delta_w + 1)\beta_y.$$

(Here we use the obvious fact that $\delta_c \leq \delta_w$.) This leads to the desired conclusion. ∎

11.3.5 Sample Complexity Considerations

Vapnik-Chervonekis theory of uniform convergence of empirical means to their true values represents a great intellectual achievement. In as much as the theory provides both *necessary as well as sufficient conditions* for the UCEM property to hold, the only room for improvement in the theory is that of obtaining less conservative estimates for the sample complexity, that is, the number of randomly generated samples that are required to achieve a specified level of accuracy and confidence. In this section, we study the issue of sample complexity for both the general UCEM problem as well as the specific problem studied in this paper of minimizing an expected-value type of objective function. In particular, the sample complexity bounds obtained from VC theory are compared with those obtained from the older Hoeffding's inequality. The conclusion is that, in most practical situations, the bounds based on VC theory *as they are at present* lead to much larger numbers of samples than those based on Hoeffding's inequality. However, the bounds obtained from Hoeffding's inequality are in some sense the "best possible," whereas there is considerable scope for improvement in the bounds based on VC theory. Hence it is possible that as the latter bounds improve, they may eventually prove to be less conservative than the bounds based on Hoeffding's inequality. Thus it clearly behoves the statistical learning theory research community to continue studying the issue of sample complexity using VC theory.

Let us begin by studying the problem of uniform convergence of empirical means (UCEM). In the interests of simplicity, let us restrict attention to binary-valued functions; this restriction does not detract from the generality of the conclusions drawn below. Given a probability space (X, P_X) and a measurable function $f : X \to [0,1]$, Hoeffding's inequality states that

$$P^m\{\mathbf{x} \in X^m : |\hat{E}(f; \mathbf{x}) - E_{P_X}(f)| > \epsilon\} \leq 2e^{-2m\epsilon^2}.$$

Note that the above bound is independent of both the function f and the probability measure P_X. In this sense, the above bound is the "best possible," because it is possible to choose P_X and f in such a way that the above bound holds exactly; see [129]. Next, suppose $f_i : X \to [0,1]$, $i = 1, \ldots, n$ are measurable functions, and define

$$S_i(m, \epsilon) := \{\mathbf{x} \in X^m : |\hat{E}(f_i; \mathbf{x}) - E_{P_X}(f_i)| > \epsilon\}, \ i = 1, \ldots, n,$$

$$S(m, \epsilon) := \{\mathbf{x} \in X^m : \max_{1 \leq i \leq n} |\hat{E}(f_i; \mathbf{x}) - E_{P_X}(f_i)| > \epsilon\}.$$

Then

$$S(m, \epsilon) = \bigcup_{i=1}^{n} S_i(m, \epsilon),$$

and it follows from the subadditivity of P_X and Hoeffding's inequality that

$$P_X^m[S(m, \epsilon)] \leq 2ne^{-2m\epsilon^2}.$$

Moreover, it can always be arranged by suitable choice of the functions f_i that the sets $S_i(m, \epsilon)$ are pairwise disjoint, so that the above bound is also the "best possible." This statement presupposes that the family of functions $\{f_1, \ldots, f_n\}$ does not have any "structure." VC theory is a way of (i) taking advantage of the "structure" if any of the family of functions $\{f_1, \ldots, f_n\}$, and (ii) extending the above bound to the case of an *infinite* family of functions, in which case the above reasoning breaks down.

Next, let us compare the sample complexity estimates obtained through Hoeffding's inequality and VC theory, in the case where the underlying family of functions is finite. Suppose an accuracy parameter $\epsilon > 0$ and a confidence parameter $\delta > 0$ are specified. The aim is to estimate an integer $m_0(\epsilon, \delta)$ to ensure that

11.3 VC-Dimension Estimates for Problems in Robust Controller Synthesis 447

$$P^m\{\mathbf{x} \in X^m : \max_{1 \leq i \leq n} |\hat{E}(f_i; \mathbf{x}) - E_{P_X}(f_i)| > \epsilon\} \leq \delta, \ \forall m \geq m_0(\epsilon, \delta).$$

Suppose Hoeffding's inequality is used to estimate the left side. Then it is enough to choose m_0 large enough to ensure that

$$2ne^{-2m_0\epsilon^2} \leq \delta,$$

or

$$m_{0,\text{Hoeff}}(\epsilon, \delta) = \frac{\ln(2n/\delta)}{2\epsilon^2}. \tag{11.3.5}$$

It is important to note that the integer n, corresponding to the number of functions, appears *inside* the $\ln(\cdot)$. Now let d denote the VC-dimension of the family of functions $\{f_1, \ldots, f_n\}$. Then it follows from (7.1.1) that

$$m_{0,\text{VC}}(\epsilon, \delta) = \max\left\{\frac{16}{\epsilon^2} \ln \frac{4}{\delta}, \frac{32d}{\epsilon^2} \ln \frac{32e}{\epsilon^2}\right\}. \tag{11.3.6}$$

To facilitate comparison, let us replace $2n/\delta$ by the larger term $4n/\delta$, so that

$$m_{0,\text{Hoeff}}(\epsilon, \delta) = \frac{\ln(4n/\delta)}{2\epsilon^2} = \frac{\ln n + \ln(4/\delta)}{2\epsilon^2}.$$

Then a simple computation shows that

$$m_{0,\text{VC}}(\epsilon, \delta) \leq m_{0,\text{Hoeff}}(\epsilon, \delta) \Rightarrow \frac{16\ln(4/\delta)}{\epsilon^2} \leq \frac{\ln n + \ln(4/\delta)}{2\epsilon^2}$$

$$\Rightarrow \frac{31\ln(4/\delta)}{2\epsilon^2} \leq \frac{\ln n}{2\epsilon^2} \Rightarrow \ln n \geq 31\ln(4/\delta) \Rightarrow n \geq \left(\frac{4}{\delta}\right)^{31}.$$

It is obvious that, for any reasonable value of δ, the corresponding value of n is astronomically large. Thus, in any practical situation, one is always better off using the Hoeffding bound (11.3.5) instead of the VC-bound (11.3.6).

Now let us study the specific problem forming the subject of the present section, namely, the minimization of an expected-value type of objective function. The above computation can be modified to compare the sample complexities of Algorithms 11.5 and 11.6. Suppose an accuracy $\epsilon > 0$, a confidence $\delta > 0$, and a level $\alpha > 0$ are specified. Then, in both algorithms, we have

$$n = \frac{\lg(2/\delta)}{\lg[1/(1-\alpha)]} \approx \frac{2/\delta}{\alpha}.$$

In Algorithm 11.5, we have

$$m_{\text{Hoeff}} = \frac{1}{2\epsilon^2} \ln \frac{4n}{\delta} \approx \frac{\ln(1/\alpha) + \ln(4/\delta) + \ln\ln(2/\delta)}{2\epsilon^2}.$$

In Algorithm 11.6, we have, as before

$$m_{VC} = \max\left\{\frac{16}{\epsilon^2}\ln\frac{4}{\delta}, \frac{32d}{\epsilon^2}\ln\frac{32e}{\epsilon^2}\right\}.$$

If we ignore the $\ln\ln(2/\delta)$ term in m_{Hoeff} as being insignificantly small, then

$$m_{VC} \leq m_{Hoeff} \Rightarrow \ln(1/\alpha) \geq 31\ln(4/\delta) \Rightarrow \alpha \leq \left(\frac{\delta}{4}\right)^{31}.$$

Again, for any reasonable values of α and δ, the above inequality will not hold, which implies that in practical situations, one is better off using Algorithm 11.5 instead of Algorithm 11.6.

However, this does not mean that all of the VC-dimension estimates derived in the preceding sections are of mere academic interest. The conservatism of the sample complexity estimates in (11.3.6) is well known in the statistical learning theory research community, and many researchers have attempted to improve these bounds. There can be two possible sources of conservatism in the estimates for m, namely: (i) the estimate of the VC-dimension d as given in Theorem 11.1, and (ii) the estimate of the sample complexity m for a given VC-dimension d, as given in (11.3.6). While the bound of Theorem 11.1 is, as of now, not known to be the best possible, it would not be surprising if this were to be so. Thus point (i) above is unlikely to be the source of conservatism in the sample complexity estimate. On the other hand, there is considerable room for improvement in the estimate (11.3.6). In [108] it is shown that under suitable conditions, we have the improved estimate

$$m = \max\left\{\frac{8}{\epsilon}, 5.170d + 2.0\log_2\frac{4C}{\delta}\right\},$$

where C is a "smoothness constant" that is specific to the problem at hand. It is not *a priori* clear from [108] how the constant C is to be estimated in practice, but if the above bound were indeed to be applicable to the particular problem at hand, then the value of m is *considerably* reduced. For instance, in [203], the problem of designing a first-order controller for the longitudinal axis of an aircraft is studied (see the next subsection for more details). In this instance, it turns out that $d = 118$ using Theorem 11.3. Setting $\alpha = 0.1, \epsilon = 0.1, \delta = 0.01$ leads to $n = 51$. Applying the bound of (11.3.6) leads to the estimate $m = 2,996,647$, which is clearly unrealistically large. On the other hand, using the above estimate and assuming that the "smoothness constant" C equals 1 leads to $m = 620$, which is quite a reasonable number. Note that in the present case, the sample complexity estimate of Algorithm 11.5 gives $m_{Hoeff} \approx 500$, which is comparable to the VC-type of bound *provided the bound of [108] holds*. However, it is not known at present whether this particular bound is applicable to the class of problems studied here, or even if it does, how one goes about estimating the smoothness constant C. Clearly this issue merits further investigation.

11.3.6 Robust Controller Design Using Randomized Algorithms: An Example

In this subsection, a "real life" example of the use of randomized algorithms is presented, namely the design of an inner loop controller for the longitudinal axis of an aircraft. Complete details of the example can be found in [203].[4]

The problem is to minimize the H_∞-norm of the weighted sensitivity function for the inner loop as far as possible while achieving good handling qualities by means of a prefilter in the stick path. The design of the prefilter is not discussed here.

The plant: A typical linearized model (short period approximation) for the longitudinal axis of an aircraft is given by

$$\dot{x} = Ax + Bu$$
$$y = Cx$$

where

$$x = y = \begin{bmatrix} \alpha \\ q \end{bmatrix}$$

$$A = \begin{bmatrix} Z_\alpha & 1 - Z_q \\ M_\alpha & M_q \end{bmatrix}, \quad B = \begin{bmatrix} Z_{\delta e} \\ M_{\delta e} \end{bmatrix}, \quad K = \begin{bmatrix} 1 & 0 \\ 0 & 1 \end{bmatrix}.$$

The derivatives at a flight condition are available from wind tunnel experiments in the form of Gaussian distributions as given below :

Z_α : mean = -0.9381, standard deviation = 0.0736
Z_q : mean = 0.0424, standard deviation = 0.0035
M_α : mean = 1.6630, standard deviation = 0.1385
M_q : mean = -0.8120, standard deviation = 0.0676
$Z_{\delta e}$: mean = -0.3765, standard deviation = 0.0314
$M_{\delta e}$: mean = -10.8791, standard deviation = 3.4695.

Thus, in the present instance, the parameter vector x consists of the above six variables, and $X = \mathbb{R}^6$. The probability measure $\mathcal{P}_X$ is the product of the above six individual Gaussian probability measures. Let the symbol $\mathcal{G}$ denote the corresponding collection of plants $\{G(x), x \in X\}$.

The Problem Formulation: Let G_0 denote the nominal plant, and let $\mathcal{G}$ denote the collection of all the plants generated by the distribution above. Typical models for the various hardware elements such as the sensors, actuators, structural filters, etc. are lumped together as a second-order transfer function at the input to the plant, as follows:

$$HW(s) = \frac{0.000697s^2 - 0.0397s + 1}{0.000867s^2 + 0.0591s + 1}$$

---

[4] The example in this section is due to Drs. Vijay V. Patel and Girish S. Deodhare of the Centre for Artificial Intelligence & Robotics, Bangalore, India.

(See Figure 11.6). The objective is to design a controller that minimizes the weighted sensitivity function at the output for the set $\mathcal{G}$ to ensure good disturbance rejection, while ensuring that a modified complementary sensitivity function at the input meets a certain bound for the nominal plant G_0, as follows:

$$\min \| W(I+GK)^{-1} \|_\infty \text{ subject to } \left\| \frac{0.75 K G_0}{1 + 1.25 K G_0} \right\|_\infty \leq 1,$$

where

$$W(s) = \begin{bmatrix} \frac{2.8*6.28*31.4}{(s+6.28)(s+31.4)} & 0 \\ 0 & \frac{2.8*6.28*3.14}{(s+6.28)(s+31.4)} \end{bmatrix}.$$

The bound on the modified complementary sensitivity ensures that the controller has a $\pm 6dB$ gain margin and a ± 35 degrees phase margin at the input to the nominal plant as required by the MIL Specs.

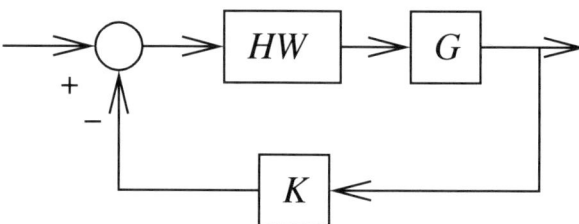

Fig. 11.6. Feedback System

The choice of the plant G in the minimization above is not obvious. If one chooses $G = G_0$, then one is guaranteeing only nominal performance with robust stabilization. However, when one perturbs a "good" nominal design with a 6 dB increase in the plant gain, the closed-loop response is almost oscillatory. Such behaviour is not possible with a structured singular value minimization. Therefore another option is to minimize the structured singular value and thereby guarantee robust performance over the set defined by the uncertainty specification of the gain and phase margin. This is a superset of $\mathcal{G}$, the set of models determined by the Gaussian distributions on the derivatives. As a result, the minimization of the structured singular value is overly conservative and leads to a significant sacrifice in the performance at the nominal plant condition G_0.

An H_∞-optimal design can overcome this problem to some extent. However, in the mixed sensitivity problem formulation, because of the nature of the modified complementary sensitivity function, one can optimize the performance only for $1.25 G_0$ and not G_0. (Note that this problem does not arise if one is designing for robust performance.) Moreover, no guarantee can be given regarding the performance of the controller for the set of plants $\mathcal{G}$.

11.3 VC-Dimension Estimates for Problems in Robust Controller Synthesis

For the purposes of comparison, an H_∞ design was performed for the above problem *without* taking into account the constraint on the modified sensitivity function, and an optimal eighth order controller was obtained.

Now suppose it is desired to design a *first-order controller* to satisfy the above design criteria. One possibility is to do an order reduction of the above eighth order H_∞-optimal controller to a first order controller, using a standard method such as Hankel norm minimization. In this case, it turns out that, if the order is reduced below 5, such a reduced-order controller does not even stabilize the nominal plant G_0. Thus, if we wish to find a good first-order controller, we are forced to look for alternate approaches.

Design using randomized algorithms: A formulation for the problem above using randomized algorithms is given below. Define the cost function to be minimized as follows:

$$\Psi(y) = \max\{\psi_1(y), \psi_2(y)\},$$

where

$$\psi_1(y) = \begin{cases} 1, & \text{if } \|\frac{0.75KG_0}{1+1.25KG_0}\|_\infty > 1 \\ 0 & \text{otherwise} \end{cases}$$

and

$$\psi_2(y) = E_{P_X}(\zeta(x,y)),$$

where

$$\zeta(x,y) = \begin{cases} 1 & \text{if } (G(x), K(y)) \text{ is unstable,} \\ \frac{\|W(I+G(x)K(y))^{-1}\|_\infty}{1+\|W(I+G(x)K(y))^{-1}\|_\infty} & \text{otherwise.} \end{cases}$$

Note that the definition of the cost function assures that, if a controller K fails to satisfy the constraint on the modified complementary sensitivity function, then the corresponding value of the objective function is automatically set to 1. Thus the role of the quantity ψ_1 is simply to guarantee that the controller generated by the randomized algorithm meets the gain and phase requirements for all the plants in $\mathcal{G}$ while providing reasonably good performance at the nominal plant condition.

The controller is assumed to have the form

$$K = \left[K_a \quad K_q \frac{(1+\tau_1 s)}{(1+\tau_2 s)}\right],$$

where the four variable parameters are chosen to be in the following ranges.

$$K_a \in [0, 2], \quad K_q \in [0, 1], \quad \tau_1 \in [0.01, 0.1], \quad \tau_2 \in [0.01, 0.1].$$

These limits come from practical considerations and previous experience. $\psi_1(y)$ does not figure in the calculation of the VC-dimensions below since it does not involve any empirical means.

Let the plant be represented by the polynomial factorization

452   11. Applications to Control Systems

$$G(x) = \begin{bmatrix} n_{G1}(x)/d_G(x) \\ n_{G2}(x)/d_G(x), \end{bmatrix}$$

and the controller be given by the polynomial factorization

$$K(y) = \begin{bmatrix} \dfrac{n_{K1}(y)}{d_K(y)} & \dfrac{n_{K2}(y)}{d_K(y)} \end{bmatrix}.$$

Then we can apply Theorem 11.3, after accounting for the fact that the H_∞-norm of a 2×2 transfer function needs to be evaluated. With this modification, the bound (11.3.4) becomes

$$\text{P-dim}(G) \leq 2l \log_2(4ert).$$

where $r = 4\beta_y n_w$, $t = 2n_c + 4n_w + 5$. In the present case, $\beta_y = 2$, $n_w = 7$, $n_c = 5$ and $l = 4$. This gives the bound on the P- dimension as 118.

Procedure:

1. Select m plants according to the Gaussian distributions where

$$m \geq \frac{32}{\epsilon^2}\left[\ln\frac{8}{\delta} + d\left(\ln\frac{16e}{\epsilon} + \ln\ln\frac{16e}{\epsilon}\right)\right].$$

   With $d = 118$, $\epsilon = 0.1$, $\delta = 0.01$, the bound on m evaluates to 2,619,047 plants.
2. Select n controllers with a uniform distribution where

$$n \geq \frac{lg(2/\delta)}{lg[1/(1-\alpha)]}.$$

   For $\alpha = 0.1$, this evaluates to 51 controllers.
3. Calculate the cost function for each controller with all the plants and select the minimum value. This is the optimal controller.

Design Using Randomzation: The above procedure was applied with $m = 200$ randomly generated plants and $n = 40$ controllers. Note that, while the number n of controllers is quite close to the number specified by the theory, the number m of plants is many orders of magnitude less than the theory specifies. The results are presented here and a comparison is made with the eighth-order H_∞ controller designed for nominal plant condition. The comparison is not really fair since the orders of the two controllers are quite different. Indeed, as mentioned above, the H_∞-optimal controller cannot be manipulated to yield a stabilizing first-order controller. However, in spite of being only of first order, the controller generated using the randomized approach performs quite well. In particular, the H_∞-norm of the weighted sensitivity function using the controller generated by the randomized approach is only about 7% worse than that using the H_∞-optimal controller.

Moreover, the randomized controller guarantees that the gain margin and phase margin conditions are met, while the H_∞-optimal controller does not.

Results:

Controller Type	$\| W(I + G_0K)^{-1} \|_\infty$	$\Psi(y_{\text{opt}})$
H_∞ Controller	2.9457	0.7477
Randomized Design	3.1570	0.7684

In summary, the above design example shows that randomized algorithms can be used effectively in a real-life example, and that it is possible to achive success using a far smaller number of samples than indicated by VC-dimension estimates.

11.4 A Learning Theory Approach to System Identification

In this section, the problem of system identification is formulated as a problem in statistical learning theory. By doing so, it is possible to derive *quantitative estimates* of the convergence rate of an identification algorithm, something that is not customarily done in the identification literature. Moreover, by appealing to the results in Section 9.3, it can be shown that in *any* situation where the family of models is parametrized by elements of a separable metric space, it is always possible to learn with prior information.

11.4.1 Problem Formulation

The aim of system identification is to fit given data, usually supplied in the form of a time series, with models from within a given model class. Let us denote the time series by $\{(y_t, u_t)\}_{t=-\infty}^{\infty}$, where u_t and y_t denote respectively the input and output of the system at time t. Let us denote by $\{h(\theta), \theta \in \Theta\}$ the family of input-output models that are to be used to fit this time series. The notation is now made precise.

For the class of systems under study, the output set is some $Y \subseteq \mathbb{R}^k$, while the input set is some $U \subseteq \mathbb{R}^\ell$ for some k and ℓ. To avoid technicalities, let us suppose that the inputs are restricted to belong to a *bounded* set U; this assumption ensures that any random variable assuming values in U has bounded moments of all orders. There is also a "loss function" $\ell : Y \times Y \to [0, 1]$ which is used to measure how well the predicted output matches the actual output.

To set up the time series that forms the input to identification, let us define $\mathcal{U} := \prod_{-\infty}^{\infty} U$, and define $\mathcal{Y}$ analogously. Equip the doubly infinite

454   11. Applications to Control Systems

Cartesian product $\mathcal{Y} \times \mathcal{U} := \prod_{-\infty}^{\infty}(Y \times U)$ with the product Borel σ-algebra, and call it $\mathcal{S}^{\infty}$. Next, introduce a probability measure $\tilde{P}_{\mathbf{y},\mathbf{u}}$ on the measurable space $(\mathcal{Y} \times \mathcal{U}, \mathcal{S}^{\infty})$. Following our earlier practice, let us define a "stochastic process" $\{(y_t, u_t)\}_{t=-\infty}^{\infty}$ as a measurable map from $(\mathcal{Y} \times \mathcal{U}, \mathcal{S}^{\infty}, \tilde{P}_{\mathbf{y},\mathbf{u}})$ into $\mathcal{Y} \times \mathcal{U}$. Let the coordinate random variables (y_t, u_t) be thought of as the components of the time series at time t, and let us assume that the time series is stationary, i.e. that the probability measure $\tilde{P}_{\mathbf{y},\mathbf{u}}$ is shift-invariant. Let $\tilde{P}_{\mathbf{y},\mathbf{u}}$ denote the one-dimensional marginal probability associated with $\tilde{P}_{\mathbf{y},\mathbf{u}}$ on Y, and note that $\tilde{P}_{\mathbf{y},\mathbf{u}}$ is a probability measure on the set $Y \times U$. Let $U_{-\infty}^0$ denote the one-sided infinite Cartesian product $U_{-\infty}^0 := \prod_{-\infty}^{0} U$, and for a given two-sided infinite sequence $\mathbf{u} \in \mathcal{U}$, define

$$\mathbf{u}_t := (u_{t-1}, u_{t-2}, u_{t-3}, \ldots) \in U_{-\infty}^0.$$

Thus $\mathbf{u}_t$ denotes the "infinite past" of the input sequence at time t. With this preliminary notation, we can set up the problem under study.

The input to the identification process is a time series $\{(y_t, u_t)\}_{t \geq 1}$ generated through a stochastic process, as described above. Thus, while it is assumed that stochastic process stretches into the infinite past, the identification algorithm has a definite starting point, which is taken as time $t = 0$. To fit this time series, we use a family of models $\{h(\theta), \theta \in \Theta\}$, where each $h(\theta)$ denotes an input-output mapping from $U_{-\infty}^0$ to Y, and the parameter θ captures the variations in the model family. Thus for the system parametrized by θ, the output at time t in response to the input sequence $\mathbf{u} \in \mathcal{U}$ is given by $h(\theta) \cdot \mathbf{u}_t$. Note that this definition automatically guarantees that each system in the family of models is time-invariant.

For each parameter $\theta \in \Theta$, define the objective function

$$J(\theta) := E[\ell(y_t, h(\theta) \cdot \mathbf{u}_t), \tilde{P}_{\mathbf{y},\mathbf{u}}]. \tag{11.4.1}$$

Thus $J(\theta)$ is the expected value of the loss incurred by using the model output $h(\theta) \cdot \mathbf{u}_t$ to predict the actual output y_t. Note that, since the only value of $\mathbf{y}$ that appears within the expected value is y_t, we can actually replace the measure $\tilde{P}_{\mathbf{y},\mathbf{u}}$ by $\tilde{P}_{y,\mathbf{u}}$. In other words, we can also write

$$J(\theta) := E[\ell(y_t, h(\theta) \cdot \mathbf{u}_t), \tilde{P}_{y,\mathbf{u}}].$$

Thus the expectation is taken with respect to the 'one-dimensional' marginal measure $\tilde{P}_{y,\mathbf{u}}$ on $Y \times \mathcal{U}$.

One of the most commonly used loss functions is the squared error; thus

$$\ell(y, z) := \| y - z \|^2,$$

where $\| \cdot \|$ is the usual Euclidean or ℓ_2-norm on $\mathbb{R}^k$. In this case $J(\theta)$ is the expected value of the mean squared prediction error when the map $h(\theta)$ is used to predict y_t. Note that, by the assumption of stationarity, the quantity on the right side of (11.4.1) is independent of t.

11.4 A Learning Theory Approach to System Identification

Now we are in a position to state the problem under study.

The Identification Problem: Determine a $\theta \in \Theta$ that minimizes the error measure $J(\theta)$.

Suppose the measured output y_t corresponds to a noise-corrupted output of a 'true' system f_{true}, and that ℓ is the squared error, as above. In such a case, the problem formulation becomes the following: Suppose the input sequence $\{u_t\}_{-\infty}^{\infty}$ is i.i.d. according to some law P, and that $\{\eta_t\}_{-\infty}^{\infty}$ is a measurement noise sequence that is zero mean and i.i.d. with law Q. Suppose in addition that u_i, η_j are independent for each i, j. Now suppose that

$$y_t = f_{\text{true}} \cdot \mathbf{u}_t + \eta_t, \quad \forall t. \tag{11.4.2}$$

In such a case, the expected value in (11.4.1) can be expressed in terms of the probability measure $Q \times P^{\infty}$, and becomes

$$\begin{aligned} J(\theta) &= E[\|\, (f_{\text{true}} - h(\theta)) \cdot \mathbf{u}_t + \eta_t \,\|^2, Q \times P^{\infty}] \quad (11.4.3) \\ &= E[\|\, \tilde{h}(\theta) \cdot \mathbf{u}_t \,\|^2, P^{\infty}] + E[\|\, \eta \,\|^2, Q], \end{aligned}$$

where $\tilde{h}(\theta) := h(\theta) - f_{\text{true}}$. Since the second term is independent of θ, we effectively minimize only the first term. In other words, by minimizing $J(\theta)$ with respect to θ, we will find the best approximation to the true system f_{true} in the model family $\{h(\theta), \theta \in \Theta\}$. Note that it is *not* assumed that the true system f_{true} belongs to $\{h(\theta), \theta \in \Theta\}$.

11.4.2 A General Result

One can divide the main challenges of system identification into three successively stronger questions, as follows: As more and more data is provided to the identification algorithm:

1. Does the estimation error between the outputs of the identified model and the actual time series approach the minimum possible estimation error achievable by any model within the given model class? In other words, if θ_t denotes the parameter estimate at time t, does $J(\theta_t)$ approach $J^* := \inf_{\theta \in \Theta} J(\theta)$?
2. Assuming that the data is generated by a "true" system whose output is corrupted by measurement noise, does the identified model converge to the best possible approximation of the "true" system within the model class? In other words, suppose we define some kind of metric distance ρ between pairs of input-output maps. Does $\rho[f_{\text{true}}, h(\theta_t)]$ approach the quantity $\inf_{\theta \in \Theta} \rho[f_{\text{true}}, h(\theta)]$?
3. Assuming that the the true system belongs to the model class, does the estimated model converge to the true system? In other words, suppose $f_{\text{true}} = h(\theta_{\text{true}})$ for some "true" parameter vector θ_{true}. Does $\theta_t \to \theta_{\text{true}}$ as $t \to \infty$?

From a technical standpoint, Questions 2 and 3 are easier to answer than Question 1. Since identification is carried out recursively, the output of the identification algorithm is a sequence of estimates $\{\theta_t\}_{t\geq 1}$, or equivalently, a sequence of estimated models $\{h(\theta_t)\}_{t\geq 1}$. Traditionally a positive answer to Question 2 is assured by assuming that Θ is a *compact* set, which in turn ensures that the sequence $\{\theta_t\}$ contains a convergent subsequence. Alternatively, it is possible to use some kind of "regularization" whereby the objective function $J(\theta)$ is augmented by an additional term $\| \theta \|^2$. Adding this term to $J(\theta)$ effectively ensures that all iterations θ_t are contained within a sphere of finite radius. Either way, the aim is to ensure that $\{\theta_t\}$ contains a convergent subsequence. If the answer to Question 1 is "yes," and if θ^* is a limit point of the sequence, then by appealing some continuity arguments it can be shown that $J(\theta^*) = J^*$. In turn this implies that the expected value of the loss function $\ell(f_{\text{true}}, h(\theta^*))$ equals J^*. Suppose now that the loss function ℓ is chosen in such a way that $\ell(f_{\text{true}}, h(\theta^*))$ measures a distance between f_{true} and $h(\theta^*)$. For instance, in case $\ell(y, z) = \| y - z \|^2$, then $J(\theta_t)$ is the square of a metric distance between f_{true} and $h(\theta_t)$. In this case, it readily follows that $h(\theta^*)$ is the "best possible" fit to the true system f_{true} with respect to this metric distance. Coming now to Question 3, suppose θ_{true} is the parameter of the "true" model, and let $h(\theta_{\text{true}})$ denote the "true" system. Suppose θ^* is a limit point of the sequence $\{\theta_t\}$. The traditional way to ensure that $\theta_{\text{true}} = \theta^*$ is to assume that the input to the true system is "persistingly exciting" or "sufficiently rich," so that the only way for $h(\theta^*)$ to match the performance of $h(\theta_{\text{true}})$ is to have $\theta^* = \theta_{\text{true}}$.

With this background, the emphasis in this section is on providing an affirmative answer to Question 1. A sufficient condition for this is given in Theorem 11.5 below. Observe that the probability measure $\tilde{P}_{\mathbf{y},\mathbf{u}}$ in (11.4.1) is in general *unknown*. This is because $\tilde{P}_{\mathbf{y},\mathbf{u}}$ often corresponds to the "true but unknown" system, as in (11.4.2). Thus it is not reasonable to assume that $\tilde{P}_{\mathbf{y},\mathbf{u}}$ is known. In turn, this implies that in general it is *not possible* to compute the objective function $J(\theta)$ in (11.4.1) directly, for a given choice of the parameter vector θ. An indirect method must therefore be found to minimize $J(\theta)$. A natural algorithm is the so-called "minimum empirical cost" algorithm described next. For each $t \geq 1$ and each $\theta \in \Theta$, define the empirical error

$$\hat{J}_t(\theta) := \frac{1}{t} \sum_{i=1}^{t} \ell[y_t, h(\theta) \cdot \mathbf{u}_t].$$

Note that, unlike $J(\theta)$, the quantity $\hat{J}_t(\theta)$ can indeed be computed on the basis of the data available at time t. At time t, choose θ_t^* so as to minimize $\hat{J}_t(\theta)$; that is,

$$\theta_t^* = \operatorname*{Arg\,min}_{\theta \in \Theta} \hat{J}_t(\theta).$$

For this algorithm, we have the following result.

11.4 A Learning Theory Approach to System Identification

Theorem 11.5. *Let*
$$J^* := \inf_{\theta \in \Theta} J(\theta).$$

Define the quantity
$$q(t, \epsilon) := \tilde{P}_{\mathbf{y},\mathbf{u}}\{\sup_{\theta \in \Theta} |\hat{J}_t(\theta) - J(\theta)| > \epsilon\}. \quad (11.4.4)$$

Suppose it is the case that $q(t, \epsilon) \to 0$ as $t \to \infty$. Then for the minimum empirical cost algorithm we have
$$\tilde{P}_{\mathbf{y},\mathbf{u}}\{\hat{J}_t(\theta_t^*) > J^* + \epsilon\} \to 0 \text{ as } t \to \infty.$$

In particular, given any $\epsilon, \delta > 0$, choose an integer $t_0(\epsilon, \delta)$ such that
$$q(t, \epsilon/3) \leq \delta \; \forall t \geq t_0(\epsilon, \delta). \quad (11.4.5)$$

Then
$$\tilde{P}_{\mathbf{y},\mathbf{u}}\{\hat{J}_t(\theta_t^*) > J^* + \epsilon\} \leq \delta \; \forall t \geq t_0(\epsilon, \delta).$$

Remark: The theorem states that if the family of error measures $\{J(\theta), \theta \in \Theta\}$ has the UCEM property, then the natural algorithm of choosing θ_t so as to minimize the empirical estimate $\hat{J}(\theta)$ at time t is "asymptotically optimal."

Note that the result given in Theorem 11.5 is not by any means the most general possible. In particular, it is possible to show that if θ_t is chosen so as to "nearly" minimize the empirical error "most of the time," then the minimum empirical cost algorithm is still asymptotically optimal.

Proof. Suppose $q(t, \epsilon) \to 0$ as $t \to \infty$. Given $\epsilon, \delta > 0$, choose t_0 large enough that (11.4.5) holds. Then
$$\tilde{P}_{\mathbf{y},\mathbf{u}}\{\sup_{\theta \in \Theta} |\hat{J}_t(\theta) - J(\theta)| > \epsilon/3\} < \delta \; \forall t \geq t_0(\epsilon, \delta). \quad (11.4.6)$$

Select a $\theta_\epsilon \in \Theta$ such that $J(\theta_\epsilon) \leq J^* + \epsilon/3$. Such a θ_ϵ exists in view of the definition of J^*. Then, in view of (11.4.6), we can say with confidence $1 - \delta$ that
$$\hat{J}(\theta_t) \geq J(\theta_t) - \epsilon/3, \text{ and } \hat{J}(\theta_\epsilon) \leq J(\theta_\epsilon) + \epsilon/3.$$

By definition,
$$\hat{J}(\theta_t) \leq \hat{J}(\theta_\epsilon).$$

Combining these two inequalities shows that
$$J(\theta_t) \leq \hat{J}(\theta_t) + \epsilon/3 \leq \hat{J}(\theta_\epsilon) + \epsilon/3 \leq J(\theta_\epsilon) + 2\epsilon/3 \leq J^* + 2\epsilon/3 + \epsilon/3 = J^* + \epsilon.$$

This statement holds with confidence $1 - \delta$. ∎

Thus the sample complexity of ensuring that $J(\theta_t) \leq J^* + \epsilon$ is at most equal to the sample complexity of $q(m, \epsilon/3)$. This naturally brings up the question as to what kinds of families $\{h(\theta), \theta \in \Theta\}$ have this particular UCEM property, and what their sample complexities are like. These questions are addressed in the next subsection.

11.4.3 Sufficient Conditions for the UCEM Property

In a seminal paper [116], Ljung has shown that indeed Question 1 can be answered in the affirmative provided empirical estimates of the performance of each model $h(\theta)$ converge *uniformly* to the corresponding true performance, where the uniformity is with respect to $\theta \in \Theta$. Thus he was among the first to realize the importance of the UCEM property in establishing the convergence properties of identification algorithms. He also showed that this particular uniform convergence property does hold, provided two assumptions are satisfied, namely:

- The model class consists of uniformly exponentially stable systems, and
- The parameter θ enters the description of the model $h(\theta)$ in a "differentiable" manner. Coupled with the assumption that Θ is a compact set, this assumption implies that various quantities have bounded gradients with respect to θ.

It is interesting to note that Ljung's UCEM result predates that of Vapnik and Chervonenkis for real-valued functions as found in [196]. Some related work can be found in [39, 40].

Now it turns out that the assumptions employed in [116] in fact guarantee the finiteness of the VC-dimension of the family of cost functions associated with the model family $\{h(\theta), \theta \in \Theta\}$. See for example [80]. Thus in principle it is possible to derive UCEM results very similar to (if not exactly identical to) those in [116] using the results of Chapter 10. However, it is well-known that the estimates of the (finite) VC-dimension based on differentiability assumptions are extremely conservative. In the present subsection, we shall derive an alternate UCEM result based on more direct arguments.

We begin by listing below the assumptions regarding the family of models employed in identification, and on the time series. Recall the symbol $\tilde{h}(\theta)$ to denote the function $\mathbf{u}_t \mapsto (f_{\text{true}} - h(\theta)) \cdot \mathbf{u}_t$, which maps $U_{-\infty}^0$ into $\mathbb{R}^k$. Define the collection of functions $\mathcal{G}$ mapping $\mathcal{U}$ into $\mathbb{R}$ as follows:

$$g(\theta) := \mathbf{u}_0 \mapsto \| (f - h(\theta)) \cdot \mathbf{u}_0 \|^2 : \mathcal{U} \to \mathbb{R},$$

$$\mathcal{G} := \{g(\theta) : \theta \in \Theta\}.$$

Now the various assumptions are listed.

A1. There exists a constant M such that

$$|g(\theta) \cdot \mathbf{u}_0| \leq M, \; \forall \theta \in \Theta, \mathbf{u} \in \mathcal{U}.$$

This assumption can be satisfied, for example, by assuming that the true system and each system in the family $\{h(\theta), \theta \in \Theta\}$ is BIBO stable (with an upper bound on the gain, independent of θ), and that the set U is bounded (so that $\{u_t\}$ is a bounded stochastic process).

A2. For each integer $k \geq 1$, define

$$g_k(\theta) \cdot \mathbf{u}_t := g(\theta) \cdot (u_{t-1}, u_{t-2}, \ldots, u_{t-k}, 0, 0, \ldots).$$

With this notation, define

$$\mu_k := \sup_{\mathbf{u} \in \mathcal{U}} \sup_{\theta \in \Theta} |(g(\theta) - g_k(\theta)) \cdot u_0|.$$

Then the assumption is that μ_k is finite for each k and approaches zero as $k \to \infty$. This assumption essentially means that each of the systems in the model family has decaying memory (in the sense that the effect of the values of the input in the distant past on the current output becomes negligibly small). In the SISO case, this assumption is satisfied, for example, if
– Each of the models $h(\theta)$ is a linear ARMA model of the form

$$y_t = \sum_{i=1}^{l} a_i(\theta) u_{t-i} + b_i(\theta) y_{t-i},$$

– The characteristic polynomials

$$\phi(\theta, z) := z^{l+1} - \sum_{i=1}^{l} b_i(\theta) z^{l-i}$$

all have their zeros inside a circle of radius $\rho < 1$, where ρ is independent of θ.
– The numbers $a_i(\theta)$ are uniformly bounded with respect to θ.
The extension of the above condition to MIMO systems is straightforward and is left to the reader.

A3. Consider the collection of maps $\mathcal{G}_k = \{g_k(\theta) : \theta \in \Theta\}$, viewed as maps from U^k into $\mathbb{R}$. For each k, the family $\mathcal{G}_k$ has finite P-dimension, bounded by $d(k)$. (See [200], Chapter 4 for a definition of the P-dimension.)

Now we can state the main theorem.

Theorem 11.6. *Define the quantity $q(t, \epsilon)$ as in (11.4.4) and suppose Assumptions A1 through A3 are satisfied. Given un $\epsilon > 0$, choose $k(\epsilon)$ large enough that $\mu_k \leq \epsilon/4$ for all $k \geq k(\epsilon)$. Then for all $t \geq k(\epsilon)$ we have*

$$q(t, \epsilon) \leq 8k(\epsilon) \left(\frac{32e}{\epsilon} \ln \frac{32e}{\epsilon}\right)^{d(k(\epsilon))} \exp(-\lfloor t/k(\epsilon) \rfloor \epsilon^2/512M^2), \quad (11.4.7)$$

where $\lfloor t/k(\epsilon) \rfloor$ denotes the largest integer part of $t/k(\epsilon)$.

Remark: From Theorem 11.5, it is also possible to quantify the rate of convergence of the performance of the estimated model to the optimal performance.

460   11. Applications to Control Systems

Proof. Write $g(\theta) = g_k(\theta) + (g(\theta) - g_k(\theta))$, and define

$$q_1^k(t,\epsilon) := \Pr\{\sup_{\theta \in \Theta} \left|\frac{1}{t}\sum_{i=1}^{t} g_k(\theta) \cdot \mathbf{u}_i - E[g_k(\theta) \cdot \mathbf{u}_i, \tilde{P}]\right| > \epsilon\},$$

$$q_2^k(t,\epsilon) := \Pr\{\sup_{\theta \in \Theta} \left|\frac{1}{t}\sum_{i=1}^{t} (g(\theta) - g_k(\theta)) \cdot \mathbf{u}_i - E[(g(\theta) - g_k(\theta)) \cdot \mathbf{u}_i, \tilde{P}]\right| > \epsilon\},$$

where $\tilde{P}$ is used as a shorthand for $\tilde{P}_{\mathbf{y},\mathbf{u}}$. Then it is easy to see that

$$q(t,\epsilon) \leq q_1^k(t,\epsilon/2) + q_2^k(t,\epsilon/2). \tag{11.4.8}$$

Now observe that if k is sufficiently large that $\mu_k \leq \epsilon/4$, then $q_2^k(t,\epsilon) = 0$. This is because, if $|(g(\theta) - g_k(\theta)) \cdot \mathbf{u}_i|$ is always smaller than $\epsilon/4$, then its expected value is also smaller than $\epsilon/4$, so that their difference can be at most equal to $\epsilon/2$. Since this is true for all $\mathbf{u}$ and all θ, the above observation follows. Thus it follows that if $k(\epsilon)$ is chosen large enough that $\mu_k \leq \epsilon/4$ for all $k \geq k(\epsilon)$, then

$$q(t,\epsilon) \leq q_1^{k(\epsilon)}(t,\epsilon/2) \; \forall t \geq k(\epsilon), \; \forall \epsilon. \tag{11.4.9}$$

Hence the rest of the proof consists of estimating $q_1^{k(\epsilon)}(t,\epsilon)$ when $t \geq k(\epsilon)$.

From here onwards, let us replace $k(\epsilon)$ by k in the interests of notational clarity. When $t \geq k$, define $l := \lfloor t/k \rfloor$, and $r = t - kl$. Partition $\{1, \ldots, t\}$ into k intervals, as follows.

$$I_j := \{i, i+k, \ldots, i+lk\} \text{ for } 1 \leq j \leq r, \text{ and}$$

$$I_j := \{i, i+k, \ldots, i+(l-1)k\} \text{ for } r+1 \leq j \leq k.$$

Then we can write

$$\frac{1}{t}\sum_{i=1}^{t} g_k(\theta) \cdot \mathbf{u}_i = \frac{1}{t}\sum_{j=1}^{k}\sum_{i \in I_j} g_k(\theta) \cdot \mathbf{u}_i.$$

Now define

$$\alpha_j := \frac{1}{l+1}\left|\sum_{i \in I_j}\left(g_k(\theta) \cdot \mathbf{u}_i - E[g_k(\theta) \cdot \mathbf{u}_i, \tilde{P}]\right)\right|, \; 1 \leq j \leq r, \text{ and}$$

$$\alpha_j := \frac{1}{l}\left|\sum_{i \in I_j}\left(g_k(\theta) \cdot \mathbf{u}_i - E[g_k(\theta) \cdot \mathbf{u}_i, \tilde{P}]\right)\right|, \; r+1 \leq j \leq k.$$

Then, noting that $E[g_k(\theta) \cdot \mathbf{u}_i, \tilde{P}]$ is independent of i due to the stationarity assumption, we get

11.4 A Learning Theory Approach to System Identification

$$\left| \frac{1}{t} \sum_{i=1}^{t} g_k(\theta) \cdot \mathbf{u}_i - E[g_k(\theta) \cdot \mathbf{u}_i, \tilde{P}] \right| \leq \left| \sum_{j=1}^{r} \frac{l+1}{t} \alpha_j + \sum_{j=r+1}^{k} \frac{l}{t} \alpha_j \right|.$$

It follows that if $\alpha_j \leq \epsilon$ for each j, then the left side of the inequality is also less than ϵ. So the following containment of events holds:

$$\left\{ \sup_{\theta \in \Theta} \left| \frac{1}{t} (g_k \cdot \mathbf{u}_i - E[g_k \cdot \mathbf{u}_i, \tilde{P}]) \right| > \epsilon \right\} \subseteq \bigcup_{j=1}^{k} \{\alpha_j > \epsilon\}.$$

Hence

$$q_1^k(t,\epsilon) \leq \sum_{j=1}^{k} \tilde{P}\{\alpha_j > \epsilon\}. \tag{11.4.10}$$

Now note that each $g_k \cdot \mathbf{u}_i$ depends on only u_{i-1} through u_{i-k}. Hence, in the summation defining each of the α_j, the various quantities being summed are independent. Since it is assumed that the family $\{g_k(\theta), \theta \in \Theta\}$ has finite P-dimension $d(k)$, the quantity $q_1^k(t,\epsilon)$ can be estimated using (7.1.1). A small adjustment is necessary, however. The hypotheses of Theorem 7.1 assume that all the functions under study assume values in the interval $[0,1]$, whereas in the present instance the functions $h(\theta) \cdot \mathbf{u}_i$ all assume values in the interval $[-M, M]$. Thus the range of values now has width $2M$ instead of one. With this adjustment, (7.1.1) implies that

$$\Pr\{\alpha_j > \epsilon\} \leq 8 \left(\frac{16e}{\epsilon} \ln \frac{16e}{\epsilon} \right)^{d(k)} \exp(-(l+1)^2 \epsilon^2 / 128 M^2), \text{ for } 1 \leq j \leq r,$$

and

$$\Pr\{\alpha_j > \epsilon\} \leq 8 \left(\frac{16e}{\epsilon} \ln \frac{16e}{\epsilon} \right)^{d(k)} \exp(-l^2 \epsilon^2 / 128 M^2), \text{ for } r+1 \leq j \leq k.$$

Since $\exp(-(l+1)^2) < \exp(-l^2)$, the $l+1$ term can be replaced by l in the first inequality as well. Substituting these estimates into (11.4.10) yields the desired estimate

$$q_1^k(t,\epsilon) \leq 8k \left(\frac{16c}{\epsilon} \ln \frac{16e}{\epsilon} \right)^{d(k)} \exp(-l^2 \epsilon^2 / 128 M^2). \tag{11.1.11}$$

Finally, the conclusion (11.4.7) is obtained by replacing ϵ by $\epsilon/2$ in the above expression, and then applying (11.4.9). ∎

11.4.4 Bounds on the P-Dimension

In order for the estimate in Theorem 11.6 to be useful, it is necessary for us to derive an estimate for the P-dimension of the family of functions defined by

462   11. Applications to Control Systems

$$\mathcal{G}_k := \{g_k(\theta) : \theta \in \Theta\}, \tag{11.4.12}$$

where $g_k(\theta) : U^k \to \mathbb{R}$ is defined by

$$g_k(\theta)(\mathbf{u}) := \| (f - h(\theta)) \cdot \mathbf{u}_k \|^2,$$

where

$$\mathbf{u}_k := (\ldots, 0, u_k, u_{k-1}, \ldots, u_1, 0, 0, \ldots).$$

Note that, in the interests of convenience, we have denoted the infinite sequence with only k nonzero elements as $u_k, \ldots, u_1$ rather than $u_0, \ldots, u_{1-k}$ as done earlier. Clearly this makes no difference. In this subsection, we state and prove such an estimate for the commonly occurring case where each system model $h(\theta)$ is an ARMA model where the parameter θ enters linearly. Specifically, it is supposed that the model $h(\theta)$ is described by

$$x_{t+1} = \sum_{i=1}^{l} \theta_i \, \phi_i(x_t, u_t), \quad y_t = x_t, \tag{11.4.13}$$

where $\theta = (\theta_1, \ldots, \theta_l) \in \Theta \subseteq \mathbb{R}^l$, and each $\phi_i(\cdot, \cdot)$ is a polynomial of degree no larger than r in the components of x_t, u_t.

Theorem 11.7. *With the above assumptions, we have that*

$$\begin{aligned} \text{P-dim}(\mathcal{G}_k) &\leq 9l + 2l \lg[2(r^{k+1} - 1)/(r-1)] \\ &\approx 9l + 2lk \lg(2r) \text{ if } r > 1. \end{aligned} \tag{11.4.14}$$

In case $r = 1$ so that each system is linear, the above bound can be simplified to

$$\text{P-dim}(\mathcal{G}_k) \leq 9l + 2l \lg(2k). \tag{11.4.15}$$

Remark: It is interesting to note that the above estimate is *linear* in both the number of parameters l and the duration k of the input sequence $\mathbf{u}$, but is only logarithmic in the degree of the polynomials ϕ_i. In the practically important case of linear ARMA models, even k appears inside the logarithm.

Proof. For each function $g_k(\theta) : U^k \to \mathbb{R}$ defined as in (11.4.12), define an associated function $g'_k : U^k \times [0,1] \to \{0,1\}$ as follows:

$$g'_k(\theta)(\mathbf{u}, c) := \eta[g_k(\theta)(\mathbf{u}) - c],$$

where $\eta(\cdot)$ is the Heaviside or "step" function. Then it follows from Lemma 4.1 that

$$\text{P-dim}(\mathcal{G}_k) = \text{VC-dim}(\mathcal{G}'_k).$$

Next, to estimate VC-dim($\mathcal{G}'_k$), we use Corollary 10.2, which states that, if the condition $\eta[g_k(\theta)\mathbf{u} - c] = 1$ can be stated as a Boolean formula involving s polynomial inequalities, each of degree no larger than d, then

11.4 A Learning Theory Approach to System Identification

$$\text{VC-dim}(\mathcal{G}'_k) \leq 2l \lg(4eds). \quad (11.4.16)$$

Thus the proof consists of showing that the conditions needed to apply this bound hold, and of estimating the constants d and s.

Towards this end, let us back-substitute repeatedly into the ARMA model (11.4.13) to express the inequality

$$\| (f - h(\theta))\mathbf{u}_k \|^2 - c < 0$$

as a polynomial inequality in $\mathbf{u}$ and the θ-parameters. To begin with, we have

$$\begin{aligned} x_{k+1} &= \sum_{i=1}^{l} \theta_i \, \phi_i(x_k, u_k) \\ &= \sum_{i=1}^{l} \theta_i \phi_i \left(\sum_{j=1}^{l} \theta_j \, \phi_j(x_{k-1}, u_{k-1}) \right) \\ &= \ldots \quad (11.4.17) \end{aligned}$$

Thus each time one of the functions ϕ_i is applied to its argument, the degree with respect to any of the θ_j goes up by a factor of r. In other words, the total degree of x_{k+1} with respect to each of the θ_j is no larger than $1 + r + r^2 + \ldots + r^k = (r^{k+1} - 1)/(r - 1)$. If $r = 1$, then the degree is simply k. Next, we can write

$$\| x_{k+1} \|^2 - c < 0 \iff x'_{k+1} x_{k+1} - c < 0.$$

This is a single polynomial inequality. Moreover, the degree of this polynomial in the components of θ is at most $2(r^{k+1} - 1)/(r - 1)$ if $r > 1$, and $2k$ if $r = 1$. Thus we can apply the bound (11.4.16) with and $s = 1$, and

$$d = \begin{cases} \frac{2(r^{k+1}-1)}{r-1} & \text{if } r > 1, \\ 2k & \text{if } r = 1. \end{cases}$$

The desired estimate now follows on noting that $\lg e < 1.5$, so that $\lg(8e) < 4.5$. ■

Notes and References The probabilistic approach to robust control systems can perhaps be dated back to the work of Stengel and his coworkers [165], [128]. Recent interest in probabilistic methods can perhaps be said to have been spurred by the NP-hardness results of Poljak-Rohn [161], Nemirovskii [150] and other related work. The material in Section 11.1 on the randomized approach to robustness analysis is taken from the work of Khargonekar-Tikku [101] and Tempo et al. [181]. The material in Sections 11.2 and 11.3 can be found in [202]. There is a vast literature on the topic of system identification. See for example the book by Ljung [117] for a comprehensive treatment. Many new discoveries are constantly being made in this ever-expanding subject. The idea of treating system identification as a learning problem is a natural one, and is pursued in [42, 211, 44] for example. The present results are from [206].

12. Some Open Problems

In the first edition of this book, a total of sixteen open problems have been listed that appeared both interesting and important at that time. Since that time, some of these open problems were solved, but the rest remain open. Consequently, in the present chapter, the original sixteen open problems are retained, and are discussed in that order. Wherever the problem has been solved, the solution is briefly summarized. In case the problem still remains open, the original motivating discussion is retained.

Problem 12.1. *Is it possible to dispense with the assumption that the learning inputs are independent? Is it possible to introduce explicitly a notion of "time" into the learning problem formulation?* (cf. Chapters 2, 3 and 11). A substantial amount of progress has been made in this direction since the publication of the first edition. The notion of "mixing" stochastic processes provides an excellent setting for dispensing with the assumption that the samples are independent. Specifically, the results in Section 3.4 show that both the UCEM property and PAC learnability are preserved when the i.i.d. input process is replaced by a β-mixing process. The results in Section 3.5 provide some natural sufficient conditions for the state sequence of a Markov process and the output sequence of a hidden Markov model to be β-mixing.

However, there is still an interesting open problem, namely: *Is the UCEM property preserved when an i.i.d. input sequence is replaced by an α-**mixing** sequence?*

Problem 12.2. *Are the properties of uniform convergence of empirical means, and learnability preserved when the family of probability measures is replaced by its closure?* (cf. Chapter 3.) It is now known that the answer to both of these questions is in the affirmative.

Problem 12.3. *Is there a notion analogous to Probably Uniformly Approximately Correctness (PUAC) in the case of model-free learning?* (cf. Chapter 3). This problem still remains open, so the original motivation is retained.

Recall that in the above-cited chapter, we defined the efficacy of a learning algorithm in terms of various quantities. In the case where the learning data consists of labelled samples of the form

$$(x_1, f(x_1)), \ldots, (x_m, f(x_m)),$$

where f is the unknown target function, we used the symbol $h_m(f;\mathbf{x})$ to denote the hypothesis generated by the algorithm when the target function is f and the multisample is $\mathbf{x}$. Then we defined the quantities (cf. (3.2.4))

$$r(m,\epsilon) := \sup_{P \in \mathcal{P}} \sup_{f \in \mathcal{F}} P^m\{\mathbf{x} \in X^m : d_P[f, h_m(f;\mathbf{x})] > \epsilon\},$$

and (cf. (3.2.13))

$$s(m,\epsilon) := \sup_{P \in \mathcal{P}} P^m\{\mathbf{x} \in X^m : \sup_{f \in \mathcal{F}} d_P[f, h_m(f;\mathbf{x})] > \epsilon\}.$$

The main difference is that the supremum with respect to $f \in \mathcal{F}$ is taken *inside* the probability P^m in the case of $s(m,\epsilon)$. An algorithm was defined to be probably approximately correct (PAC) if $r(m,\epsilon) \to 0$ as $m \to \infty$, and was defined to be probably *uniformly* approximately correct (PUAC) if $s(m,\epsilon) \to 0$ as $m \to \infty$. It is clear that PUAC is a stronger property than PAC. Moreover, an algorithm can be PAC without being PUAC; see Example 3.9.

The main reason for introducing the PUAC property is this: In all the literature preceding this book, a family $\mathcal{F}$ having finite P-dimension (or a concept class $\mathcal{C}$ having finite VC-dimension) has always been presented as a sufficient condition for the function (or concept) class to be *PAC learnable*. However, a perusal of Theorem 7.7 shows that these conditions are actually enough to imply the much stronger conclusion that the function (or concept) class is *PUAC learnable*. Moreover, in the case of distribution-free concept learning, the finiteness of the VC-dimension is also *necessary* for PAC learnability, whence it is also necessary for PUAC learnability. Thus, in some sense, PUAC instead of PAC "comes for free." The only price to be paid for concluding PUAC instead of PAC is a *very* slight increase in sample complexity; compare the bounds in (7.2.2) and (7.2.4). Thus, in the case where the learning problem is that of trying to fit a hypothesis to an unknown target function, the PUAC property is at least as interesting and natural as the PAC property. Moreover, in the case where learning takes place under a fixed probability measure, the property of an algorithm being PUAC can be naturally interpreted in terms of the convergence of the stochastic process

$$\mathbf{x}^* \mapsto P^m\{\mathbf{x} \in X^m : \sup_{f \in \mathcal{F}} d_P[f, h_m(f;\mathbf{x})] > \epsilon\},$$

to zero, while the PAC-ness of an algorithm does not have such a natural interpretation.

All this leads to the problem posed above, namely: in the model-free learning problem, what is an appropriate analog of an algorithm being PUAC?

It has been said above that "PUAC learnability comes for free." The basis for this statement is that the exponents in the bounds for $s(m,\epsilon)$ in (7.2.3) and $r(m,\epsilon)$ in (7.2.1) are the same. Note that $s(m,\epsilon)$ is pertinent to PUAC learnability, while $r(m,\epsilon)$ is pertinent to PAC learnability. However, thanks to the ingenious arguments of Blumer *et al.*, the exponent in the bound for $r(m,\epsilon)$ in the case of *concept* learning changes to $-m\epsilon/2$, which means that the bound decays more quickly. Thus one can ask whether the bound (7.2.3) for the quantity $s(m,\epsilon)$ in the case of function learning can be improved in the case of *concept* learning.

Problem 12.4. *What are necessary and/or sufficient conditions for a function or concept class to be PUAC learnable? Are there examples of a function or concept class that is PAC learnable but not PUAC learnable?* (cf. Chapter 6). This problem has been solved by Barbara Hammer in [79]. Specifically, she has shown that consistent PUAC learnability and PUAC learnability are equivalent. Her result is reproduced here as Theorem 6.5.

Problem 12.5. *Is it possible to compute the limiting constant of the maximum disparity between empirical means and true means?* (cf. Chapter 5). This problem is still open.

As always, let $\mathcal{F}$ be a given family of functions mapping X into $[0,1]$. For each sequence $\mathbf{x}^* \in X^\infty$, each integer m, and each $f \in \mathcal{F}$, define

$$\hat{E}_m(f;\mathbf{x}^*) := \frac{1}{m}\sum_{i=1}^{m} f(x_i)$$

to be the empirical mean of f based on the first m elements of $\mathbf{x}^*$. Now define the stochastic process (cf. (5.2.1))

$$a_m(\mathbf{x}^*) := \sup_{f \in \mathcal{F}} |\hat{E}_m(f;\mathbf{x}^*) - E_P(f)|.$$

Then it is shown in Lemma 5.1 that this stochastic process converges almost surely to a constant, call it $c_0(\mathcal{F})$, as $m \to \infty$. This constant $c_0(\mathcal{F})$ can be thought of as the maximum disparity between empirical means and true means. The property of uniform convergence of empirical means (UCEM) corresponds to $c_0(\mathcal{F})$ being equal to zero, and a necessary and sufficient condition for $\mathcal{F}$ to have this property is stated in Theorem 5.3. But what happens if the condition of Theorem 5.3 does not hold? One can of course conclude that $c_0(\mathcal{F}) > 0$, but cannot infer the *actual value* of $c_0(\mathcal{F})$. In other words, Theorem 5.3 does not help us in determining *how far away* the family $\mathcal{F}$ is from having the UCEM property. This leads to the problem posed here.

In particular, it may be possible at least to bracket $c_0(\mathcal{F})$ in terms of another constant that is also characteristic of the family $\mathcal{F}$. To define this constant, let us define another stochastic process

$$b_m(\mathbf{x}^*) := \frac{\lg L(\epsilon, \mathcal{F}|_\mathbf{x}, \|\cdot\|_\infty)}{m},$$

where, as in Section 5.3, $\mathcal{F}|_\mathbf{x} \subseteq [0,1]^m$ denotes the set of vectors in $[0,1]^m$ that correspond to values of functions in $\mathcal{F}$ evaluated at $x_1, \ldots, x_m$. Now, as pointed out in Lemma 5.2, the stochastic process $\{b_m\}$ also converges almost surely to some constant as $m \to \infty$. Let $c_1(\mathcal{F})$ denote this constant. It would be interesting to study the relationship, if any, between $c_0(\mathcal{F})$ and $c_1(\mathcal{F})$.

In the case of empirical *probabilities* rather than empirical means, there is yet a third constant that one can define. Suppose $\mathcal{F}$ consists of *binary-valued* functions, and given a sequence $\mathbf{x}^* \in X^\infty$, let $\mathbf{x}_m \in X^m$ denote the vector consisting of its first m components. As in Section 5.3, let $d(\mathbf{x}_m)$ denote the VC-dimension of $\mathcal{F}$ restricted to the set $S_m := \{x_1, \ldots, x_m\}$. Then Theorem 5.4 gives an alternate necessary and sufficient condition for $\mathcal{F}$ to have the UCEM property. Now it is possible to define yet a third stochastic process, namely

$$g_m(\mathbf{x}^*) := \frac{d(\mathbf{x}_m)}{m}.$$

Then, as pointed out in Lemma 5.10, the stochastic process $\{g_m\}$ also converges almost surely to a constant, which can be denoted by $c_2(\mathcal{F})$ (with the caveat that $c_2(\mathcal{F})$ is defined *only* for families of *binary-values* functions, whereas c_0 and c_1 are defined even otherwise). In the proof of Theorem 5.4 it is shown that

$$c_1(\mathcal{F}) \leq \phi(c_2(\mathcal{F})),$$

where $\phi(x) = x(1 - \ln x)$. But the relationship between $c_0(\mathcal{F})$ and $c_2(\mathcal{F})$ is worth exploring.

Problem 12.6. *How can one reconcile the fact that in distribution-free learning, every learnable concept class is also "polynomially" learnable, whereas this might not be so in fixed-distribution learning?* (cf. Chapters 6 and 7). The problem is still open.

In the case of distribution-free learning of concept classes, Theorem 7.8 shows that there are only two possibilities:

1. $\mathcal{C}$ has infinite VC-dimension, in which case $\mathcal{C}$ is not PAC learnable at all.
2. $\mathcal{C}$ has finite VC-dimension, in which case $\mathcal{C}$ is not only PAC learnable, but the sample complexity $m_0(\epsilon, \delta)$ is $O(1/\epsilon + \ln(1/\delta))$. Let us call such a concept class "polynomially learnable."

In other words, there is no "intermediate" possibility of a concept class being learnable, but having a sample complexity that is superpolynomial in $1/\epsilon$.

In the case of fixed-distribution learning, the situation is not so clear. Of course it is very easy to construct *algorithms* whose sample complexity is superpolynomial in $1/\epsilon$. For example, one can construct a concept class $\mathcal{C}$ and a probability measure P such that $\mathcal{C}$ has the UCEP property with respect to P, but where the convergence of the empirical probabilities to their true values is arbitrarily slow ([191], p. 52). Then a consistent algorithm that always picks a hypothesis that is at the "edge" of the version space (i.e., the set of all hypotheses that are consistent with the labelled samples) would also have arbitrarily slow rate of convergence. But this is an artifice that reflects poorly on the algorithm and not on the concept class. The question being asked is something else, namely: Is there a concept class for which *every* algorithm would require a superpolynomial number of samples? The only known way of constructing such a concept class would be to appeal to Theorem 6.6, and attempt to construct a concept class whose ϵ-covering number grows *faster than any exponential in* $1/\epsilon$. It would be interesting to know whether such a concept class exists. As shown in Lemma 6.3, it is enough to find a *function class* with this property, because such an example could immediately be translated into a corresponding example of a concept class.

Let us suppose for the sake of argument that some clever person succeeds in constructing such an example. From Theorem 7.8 it is clear that any such example must have infinite VC-dimension. I would like to have an "intrinsic" explanation as to why in distribution-free learning, every learnable concept class is also forced to be polynomially learnable. Next, how far can one "push" this line of argument? Suppose $\mathcal{P}$ is a family of probabilities that contains a ball in the total variation metric ρ. From Theorem 8.8 it follows that every concept class that is learnable with respect to $\mathcal{P}$ must also be polynomially learnable (because $\mathcal{C}$ must have finite VC-dimension). Is it possible to identify other such classes of probabilities?

Problem 12.7. *Is there a sufficient condition for function learning under an intermediate family of probability measures that is weaker than Theorem 8.4?* (cf. Chapter 8.) The problem is still open.

Chapter 8 contains a "universal" necessary condition for a concept class $\mathcal{C}$ to be PAC learnable under a family of probability measures $\mathcal{P}$, namely the uniformly bounded metric entropy (UBME) condition given in Theorem 8.5. In the two "extreme" cases where $\mathcal{P}$ is a singleton or $\mathcal{P} = \mathcal{P}^*$, this necessary condition is "tight" in that it reduces to the known necessary and sufficient condition in each case. Chapter 8 also contains a universal *sufficient* condition, namely Theorem 8.4. This condition reduces to the known (necessary and) sufficient condition if $\mathcal{P}$ is a singleton set, but is *stronger* than the known necessary and sufficient condition in the other extreme case where $\mathcal{P} = \mathcal{P}^*$. Thus it is worthwhile to seek a better sufficient condition than that given in Theorem 8.4.

Problem 12.8. *Suppose $X \subseteq \mathbb{R}^k$ for some integer k, and that $\mathcal{P}$ is the set of all* nonatomic *measures on X. What are some necessary and sufficient conditions for a concept class to be PAC learnable under $\mathcal{P}$?* (cf. Chapters 7 and 8.) The problem is still open.

A perusal of Chapter 7 shows that all the proofs to the effect that finite VC-dimension is necessary for a concept class to be distribution-free PAC learnable rely crucially on being able to choose a *purely atomic measure* on the sample space X. What happens if $\mathcal{P}$ does not contain any atomic measures? Distribution-free learning is very popular among computer scientists, as a way of ensuring that no prior knowledge is assumed about the learning problem.

However, in the computer science world, invariably the sample space X is *graded*, i.e., a collection of the form $X = \{X_n\}$. Moreover, usually each X_n is a *finite set*. Now, on a finite set, *all* probability measures are purely atomic. So in the types of problems studied in the computational learning theory literature, it is not unnatural to assume that $\mathcal{P} = \mathcal{P}^*(X_n)$ for each n. On the other hand, when X is a "continuous" set such as $[0,1]^k$, it is not entirely natural to permit $\mathcal{P}$ to contain purely atomic measures; this is especially so if the presence of purely atomic measures in $\mathcal{P}$ serves solely as a device to enable us to prove theorems that we could not prove otherwise. It seems to me that letting $\mathcal{P}$ consist of all nonatomic measures meets the requirement of assuming (almost) no prior knowledge, and at the same time, results in a very interesting learning problem. Note that $\mathcal{P}$ is an example of a noncompact set with an empty interior, so that the results of Chapter 8 do not apply.

I would guess that the answer might involve some sort of topological feature of the concept class, such as not having any "isolated" concepts. Though introduced for quite a different purpose, the notion of concept classes that are "dense in themselves" [67] might play a role.

Problem 12.9. *In the case of learning problems with nonparametric uncertainty in the probability measure, what is the trade-off between sample complexity and the extent of uncertainty?* (cf. Chapter 8.) The problem is still open.

In Theorem 8.8 it is shown that if a concept class is learnable with respect to a ball $\bar{\mathcal{B}}(\lambda, P, \rho)$ in the total variation metric ρ with *nonzero* radius λ, then in fact the concept class is distribution-free learnable. This is achieved by showing that a concept class satisfies the uniformly bounded metric entropy (UBME) condition with respect to such a family of probabilities if and only if it has finite VC-dimension. An issue that is left unresolved by this theorem is the effect of the parameter λ (reflecting the "extent" of nonparametric uncertainty in the probability measure) on the sample complexity. As it stands, Theorem 8.8 shows that there is a "discontinuity" in learnability at $\lambda = 0$, since $\lambda = 0$ corresponds to fixed-distribution learning, whereas $\lambda > 0$, no matter how small, corresponds (at least so far as learnability goes) to distribution-free learning. How is this discontinuity reflected in the sample complexity estimates? Perhaps an answer to this question might shed some light on Problem 12.6.

Problem 12.10. *Is it possible to learn the underlying probability measure in a learning problem, and use this information to accelerate (even by a constant factor) the learning process?* (cf. Chapter 8.) The problem is still open.

In Chapter 6, we studied the problem of fixed-distribution learning, in which the probability measure that generates the learning samples is known ahead of time. In Chapters 7 and 8, this is not assumed to be the case. And yet, even though the learning samples $x_1, \ldots, x_m$ are ostensibly generated by an unknown probability $P \in \mathcal{P}$, in actuality they "encode" some information about P. For instance, if one were to construct an "empirical probability measure"

$$\hat{P}_m(x) := \frac{1}{m}\sum_{i=1}^{m}\delta(x-x_i),$$

where $\delta(x-x_i)$ is the Dirac atomic measure concentrated at x_i, then it follows from Sanov's theorem that $\{\hat{P}_m(\cdot)\}$ converges almost surely to the "true" probability measure P in an appropriate metric. In principle then, one would like to try something like the following strategy: Given ϵ, δ, draw a sufficiently large number m of samples such that the unknown probability measure P is localized to some ball $\mathcal{B}(\alpha, \hat{P}_m, \mu)$ with confidence $1-\delta/2$, where μ is a suitable metric on the space of probability measures on $(X, \mathcal{S})$, and α is a measure of uncertainty around the empirical probability measure $\hat{P}_m$. Then, assuming that $P \in \mathcal{B}(\alpha, \hat{P}_m, \mu)$, learn to accuracy ϵ and confidence $\delta/2$ by drawing some more samples. If μ is the total variation metric, then Lemma 8.2 can be used for the latter purpose. Unfortunately, the convergence guaranteed by Sanov's theorem is *not* in the total variation metric, but in some other metric such as the Prohorov metric, which is strictly weaker than the total variation metric unless the set X is finite. And we have no analog of Lemma 8.2 for anything other than the total variation metric.

A promising special case occurs if it is assumed that X is some subset of $\mathbb{R}$, and that every probability measure in $\mathcal{P}$ has a continuous distribution function. In this case, the well-known results of Kolmogorov-Smirnov and their subsequent generalizations due to Massart [129] show that the rate of convergence of the empirical *distribution* to the true distribution function is well-understood. In particular, define the empirical distribution function

$$\hat{r}_m(y) := \frac{1}{m}\sum_{i=1}^{m}\eta(y-x_i),$$

where $\eta(\cdot)$ is the Heaviside function defined by

$$\eta(y) := \begin{cases} 1, & \text{if } y \geq 0, \text{ and} \\ 0, & \text{if } y < 0. \end{cases}$$

Thus $\hat{r}_m(\cdot)$ is the empirical distribution function based on the multisample **x**. Let $r(\cdot)$ denote the actual distribution function of P; that is:

$$r(y) := P\{(-\infty, y]\}.$$

Then, for each $\epsilon > 0$, it is true that

$$P^m\{\mathbf{x} \in X^m : \sup_{y \in \mathbb{R}}|r(y) - \hat{r}_m(y)| > \epsilon\} \leq 2\exp(-2m\epsilon^2).$$

In other words, the empirical distribution function converges uniformly (with respect to y) to the true distribution function. Can this result be exploited in some way?

The type of learning strategy described above is not likely to speed up the "rate" of learning, since in distribution-free learning the upper bounds and lower bounds on sample complexity are quite close ($O((1/\epsilon)\ln(1/\epsilon))$ for the upper bounds and $O(1/\epsilon)$ for the lower bounds). But it might reduce the constant hidden under the O-symbol. Another interesting possibility is the following: Suppose that calls to the random number generator are much less "expensive" than calls to the oracle. In other words, it is quite cheap to generate *unlabelled* random samples, but costly to generate *labelled* samples. Then a strategy such as the above might be effective, since one could generate several unlabelled samples to "learn" the probability measure, and then generate relatively fewer labelled samples to learn the target concept or function.

However, in order for any of these ideas to pan out, it is necessary to extend Lemma 8.2 to balls in a metric compatible with the Prohorov metric.

Problem 12.11. *Can one define "local" versions of metric dimension and VC-dimension, and use them to estimate the complexity of learning a specific target concept within a given class?* (cf. Chapters 6 and 7.) (This problem is due to Sanjeev Kulkarni.) The problem is still open.

By nature, the definition of sample complexity in the standard PAC learning problem is "worst-case" with respect to the target concept to be learnt. When learning commences, the unknown target concept T could be *anywhere* within $\mathcal{C}$. However, as learning progresses, T gradually gets "localized." Is it possible to "adapt" the learning algorithm to take advantage of this localization phenomenon and thus accelerate the learning process? Again, the "rate" of learning might not be speeded up, and the acceleration might only be by a constant factor.

To illustrate what is meant, consider first the case of fixed-distribution learning. Suppose ϵ, δ are specified. The learning algorithm is itself a concatenation of several sub-algorithms. Initially, one finds an $\alpha/2$-cover of $\mathcal{C}$ with respect to the total variation metric ρ, where α is a parameter to be specified. Then the minimal empirical risk algorithm is run on this $\alpha/2$-cover to an accuracy of $\delta/2$. By drawing a suitable number of samples, one can ensure that, with a probability of at least $1 - \delta/2$, the unknown target concept T belongs to a particular α-ball, call it $\mathcal{B}(\alpha, A_1, \rho)$. Then one finds an $\alpha/4$-cover of this ball of radius α, and runs the minimal empirical risk algorithm with a confidence parameter of $\delta/4$. This further localizes T to a smaller ball $\mathcal{B}(\alpha/2, A_2, \rho)$ with a probability of at least $1 - 3\delta/4$. The process is repeated i times, until the radius of uncertainty α^{-i} is less than ϵ. Now it is clear that the sample complexity as a function of $1/\epsilon$ depends on the $\alpha/2^{i+2}$-covering number of a ball of radius $\alpha/2^i$ around T, and the rate at which this number increases as i approaches infinity. This exponent of growth can be thought of as the metric dimension of the concept class $\mathcal{C}$ *around* T. In other words, it is a "local" version of the metric dimension around T. There is no reason to suppose that the local metric dimension is the same at all $T \in \mathcal{C}$. By studying how this number varies across $\mathcal{C}$, one can try to differentiate between the sample complexity of learning different target concepts within $\mathcal{C}$. However, this argument needs to be formalized, and it needs to be established whether such an approach leads to any nontrivial sample complexity estimates. Moreover, it is not clear whether there exists an analogous notion of a "local VC-dimension" around T, and if so, how one would go about defining it.

Problem 12.12. *When is the loading problem for sigmoidal neural networks NP-hard?* (cf. Chapter 10.) (This problem is due to Eduardo Sontag.) The problem is still open.

Chapter 10 contains several examples of loading problems for neural networks that are NP-complete or NP-hard. Consider for instance the three-neuron network of Example 10.1. It was shown in [31] that if all three neurons are perceptrons, then the loading problem for this network is NP-complete. Subsequently the result was extended in [50] to the case where the first-level neurons are hard limiters. It is conjectured by Eduardo Sontag that the problem is NP-hard if the first-level neurons are standard sigmoids.

Problem 12.13. *When is neural network learning intractable as the* number *of neurons increases?* (cf. Chapter 10.) The problem is still open.

Many if not most of the negative results concerning the intractability of learning neural networks have to do with the NP-hardness of finding a hypothesis with minimum empirical error, i.e., a neural network that reproduces the training data as well as possible within the given class of networks. Usually, the NP-hardness is with respect to the *number of inputs* to the network. However, in my opinion it is much more interesting to study what happens to the complexity of finding a hypothesis with minimum empirical error as *the number of neurons* increases, while the input dimension remains constant. In [125], Maass makes this point forcefully, and proves a few preliminary results. However, a great deal more needs to be done.

Problem 12.14. *Is it possible to obtain good estimates for the metric entropy of neural networks under, for example, the uniform distribution on the input space?* (cf. Chapters 7 and 10.) The problem is still open.

In Chapter 10, the emphasis is on bounding the VC-dimension of various types of neural network architectures. Using the results of Chapter 4, these bounds on the VC-dimension can in turn be used to estimate the metric entropy of the concept class with respect to every probability measure on the input space. Now, if one fixes a specific probability on the input space, such as for example the uniform distribution, then the upper bounds thus obtained might be too conservative, compared to the actual the metric entropy of the concept class with respect to this particular probability. It might be possible to obtain less conservative estimates for the metric entropy using more "direct" methods. In turn, such bounds would lead to less conservative bounds on the sample complexity.

Problem 12.15. *What is the reduction in the sample complexity of learning a family of binary-output neural networks if membership queries are permitted?* (cf. Chapters 9 and 10.) The problem is still open.

In Chapter 9 it is shown that active learning using *arbitrary* binary queries can be substantially faster than passive learning, but this need not be so if the active learner is restricted to only *membership queries*. This is because, in a general learning problem, membership queries alone might not significantly reduce the extent of ignorance about the target concept. Specifically in the case of neural networks, it is not clear how much, if at all, membership queries help in reducing sample complexity.

Problem 12.16. *When does the class of performance indices of feedback control systems have the property of uniform convergence of empirical means?* (cf. Chapters 11 and 10.) This problem is largely solved via several results in Chapter 11.

References

1. M. A. Aizerman, E. M. Braverman and L. I. Rozonoer, "Theoretical foundations of the potential function method in pattern recognition," *Automation and Remote Control*, 25, 821-837, 1964.
2. D. Aldous and U. Vazirani, "A Markovian extension of Valiant's learning model," *Proc. 31st Annual IEEE Symp. on the Foundations of Comput. Sci.*, 392-396, 1990.
3. N. Alon, S. Ben-David, N. Cesa-Bianchi and D. Haussler, "Scale-sensitive dimensions, uniform convergence, and learnability," *Proc. 34th Annual IEEE Conf. on Foundations of Comput. Sci.*, 292-301, 1993.
4. D. Angluin, "Queries and concept learning," *Machine Learning*, 2, 319-342, 1987.
5. D. Angluin, "Computational learning theory: Survey and selected bibliography," *Proc. 24th ACM Symp. on Theory of Computing*, 351-369, 1992.
6. D. Angluin and M. Kharitonov, "When won't membership queries help?" *J. Comput. Syst. Sci.*, 50, 336-355, 1995.
7. M. Anthony and P. L. Bartlett, *Neural Network Learning: Theoretical Foundations*, Cambridge University Press, Cambridge, UK, 1999.
8. M. Anthony, P. Bartlett, Y. Ishai and J. Shawe-Taylor, "Valid generalisation from approximate interpolation," (preprint).
9. M. Anthony and N. Biggs, *Computational Learning Theory*, Cambridge University Press, Cambridge, UK, 1992.
10. M. Anthony, N. Biggs and J. Shawe-Taylor, "The learnability of formal concepts," *Proc. Third Workshop on Computational Learning Theory*, Morgan-Kaufmann, San Mateo, CA, 246-257, 1990.
11. P. Assouad, "Densité et dimension," *Ann. Inst. Fourier, Grenoble*, 33(3), 233-282, 1983.
12. K. B. Athreya and S. G. Pantula, "Mixing properties of Harris chains and autoregressive processes," *J. Appl. Probab.*, 23, 880-892, 1986.
13. B. R. Barmish, *New Tools for Robustness of Linear Systems*, MacMillan, New York, 1994.
14. A. R. Barron, "Universal approximation bounds for superpositions of a sigmoidal function," *IEEE Trans. Inf. Theory*, 39(3), 930-945, 1993.
15. P. L. Bartlett and S. R. Kulkarni, "The complexity of model classes, and smoothing of noisy data," *Proc. Conf. on Decision and Control*, 1996.
16. P. L. Bartlett, P. M. Long and R. C. Williamson, "Fat-shattering and the learnability of real-valued functions," *Proc. 7th ACM Conf. on Computational Learning Theory*, 299-310, 1994.
17. P. L. Bartlett, P. M. Long and R. C. Williamson, "Fat-shattering and the learnability of real-valued functions," *J. Comput. Syst. Sci.*, 52(3), 534-452, 1996.

18. P. L. Bartlett and R. C. Williamson, "Investigating the distribution assumption in the pac learning model," *Proc. Fourth Annual Workshop on Computational Learning Theory*, Morgan-Kaufmann, San Mateo, CA, 24-32, 1991.
19. E. Baum and D. Haussler, "What size net gives valid generalization?" *Neural Computation*, 1(1), 151-160, 1989.
20. S. Ben-David, N. Cesa-Bianchi, D. Haussler and P. M. Long, "Characterizations of learnability for classes of $\{0,\ldots,n\}$-valued functions," *J. Comput. Syst. Sci*, 50, 74-86, 1995.
21. S. Ben-David and M. Lindenbaum, "Localization vs. identification of semi-algebraic sets," *Proc. Sixth ACM Workshop on Computational Learning Theory*, 327-336, 1993.
22. G. M. Benedek and A. Itai, "Learnability by fixed distributions," *Proc. First Workshop on Computational Learning Theory*, Morgan-Kaufmann, San Mateo, CA, 80-90, 1988.
23. G. M. Benedek and A. Itai, "Learnability with respect to fixed distributions," *Theoretical Computer Science*, 86(2), 377-390, 1991.
24. G. M. Benedek and A. Itai, "Dominating distributions and learnability," *Proc. Fifth Workshop on Computational Learning Theory*, ACM, 253-264, 1992.
25. P. Billingsley, *Probability and Measure*, Wiley, New York, 1986.
26. P. Billingsley, *Probability and Measure*, (Third Edition), Wiley, New York, 1995.
27. V. Blondel and J. N. Tsitsiklis, "NP-hardness of some linear control design problems," *SIAM J. Control and Optim.*, 35(6), pp. 2118-2127, 1997.
28. V. Blondel and J. N. Tsitsiklis, "A survey of computational complexity results in systems and control," *Automatica*, 45(9), pp. 1249-1274, 2000.
29. A. Blum, A. Frieze, R. Kannan and S. Vempala, "A polynomial-time algorithm for learning noisy linear threshold elements," *Algorithmica*, 22(1), 35-52, 1997.
30. A. Blum and R. Kannan, "Learning an intersection of k halfspaces over a uniform distribution," *Proc. 34th Annual IEEE Symp. on Foundations of Comput. Sci*, 312-320, 1993.
31. A. Blum and R. L. Rivest, "Training a 3-node neural network is NP-complete," *Proc. First Workshop on Computational Learning Theory*, Morgan-Kaufmann, San Mateo, CA, 9-18, 1988.
32. A. Blumer, A. Ehrenfeucht, D. Haussler and M. Warmuth, "Learnability and the Vapnik-Chervonenkis dimension," *J. ACM*, 36(4), 929-965, 1989.
33. S. Boyd, V. Balakrishnan and P. Kabamba, "A bisection method for computing the $\mathbf{H}_\infty$ norm of a transfer matrix and related problems," *Math. of Control, Signals and Systems*, 2(3), 207-219, 1989.
34. R. Braatz, P. Young, J. Doyle and M. Morari, "Computational complexity of the μ calculation," *IEEE Trans. Autom. Control*, 39, pp. 1000-1002, 1994.
35. L. Breiman, *Probability*, Addison-Wesley, Reading, MA, 1968.
36. L. Breiman, "Hinging hyperplanes for regression, classification and function approximation," *IEEE Trans. Inf. Theory*, 39(3), 999–1013, 1993.
37. K. L. Buescher and P. R. Kumar, "Learning by canonical smooth estimation, Part I: Simultaneous estimation," *IEEE Trans. Autom. Control*, 42(4), 545-556, April 1996.
38. K. L. Buescher and P. R. Kumar, "Learning by canonical smooth estimation, Part II: Learning and choice of model complexity," *IEEE Trans. Autom. Control*, 42(4), 557-569, April 1996.
39. P. E. Caines, "Prediction error identification methods for stationary stochastic processes," *IEEE Trans. Autom. Control*, AC-21(4), 500-505, Aug. 1976.
40. P. E. Caines, "Stationary linear and nonlinear system identification and predictor set completeness," *IEEE Trans. Autom. Control*, AC-23(4), 583-594, Aug. 1978.

41. M. Campi, "Decision-directed learning in a Bayesian framework," (preprint).
42. M. Campi and P. R. Kumar, "Learning dynamical systems in a stationary environment," *Proc. Conf. on Decision and Control*, Kobe, Japan, 2308-2311, Dec. 1996.
43. M. C. Campi and M. Vidyasagar, "Learning with prior information," *IEEE Trans. Autom. Control*, AC-46(11), 1682-1695, Nov. 2001.
44. M. C. Campi and E. Weyer, "Finite sample properties of system identification methods," *IEEE Trans. Autom. Control*, to appear.
45. H. Chernoff, "A measure of asymptotic efficiency for tests of a hypothesis based on the sum of observations," *Ann. Math. Stat.*, 23, 493-507, 1952.
46. C. Cortes and V. N. Vapnik, "Support vector networks," *Machine Learning*, 20, 273-295, 1997.
47. T. M. Cover, "Capacity problems for linear machines," in *Pattern Recognition*, L. Kanal (Editor), Thompson Book Co., 283-289, 1968.
48. G. E. Coxson and C. DeMarco "The computational complexity of approximating the minimal perturbation scaling to achieve instability in an interval matrix," *Math. of Control, Signals and Systems*, 7, 279-291, 1994.
49. N. Cristianini and J. Shawe-Taylor, *Support Vector Machines*, Cambridge University Press, Cambridge, UK, 2000.
50. B. Dasgupta, H. T. Siegelmann and E. D. Sontag, "On the intractability of loading neural networks," in *Theoretical Advances in Neural Computation and Learning*, V. P. Roychowduhry, K. Y. Siu and A. Orlitsky (Editors), Kluwer, Boston, 357-389, 1994.
51. B. Dasgupta, H. T. Siegelmann and E. D. Sontag, "On the complexity of training neural networks with continuous activation functions," *IEEE Trans. Neural Networks*, 6, 1490-1504, 1995.
52. B. Dasgupta and E. D. Sontag, "Sample complexity for learning recurrent perceptron mappings," summary in *Advances in Neural Information Processing*, 8, MIT Press, Cambridge, MA, 204-210, 1996.
53. B. Dasgupta and E. D. Sontag, "Sample complexity for learning recurrent perceptron mappings," *IEEE Trans. Info. Theory*, 42, 1479-1487, 1996.
54. A. Dembo and O. Zeitoni, *Large Deviations Techniques and Applications*, Springer-Verlag, New York, 1993.
55. L. Devroye and L. Györfi, *Nonparametric Density Estimation: An L_1 view*, Wiley, New York, 1985.
56. L. Devroye, L. Gyorfi and G. Lugosi, *A Probabilistic Theory of Pattern Recognition*, Springer, 1996.
57. J. Doyle, "Analysis of feedback systems with structured uncertainties," *Proc. IEEE*, 129, 242-250, 1982.
58. J. Doyle, K. Glover, P. P. Khargonekar and B. A. Francis, "State space solutions to standard H_2 and H_∞ control problems," *IEEE Trans. Autom. Control*, 34(8), 831-847, 1989.
59. J. Doyle and G. Stein, "Multivariable feedback design: Concepts for a classical/modern synthesis," *IEEE Trans. Autom. Control*, 26(1), 4-16, Feb. 1981.
60. R. O. Duda and P. E. Hart, *Pattern Classification and Scene Analysis*, Wiley, 1973.
61. R. M. Dudley, "Central limit theorems for empirical measures" *Ann. Probab.*, 6(6), 899-929, 1978.
62. R. M. Dudley, *A Course on Empirical Processes*, in Lecture Notes in Mathematics, No. 1097, 1-142, Springer-Verlag, New York, 1984.
63. R. M. Dudley, "Universal Donsker classes and metric entropy," *Ann. Probab.*, 15(4), 1306-1326, 1987.

64. R. M. Dudley, S. R. Kulkarni, T. J. Richardson and O. Zeitouni, "A metric entropy bound is not sufficient for learnability," *IEEE Trans. Information Theory*, 40, 883-885, 1994.
65. N. Dunford and J. T. Schwartz, *Linear Operators: Part I*, Interscience, New York, 1959.
66. A. Ehrenfeucht, D. Haussler, M. Kearns and L. Valiant, "A general lower bound on the number of examples needed for learning," *Proc. First Workshop on Computational Learning Theory*, Morgan-Kaufmann, San Mateo, CA, 139-154, 1988; also *Information and Computation*, 82, 247-261, 1989.
67. B. Eisenberg and R. L. Rivest, "On the Sample Complexity of Pac-Learning Using Random and Chosen Examples," *Proc. Third Annual Workshop on Computational Learning Theory*, Morgan-Kaufmann, San Mateo, CA, 154-162, 1990.
68. W. Feller, *An Introduction to Probability Theory and Its Applications*, (Second Edition), Wiley, New York, 1957.
69. T. L. Fine, *Feedforward Neural Network Methodology*, Springer-Verlag, New York, 1999.
70. B. A. Francis, *A Course in H^∞-Control Theory*, in *Lecture Notes in Control and Information Sciences*, Vol. 88, Springer-Verlag, New York, 1988.
71. D. Gamarnik, "Extension of the PAC framework to finite and countable Markov chains," *Proc. Twelfth Annual Conf. on Computational Learning Theory*, 1999.
72. F. R. Gantmacher, *Matrix Theory*, Volume II, Chelsea, New York, 1959.
73. M. R. Garey and D. S. Johnson, *Computers and Intractability: A Guide to the Theory of NP-Completeness*, W. H. Freeman, New York, 1979.
74. B. V. Gnedenko, *Theory of Probability*, (Fourth Edition), Chelsea, New York, 1968.
75. P. Goldberg and M. Jerrum, "Bounding the Vapnik-Chervonenkis dimension of concept classes parametrized by real numbers," *Proc. 6th ACM Workshop on Computational Learning Theory*, 361-369, 1993.
76. P. Goldberg and M. Jerrum, "Bounding the Vapnik-Chervonenkis dimension of concept classes parametrized by real numbers," *Machine Learning*, 18, 131-148, 1995.
77. P. R. Halmos, *Measure Theory*, Van Nostrand, 1950.
78. P. Hall and C. C. Heyde, *Martingale Limit Theory and Its Application*, Academic Press, New York, 1980.
79. B. Hammer, "Learning recursive data," *Math. of Control, Signals and Systems*, 12(1), 62-79, 1999.
80. D. Haussler, "Decision theoretic generalizations of the PAC model for neural net and other learning applications," *Information and Computation*, 100, 78-150, 1992.
81. D. Haussler, M. Kearns, N. Littlestone and M. K. Warmuth, "Equivalence of models for polynomial learnability," *Proc. First Workshop on Computational Learning Theory*, Morgan-Kaufmann, San Mateo, CA, 42-55, 1988.
82. D. Haussler, M. Kearns, N. Littlestone and M. K. Warmuth, "Equivalence of models for polynomial learnability," *Information and Computation*, 95, 129-161, 1991.
83. D. Haussler, M. Kearns, M. Opper and R. Schapire, "Estimating average-case learning curves using Bayesian, statistical physics and VC-dimension models," *Advances in Neural Information Processing*, 855-862, 1992.
84. D. Haussler, M. Kearns and R. Schapire, "Bounds on the sample complexity of Bayesian learning using information theory and the VC dimension," *Proc. Fourth Workshop on Computational Learning Theory*, Morgan-Kaufmann, San Mateo, CA, 61-74, 1991.

85. D. Haussler, N. Littlestone and M. K. Warmuth, "Predicting $\{0,1\}$-functions on randomly drawn points," *Proc. First Workshop on Computational Learning Theory*, Morgan-Kaufmann, San Mateo, CA, 280-296, 1988.
86. W. Hoeffding, "Probability inequalities for sums of bounded random variables," *J. Amer. Statist. Assoc.* 58, 13-30, 1963.
87. K-U. Höffgen, H-U. Simon and K. S. Van Horn, "Robust trainability of single neurons," *J. Comput. Syst. Sci.*, 50(1), 114-125, 1995.
88. I. A. Ibragimov, "Some limit theorems for stationary processes," *Theory Probab. Appl.*, 7, 349-382, 1962.
89. J. S. Judd, *Neural Network Design and the Complexity of Learning*, MIT Press, Cambridge, MA, 1990.
90. E. I. Jury, *Inners and Stability of Dynamical Systems*, John Wiley, New York, 1977.
91. T. Kailath, *Linear Systems*, Prentice-Hall, Englewood Cliffs, NJ, 1979.
92. L. V. Kantorovich and G. P. Akilov, *Functional Analysis*, (Second Edition), Pergamon Press, New York, 1982.
93. R. L. Karandikar and M. Vidyasagar, "Rates of convergence of empirical means under mixing processes," *Stat. and Probab. Letters*, (to appear).
94. N. Karmarkar, "A new polynomial-time algorithm for linear programming," *Combinatorica*, 4(4), 373-395, 1984.
95. M. Karpinski and A. J. Macintyre, "Polynomial bounds for VC dimension of sigmoidal neural networks," *Proc. 27th ACM Symp. Theory of Computing*, 200-208, 1995.
96. M. Karpinski and A. J. Macintyre, "Polynomial bounds for VC dimension of sigmoidal and general Pfaffian neural networks," *J. Comput. Syst. Sci.*, 54, 169-176, 1997.
97. M. Kearns, M. Li, L. Pitt and L. Valiant, "On the learnability of Boolean formulae," *19th ACM Symp. on the Theory of Computing*, 285-295, 1987.
98. M. Kearns and R. E. Schapire, "Efficient distribution-free learning of probabilistic concepts," *J. Comput. Syst. Sci*, 48, 464-497, 1994.
99. M. Kearns and U. Vazirani, *Introduction to Computational Learning Theory*, MIT Press, Cambridge, MA, 1994.
100. J. L. Kelley, *General Topology*, Van Nostrand, Princeton, NJ, 1955.
101. P. P. Khargonekar and A. Tikku, "Randomized algorithms for robust control analysis have polynomial complexity," *Proc. Conf. on Decision and Control*, 1996.
102. M. Kharitonov, "Cyrptographic lower bounds for learnability of Boolean functions on the uniform distribution," *J. Comput. Syst. Sci.*, 50, 600-610, 1995.
103. A. G. Khovanski, *Fewnomials*, American Mathematical Society, Providence, RI, 1991.
104. J. F. C. Kingman, "The ergodic theory of subadditive stochastic processes," *J. Royal Stat. Soc., Ser. B*, 30, 499-510, 1968.
105. J. F. C. Kingman, "Subadditive ergodic theory," *Ann. Probab.*, 1, 883-909, 1973.
106. P. Koiran and E. D. Sontag, "Neural networks with quadratic VC dimension," *J. Comput. Syst. Sci.*, 54, 190-198, 1997.
107. A. N. Kolmogorov and V. M. Tikhomirov, "ϵ-Entropy and ϵ-capacity of sets in functional spaces," *Amer. Math. Soc. Transl.* 17, 277-364, 1961.
108. A. Kowalczyk, H. Ferra and J. Szymanski, "Combining statistical physics with VC-bounds on generalisation in learning systems," *Proc. Sixth Australian Conf. on Neural Networks (ACNN'95)*, pp. 41-44, Sydney, 1995.

109. S. R. Kulkarni, "On metric entropy, Vapnik-Chervonenkis dimension and learnability for a class of distributions," Center for Intelligent Control Systems, Laboratory for Information and Decision Systems, M.I.T., Report No. P-1910, 1989.
110. S. R. Kulkarni, "A review of some extensions to the PAC learning model," *Proc. of Silver Jubilee Workshop on Computing and Intelligent Control*, Bangalore, India, 1993.
111. S. R. Kulkarni, S. K. Mitter, J. N. Tsitsiklis, "Active learning using arbitrary binary valued queries," *Machine Learning*, 11, 23-35, 1993.
112. S. R. Kulkarni and M. Vidyasagar, "Learning decision rules under a family of probability measures," *IEEE Trans. Info. Theory*, IT-43(1), 154-166, January 1997.
113. M. C. Laskowski, "Vapnik-Chervonenkis classes of definable sets," *J. London Math. Soc.*, 45(2), 377-384, 1992.
114. A. Levy, P. Fraenkel and Y. Bar-Hillel, *Foundations of Set Theory*, Elsevier Science, Amsterdam, 1973.
115. M. Linial, Y. Mansour and R. L. Rivest, "Results on learnability and the Vapnik-Chervonenkis dimension," *29th Annual IEEE Symp. on Foundations of Comput. Sci.*, 120-129, 1988; also *Proc. First Workshop on Computational Learning Theory*, Morgan-Kaufmann, San Mateo, CA, 56-68, 1988 and *Information and Computation*, 90(1), 33-49, 1989.
116. L. Ljung, "Convergence analysis of parametric identification methods," *IEEE Trans. Autom. Control*, AC-23(5), 770-783, Oct. 1978.
117. L. Ljung, *System Identification: Theory for the User*, Prentice-Hall, Englewood Cliffs, NJ, 1987.
118. L. Ljung, *System Identification: Theory for the User*, (Second Edition), Prentice-Hall, Englewood Cliffs, NJ, 1999.
119. M. Loève, *Probability Theory*, Vol. I, Van Nostrand, Princeton, NJ, 1963.
120. O. B. Lupanov, "Circuits using threshold elements," *Soviet Physics Doklady*, 17(2), 91-93, Aug. 1972.
121. W. Maass, "Bounds for the computational power and learning complexity of analog neural nets," *Proc. 25th ACM Symp. Theory of Computing*, 335-344, 1993.
122. W. Maass, "Neural nets with superlinear VC-dimension," *Neural Computation*, 6, 875-882, 1994.
123. W. Maass, "Perspectives of current research about the complexity of learning on neural nets," in *Theoretical Advances in Neural Computation and Learning*, V. P. Roychowduhry, K. Y. Siu and A. Orlitsky (Editors), Kluwer, Boston, 295-336, 1994.
124. W. Maass, "Vapnik-Chervonenkis dimension of neural nets," in *Handbook of Brain Theory and Neural Networks*, M. Arbib, (Editor), 1000-1003, 1995.
125. W. Maass, "Agnostic PAC learning of functions on analog neural nets," *Neural Computation*, 7(5), 1054-1078, Sept. 1995.
126. A. J. Macintyre and E. D. Sontag, "Finiteness results for sigmoidal neural networks," *Proc. 25th ACM Symp. Theory of Computing*, 325-334, 1993.
127. R. Mañé, *Ergodic Theory and Differentiable Dynamics*, Springer-Verlag, New York, 1987.
128. C. Marrison and R. Stengel, "The use of random search and genetic algorithms to optimize stochastic robustness functions," *Proc. Amer. Control Conf.*, Baltimore, MD, 1484-1489, 1994.
129. P. Massart "The tight constant in the Dvoretzky-Kiefer-Wolfowitz inequality," *Ann. Probab.*, Vol. 18, No. 3, 1269-1283, 1990.

130. J. L. McCulloch and W. Pitts, "A logical calculus of ideas immanent in nervous activity," *Bull. Math. Biophysics*, 5, 115-133, 1943.
131. N. Meggido, "On the complexity of polynomial separability," *Discrete Computational Geometry*, 3, 325-337, 1988.
132. A. Megretski, "On the gap between structured singular values and their upper bounds," *Proc. IEEE Conf. Decision and Control*, 3461-3462, 1993.
133. R. Meir, "Structural risk minimization: A case study," *Neural Computation*, 7, 144-157, 1995.
134. R. Meir, "Nonparametric time series prediction through adaptive model selection," *Machine Learning*, 39(1), 5-34, Apr. 2000.
135. S. P. Meyn and R. L. Tweedie, *Markov Chains and Stochastic Stability*, Springer-Verlag, London, 1993.
136. J. W. Milnor, "On the Betti numbers of real varieties," *Proc. Amer. Math. Soc.*, 15, 275-280, 1964.
137. M. Minsky and S. Papert, *Perceptrons: An Introduction to Computational Geometry*, MIT Press, Cambridge, MA, 1969.
138. D. S. Modha and E. Masry, "Minimum complexity regression estimation with weakly dependent observations," *IEEE Trans. Info. Theory*, 42(6), 2133-2145, November 1996.
139. D. S. Modha and E. Masry, "Memory-universal prediction of stationary random processes," *IEEE Trans. Info. Theory*, 44(1), 117-133, Jan. 1998.
140. A. Mokkadem, "Mixing properties of ARMA sequences," *Stoch. Process. and Appl.*, 29, 309-315, 1988.
141. A. Mokkadem, "Propriétés de mélange des processus autorégressifs polynomiaux," *Ann. Inst. Henri Poincaré*, 26(2), 219-260, 1990.
142. R. Motwani and P. Raghavan, *Randomized Algorithms*, Cambridge U. Press, Cambridge, 1995.
143. K. Najarian, G. A. Dumont, M. S. Davies and N. E. Heckman, "PAC learning in non-linear FIR models," *Int. J. Adaptive Control and Signal Process.*, 15, 37-52, 2001.
144. B. K. Natarajan, "On learning Boolean Functions," *19th ACM Symp. on the Theory of Computing*, 296-304, 1987.
145. B. K. Natarajan, "Learning over families of distributions," *Proc. First Workshop on Computational Learning Theory*, Morgan-Kaufmann, San Mateo, CA, 408-409, 1988.
146. B. K. Natarajan, "On learning sets and functions," *Machine Learning*, 4(1), 67-97, 1989.
147. B. K. Natarajan, *Machine Learning: A Theoretical Approach*, Morgan-Kaufmann, San Mateo, CA, 1991.
148. B. K. Natarajan, "Probably approximate learning of sets and functions," *SIAM J. Computing*, 20(2), 328-351, 1991.
149. B. K. Natarajan, "Probably approximate learning over classes of distributions," *SIAM J. Computing*, 21(3), 438-449, 1992.
150. A. Nemirovskii, "Several NP-hard problems arising in robust stability analysis," *Math. of Control, Signals, and Systems*, 6(2), 99-105, 1993.
151. G. C. Newton (Jr), L. A. Gould and J. F. Kaiser, *Analytic Design of Linear Feedback Controls*, John Wiley, New York, 1967.
152. A. Nobel, *On Uniform Laws of Averages*, Ph.D. thesis, Dept. of Statistics, Standord University, 1992.
153. A. Nobel and A. Dembo, "A note on uniform laws of averages for dependent processes," *Stat. & Probab. Letters*, 17, 169-172, 1993.
154. O. Oleinik and I. Petrovsky, "On the topology of real algebraic surfaces," *Izv. Akad. Nauk SSSR*, 13, 389-402, 1949.

155. M. Opper and D. Haussler, "Calculation of the learning curve of Bayes optimal classification algorithm for learning a perceptron with noise," *Proc. 4th Conf. on Learning Theory*, 75-87, 1991.
156. A. Packard and J. Doyle, "The complex structured singular value," *Automatica*, 29, 71-110, 1993.
157. C. Papadimitrou, *Computational Complexity*, Addison-Wesley, Reading, MA, USA, 1994.
158. J. M. Parrondo and C. van den Broeck, "Vapnik-Chervonenkis bounds for generalization," *J. Phys. A*, 26, 2211-2223, 1993.
159. K. R. Parthasarathy, *Probability Measures on Metric Spaces*, Academic Press, New York, 1967.
160. L. Pitt and L. G. Valiant, "Computational limits on learning from examples," *J. ACM*, 35(4), 965-984, 1988.
161. S. Poljak and J. Rohn, "Checking robust nonsingularity is NP-hard," *Math. Control, Signals, and Systems*, 6(1), 1-9, 1993.
162. D. Pollard, *Convergence of Stochastic Processes*, Springer-Verlag, 1984.
163. D. Pollard, *Empirical Processes: Theory and Applications*, NSF-CBMS Regional Conference Series in Probability and Statistics, Institute of Mathematical Statistics, Volume 2, 1990.
164. R. Ranga Rao, "Relations between weak and uniform convergence of measures with applications," *Ann. Math. Stat.*, 33, 659-680, 1962.
165. L. R. Ray and R. F. Stengel, "Stochastic robustness of linear time-invariant control systems," *IEEE Trans. Autom. Control*, 36, 82-87, 1991.
166. J. M. Rojas, "Some speed-ups and speed limits in real algebraic geometry," *J. Complexity*, (FoCM 1999 Special Issue), 16(3), 552-571, 2000.
167. F. Rosenblatt, *Principles of Neurodynamics: Perceptrons and the Theory of Brain Mechanisms*, Spartan Press, Washington, 1962.
168. D. E. Rumelhart and J. L. McClelland, *Parallel Distributed Processing: Exploration in the Microstructure of Cognition*, Vol. I, MIT Press, Cambridge, MA, 1986.
169. D. E. Rumelhart and J. L. McClelland, *Parallel Distributed Processing: Exploration in the Microstructure of Cognition*, Vol. II, MIT Press, Cambridge, MA, 1986.
170. A. Sard, "The measure of the critical points of differentiable maps," *Bull. Amer. Math. Soc.*, 48, 883-890, 1942.
171. N. Sauer, "On the densities of families of sets," *J. Comb. Theory*, Ser. A, 13, 145-147, 1972.
172. B. Schölkopf, C. J. C. Burges and A. J. Smola, *Advances in Kernel Methods – Support Vector Learning*, MIT Press, Cambridge, MA, 1999.
173. S. Shelah, "Stability, the f.c.p, and superstability; model theoretic properties of formulas in first-order theory," *Annals of Math. Logic*, 3, 271-362, 1971.
174. S. Shelah, "A combinatorial problem: stability and order for models and theories in infinitary languages," *Pacific J. Math.*, 41, 247-261, 1972.
175. E. D. Sontag, "Feedforward nets for interpolation and classification," *J. Comput. Syst. Sci.*, 45(1), 20-48, 1992.
176. E. D. Sontag, "Neural networks for control," in *Essays on Control: Perspectives in the Theory and Applications*, H. L. Trentleman and J. C. Willems (Editors), Birkhauser, Boston, 339-380, 1993.
177. E. D. Sontag, "Critical points for least-squares problems involving certain analytic functions, with applications to sigmoidal nets," *Advances in Computational Mathematics*, 5, 245-268, 1996.
178. E. D. Sontag, "Shattering all sets of k points in general position requires $(k-1)/2$ parameters," *Neural Computation*, 9, 337-348, 1997.

179. J. M. Steele, "Empirical discrepancies and subadditive processes," *Ann. Probab.*, 6, 118-127, 1978.
180. G. Stengle and J. E. Yukich, "Some new Vapnik-Chervonenkis classes," *Ann. Stat.*, 17(4), 1441-1446, 1989.
181. R. Tempo, E. W. Bai and F. Dabbene, "Probabilistic robustness analysis: Explicit bounds for the minimum number of sampling points," *Systems and Control Letters*, 30, 237-242, 1997.
182. R. Thom, "Sur l'homologie des variété algébriques réelles," in *Differential and Combitorial Topology*, S. Cairns (Ed.), Princeton University Press, Princeton, NJ, USA, 1965.
183. V. M. Tikhomirov, "Kolmogorov's work on ϵ-entropy of functional classes and the superposition of functions," *Russian Math. Surveys*, k8, 51-75, 1963.
184. O. Toker and H. Özbay, "Complexity issues in robust stability of linear delay-differential systems," *Math. Control, Signals and Syst.*, 9, 386-400, 1996.
185. S. Treil, "The gap between complex structured singular value μ and its upper bound is infinite," *IEEE Trans. Autom. Control*, (to appear).
186. A. W. van der Vaart and J. A. Wellner, *Weak Convergence and Empirical Processes*, Springer-Verlag, Heidelberg, 1996.
187. L. G. Valiant, "A theory of the learnable," *Commun. ACM*, 27(11), 1134-1142, 1984.
188. L. van den Dries, A. Macintyre and D. Marker, "The elementary theory of restricted analytic fields with exponentiation," *Anal. Math.*, 140, 183-205, 1994.
189. L. van den Dries and L. Miller, "On the real exponential field with restricted analytic functions," *Israel J. Math*, 85, 19-56, 1994.
190. V. N. Vapnik, *Estimation of Dependences Based on Empirical Data*, Springer-Verlag, 1982.
191. V. N. Vapnik, *The Nature of Statistical Learning Theory*, Springer-Verlag, New York, 1995.
192. V. N. Vapnik, *Statistical Learning Theory*, Wiley, New York, 1998.
193. V. N. Vapnik and A. Ya. Chervonenkis, "Uniform convergence of the frequencies of occurence of events to their probabilities," *Soviet Math. Doklady*, 9, 915-918, 1968.
194. V. N. Vapnik and A. Ya. Chervonenkis, "On the uniform convergence of relative frequencies to their probabilities," *Theory of Probab. Appl.* 16(2), 264-280, 1971.
195. V. N. Vapnik and A. Ya. Chervonenkis, *Theory of Pattern Recognition*, (in Russian), Nauka, Moscow, 1974.
196. V. N. Vapnik and A. Ya. Chervonenkis, "Necessary and and sufficient conditions for the uniform convergence of means to their expectations," *Theory of Probab. Appl.*, 26(3), 532-553, 1981.
197. V. N. Vapnik and A. Ya. Chervonenkis, "The necessary and sufficient conditions for consistency of the method od empirical risk minimization," *Pattern Recognition and Image Analysis*, 1(3), 284-305, 1991.
198. M. Vidyasagar, *Control System Synthesis: A Factorization Approach*, MIT Press, Cambridge, MA, 1985.
199. M. Vidyasagar, *Nonlinear Systems Analysis*, (Second Edition), Prentice-Hall, New York, 1993.
200. M. Vidyasagar *A Theory of Learning and Generalization: With Applications to Neural Networks and Control Systems*, Springer-Verlag, London, 1997.
201. M. Vidyasagar "Statistical Learning Theory and Its Applications to Randomized Algorithms for Robust Controller Synthesis," Semi-Plenary Lecture, *European Control Conference*, Brussels, Belgium, (G. Basten and M. Gevers, Eds.), 161-189, 1997.

202. M. Vidyasagar, "Randomized algorithms for robust controller synthesis using statistical learning theory," *Automatica*, 37, 1515-1528, 2001.
203. M. Vidyasagar, "Randomized algorithms for robust controller synthesis using statistical learning theory: A tutorial overview," *European J. Control*, 7(2-3), 287-310, 2001.
204. M. Vidyasagar, V. Balaji and B. Hammer, "Closure properties of uniform convergence of empirical means and PAC learnability under a family of probability measures," *Systems & Control Letters*, 42, 151-157, 2001.
205. M. Vidyasagar and V. Blondel, "Probabilistic solutions to some NP-hard matrix problems," *Automatica*, 37, 1397-1405, 2001.
206. M. Vidyasagar and R. L. Karandikar, "A learning theory approach to system identification and stochastic adaptive control," *IFAC Symp. on Adaptation and Learning*, Como, Italy, Aug. 2001.
207. M. Vidyasagar and R. L. Karandikar, "System identification: A learning theory approach," *Proc. IEEE Conf. on Decision and Control*, Orlando, FL, 2001-2006, Dec. 2001.
208. M. Vidyasagar and H. Kimura, "Robust controllers for uncertain multivariable linear systems," *Automatica*, 22(1), 85-94, January 1986.
209. H. E. Warren, "Lower bounds for approximation by nonlinear manifolds," *Trans. AMS*, 133, 167-178, Aug. 1968.
210. R. S. Wenocur and R. M. Dudley, "Some special Vapnik-Chervonenkis Classes," *Discrete Mathematics*, 33, 313-318, 1981.
211. E. Weyer, "Finite sample properties of system identification of ARX models under mixing conditions," *Automatica*, 36(9), 1291-1299, Sept. 2000.
212. B. Yu, "Rates of convergence of empirical processes for mixing sequences," *Annals of Probab.*, 22(1), 94-116, 1994.

Index

A

Accuracy parameter 57
Activation function 366
Active learning 326
- distribution-free 332
- fixed-distribution 329
Agnostic learning 75
Algorithm 55
- almost surely consistent 212
- asymptotically consistent 213
- consistent 210
- efficient 314
- minimal empirical risk 216
Alpha-mixing 36
Arzela-Ascoli theorem 198
ASCEM property
- definition 52
- equivalence to UCEM property 154
ASCEP property 46
ASEC learnability 75
Asymptotically consistent algorithm 213
Atomic formula 396
Averaged l_1-norm 138
Axis-parallel rectangles 58

B

Bayesian algorithm 354
Bernoulli process 24
Bernoulli trials 24
Beta-mixing 36
Boolean formula 319
Boolean functions 141
- UCEP property 195
- VC-dimension bounds 141
Boolean variables 318
Borel-Cantelli lemma 30

C

Chernoff bound
- additive form 24
- multiplicative form 25
Chernoff-Okamoto bound 26
Compact set 16
Complete regularity 36
Concept class 55
Concept learning 55
Confidence parameter 57
Conjunctive normal form 319
Connected component 397
Consistent algorithm 210
Consistent algorithm, existence 211
Consistent learnability 224
- conditions for 227
Consistent PUAC learnability 226
- conditions for 226
Convergence
- almost sure 30
- in probability 30
Cover 14
- external 14
- minimal 14
- proper 15
Covering number 14
- bounds involving F-dimension 139
- bounds involving P-dimension 132
- external 14
- right continuity 16
Cylinder set 29

D

Disjunctive normal form 319
Distribution function 21
- joint 21

E

Efficient algorithm 314
Efficient learnability 315
$\hat{E}(f; \mathbf{x})$ 50
Empirical estimation of supremum 426
Empirical mean 51
– almost sure convergence 52
– one-sided convergence 112
– uniform convergence 51
Empirical probability 43
– almost sure convergence 46
– uniform convergence 45
Empirical risk 82
Expected value 21

F

Finite metric entropy condition 216
– implies PAC learnability 217
– implies PAC learnability 219
– is implied by PAC learnability 236
– is not implied by PAC learnability 240
First-order logic 390
– formula 391
– term 391
Function class 55
Function learning 64

G

Generalization
– by neural networks 61
– Impossibility of perfect 63
Gibbsian algorithm 354
Glivenko-Cantelli lemma 46
Graph colouring problem 323
Growth function 124
– of iterated families 141

H

Heaviside function 61
Hoeffding's inequality 26
Hypothesis class 70

I

I.i.d. 30
Independence 22
Indicator function 43
Inequality
– Chernoff's 24, 25
– Chernoff-Okamoto 26
– Hoeffding's 26
– Jensen's 191, 358, 362
– Markov's 27
Interval matrix 426
Invariant measure 101

J

Jensen's inequality 191
Jensen's inequality 358
Jensen's inequality 362

K

k-CNF formula 321
k-DNF formula 322
Kolmogorov's 0-1 law 154
k-term CNF 322
k-term DNF 325

L

Labelled multisample 55
Learnability
– efficient 315
– with prior information 335
Literal 318
Loading problem
– definition 377
– intractable example 377
– tractable example 379
Loss function 77
– UCEMUP property 200

M

Marginal probability 22
Markov chain 100
– Geometric ergodicity 101
Markov's inequality 27
Measurable function 17
Measurable space 17
Measure 17
Metric 13
Metric entropy 132
– uniform boundedness condition 292
Minimal empirical risk algorithm 216
– conditions to be PAC 217
– conditions to be PAC 219
– sample complexity 221
Mixing

– α-coefficient 34
– β-coefficient 34
– ϕ-coefficient 34
Model-free learnability
– definition 78
– relationship to UCEM property 81
Model-free learning 75
– under a fixed distribution 242
Model theory of real numbers 390
Monomial 320
– algorithm for learning 320
Monte Carlo simulation 5

N

Neural network 367
– architecture 368
– depth 368
– loading problem 377
– timed 369
Neuron model 366
NMER algorithm 82
Normal form
– conjunctive 319
– disjunctive 319
NP-complete problem 318
NP-hard problem 318

O

Oracle 55
– noisy 79
Order-minimality 413
– sufficient condition 414
Over-fitting 416

P

PAC algorithm
– definition 56
– to a fixed accuracy 56
Packing number 15
– bounds involving F-dimension 139
– bounds involving P-dimension 133
– right continuity 16
PAC learnability 56
Passive learning 326
P-dimension 120
– relationship to VC-dimension 374
Percerptron 61, 366
Pfaffian chain 410
$\hat{P}(A;\mathbf{x})$ 45
Phi-mixing 36

Precompact set 16
Probability measure 17
Probability space 17
Pseudo-dimension 120
Pseudometric 13
P-shattering 120
PUAC learnability 71

Q

$q(m,\epsilon,P)$ 45
Quasicube 165
Query model in active learning 328
– arbitrary binary query 328
– membership query 328

R

Randomized algorithms
– A real life example 449
– for robust control 429
Random variable 21
Regular value 402
Restricted analytic function 414
Risk function 77
$r(m,\epsilon)$ 56
Robust stabilization 422
– NP-hardness 426
– probabilistic algorithms 427
– tractability 423
– VC-dimension estimates 439

S

Sample complexity
– active learning 330
– definition 57
– distribution-free concept learning 269
– distribution-free function learning 264
– fixed-distribution concept learning 219
– fixed-distribution function learning 217
– intermediate families of probabilities 299
– universal lower bound 274
Sard's theorem 403
Sauer's lemma 124
Separated set 15
– maximal 15
Shaping function 366
Shattering 115

Shrinking width property 226
- equivalence to consistent PUAC learnability 227

σ-algebra 17
- Borel 17
- Generated by a random variable 22

Sigmoidal function 367
Sign vector 397
Standard sigmoid 367
Standard sigmoidal neuron 65
Stationary distribution 101
Step function 61
Stochastic process 29
- canonical representation 30
- stationary 30
Strong law of large numbers 33
Strong regularity 36
Structural risk 415
Structured singular value 425
Subadditive process 153
- convergence properties 153
Symmetric difference 18
System identification 453
- learning theory approach 453
- bounds on the P-dimension 461

T

Testing probability 66
Totally bounded set 16
- relationship to compactness 16
Total variation metric 19
Training probability 66
Type 1 near minimum 433
Type 2 near minimum 433
Type 3 near minimum 434

U

UBME condition
- definition 292
- necessity 292
UBME condition
- nonsufficiency 293
- sufficiency 297
UCEM property
- conditions 156
- definition 51
- equivalence to ASCEM property 154
UCEMUP property
- conditions 158
- definition 52
- loss functions 262

UCEP property 45
UCEPUP property
- conditions for 158
- definition 52
- distribution-free 259
Uniform convergence
- of empirical distances 199
- of empirical means 156
- continuous operations 196
- families of loss functions 200
- Boolean operations 195
- of convex sets 159
Uniform law of large numbers 45
Uniform Regularity 36

V

Valuation 319
Vapnik-Chervonenkis dimension 115
- of axis-parallel rectangles 118
- of convex sets 119
- of finite sets 115
- of half-planes 117
- dual 394
- relationship to P-dimension 374

W

Weighted H_∞-norm minimization
- VC-dimension estimates 442
Weighted H_2-norm minimization
- VC-dimension estimates 444